W9-CRY-388

Nonlinear Optics

Second Edition

Nonlinear Optics

Second Edition

Robert W. Boyd

The Institute of Optics
University of Rochester
Rochester, New York USA

ACADEMIC PRESS

An Imprint of Elsevier

Amsterdam   Boston   London   New York   Oxford   Paris   San Diego
San Francisco   Singapore   Sydney   Tokyo

ACADEMIC PRESS
An Imprint of Elsevier
525 B Street, Suite 1900, San Diego, CA 92101-4495, USA
http://www.academicpress.com

Academic Press
An Imprint of Elsevier
84 Theobald's Road, London WC1X 8RR, UK
http://www.academicpress.com

Library of Congress Control Number: 2002115608
ISBN- 13: 978-0-12-121682-5   ISBN-10: 0-12-121682-9

PRINTED IN THE UNITED STATES OF AMERICA
    06   07   08   9   8   7   6   5

for Diane, Jessica, and John

Contents

Contents   xi

Appendices

Index   571

Preface to the Second Edition

In the 10 years since the publication of the first edition of this book, the field of nonlinear optics has continued to achieve new advances both in fundamental physics and in practical applications. Moreover, the author's fascination with this subject has held firm over this time inerval. The present work extends the treatment of the first edition by including a considerable body of additional material and by making numerous small improvements in the presentation of the material included in the first edition.

The primary differences between the first and second editions are as follows.

Two additional sections have been added to Chapter 1, which deals with the nonlinear optical susceptibility. Section 1.6 deals with time-domain descriptions of optical nonlinearities, and Section 1.7 deals with Kramers–Kronig relations in nonlinear optics. In addition, a description of the symmetry properties of gallium arsenide has been added to Section 1.5.

Three sections have been added to Chapter 2, which treats wave-equation descriptions of nonlinear optical interactions. Section 2.8 treats optical parametric oscillators, Section 2.9 treats quasi-phase-matching, and Section 2.11 treats nonlinear optical surface interactions.

Two sections have been added to Chapter 4, which deals with the intensity-dependent refractive index. Section 4.5 treats thermal nonlinearities, and Section 4.6 treats semiconductor nonlinearities.

Chapter 5 is an entirely new chapter dealing with the molecular origin of the nonlinear optical response. (Consequently the chapter numbers of all the following chapters are one greater than those of the first edition.) This chapter treats electronic nonlinearities in the static approximation, semiempirical models of the nonlinear susceptibility, the nonlinear response of conjugated polymers, the bond charge model of optical nonlinearities, nonlinear optics of chiral materials, and nonlinear optics of liquid crystals.

In Chapter 7 on processes resulting from the intensity-dependent refractive index, the section on self-action effects (now Section 7.1) has been significantly expanded. In addition, a description of optical switcing has been included in Section 7.3, now entitled optical bistability and optical switching.

In Chapter 9, which deals with stimulated Brillouin scattering, a discussion of transient effects has been included.

Chapter 12 is an entirely new chapter dealing with optical damage and multiphoton absorption. Chapter 13 is an entirely new chapter dealing with ultrafast and intense-field nonlinear optics.

The Appendices have been expanded to include a treatment of the Gaussian system of units. In addition, many additional homework problems and literature references have been added.

I would like to take this opportunity to thank my many colleagues who have given me advice and suggestions regarding the writing of this book. In addition to the individuals mentioned in the preface to the first edition, I would like to thank G. S. Agarwal, P. Agostini, G. P. Agrawal, M. D. Feit, A. L. Gaeta, D. J. Gauthier, L. V. Hau, F. Kajzar, M. Kauranen, S. G. Lukishova, A. C. Melissinos, Q-H. Park, M. Saffman, B. W. Shore, D. D. Smith, I. A. Walmsley, G. W. Wicks, and Z. Zyss. I especially wish to thank M. Kauranen and A. L. Gaeta for suggesting additional homework problems and to thank A. L. Gaeta for advice on the preparation of Section 13.2.

Preface to the First Edition

Nonlinear optics is the study of the interaction of intense laser light with matter. This book is a textbook on nonlinear optics at the level of a beginning graduate student. The intent of the book is to provide an introduction to the field of nonlinear optics that stresses fundamental concepts and that enables the student to go on to perform independent research in this field. The author has successfully used a preliminary version of this book in his course at the University of Rochester, which is typically attended by students ranging from seniors to advanced PhD students from disciplines that include optics, physics, chemistry, electrical engineering, mechanical engineering, and chemical engineering. This book could be used in graduate courses in the areas of nonlinear optics, quantum optics, quantum electronics, laser physics, electrooptics, and modern optics. By deleting some of the more difficult sections, this book would also be suitable for use by advanced undergraduates. On the other hand, some of the material in the book is rather advanced and would be suitable for senior graduate students and research scientists.

The field of nonlinear optics is now thirty years old, if we take its beginnings to be the observation of second-harmonic generation by Franken and coworkers in 1961. Interest in this field has grown continuously since its beginnings, and the field of nonlinear optics now ranges from fundamental studies of the interaction of light with matter to applications such as laser frequency conversion and optical switching. In fact, the field of nonlinear optics has grown so enormously that it is not possible for one book to cover all of the topics of current interest. In addition, since I want this book to be accessible to beginning graduate students, I have attempted to treat the topics that are covered in a reasonably self-contained manner. This consideration also restricts the number of topics that can be treated. My strategy in deciding what topics to include has been to stress the fundamental aspects of nonlinear optics, and to include applications and experimental results only as necessary to illustrate these fundamental issues. Many of the specific topics that I have chosen to include are those of particular historical value.

Nonlinear optics is notationally very complicated, and unfortunately much of the notational complication is unavoidable. Because the notational aspects of nonlinear optics have historically been very confusing, considerable effort is made, especially in the early chapters, to explain the notational conventions. The book uses primarily the Gaussian system of units, both to establish a connection with the historical papers of nonlinear optics, most of which were written using the Gaussian system, and also because the author believes that the laws of electromagnetism are more physically transparent when written in this system. At several places in the text (see especially the appendices at the end of the book), tables are provided to facilitate conversion to other systems of units.

The book is organized as follows: Chapter 1 presents an introduction to the field of nonlinear optics from the perspective of the nonlinear susceptibility. The nonlinear susceptibility is a quantity that is used to determine the nonlinear polarization of a material medium in terms of the strength of an applied optical-frequency electric field. It thus provides a framework for describing nonlinear optical phenomena. Chapter 2 continues the description of nonlinear optics by describing the propagation of light waves through nonlinear optical media by means of the optical wave equation. This chapter introduces the important concept of phase matching and presents detailed descriptions of the important nonlinear optical phenomena of second-harmonic generation and sum- and difference-frequency generation. Chapter 3 concludes the introductory portion of the book by presenting a description of the quantum mechanical theory of the nonlinear optical susceptibility. Simplified expressions for the nonlinear susceptibility are first derived through use of the Schrödinger equation, and then more accurate expressions are derived through use of the density matrix equations of motion. The density matrix formalism is itself developed in considerable detail in this chapter in order to render this important discussion accessible to the beginning student.

Chapters 4 through 6 deal with properties and applications of the nonlinear refractive index. Chapter 4 introduces the topic of the nonlinear refractive index. Properties, including tensor properties, of the nonlinear refractive index are discussed in detail, and physical processes that lead to the nonlinear refractive index, such as nonresonant electronic polarization and molecular orientation, are described. Chapter 5 is devoted to a description of nonlinearities in the refractive index resulting from the response of two-level atoms. Related topics that are discussed in this chapter include saturation, power broadening, optical Stark shifts, Rabi oscillations, and dressed atomic states. Chapter 6 deals with applications of the nonlinear refractive index. Topics that are included are optical phase conjugation, self focusing, optical bistability, two-beam coupling, pulse propagation, and the formation of optical solitons.

Chapters 7 through 9 deal with spontaneous and stimulated light scattering and the related topic of acoustooptics. Chapter 7 introduces this area by presenting a description of theories of spontaneous light scattering and by describing the important practical topic of acousto-optics. Chapter 8 presents a description of stimulated Brillouin and stimulated Rayleigh scattering. These topics are related in that they both entail the scattering of light from material disturbances that can be described in terms of the standard thermodynamic variables of pressure and entropy. Also included in this chapter is a description of phase conjugation by stimulated Brillouin scattering and a theoretical description of stimulated Brillouin scattering in gases. Chapter 9 presents a description of stimulated Raman and stimulated Rayleigh-wing scattering. These processes are related in that they entail the scattering of light from disturbances associated with the positions of atoms within a molecule.

The book concludes with Chapter 10, which treats the electrooptic and photorefractive effects. The chapter begins with a description of the electrooptic effect and describes how this effect can be used to fabricate light modulators. The chapter then presents a description of the photorefractive effect, which is a nonlinear optical interaction that results from the electrooptic effect. The use of the photorefractive effect in two-beam coupling and in four-wave mixing is also described.

The author wishes to acknowledge his deep appreciation for discussions of the material in this book with his graduate students at the University of Rochester. He is sure that he has learned as much from them as they have from him. He also gratefully acknowledges discussions with numerous other professional colleagues, including N. Bloembergen, D. Chemla, R. Y. Chiao, J. H. Eberly, C. Flytzanis, J. Goldhar, G. Grynberg, J. H. Haus, R. W. Hellwarth, K. R. MacDonald, S. Mukamel, P. Narum, M. G. Raymer, J. E. Sipe, C. R. Stroud, Jr., C. H. Townes, H. Winful, and B. Ya. Zel'dovich. In addition, the assistance of J. J. Maki and A. Gamliel in the preparation of the figures is gratefully acknowledged.

Chapter 1

The Nonlinear Optical Susceptibility

1.1. Introduction to Nonlinear Optics

Nonlinear optics is the study of phenomena that occur as a consequence of the modification of the optical properties of a material system by the presence of light. Typically, only laser light is sufficiently intense to modify the optical properties of a material system. In fact, the beginning of the field of nonlinear optics is often taken to be the discovery of second-harmonic generation by Franken *et al.* in 1961, shortly after the demonstration of the first working laser by Maiman in 1960. Nonlinear optical phenomena are "nonlinear" in the sense that they occur when the response of a material system to an applied optical field depends in a nonlinear manner upon the strength of the optical field. For example, second-harmonic generation occurs as a result of the part of the atomic response that depends quadratically on the strength of the applied optical field. Consequently, the intensity of the light generated at the second-harmonic frequency tends to increase as the square of the intensity of the applied laser light.

In order to describe more precisely what we mean by an optical nonlinearity, let us consider how the dipole moment per unit volume, or polarization $\tilde{P}(t)$, of a material system depends upon the strength $\tilde{E}(t)$ of the applied optical field.[*] In the case of conventional (i.e., linear) optics, the induced polarization depends linearly upon the electric field strength in a manner that can often be

---

[*] Throughout the text, we use the tilde to denote a quantity that varies rapidly in time. Constant quantities, slowly varying quantities, and Fourier amplitudes are written without the tilde. See, for example, Eq. (1.2.1).

described by the relationship

$$\tilde{P}(t) = \chi^{(1)}\tilde{E}(t), \qquad (1.1.1)$$

where the constant of proportionality $\chi^{(1)}$ is known as the linear susceptibility. In nonlinear optics, the optical response can often be described by generalizing Eq. (1.1.1) by expressing the polarization $\tilde{P}(t)$ as a power series in the field strength $\tilde{E}(t)$ as

$$\tilde{P}(t) = \chi^{(1)}\tilde{E}(t) + \chi^{(2)}\tilde{E}^2(t) + \chi^{(3)}\tilde{E}^3(t) + \cdots$$
$$\equiv \tilde{P}^{(1)}(t) + \tilde{P}^{(2)}(t) + \tilde{P}^{(3)}(t) + \cdots. \qquad (1.1.2)$$

The quantities $\chi^{(2)}$ and $\chi^{(3)}$ are known as the second- and third-order nonlinear optical susceptibilities, respectively. For simplicity, we have taken the fields $\tilde{P}(t)$ and $\tilde{E}(t)$ to be scalar quantities in writing Eqs. (1.1.1) and (1.1.2). In Section 1.3 we show how to treat the vector nature of the fields; in such a case $\chi^{(1)}$ becomes a second-rank tensor, $\chi^{(2)}$ becomes a third-rank tensor, etc. In writing Eqs. (1.1.1) and (1.1.2) in the form shown, we have also assumed that the polarization at time t depends only on the instantaneous value of the electric field strength. The assumption that the medium responds instantaneously also implies (through the Kramers–Kronig relations)* that the medium must be lossless and dispersionless. We shall also see in Section 1.3 how to generalize these equations for the case of a medium with dispersion and loss. In general, the nonlinear susceptibilities depend on the frequencies of the applied fields, but under our present assumption of instantaneous response we take them to be constants.

We shall refer to $\tilde{P}^{(2)}(t) = \chi^{(2)}\tilde{E}(t)^2$ as the second-order nonlinear polarization and to $\tilde{P}^{(3)}(t) = \chi^{(3)}\tilde{E}(t)^3$ as the third-order nonlinear polarization. We shall see later in this section that the physical processes that occur as a result of the second-order polarization $\tilde{P}^{(2)}$ are distinct from those that occur as a result of the third-order polarization $\tilde{P}^{(3)}$. In addition, we shall show in Section 1.5 that second-order nonlinear optical interactions can occur only in noncentrosymmetric crystals, that is, in crystals that do not display inversion symmetry. Since liquids, gases, amorphous solids (such as glass), and even many crystals do display inversion symmetry, $\chi^{(2)}$ vanishes identically for such media, and consequently they cannot produce second-order nonlinear optical interactions. On the other hand, third-order nonlinear optical interactions

---

* See, for example, Loudon (1973) Chapter 4 or the discussion in Section 1.7 of the present book for a discussion of the Kramers–Kronig relations.

(i.e., those described by a $\chi^{(3)}$ susceptibility) can occur both for centrosymmetric and noncentrosymmetric media.

We shall see in later sections of this book how to calculate the values of the nonlinear susceptibilities for various physical mechanisms that lead to optical nonlinearities. For the present, we shall make a simple order-of-magnitude estimate of the size of these quantities for the common case in which the nonlinearity is electronic in origin (see, for instance, Armstrong *et al.*, 1962). One might expect that the lowest-order correction term $\tilde{P}^{(2)}$ would be comparable to the linear response $\tilde{P}^{(1)}$ when the amplitude of the applied field $\tilde{E}$ is of the order of the characteristic atomic electric field strength $E_{at} = e/a_0^2$, where $-e$ is the charge of the electron and $a_0 = \hbar^2/me^2$ is the Bohr radius of the hydrogen atom (here $\hbar$ is Planck's constant divided by 2π, and m is the mass of the electron). Numerically, we find that $E_{at} = 2 \times 10^7$ statvolt/cm.* We thus expect that under conditions of nonresonant excitation the second-order susceptibility $\chi^{(2)}$ will be of the order of $\chi^{(1)}/E_{at}$. For condensed matter $\chi^{(1)}$ is of the order of unity, and we hence expect that $\chi^{(2)}$ will be of the order of $1/E_{at}$, or that

$$\chi^{(2)} \simeq 5 \times 10^{-8} \frac{cm}{statvolt} = 5 \times 10^{-8} (cm^3/erg)^{1/2} = 5 \times 10^{-8} \text{ esu.} \quad (1.1.3)$$

Similarly, we expect $\chi^{(3)}$ to be of the order of $\chi^{(1)}/E_{at}^2$, which for condensed matter is of the order of

$$\chi^{(3)} \simeq 3 \times 10^{-15} \frac{cm^2}{statvolt^2} = 3 \times 10^{-15} cm^3/erg = 3 \times 10^{-15} \text{ esu.} \quad (1.1.4)$$

These predictions are in fact quite accurate, as one can see by comparing these values with actual measured values of $\chi^{(2)}$ (see for instance Table 1.5.3) and $\chi^{(3)}$ (see for instance Table 4.3.1). For certain purposes, it is useful to express the second- and third-order susceptibilities in terms of fundamental physical constants. Noting that the number density N of condensed matter is of the order of $(a_0)^{-3}$, we find that $\chi^{(2)} \simeq \hbar^4/m^2 e^5$ and $\chi^{(3)} \simeq \hbar^8/m^4 e^{10}$. See Boyd (1999) for further details.

The most common procedure for describing nonlinear optical phenomena is based on expressing the polarization $\tilde{P}(t)$ in terms of the applied electric field strength $\tilde{E}(t)$, as we have done in Eq. (1.1.2). The reason why the polarization plays a key role in the description of nonlinear optical phenomena is that a

---

* Except where otherwise noted, we use the gaussian system of units in this book. Note that in the scientific literature the units of an electrical quantity expressed in the gaussian system are often not given explicitly, but rather are simply said to be stated in electrostatic units (esu). As an example, in the present instance one would say that $E_{at} = 2 \times 10^7$ esu. See also the discussion in the appendix to this book on the conversion between the systems of units.

time-varying polarization can act as the source of new components of the electromagnetic field. For example, we shall see in Section 2.1 that the wave equation in nonlinear optical media often has the form

$$\nabla^2 \tilde{E} - \frac{n^2}{c^2} \frac{\partial^2 \tilde{E}}{\partial t^2} = \frac{4\pi}{c^2} \frac{\partial^2 \tilde{P}^{\mathrm{NL}}}{\partial t^2},$$ (1.1.5)

where n is the usual linear refractive index and c is the speed of light in vacuum. We can interpret this expression as an inhomogeneous wave equation in which the polarization $\tilde{P}^{\mathrm{NL}}$ associated with the nonlinear response drives the electric field $\tilde{E}$. This equation expresses the fact that, whenever $\partial^2 \tilde{P}^{\mathrm{NL}} / \partial t^2$ is nonzero, charges are being accelerated, and according to Larmor's theorem from electromagnetism accelerated charges generate electromagnetic radiation.

It should be noted that the power series expansion expressed by Eq. (1.1.2) need not necessarily converge. In such circumstances the relationship between the material response and the applied electric field amplitude must be expressed using different procedures. One such example is that under resonant excitation of an atomic system, an appreciable fraction of the atoms can be removed from the ground state. Saturation effects of this sort can be described by procedures developed in Chapter 6. Even under nonresonant conditions, Eq. (1.1.2) loses its validity if the applied laser field strength becomes comparable to the characteristic atomic field strength E_{at}, because of strong photoionization that can occur under these conditions. For future reference, we note that the laser intensity associated with a peak field strength of E_{at} is given by

$$I_{\mathrm{at}} = \frac{c}{8\pi} E_{\mathrm{at}}^2 = 5 \times 10^{23} \ \mathrm{erg/cm^2 s} = 5 \times 10^{16} \ \mathrm{W/cm^2}.$$ (1.1.6)

We shall see in later sections of this book how nonlinear optical processes display qualitatively distinct features when excited by such super-intense fields.

1.2. Descriptions of Nonlinear Optical Interactions

In the present section, we present brief qualitative descriptions of a number of nonlinear optical interactions. In addition, for those processes that can occur in a lossless medium, we indicate how they can be described in terms of the nonlinear contributions to the polarization described by Eq. (1.1.2).* Our motivation is to provide the reader with an indication of the variety of nonlinear optical phenomena that can occur. These interactions are described in greater detail in later sections of this book. In this section we also introduce some notational conventions and some of the basic concepts of nonlinear optics.

---

* Recall that Eq. (1.1.2) is valid only for a medium that is lossless and dispersionless.

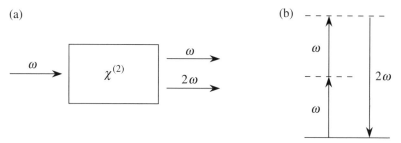

FIGURE 1.2.1 (a) Geometry of second-harmonic generation. (b) Energy-level diagram describing second-harmonic generation.

Second-Harmonic Generation

As an example of a nonlinear optical interaction, let us consider the process of second-harmonic generation, which is illustrated schematically in Fig. 1.2.1. Here a laser beam whose electric field strength is represented as

$$\tilde{E}(t) = Ee^{-i\omega t} + \text{c.c.} \tag{1.2.1}$$

is incident upon a crystal for which the second-order susceptibility $\chi^{(2)}$ is nonzero. The nonlinear polarization that is created in such a crystal is given according to Eq. (1.1.2) as $\tilde{P}^{(2)}(t) = \chi^{(2)}\tilde{E}^2(t)$ or as

$$\tilde{P}^{(2)}(t) = 2\chi^{(2)}EE^* + \left(\chi^{(2)}E^2e^{-2\omega t} + \text{c.c.}\right). \tag{1.2.2}$$

We see that the second-order polarization consists of a contribution at zero frequency (the first term) and a contribution at frequency 2ω (the second term). According to the driven wave equation (1.1.5), this latter contribution can lead to the generation of radiation at the second-harmonic frequency. Note that the first contribution in Eq. (1.2.2) does not lead to the generation of electromagnetic radiation (because its second time derivative vanishes); it leads to a process known as optical rectification in which a static electric field is created within the nonlinear crystal.

Under proper experimental conditions, the process of second-harmonic generation can be so efficient that nearly all of the power in the incident radiation at frequency ω is converted to radiation at the second-harmonic frequency 2ω. One common use of second-harmonic generation is to convert the output of a fixed-frequency laser to a different spectral region. For example, the Nd:YAG laser operates in the near infrared at a wavelength of 1.06 μm. Second-harmonic generation is routinely used to convert the wavelength of the radiation to 0.53 μm, in the middle of the visible spectrum.

Second-harmonic generation can be visualized by considering the interaction in terms of the exchange of photons between the various frequency components of the field. According to this picture, which is illustrated in part (b) of Fig. 1.2.1, two photons of frequency ω are destroyed and a photon of frequency 2ω is simultaneously created in a single quantum-mechanical process. The solid line in the figure represents the atomic ground state, and the dashed lines represent what are known as virtual levels. These levels are not energy eigenlevels of the free atom, but rather represent the combined energy of one of the energy eigenstates of the atom and of one or more photons of the radiation field.

The theory of second-harmonic generation is developed more fully in Section 2.6.

Sum- and Difference-Frequency Generation

Let us next consider the circumstance in which the optical field incident upon a nonlinear optical medium characterized by a nonlinear susceptibility $\chi^{(2)}$ consists of two distinct frequency components, which we represent in the form

$$\tilde{E}(t) = E_1 e^{-i\omega_1 t} + E_2 e^{-i\omega_2 t} + \text{c.c.} \tag{1.2.3}$$

Then, assuming as in Eq. (1.1.2) that the second-order contribution to the nonlinear polarization is of the form

$$\tilde{P}^{(2)}(t) = \chi^{(2)} \tilde{E}(t)^2, \tag{1.2.4}$$

we find that the nonlinear polarization is given by

$$\tilde{P}^{(2)}(t) = \chi^{(2)} \big[E_1^2 e^{-2i\omega_1 t} + E_2^2 e^{-2i\omega_2 t} + 2E_1 E_2 e^{-i(\omega_1 + \omega_2)t}$$
$$+ 2E_1 E_2^* e^{-i(\omega_1 - \omega_2)t} + \text{c.c.} \big] + 2\chi^{(2)} [E_1 E_1^* + E_2 E_2^*]. \tag{1.2.5}$$

It is convenient to express this result using the notation

$$\tilde{P}^{(2)}(t) = \sum_n P(\omega_n) e^{-i\omega_n t}, \tag{1.2.6}$$

where the summation extends over positive and negative frequencies ω_n. The complex amplitudes of the various frequency components of the nonlinear polarization are hence given by

$$P(2\omega_1) = \chi^{(2)} E_1^2 \quad \text{(SHG)},$$
$$P(2\omega_2) = \chi^{(2)} E_2^2 \quad \text{(SHG)},$$
$$P(\omega_1 + \omega_2) = 2\chi^{(2)} E_1 E_2 \quad \text{(SFG)}, \tag{1.2.7}$$
$$P(\omega_1 - \omega_2) = 2\chi^{(2)} E_1 E_2^* \quad \text{(DFG)},$$
$$P(0) = 2\chi^{(2)} (E_1 E_1^* + E_2 E_2^*) \quad \text{(OR)}.$$

Here we have labeled each expression by the name of the physical process that it describes, such as second-harmonic generation (SHG), sum-frequency generation (SFG), difference-frequency generation (DFG), and optical rectification (OR). Note that, in accordance with our complex notation, there is also a response at the negative of each of the nonzero frequencies given above:

$$P(-2\omega_1) = \chi^{(2)} E_1^{*2}, \qquad P(-2\omega_2) = \chi^{(2)} E_2^{*2},$$

$$P(-\omega_1 - \omega_2) = 2\chi^{(2)} E_1^* E_2^*, \qquad P(\omega_2 - \omega_1) = 2\chi^{(2)} E_2 E_1^*. \tag{1.2.8}$$

However, since each of these quantities is simply the complex conjugate of one of the quantities given in Eq. (1.2.7), it is not necessary to take explicit account of both the positive and negative frequency components.*

We see from Eq. (1.2.7) that four different nonzero frequency components are present in the nonlinear polarization. However, typically no more than one of these frequency components will be present with any appreciable intensity in the radiation generated by the nonlinear optical interaction. The reason for this behavior is that the nonlinear polarization can efficiently produce an output signal only if a certain phase-matching condition (which is discussed in detail in Section 2.7) is satisfied, and usually this condition cannot be satisfied for more than one frequency component of the nonlinear polarization. Operationally, one often chooses which frequency component will be radiated by properly selecting the polarization of the input radiation and orientation of the nonlinear crystal.

Sum-Frequency Generation

Let us now consider the process of sum-frequency generation, which is illustrated in Fig. 1.2.2. According to Eq. (1.2.7), the complex amplitude of the

---

* Not all workers in nonlinear optics use our convention that the fields and polarizations are given by Eqs. (1.2.3) and (1.2.6). Another common convention is to define the field amplitudes according to

$$\tilde{E}(t) = \tfrac{1}{2}(E_1' e^{-i\omega_1 t} + E_2' e^{-i\omega_2 t} + \text{c.c.}),$$

$$\tilde{P}^2(t) = \tfrac{1}{2} \sum_n P'(\omega_n) e^{i\omega_n t},$$

where in the second expression the summation extends over all positive and negative frequencies. Using this convention, one finds that

$$P'(2\omega_1) = \tfrac{1}{2}\chi^{(2)} E_1'^2, \qquad\qquad P'(2\omega_2) = \tfrac{1}{2}\chi^{(2)} E_2'^2,$$

$$P'(\omega_1 + \omega_2) = \chi^{(2)} E_1' E_2', \qquad\qquad P'(\omega_1 - \omega_2) = \chi^{(2)} E_1' E_2'^*,$$

$$P'(0) = \chi^{(2)}(E_1' E_1'^* + E_2' E_2'^*).$$

Note that these expressions differ from Eqs. (1.2.7) by factors of 1/2.

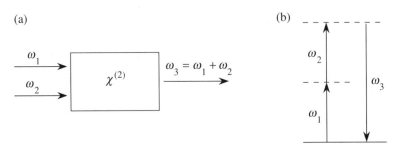

FIGURE 1.2.2 Sum-frequency generation. (a) Geometry of the interaction. (b) Energy-level description.

nonlinear polarization describing this process is given by the expression

$$P(\omega_1 + \omega_2) = 2\chi^{(2)} E_1 E_2. \tag{1.2.9}$$

In many ways the process of sum-frequency generation is analogous to that of second-harmonic generation, except that in sum-frequency generation the two input waves are at different frequencies. One application of sum-frequency generation is to produce tunable radiation in the ultraviolet spectral region by choosing one of the input waves to be the output of a fixed-frequency visible laser and the other to be the output of a frequency-tunable visible laser. The theory of sum-frequency generation is developed more fully in Sections 2.2 and 2.4.

Difference-Frequency Generation

The process of difference-frequency generation is described by a nonlinear polarization of the form

$$P(\omega_1 - \omega_2) = 2\chi^{(2)} E_1 E_2^* \tag{1.2.10}$$

and is illustrated in Fig. 1.2.3. Here the frequency of the generated wave is the difference of those of the applied fields. Difference-frequency generation can be used to produce tunable infrared radiation by mixing the output of a frequency-tunable visible laser with that of a fixed-frequency visible laser.

Superficially, difference-frequency generation and sum-frequency generation appear to be very similar processes. However, an important difference between the two processes can be deduced from the description of difference-frequency generation in terms of a photon energy-level diagram (part (b) of Fig. 1.2.3). We see that conservation of energy requires that for every photon that is created at the difference frequency $\omega_3 = \omega_1 - \omega_2$, a photon at the higher input frequency (ω_1) must be destroyed and a photon at the lower input

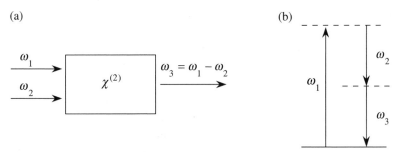

FIGURE 1.2.3   Difference-frequency generation. (a) Geometry of the interaction. (b) Energy-level description.

frequency (ω_2) must be created. Thus, the lower-frequency input field is amplified by the process of difference-frequency generation. For this reason, the process of difference-frequency generation is also known as optical parametric amplification. According to the photon energy-level description of difference-frequency generation, the atom first absorbs a photon of frequency ω_1 and jumps to the highest virtual level. This level decays by a two-photon emission process that is stimulated by the presence of the ω_2 field, which is already present. Two-photon emission can occur even if the ω_2 field is not applied. The generated fields in such a case are very much weaker, since they are created by *spontaneous* two-photon emission from a virtual level. This process is known as parametric fluorescence and has been observed experimentally (Harris *et al.*, 1967; Byer and Harris, 1968).

The theory of difference-frequency generation is developed more fully in Section 2.5.

Optical Parametric Oscillation

We have just seen that in the process of difference-frequency generation the presence of radiation at frequency ω_2 or ω_3 can stimulate the emission of additional photons at these frequencies. If the nonlinear crystal used in this process is placed inside an optical resonator, as shown in Fig. 1.2.4, the ω_2

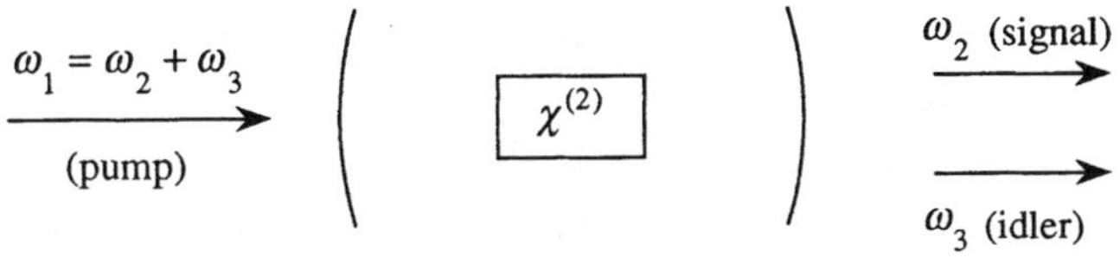

$$\omega_1 = \omega_2 + \omega_3$$
(pump)

$\chi^{(2)}$

ω_2 (signal)

ω_3 (idler)

FIGURE 1.2.4   The optical parametric oscillator. The cavity end mirrors have high reflectivities at frequencies ω_2 and/or ω_3.

and/or ω_3 fields can build up to large values. Such a device is known as an optical parametric oscillator. Optical parametric oscillators are frequently used at infrared wavelengths, where other sources of tunable radiation are not readily available. Such a device is tunable because any frequency ω_2 that is smaller than ω_1 can satisfy the condition $\omega_2 + \omega_3 = \omega_1$ for some frequency ω_3. In practice, one controls the output frequency of an optical parametric oscillator by adjusting the phase-matching condition, as discussed in Section 2.7. The applied field frequency ω_1 is often called the pump frequency, the desired output frequency is called the signal frequency, and the other, unwanted, output frequency is called the idler frequency.

Third-Order Polarization

We next consider the third-order contribution to the nonlinear polarization

$$\tilde{P}^{(3)}(t) = \chi^{(3)} \tilde{E}(t)^3. \tag{1.2.11}$$

For the general case in which the field $\tilde{E}(t)$ is made up of several different frequency components, the expression for $\tilde{P}^{(3)}(t)$ is very complicated. For this reason, we first consider the simple case in which the applied field is monochromatic and is given by

$$\tilde{E}(t) = \mathscr{E}\cos \omega t. \tag{1.2.12}$$

Then, through use of the identity $\cos^3 \omega t = \frac{1}{4}\cos 3\omega t + \frac{3}{4}\cos \omega t$, the nonlinear polarization can be expressed as

$$\tilde{P}^{(3)}(t) = \frac{1}{4}\chi^{(3)}\mathscr{E}^3 \cos 3\omega t + \frac{3}{4}\chi^{(3)}\mathscr{E}^3 \cos \omega t. \tag{1.2.13}$$

The significance of each of the two terms in this expression is described briefly below.

Third-Harmonic Generation

The first term in Eq. (1.2.13) describes a response at frequency 3ω that is due to an applied field at frequency ω. This term leads to the process of third-harmonic generation, which is illustrated in Fig. 1.2.5. According to the photon description of this process, shown in part (b) of the figure, three photons of frequency ω are destroyed and one photon of frequency 3ω is created in each elementary event.

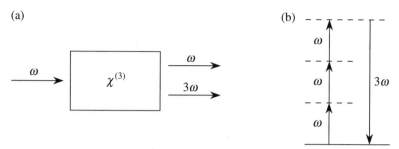

FIGURE 1.2.5 Third-harmonic generation. (a) Geometry of the interaction. (b) Energy-level description.

Intensity-Dependent Refractive Index

The second term in Eq. (1.2.13) describes a nonlinear contribution to the polarization at the frequency of the incident field; this term hence leads to a nonlinear contribution to the refractive index experienced by a wave at frequency ω. We shall see in Section 4.1 that the refractive index in the presence of this type of nonlinearity can be represented as

$$n = n_0 + n_2 I \tag{1.2.14a}$$

where n_0 is the usual (i.e., linear or low-intensity) refractive index, where

$$n_2 = \frac{12\pi^2}{n_0^2 c} \chi^{(3)} \tag{1.2.14b}$$

is an optical constant that characterizes the strength of the optical nonlinearity, and where $I = (n_0 c / 8\pi) \mathcal{E}^2$ is the intensity of the incident wave.

Self-Focusing. One of the processes that can occur as a result of the intensity-dependent refractive index is self-focusing, which is illustrated in Fig. 1.2.6. This process can occur when a beam of light having a nonuniform transverse intensity distribution propagates through a material in which n_2 is positive. Under these conditions, the material effectively acts as a positive lens, which causes the rays to curve toward each other. This process is of great practical importance because the intensity at the focal spot of the self-focused beam is usually sufficiently large to lead to optical damage of the material. The process of self-focusing is described in greater detail in Section 7.1.

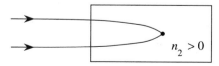

FIGURE 1.2.6 Self-focusing of light.

Third-Order Polarization (General Case)

Let us next examine the form of the nonlinear polarization

$$\tilde{P}^{(3)}(t) = \chi^{(3)} \tilde{E}(t)^3 \tag{1.2.15a}$$

induced by an applied field that consists of three frequency components:

$$\tilde{E}(t) = E_1 e^{-i\omega_1 t} + E_2 e^{-i\omega_2 t} + E_3 e^{-i\omega_3 t} + \text{c.c.} \tag{1.2.15b}$$

When we calculate $\tilde{E}(t)^3$, we find that the resulting expression contains 44 different frequency components, if we consider positive and negative frequencies to be distinct. Explicitly, these frequencies are

$$\omega_1, \omega_2, \omega_3, 3\omega_1, 3\omega_2, 3\omega_3, (\omega_1 + \omega_2 + \omega_3), (\omega_1 + \omega_2 - \omega_3),$$

$$(\omega_1 + \omega_3 - \omega_2), (\omega_2 + \omega_3 - \omega_1), (2\omega_1 \pm \omega_2), (2\omega_1 \pm \omega_3), (2\omega_2 \pm \omega_1),$$

$$(2\omega_2 \pm \omega_3), (2\omega_3 \pm \omega_1), (2\omega_3 \pm \omega_2),$$

and the negative of each. Again representing the nonlinear polarization as

$$\tilde{P}^{(3)}(t) = \sum_n P(\omega_n) e^{-i\omega_n t}, \tag{1.2.16}$$

we can write the complex amplitudes of the nonlinear polarization for the positive frequencies as

$$P(\omega_1) = \chi^{(3)}(3E_1 E_1^* + 6E_2 E_2^* + 6E_3 E_3^*)E_1,$$

$$P(\omega_2) = \chi^{(3)}(6E_1 E_1^* + 3E_2 E_2^* + 6E_3 E_3^*)E_2,$$

$$P(\omega_3) = \chi^{(3)}(6E_1 E_1^* + 6E_2 E_2^* + 3E_3 E_3^*)E_3,$$

$$P(3\omega_1) = \chi^{(3)} E_1^3, \qquad P(3\omega_2) = \chi^{(3)} E_2^3, \qquad P(3\omega_3) = \chi^{(3)} E_3^3,$$

$$P(\omega_1 + \omega_2 + \omega_3) = 6\chi^{(3)} E_1 E_2 E_3, \quad P(\omega_1 + \omega_2 - \omega_3) = 6\chi^{(3)} E_1 E_2 E_3^*,$$

$$P(\omega_1 + \omega_3 - \omega_2) = 6\chi^{(3)} E_1 E_3 E_2^*, \quad P(\omega_2 + \omega_3 - \omega_1) = 6\chi^{(3)} E_2 E_3 E_1^*,$$

$$P(2\omega_1 + \omega_2) = 3\chi^{(3)} E_1^2 E_2, \qquad P(2\omega_1 + \omega_3) = 3\chi^{(3)} E_1^2 E_3,$$

$$P(2\omega_2 + \omega_1) = 3\chi^{(3)} E_2^2 E_1, \qquad P(2\omega_2 + \omega_3) = 3\chi^{(3)} E_2^2 E_3,$$

$$P(2\omega_3 + \omega_1) = 3\chi^{(3)} E_3^2 E_1, \qquad P(2\omega_3 + \omega_2) = 3\chi^{(3)} E_3^2 E_2,$$

$$P(2\omega_1 - \omega_2) = 3\chi^{(3)} E_1^2 E_2^*, \qquad P(2\omega_1 - \omega_3) = 3\chi^{(3)} E_1^2 E_3^*,$$

$$P(2\omega_2 - \omega_1) = 3\chi^{(3)} E_2^2 E_1^*, \qquad P(2\omega_2 - \omega_3) = 3\chi^{(3)} E_2^2 E_3^*,$$

$$P(2\omega_3 - \omega_1) = 3\chi^{(3)} E_3^2 E_1^*, \qquad P(2\omega_3 - \omega_2) = 3\chi^{(3)} E_3^2 E_2^*$$

$$\tag{1.2.17}$$

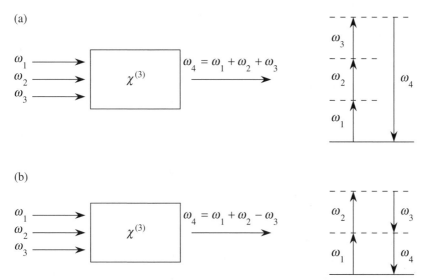

FIGURE 1.2.7   Two of the possible mixing processes described by Eq. (1.2.17) that can occur when three input waves interact in a medium characterized by a $\chi^{(3)}$ susceptibility.

We have displayed these expressions in complete detail because it is very instructive to study their form. In each case the frequency argument of P is equal to the sum of the frequencies associated with the field amplitudes appearing on the right-hand side of the equation, if we adopt the convention that a negative frequency is to be associated with a field amplitude that appears as a complex conjugate. Also, the numerical factor (1, 3, or 6) that appears in each term on the right-hand side of each equation is equal to the number of distinct permutations of the field frequencies that contribute to that term.

Some of the nonlinear optical mixing processes described by Eq. (1.2.17) are illustrated in Fig. 1.2.7.

Parametric versus Nonparametric Process

All of the processes described thus far in this chapter are examples of what are known as parametric processes. The origin of this terminology is obscure, but the word parametric has come to denote a process in which the initial and final quantum-mechanical states of the system are identical. Consequently, in a parametric process population can be removed from the ground state only for those brief intervals of time when it resides in a virtual level. According to the uncertainty principle, population can reside in a virtual level for a time interval of the order of $\hbar/\delta E$, where δE is the energy difference between the virtual level and the nearest real level. Conversely, processes that do involve the transfer of population from one real level to another are known as non-parametric processes. The processes that we describe in the remainder of the present section are all examples of nonparametric processes.

One difference between parametric and nonparametric processes is that parametric processes can always be described by a real susceptibility; conversely, nonparametric processes are described by a complex susceptibility by means of a procedure described in the following section, Section 1.3. Another difference is that photon energy is always conserved in a parametric process; photon energy need not be conserved in a nonparametric process, because energy can be transferred to or from the material medium. For this reason, photon energy level diagrams of the sort shown in Figs. 1.2.1, 1.2.2, 1.2.3, 1.2.5, and 1.2.7 to describe parametric processes play a less definitive role in describing non-parametric processes.

As a simple example of the distinction between parametric and nonparametric processes, we consider the case of the usual (linear) index of refraction. The real part of the refractive index is a consequence of parametric processes, whereas its imaginary part is a consequence of nonparametric processes, since the imaginary part of the refractive index describes the absorption of radiation, which results from the transfer of population from the atomic ground state to an excited state.

Saturable Absorption

One example of a nonparametric nonlinear optical process is saturable absorption. Many material systems have the property that their absorption coefficient decreases when measured using high laser intensity. Often the dependence of the measured absorption coefficient α on the intensity I of the incident laser radiation is given by the expression[*]

$$\alpha = \frac{\alpha_0}{1 + I/I_s},$$
(1.2.18)

where α_0 is the low-intensity absorption coefficient, and I_s is a parameter known as the saturation intensity.

Optical Bistability.   One consequence of saturable absorption is optical bistability. One way of forming a bistable optical device is to place a saturable absorber inside a Fabry–Perot resonator, as illustrated in Fig. 1.2.8. As the

---

[*] This form is valid, for instance, for the case of homogeneous broadening of a simple atomic transition.

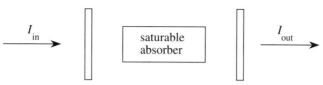

FIGURE 1.2.8   Bistable optical device.

input intensity is increased, the field inside the cavity also increases, lowering the absorption that the field experiences and thus increasing the field intensity still further. If the intensity of the incident field is subsequently lowered, the field inside the cavity tends to remain large because the absorption of the material system has already been reduced. A plot of the input-versus-output characteristics thus looks qualitatively like that shown in Fig. 1.2.9. Note that over an appreciable range of input intensities more than one output intensity is possible. The process of optical bistability is described in greater detail in Section 7.3.

Two-Photon Absorption

In the process of two-photon absorption, which is illustrated in Fig. 1.2.10, an atom makes a transition from its ground state to an excited state by the simultaneous absorption of two laser photons. The absorption cross section σ describing this process increases linearly with laser intensity according to the relation

$$\sigma = \sigma^{(2)} I, \tag{1.2.19}$$

where $\sigma^{(2)}$ is a coefficient that describes two-photon absorption. (Recall that in conventional, linear optics the absorption cross section σ is a constant.) Consequently, the atomic transition rate R due to two-photon absorption scales

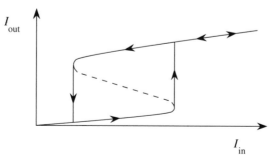

FIGURE 1.2.9   Typical input-versus-output characteristics of a bistable optical device.

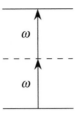

FIGURE 1.2.10   Two-photon absorption.

as the square of the laser intensity, since $R = \sigma I / \hbar \omega$, or as

$$R = \frac{\sigma^{(2)} I^2}{\hbar \omega}. \tag{1.2.20}$$

Two-photon absorption is a useful spectroscopic tool for determining the positions of energy levels that are not connected to the atomic ground state by a one-photon transition. Two-photon absorption was first observed experimentally by Kaiser and Garrett (1961).

Stimulated Raman Scattering

In stimulated Raman scattering, which is illustrated in Fig. 1.2.11, a photon of frequency ω is annihilated and a photon at the Stokes shifted frequency $\omega_s = \omega - \omega_v$ is created, leaving the molecule (or atom) in an excited state with energy $\hbar \omega_v$. The excitation energy is referred to as ω_v because stimulated Raman scattering was first studied in molecular systems, where $\hbar \omega_v$ corresponds to a vibrational energy. The efficiency of this process can be quite large, with often 10% or more of the power of the incident light being converted to the Stokes frequency. In contrast, the efficiency of normal or spontaneous Raman scattering is typically many orders of magnitude smaller. Stimulated Raman scattering is described more fully in Chapter 9.

Other stimulated scattering processes such as stimulated Brillouin scattering and stimulated Rayleigh scattering also occur and are described more fully in Chapter 8.

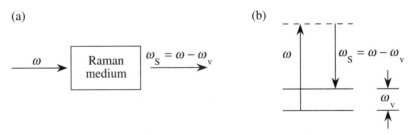

FIGURE 1.2.11   Stimulated Raman scattering.

1.3.   Formal Definition of the Nonlinear Susceptibility

Nonlinear optical interactions can be described in terms of the nonlinear polarization given by Eq. (1.1.2) only for a material system that is lossless and dispersionless. In the present section, we consider the more general case of a material with dispersion and/or loss. In this general case the nonlinear susceptibility becomes a complex quantity relating the complex amplitudes of the electric field and polarization.

We assume that we can represent the electric field vector of the optical wave as the discrete sum of a number of frequency components as

$$\tilde{\mathbf{E}}(\mathbf{r}, t) = \sideset{}{'}\sum_{n} \tilde{\mathbf{E}}_n(\mathbf{r}, t). \tag{1.3.1}$$

The prime on the summation sign of Eq. (1.3.1) indicates that the summation is to be taken over positive frequencies only. It is often convenient to represent $\tilde{\mathbf{E}}_n(\mathbf{r}, t)$ as the sum of its positive- and negative-frequency parts as

$$\tilde{\mathbf{E}}_n = \tilde{\mathbf{E}}_n^{(+)} + \tilde{\mathbf{E}}_n^{(-)}, \tag{1.3.2}$$

where

$$\tilde{\mathbf{E}}_n^{(+)} = \mathbf{E}_n e^{-i\omega_n t} \tag{1.3.3a}$$

and

$$\tilde{\mathbf{E}}_n^{(-)} = \tilde{\mathbf{E}}_n^{(+)*} = \mathbf{E}_n^* e^{i\omega_n t}. \tag{1.3.3b}$$

By requiring $\tilde{\mathbf{E}}_n^{(-)}$ to be the complex conjugate of $\tilde{\mathbf{E}}_n^{(+)}$ we are assured that the quantity of $\tilde{\mathbf{E}}(\mathbf{r}, t)$ of Eq. (1.3.1) will be real, as it must be in order to represent a physical field. It is also convenient to define the spatially slowly varying field amplitude $\mathbf{A}_n$ by means of the relation

$$\mathbf{E}_n = \mathbf{A}_n e^{i\mathbf{k}_n \cdot \mathbf{r}}. \tag{1.3.4}$$

The total electric field of Eq. (1.3.1) can thus be represented in terms of these field amplitudes by either of the expressions

$$\tilde{\mathbf{E}}(\mathbf{r}, t) = \sideset{}{'}\sum_{n} \mathbf{E}_n e^{-i\omega_n t} + \text{c.c.}$$

$$= \sideset{}{'}\sum_{n} \mathbf{A}_n e^{(i\mathbf{k}_n \cdot \mathbf{r} - \omega_n t)} + \text{c.c.} \tag{1.3.5}$$

On occasion, we shall express these field amplitudes using the alternative notation

$$\mathbf{E}_n = \mathbf{E}(\omega_n) \quad \text{and} \quad \mathbf{A}_n = \mathbf{A}(\omega_n). \tag{1.3.6}$$

In terms of this new notation, the reality condition of Eq. (1.3.3b) becomes

$$\mathbf{E}(-\omega_n) = \mathbf{E}(\omega_n)^* \quad \text{or} \quad \mathbf{A}(-\omega_n) = \mathbf{A}(\omega_n)^*. \qquad (1.3.7)$$

Using this new notation, we can write the total field in the more compact form

$$\tilde{\mathbf{E}}(\mathbf{r}, t) = \sum_n \mathbf{E}(\omega_n)e^{-i\omega_n t}$$
$$= \sum_n \mathbf{A}(\omega_n)e^{i(\mathbf{k}_n \cdot \mathbf{r} - \omega_n t)}, \qquad (1.3.8)$$

where the unprimed summation symbol denotes a summation over all frequencies, both positive and negative.

Note that according to our definition of field amplitude, the field given by

$$\tilde{\mathbf{E}}(\mathbf{r}, t) = \mathscr{E} \cos(\mathbf{k} \cdot \mathbf{r} - \omega t) \qquad (1.3.9)$$

is represented by the complex field amplitudes

$$\mathbf{E}(\omega) = \tfrac{1}{2}\mathscr{E}e^{i\mathbf{k}\cdot\mathbf{r}}, \qquad \mathbf{E}(-\omega) = \tfrac{1}{2}\mathscr{E}e^{-i\mathbf{k}\cdot\mathbf{r}}, \qquad (1.3.10)$$

or alternatively by the slowly varying amplitudes

$$\mathbf{A}(\omega) = \tfrac{1}{2}\mathscr{E}, \qquad \mathbf{A}(-\omega) = \tfrac{1}{2}\mathscr{E}. \qquad (1.3.11)$$

In either representation, factors of $1/2$ appear because the physical field amplitude $\mathscr{E}$ has been divided equally between the positive- and negative-frequency components.

Using a notation similar to that of Eq. (1.3.8), we can express the nonlinear polarization as

$$\tilde{\mathbf{P}}(\mathbf{r}, t) = \sum_n \mathbf{P}(\omega_n)e^{-i\omega_n t}, \qquad (1.3.12)$$

where, as before, the summation extends over all positive- and negative-frequency components.

We now define the components of the second-order susceptibility tensor $\chi_{ijk}^{(2)}$ $(\omega_n + \omega_m, \omega_n, \omega_m)$ as the constants of proportionality relating the amplitude of the nonlinear polarization to the product of field amplitudes according to

$$P_i(\omega_n + \omega_m) = \sum_{jk} \sum_{(nm)} \chi_{ijk}^{(2)}(\omega_n + \omega_m, \omega_n, \omega_m)E_j(\omega_n)E_k(\omega_m). \quad (1.3.13)$$

Here the indices ijk refer to the cartesian components of the fields. The notation (nm) indicates that, in performing the summation over n and m, the sum $\omega_n + \omega_m$ is to be held fixed, although ω_n and ω_m are each allowed to vary. Since the amplitude $E(\omega_n)$ is associated with the time dependence $\exp(-i\omega_n t)$, and the amplitude $E(\omega_m)$ is associated with the time dependence $\exp(-i\omega_m t)$, their product $E(\omega_n)E(\omega_m)$ is associated with the time dependence $\exp[-i(\omega_n + \omega_m)t]$. Hence the product $E(\omega_n)E(\omega_m)$ does in fact lead to a contribution to the nonlinear polarization oscillating at frequency $\omega_n + \omega_m$, as the notation of Eq. (1.3.13) suggests. Following convention, we have written $\chi^{(2)}$ as a function of three frequency arguments. This is technically unnecessary in that the first argument is always the sum of the other two. To emphasize this fact, the susceptibility $\chi^{(2)}(\omega_3, \omega_2, \omega_1)$ is sometimes written as $\chi^{(2)}(\omega_3; \omega_2, \omega_1)$ as a reminder that the first argument is different from the other two; or it may be written symbolically as $\chi^{(2)}(\omega_3 = \omega_2 + \omega_1)$.

Let us examine some of the consequences of the definition of the nonlinear susceptibility as given by Eq. (1.3.13) by considering two simple examples.

1. *Sum-frequency generation.* We let the input field frequencies be ω_1 and ω_2 and the sum frequency be ω_3, so that $\omega_3 = \omega_1 + \omega_2$. Then, by carrying out the summation over ω_n and ω_m in Eq. (1.3.13), we find that

$$P_i(\omega_3) = \sum_{jk} \left[\chi_{ijk}^{(2)}(\omega_3, \omega_1, \omega_2) E_j(\omega_1) E_k(\omega_2) \right.$$
$$\left. + \chi_{ijk}^{(2)}(\omega_3, \omega_2, \omega_1) E_j(\omega_2) E_k(\omega_1) \right]. \tag{1.3.14}$$

This expression can be simplified by making use of the intrinsic permutation symmetry of the nonlinear susceptibility (this symmetry is discussed in more detail in Eq. (1.5.6) below), which requires that

$$\chi_{ijk}^{(2)}(\omega_m + \omega_n, \omega_m, \omega_n) = \chi_{ikj}^{(2)}(\omega_m + \omega_n, \omega_n, \omega_m). \tag{1.3.15}$$

Through use of this relation, the expression for the nonlinear polarization becomes

$$P_i(\omega_3) = 2 \sum_{jk} \chi_{ijk}^{(2)}(\omega_3, \omega_1, \omega_2) E_j(\omega_1) E_k(\omega_2), \tag{1.3.16}$$

and for the special case in which both input fields are polarized in the x direction the polarization becomes

$$P_i(\omega_3) = 2 \chi_{ixx}^{(2)}(\omega_3, \omega_1, \omega_2) E_x(\omega_1) E_x(\omega_2). \tag{1.3.17}$$

2. *Second-harmonic generation.* We take the input frequency as ω_1 and the generated frequency as $\omega_3 = 2\omega_1$. If we again perform the summation over field frequencies in Eq. (1.3.13), we obtain

$$P_i(\omega_3) = \sum_{jk} \chi_{ijk}^{(2)}(\omega_3, \omega_1, \omega_1) E_j(\omega_1) E_k(\omega_1). \tag{1.3.18}$$

Again assuming the special case of an input field polarization along the x direction, this result becomes

$$P_i(\omega_3) = \chi_{ixx}^{(2)}(\omega_3, \omega_1, \omega_1) E_x(\omega_1)^2. \tag{1.3.19}$$

Note that a factor of two appears in Eqs. (1.3.16) and (1.3.17), which describe sum-frequency generation, but not in Eqs. (1.3.18) and (1.3.19), which describe second-harmonic generation. The fact that these expressions remain different even as ω_2 approaches ω_1 is at first sight surprising, but is a consequence of our convention that $\chi_{ijk}^{(2)}(\omega_3, \omega_1, \omega_2)$ must approach $\chi_{ijk}^{(2)}(\omega_3, \omega_1, \omega_1)$ as ω_1 approaches ω_2. Note that the expressions for $P(2\omega_2)$ and $P(\omega_1 + \omega_2)$ that apply for the case of a dispersionless nonlinear susceptibility (Eq. (1.2.7)) also differ by a factor of two. Moreover, one should expect the nonlinear polarization produced by two distinct fields to be larger than that produced by a single field (both of the same amplitude, say), because the total light intensity is larger in the former case.

In general, the summation over field frequencies ($\sum_{(nm)}$) in Eq. (1.3.13) can be performed formally to obtain the result

$$P_i(\omega_n + \omega_m) = D \sum_{jk} \chi_{ijk}^{(2)}(\omega_n + \omega_m, \omega_n, \omega_m) E_j(\omega_n) E_k(\omega_m), \quad (1.3.20)$$

where D is known as the degeneracy factor and is equal to the number of distinct permutations of the applied field frequencies ω_n and ω_m.

The expression (1.3.13) defining the second-order susceptibility can readily be generalized to higher-order interactions. In particular, the components of the third-order susceptibility are defined as the coefficients relating the amplitudes according to the expression

$$P_i(\omega_o + \omega_n + \omega_m) = \sum_{jkl} \sum_{(mno)} \chi_{ijkl}^{(3)}(\omega_0 + \omega_n + \omega_m, \omega_o, \omega_n, \omega_m)$$
$$\times E_j(\omega_o) E_k(\omega_n) E_l(\omega_m). \quad (1.3.21)$$

We can again perform the summation over m, n, and o to obtain the result

$$P_i(\omega_o + \omega_n + \omega_m) = D \sum_{jkl} \chi_{ijkl}^{(3)}(\omega_0 + \omega_n + \omega_m, \omega_o, \omega_n, \omega_m)$$
$$\times E_j(\omega_0) E_k(\omega_n) E_l(\omega_m), \quad (1.3.22)$$

where the degeneracy factor D represents the number of distinct permutations of the frequencies ω_m, ω_n, and ω_o.

1.4. Nonlinear Susceptibility of a Classical Anharmonic Oscillator

The Lorentz model of the atom, which treats the atom as a harmonic oscillator, is known to provide a very good description of the linear optical properties of atomic vapors and of nonmetallic solids. In the present section, we extend the Lorentz model by allowing the possibility of a nonlinearity in the restoring force exerted on the electron. The details of the analysis differ depending upon whether or not the medium possesses inversion symmetry.* We first treat the case of a noncentrosymmetric medium, and we find that such a medium can give rise to a second-order optical nonlinearity. We then treat the case of a medium that possesses a center of symmetry and find that the lowest-order nonlinearity that can occur in this case is a third-order nonlinear susceptibility. Our treatment is similar to that of Owyoung (1971).

The primary shortcoming of the classical model of optical nonlinearities presented here is that this model ascribes a single resonance frequency (ω_o) to each atom. In contrast, the quantum-mechanical theory of the nonlinear optical susceptibility, to be developed in Chapter 3, allows each atom to possess many energy eigenvalues and hence more than one resonance frequency. Since the present model allows for only one resonance frequency, it cannot properly describe the complete resonance nature of the nonlinear susceptibility (such as, for example, the possibility of simultaneous one- and two-photon resonances). However, it provides a good description for those cases in which all of the optical frequencies are considerably smaller than the lowest electronic resonance frequency of the material system.

Noncentrosymmetric Media

For the case of noncentrosymmetric media, we take the equation of motion of the electron coordinate $\tilde{x}$ to be the form

$$\ddot{\tilde{x}} + 2\gamma\dot{\tilde{x}} + \omega_0^2\tilde{x} + a\tilde{x}^2 = -e\tilde{E}(t)/m. \tag{1.4.1}$$

In this equation we have assumed that the applied electric field is given by $\tilde{E}(t)$, that the charge of the electron is $-e$, that there is a damping force[†] of the form $-2m\gamma\dot{\tilde{x}}$, and that the restoring force is given by

$$\tilde{F}_{\text{restoring}} = -m\omega_0^2\tilde{x} - ma\tilde{x}^2, \tag{1.4.2}$$

where a is a parameter that characterizes the strength of the nonlinearity. We obtain this form by assuming that the restoring force is a nonlinear function of the displacement of the electron from its equilibrium position and retaining the linear and quadratic terms in the Taylor series expansion of the restoring force in the displacement $\tilde{x}$. We can understand the nature of this form of the

---

* The role of symmetry in determining the nature of the nonlinear susceptibilty is discussed from a more fundamental point of view in Section 1.5. See especially the treatment leading from Eq. (1.5.31) to (1.5.35).

[†] The factor of two is introduced to make γ the dipole damping rate. 2γ is therefore the full width at half maximum in angular frequency units of the atomic absorption profile in the limit of linear response.

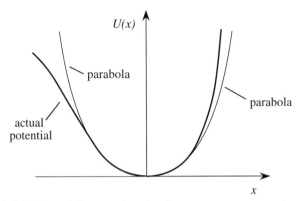

FIGURE 1.4.1   Potential energy function for a noncentrosymmetric medium.

restoring force by noting that it corresponds to a potential energy function of the form

$$U = - \int \tilde{F}_{\text{restoring}} d\tilde{x} = \tfrac{1}{2} m \omega_0^2 \tilde{x}^2 + \tfrac{1}{3} m a \tilde{x}^3. \qquad (1.4.3)$$

Here the first term corresponds to a harmonic potential and the second term corresponds to an anharmonic correction term, as illustrated in Fig. 1.4.1. This model corresponds to the physical situation of electrons in real materials, because the actual potential well that the atomic electron feels is not perfectly parabolic. The present model can describe only noncentrosymmetric media because we have assumed that the potential energy function U of Eq. (1.4.3) contains both even and odd powers of $\tilde{x}$; for a centrosymmetric medium only even powers of $\tilde{x}$ could appear, because the potential function $U(\tilde{x})$ must possess the symmetry $U(\tilde{x}) = U(-\tilde{x})$. For simplicity, we have written Eq. (1.4.1) in the scalar-field approximation; note that we cannot treat the tensor nature of the nonlinear susceptibility without making explicit assumptions regarding the symmetry properties of the material.

We assume that the applied optical field is of the form

$$\tilde{E}(t) = E_1 e^{-i\omega_1 t} + E_2 e^{-i\omega_2 t} + \text{c.c.} \qquad (1.4.4)$$

No general solution to Eq. (1.4.1) for an applied field of the form (1.4.4) is known. However, if the applied field is sufficiently weak, the nonlinear term $a\tilde{x}^2$ will be much smaller than the linear term $\omega_0^2 \tilde{x}$ for any displacement $\tilde{x}$ that can be induced by the field. Under this circumstance, Eq. (1.4.1) can be solved by means of a perturbation expansion. We use a procedure analogous to that of Rayleigh–Schrödinger perturbation theory in quantum mechanics. We replace $\tilde{E}(t)$ in Eq. (1.4.1) by $\lambda \tilde{E}(t)$, where λ is a parameter that ranges continuously between zero and one and that will be set equal to one at the end of the calculation. The expansion parameter λ thus characterizes the strength of the perturbation. Equation (1.4.1) then becomes

$$\ddot{\tilde{x}} + 2\gamma\dot{\tilde{x}} + \omega_0^2\tilde{x} + a\tilde{x}^2 = -\lambda e\tilde{E}(t)/m. \tag{1.4.5}$$

We now seek a solution to Eq. (1.4.5) in the form of a power series expansion in the strength λ of the perturbation, that is, a solution of the form

$$\tilde{x} = \lambda\tilde{x}^{(1)} + \lambda^2\tilde{x}^{(2)} + \lambda^3\tilde{x}^{(3)} + \cdots. \tag{1.4.6}$$

In order for Eq. (1.4.6) to be a solution to Eq. (1.4.5) for any value of the coupling strength λ, we require that the terms in Eq. (1.4.5) proportional to λ, λ^2, λ^3, etc., each satisfy the equation separately. We find that these terms lead respectively to the equations

$$\ddot{\tilde{x}}^{(1)} + 2\gamma\dot{\tilde{x}}^{(1)} + \omega_0^2\tilde{x}^{(1)} = -e\tilde{E}(t)/m, \tag{1.4.7a}$$

$$\ddot{\tilde{x}}^{(2)} + 2\gamma\dot{\tilde{x}}^{(2)} + \omega_0^2\tilde{x}^{(2)} + a[\tilde{x}^{(1)}]^2 = 0, \tag{1.4.7b}$$

$$\ddot{\tilde{x}}^{(3)} + 2\gamma\dot{\tilde{x}}^{(3)} + \omega_0^2\tilde{x}^{(3)} + 2a\tilde{x}^{(1)}\tilde{x}^{(2)} = 0. \tag{1.4.7c}$$

We see from Eq. (1.4.7a) that the lowest-order contribution $\tilde{x}^{(1)}$ is governed by the same equation as that of the conventional (i.e., linear) Lorentz model. Its steady-state solution is given by

$$\tilde{x}^{(1)}(t) = x^{(1)}(\omega_2)e^{-i\omega_2 t} + x^{(1)}(\omega_2)e^{-i\omega_1 t} + \text{c.c.}, \tag{1.4.8}$$

where the amplitudes $x^{(1)}(\omega_j)$ have the form

$$x^{(1)}(\omega_j) = -\frac{e}{m}\frac{E_j}{D(\omega_j)}, \tag{1.4.9}$$

where we have introduced the complex denominator function

$$D(\omega_j) = \omega_0^2 - \omega_j^2 - 2i\omega_j\gamma. \tag{1.4.10}$$

This expression for $\tilde{x}^{(1)}(t)$ is now squared and substituted into Eq. (1.4.7b), which is solved to obtain the lowest-order correction term $\tilde{x}^{(2)}$. The square of $\tilde{x}^{(1)}(t)$ contains the frequencies $\pm 2\omega_1$, $\pm 2\omega_2$, $\pm(\omega_1 + \omega_2)$, $\pm(\omega_1 - \omega_2)$, and 0. To determine the response at frequency $2\omega_1$, for instance, we must solve the equation

$$\ddot{\tilde{x}}^{(2)} + 2\gamma\dot{\tilde{x}}^{(2)} + \omega_0^2\tilde{x}^{(2)} = \frac{-a(eE_1/m)^2 e^{-2i\omega_1 t}}{D^2(\omega_1)}. \tag{1.4.11}$$

We seek a steady-state solution of the form

$$\tilde{x}^{(2)}(t) = x^{(2)}(2\omega_1)e^{-2i\omega_1 t}. \tag{1.4.12}$$

Substitution of Eq. (1.4.12) into Eq. (1.4.11) leads to the result

$$x^{(2)}(2\omega_1) = \frac{-a(e/m)^2 E_1^2}{D(2\omega_1)D^2(\omega_1)}, \tag{1.4.13}$$

where we have made use of the definition (1.4.10) of the function $D(\omega_j)$. Analogously, the amplitudes of the response at the other frequencies are found to be

$$x^{(2)}(2\omega_2) = \frac{-a(e/m)^2 E_2^2}{D(2\omega_2)D^2(\omega_2)}, \tag{1.4.14a}$$

$$x^{(2)}(\omega_1 + \omega_2) = \frac{-2a(e/m)^2 E_1 E_2}{D(\omega_1 + \omega_2)D(\omega_1)D(\omega_2)}, \tag{1.4.14b}$$

$$x^{(2)}(\omega_1 - \omega_2) = \frac{-2a(e/m)^2 E_1 E_2^*}{D(\omega_1 - \omega_2)D(\omega_1)D(-\omega_2)}, \tag{1.4.14c}$$

$$x^{(2)}(0) = \frac{-2a(e/m)^2 E_1 E_1^*}{D(0)D(\omega_1)D(-\omega_1)} + \frac{-2a(e/m)^2 E_2 E_2^*}{D(0)D(\omega_2)D(-\omega_2)}. \tag{1.4.14d}$$

We next express these results in terms of the linear ($\chi^{(1)}$) and nonlinear ($\chi^{(2)}$) susceptibilities. The linear susceptibility is defined through the relation

$$P^{(1)}(\omega_j) = \chi^{(1)}(\omega_j)E(\omega_j). \tag{1.4.15}$$

Since the linear contribution to the polarization is given by

$$P^{(1)}(\omega_j) = -Nex^{(1)}(\omega_j), \tag{1.4.16}$$

where N is the number density of atoms, we find using Eqs. (1.4.8) and (1.4.9) that the linear susceptibility is given by

$$\chi^{(1)}(\omega_j) = \frac{N(e^2/m)}{D(\omega_j)}. \tag{1.4.17}$$

The nonlinear susceptibilities are calculated in an analogous manner. The nonlinear susceptibility describing second-harmonic generation is defined by the relation

$$P^{(2)}(2\omega_1) = \chi^{(2)}(2\omega_1, \omega_1, \omega_1)E(\omega_1)^2, \tag{1.4.18}$$

where $P^{(2)}(2\omega_1)$ is the amplitude of the component of the nonlinear polarization oscillating at frequency $2\omega_1$ and is defined by the relation

$$P^{(2)}(2\omega_1) = -Nex^{(2)}(2\omega_i). \tag{1.4.19}$$

Comparison of these equations with Eq. (1.4.13) gives

$$\chi^{(2)}(2\omega_1, \omega_1, \omega_1) = \frac{N(e^3/m^2)a}{D(2\omega_1)D^2(\omega_1)}. \tag{1.4.20}$$

Through use of Eq. (1.4.17), this result can be written instead in terms of the product of linear susceptibilities as

$$\chi^{(2)}(2\omega_1, \omega_1, \omega_1) = \frac{ma}{N^2e^3}\chi^{(1)}(2\omega_1)\left[\chi^{(1)}(\omega_1)\right]^2. \tag{1.4.21}$$

The nonlinear susceptibility for second-harmonic generation of the ω_2 field is obtained trivially from Eqs. (1.4.20) and (1.4.21) through the substitution $\omega_1 \rightarrow \omega_2$.

The nonlinear susceptibility describing sum-frequency generation is obtained from the relations

$$P^{(2)}(\omega_1 + \omega_2) = 2\chi^{(2)}(\omega_1 + \omega_2, \omega_1, \omega_2)E(\omega_1)E(\omega_2) \tag{1.4.22}$$

and

$$P^{(2)}(\omega_1 + \omega_2) = -Nex^{(2)}(\omega_1 + \omega_2). \tag{1.4.23}$$

Note that in this case the relation defining the nonlinear susceptibility contains a factor of two because the two input fields are distinct, as discussed in relation to Eq. (1.3.20). By comparison of these equations with (1.4.14b), the nonlinear susceptibility is seen to be given by

$$\chi^{(2)}(\omega_1 + \omega_2, \omega_1, \omega_2) = \frac{N(e^3/m^2)a}{D(\omega_1 + \omega_2)D(\omega_1)D(\omega_2)}, \tag{1.4.24}$$

which can be expressed in terms of the product of linear susceptibilities as

$$\chi^{(2)}(\omega_1 + \omega_2, \omega_1, \omega_2) = \frac{ma}{N^2e^3}\chi^{(1)}(\omega_1 + \omega_2)\chi^{(1)}(\omega_1)\chi^{(1)}(\omega_2). \tag{1.4.25}$$

It can be seen by comparison of Eqs. (1.4.20) and (1.4.24) that, as ω_2 approaches ω_1, $\chi^{(2)}(\omega_1 + \omega_2, \omega_1, \omega_2)$ approaches $\chi^{(2)}(2\omega_1, \omega_1, \omega_1)$.

The nonlinear susceptibilities describing the other second-order processes are obtained in an analogous manner. For difference-frequency generation we find that

$$\begin{aligned}\chi^{(2)}(\omega_1 - \omega_2, \omega_1, -\omega_2) &= \frac{N(e^3/m^2)a}{D(\omega_1 - \omega_2)D(\omega_1)D(-\omega_2)} \\ &= \frac{ma}{N^2e^3}\chi^{(1)}(\omega_1 - \omega_2)\chi^{(1)}(\omega_1)\chi^{(1)}(-\omega_2),\end{aligned} \tag{1.4.26}$$

and for optical rectification we find that

$$\chi^{(2)}(0, \omega_1, -\omega_1) = \frac{N(e^3/m^2)a}{D(0)D(\omega_1)D(-\omega_1)}$$

$$= \frac{ma}{N^2e^3}\chi^{(1)}(0)\chi^{(1)}(\omega_1)\chi^{(1)}(-\omega_1).$$

(1.4.27)

The analysis just presented shows that the lowest-order nonlinear contribution to the polarization of a noncentrosymmetric material is second order in the applied field strength. This analysis can readily be extended to include higher-order effects. The solution to Eq. (1.4.7c), for example, leads to a third-order or $\chi^{(3)}$ susceptibility, and more generally terms proportional to λ^n in the expansion described by Eq. (1.4.6) lead to a $\chi^{(n)}$ susceptibility.

Miller's Rule

An empirical rule due to Miller (Miller, 1964; see also Garrett and Robinson, 1966) can be understood in terms of the calculation just presented. Miller noted that the quantity

$$\frac{\chi^{(2)}(\omega_1 + \omega_2, \omega_1, \omega_2)}{\chi^{(1)}(\omega_1 + \omega_2)\chi^{(1)}(\omega_1)\chi^{(1)}(\omega_2)}$$

(1.4.28)

is nearly constant for all noncentrosymmetric crystals. By comparison with Eq. (1.4.25), we see this quantity will be constant only if the combination

$$\frac{ma}{N^2e^3}$$

(1.4.29)

is nearly constant. In fact, the atomic number density N is nearly the same ($\sim 10^{22}$ cm^{-3}) for all condensed matter, and the parameters m and e are fundamental constants. We can estimate the size of the nonlinear coefficient a by noting that the linear and nonlinear contributions to the restoring force given by Eq. (1.4.2) would be expected to be comparable when the displacement $\tilde{x}$ of the electron from its equilibrium position is approximately equal to the size of the atom. This distance is of the order of the separation between atoms, that is, of the lattice constant d. This reasoning leads to the order-of-magnitude estimate that $m\omega_0^2 d = mad^2$ or that

$$a = \frac{\omega_0^2}{d}.$$

(1.4.30)

Since ω_0 and d are roughly the same for most solids, the quantity a would also be expected to be roughly the same for all materials where it does not vanish by reasons of symmetry.

We can also use the estimate of the nonlinear coefficient a given by Eq. (1.4.30) to make an estimate of the size of the second-order susceptibility under nonresonant conditions. If we replace $D(\omega)$ by ω_0^2 in the denominator of Eq. (1.4.24), set N equal to $1/d^3$, and set a equal to ω_0^2/d, we find that $\chi^{(2)}$ is given approximately by

$$\chi^{(2)} = \frac{e^3}{m^2 \omega_0^4 d^4}. \qquad (1.4.31)$$

Using the values $\omega_0 = 1 \times 10^{16}\,\text{rad/s}$, $d = 3\,\text{Å}$, $e = 4.8 \times 10^{-10}$ statcoulomb, and $m = 9.1 \times 10^{-28}$ g, we find that

$$\chi^{(2)} \simeq 1.6 \times 10^{-8}\ \text{esu}, \qquad (1.4.32)$$

which is in good agreement with the measured values presented in Table 1.5.3 of the next section.

Centrosymmetric Media

For the case of a centrosymmetric medium, we assume that the electronic restoring force is given not by Eq. (1.4.2) but rather by

$$\tilde{F}_{\text{restoring}} = -m\omega_0^2 \tilde{x} + mb\tilde{x}^3, \qquad (1.4.33)$$

where b is a parameter that characterizes the strength of the nonlinearity. This restoring force corresponds to the potential energy function

$$U = -\int \tilde{F}_{\text{restoring}} d\tilde{x} = \tfrac{1}{2} m\omega_0^2 \tilde{x}^2 - \tfrac{1}{4} mb\tilde{x}^4. \qquad (1.4.34)$$

This potential function is illustrated in the Fig. 1.4.2 (for the usual case in which b is positive) and is seen to be symmetric under the operation $\tilde{x} \rightarrow -\tilde{x}$,

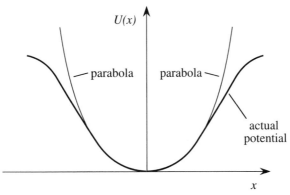

FIGURE 1.4.2   Potential energy function for a centrosymmetric medium.

which it must be for a medium that possesses a center of inversion symmetry. Note that $-mb\tilde{x}^4/4$ is simply the lowest-order correction term to the parabolic potential well described by the term $\frac{1}{2}m\omega_0^2\tilde{x}^2$. We assume that the electronic displacement $\tilde{x}$ never becomes so large that it is necessary to include higher-order terms in the potential.

We shall see below that the lowest-order nonlinear response resulting from the restoring force of Eq. (1.4.33) is a third-order contribution to the polarization, which can be described by a $\chi^{(3)}$ susceptibility. As in the case of non-centrosymmetric media, the tensor properties of this susceptibility cannot be specified unless the internal symmetries of the medium are completely known. One of the most important cases is that of a material which is isotropic (as well as being centrosymmetric). Examples of such materials are glasses and liquids. In such a case, we can take the restoring force to have the form

$$\tilde{\mathbf{F}}_{\text{restoring}} = -m\omega_0^2\tilde{\mathbf{r}} + mb(\tilde{\mathbf{r}} \cdot \tilde{\mathbf{r}})\tilde{\mathbf{r}}. \qquad (1.4.35)$$

The second contribution to the restoring force must have the form shown because it is the only form that is third-order in the displacement $\tilde{\mathbf{r}}$ and is directed in the $\tilde{\mathbf{r}}$ direction, which is the only possible direction for an isotropic medium.

The equation of motion for the electron displacement from equilibrium is thus

$$\ddot{\tilde{\mathbf{r}}} + 2\gamma\dot{\tilde{\mathbf{r}}} + \omega_0^2\tilde{\mathbf{r}} - b(\tilde{\mathbf{r}} \cdot \tilde{\mathbf{r}})\tilde{\mathbf{r}} = -e\tilde{\mathbf{E}}(t)/m. \qquad (1.4.36)$$

We assume that the applied field is given by

$$\tilde{\mathbf{E}}(t) = \mathbf{E}_1 e^{-i\omega_1 t} + \mathbf{E}_2 e^{-i\omega_2 t} + \mathbf{E}_3 e^{-i\omega_3 t} + \text{c.c.}; \qquad (1.4.37)$$

we allow the field to have three distinct frequency components because this is the most general possibility for a third-order interaction. However, the algebra becomes very tedious if all three terms are written explicitly, and hence we express the applied field as

$$\tilde{\mathbf{E}}(t) = \sum_n \mathbf{E}(\omega_n) e^{-i\omega_n t}. \qquad (1.4.38)$$

The method of solution is analogous to that used above for a noncentrosymmetric medium. We replace $\tilde{\mathbf{E}}(t)$ in Eq. (1.4.36) by $\lambda\tilde{\mathbf{E}}(t)$, where λ is a parameter that characterizes the strength of the perturbation and that is set equal to unity at the end of the calculation. We seek a solution to Eq. (1.4.36) having the form of a power series in the parameter λ:

$$\tilde{\mathbf{r}}(t) = \lambda\tilde{\mathbf{r}}^{(1)}(t) + \lambda^2\tilde{\mathbf{r}}^{(2)}(t) + \lambda^3\tilde{\mathbf{r}}^{(3)} + \cdots. \qquad (1.4.39)$$

We insert Eq. (1.4.39) into Eq. (1.4.36) and require that the terms proportional to λ^n vanish separately for each value of n. We thereby find that

$$\ddot{\tilde{\mathbf{r}}}^{(1)} + 2\gamma\dot{\tilde{\mathbf{r}}}^{(1)} + \omega_0^2\tilde{\mathbf{r}}^{(1)} = -e\tilde{\mathbf{E}}(t)/m, \quad (1.4.40a)$$

$$\ddot{\tilde{\mathbf{r}}}^{(2)} + 2\gamma\dot{\tilde{\mathbf{r}}}^{(2)} + \omega_0^2\tilde{\mathbf{r}}^{(2)} = 0, \quad (1.4.40b)$$

$$\ddot{\tilde{\mathbf{r}}}^{(3)} + 2\gamma\dot{\tilde{\mathbf{r}}}^{(3)} + \omega_0^2\tilde{\mathbf{r}}^{(3)} - b\big(\tilde{\mathbf{r}}^{(1)} \cdot \tilde{\mathbf{r}}^{(1)}\big)\tilde{\mathbf{r}}^{(1)} = 0 \quad (1.4.40c)$$

for $n = 1, 2$, and 3, respectively. Equation (1.4.40a) is simply the vector version of Eq. (1.4.7a), encountered above. Its steady-state solution is

$$\tilde{\mathbf{r}}^{(1)}(t) = \sum_n \mathbf{r}^{(1)}(\omega_n)e^{i\omega_n t}, \quad (1.4.41a)$$

where

$$\mathbf{r}^{(1)}(\omega_n) = \frac{-e\mathbf{E}(\omega_n)/m}{D(\omega_n)} \quad (1.4.41b)$$

with $D(\omega_n)$ given as before by $D(\omega_n) = \omega_0^2 - \omega_n^2 - 2i\omega_n\gamma$. Since the polarization at frequency ω_n is given by

$$\mathbf{P}^{(1)}(\omega_n) = -Ne\mathbf{r}^{(1)}(\omega_n), \quad (1.4.42)$$

we can describe the cartesian components of the polarization through the relation

$$P_i^{(1)}(\omega_n) = \sum_j \chi_{ij}^{(1)}(\omega_n)E_j(\omega_n). \quad (1.4.43a)$$

Here the linear susceptibility is given by

$$\chi_{ij}^{(1)}(\omega_n) = \chi^{(1)}(\omega_n)\delta_{ij} \quad (1.4.43b)$$

with $\chi^{(1)}(\omega_n)$ given as before by

$$\chi^{(1)}(\omega_n) = \frac{Ne^2/m}{D(\omega_n)}, \quad (1.4.43c)$$

and where δ_{ij} is the Kronecker delta, which is defined such that $\delta_{ij} = 1$ for $i = j$ and $\delta_{ij} = 0$ for $i \neq j$.

The second-order response of the system is described by Eq. (1.4.40b). Since this equation is damped but not driven, its steady-state solution vanishes, that is,

$$\tilde{\mathbf{r}}^{(2)} = 0. \quad (1.4.44)$$

To calculate the third-order response, we substitute the expression for $\tilde{\mathbf{r}}^{(1)}(t)$ given by Eq. (1.4.41a) into Eq. (1.4.40c), which becomes

$$\ddot{\tilde{\mathbf{r}}}^{(3)} + 2\gamma\dot{\tilde{\mathbf{r}}}^{(3)} + \omega_0^2\tilde{\mathbf{r}}^{(3)} = -\sum_{mnp} \frac{be^3[\mathbf{E}(\omega_m) \cdot \mathbf{E}(\omega_n)]\mathbf{E}(\omega_p)}{m^3 D(\omega_m)D(\omega_n)D(\omega_p)}$$

$$\times\, e^{-i(\omega_m+\omega_m+\omega_p)t}. \tag{1.4.45}$$

Because of the summation over m, n, and p, the right-hand side of this equation contains many different frequencies. We denote one of these frequencies by $\omega_q = \omega_m + \omega_n + \omega_p$. The solution to Eq. (1.4.45) can then be written in the form

$$\tilde{\mathbf{r}}^{(3)}(t) = \sum_q \mathbf{r}^{(3)}(\omega_q)e^{-i\omega_q t}. \tag{1.4.46}$$

We substitute Eq. (1.4.46) into Eq. (1.4.45) and find that $\mathbf{r}^{(3)}(\omega_q)$ is given by

$$\left(-\omega_q^2 - i\omega_q 2\gamma + \omega_0^2\right)\mathbf{r}^{(3)}(\omega_q) = -\sum_{(mnp)} \frac{be^3[\mathbf{E}(\omega_m) \cdot \mathbf{E}(\omega_n)]\mathbf{E}(\omega_p)}{m^3 D(\omega_m)D(\omega_n)D(\omega_p)}, \tag{1.4.47}$$

where the summation is to be carried out over frequencies ω_m, ω_n, and ω_p with the restriction that $\omega_m + \omega_n + \omega_p$ must equal ω_q. Since the coefficient of $\mathbf{r}^{(3)}(\omega_q)$ on the left-hand side is just $D(\omega_q)$, we obtain

$$\mathbf{r}^{(3)}(\omega_q) = -\sum_{(mnp)} \frac{be^3[\mathbf{E}(\omega_m) \cdot \mathbf{E}(\omega_n)]\mathbf{E}(\omega_p)}{m^3 D(\omega_q)D(\omega_m)D(\omega_n)D(\omega_p)}. \tag{1.4.48}$$

The amplitude of the polarization component oscillating at frequency ω_q then is given in terms of this amplitude by

$$\mathbf{P}^{(3)}(\omega_q) = -Ne\mathbf{r}^{(3)}(\omega_q). \tag{1.4.49}$$

We next recall the definition of the third-order nonlinear susceptibility (1.3.21),

$$P_i^{(3)}(\omega_q) = \sum_{jkl}\sum_{(mnp)} \chi_{ijkl}^{(3)}(\omega_q, \omega_m, \omega_n, \omega_p)E_j(\omega_m)E_k(\omega_n)E_l(\omega_p). \tag{1.4.50}$$

Since this equation contains a summation over the dummy variables m, n, and p, there is more than one possible choice for the expression for the nonlinear susceptibility. An obvious choice for this expression for the susceptibility, based on the way in which Eqs. (1.4.48) and (1.4.49) are written, is

$$\chi_{ijkl}^{(3)}(\omega_q, \omega_m, \omega_n, \omega_p) = \frac{Nbe^4\delta_{jk}\delta_{il}}{m^3 D(\omega_q)D(\omega_m)D(\omega_n)D(\omega_p)}. \tag{1.4.51}$$

While Eq. (1.4.51) is a perfectly adequate expression for the nonlinear susceptibility, it does not explicitly show the full symmetry of the interaction in terms of the arbitrariness of which field we call $E_j(\omega_m)$, which we call $E_k(\omega_n)$, and which we call $E_l(\omega_p)$. It is conventional to define nonlinear susceptibilities in a manner that displays this symmetry, which is known as intrinsic permutation symmetry. Since there are six possible permutations of the orders in which $E_j(\omega_m)$, $E_k(\omega_n)$, and $E_l(\omega_p)$ may be taken, we define the third-order susceptibility to be one-sixth of the sum of the six expressions analogous to Eq. (1.4.51) with the input fields taken in all possible orders. When we carry out this prescription, we find that only three distinct contributions occur, and that the resulting form for the nonlinear susceptibility is given by

$$\chi_{ijkl}^{(3)}(\omega_q, \omega_m, \omega_n, \omega_p) = \frac{Nbe^4[\delta_{ij}\delta_{kl} + \delta_{ik}\delta_{jl} + \delta_{il}\delta_{jk}]}{3m^3 D(\omega_q)D(\omega_m)D(\omega_n)D(\omega_p)}. \quad (1.4.52)$$

This expression can be rewritten in terms of the linear susceptibilities at the four different frequencies ω_q, ω_m, ω_n and ω_p by using Eq. (1.4.43c) to eliminate the resonance denominator factors $D(\omega)$. We thereby obtain

$$\chi_{ijkl}^{(3)}(\omega_q, \omega_m, \omega_n, \omega_p) = \frac{bm}{3N^3 e^4}[\chi^{(1)}(\omega_q)\chi^{(1)}(\omega_m)\chi^{(1)}(\omega_n)\chi^{(1)}(\omega_p)]$$
$$\times [\delta_{ij}\delta_{kl} + \delta_{ik}\delta_{jl} + \delta_{il}\delta_{jk}]. \quad (1.4.53)$$

We can estimate the value of the phenomenological constant b that appears in this result by means of an argument analogous to that used above (see Eq. (1.4.30)) to estimate the value of the constant a that appears in the expression for $\chi^{(2)}$. We assume that the linear and nonlinear contributions to the restoring force given by Eq. (1.4.33) will become comparable in magnitude when the displacement $\tilde{x}$ is comparable to the atomic dimension d, that is, when $m\omega_0^2 d = mbd^3$, which implies that

$$b = \frac{\omega_0^2}{d^2}. \quad (1.4.54)$$

Using this expression for b, we can now estimate the value of the nonlinear susceptibility. For the case of nonresonant excitation, $D(\omega)$ is approximately equal to ω_0^2, and hence from Eq. (1.4.52) we obtain

$$\chi^{(3)} \simeq \frac{Nbe^4}{m^3\omega_0^8} = \frac{e^4}{m^3\omega_0^6 d^5}. \quad (1.4.55)$$

Taking $d = 3$ Å and $\omega_0 = 7 \times 10^{15}$ rad/sec, we obtain

$$\chi^{(3)} \simeq 3 \times 10^{-14} \frac{\text{cm}^2}{\text{statvolt}^2}. \quad (1.4.56)$$

We shall see in Chapter 4 that this value is typical of the nonlinear susceptibility of many materials.

1.5. Properties of the Nonlinear Susceptibility

In this section we study some of the formal symmetry properties of the non-linear susceptibility. Let us first see why it is important that we understand these symmetry properties. We consider the mutual interaction of three waves of frequencies ω_1, ω_2, and $\omega_3 = \omega_1 + \omega_2$. A complete description of the interaction of these waves requires that we know the nonlinear polarizations $\mathbf{P}(\omega_i)$ influencing each of them. Since these quantities are given in general (see also Eq. (1.3.13)) by the expression

$$P_i(\omega_n + \omega_m) = \sum_{jk} \sum_{(nm)} \chi_{ijk}^{(2)}(\omega_n + \omega_m, \omega_n, \omega_m) E_j(\omega_n) E_k(\omega_m), \quad (1.5.1)$$

we therefore need to determine the six tensors

$$\chi_{ijk}^{(2)}(\omega_1, \omega_3, -\omega_2), \quad \chi_{ijk}^{(2)}(\omega_1, -\omega_2, \omega_3), \quad \chi_{ijk}^{(2)}(\omega_2, \omega_3, -\omega_1),$$

$$\chi_{ijk}^{(2)}(\omega_2, -\omega_1, \omega_3), \quad \chi_{ijk}^{(2)}(\omega_3, \omega_1, \omega_2), \quad \text{and} \quad \chi_{ijk}^{(2)}(\omega_3, \omega_2, \omega_1)$$

and six additional tensors in which each frequency is replaced by its negative. In these expressions, the indices i, j, and k can independently take on the values x, y, and z. Since each of these 12 tensors thus consists of 27 cartesian components, as many as 324 different (complex) numbers need to be specified in order to describe the interaction.

Fortunately, there are a number of restrictions resulting from symmetries that relate the various components of $\chi^{(2)}$, and hence far fewer than 324 numbers are usually needed to describe the nonlinear coupling. In this section, we study some of these formal properties of the nonlinear susceptibility. The discussion will deal primarily with the second-order $\chi^{(2)}$ susceptibility, but can readily be extended to $\chi^{(3)}$ and higher-order susceptibilities.

Reality of the Fields

Recall that the nonlinear polarization describing the sum-frequency response to input fields at frequencies ω_n and ω_m has been represented as

$$\tilde{P}_i(\mathbf{r}, t) = P_i(\omega_n + \omega_m)e^{-i(\omega_n+\omega_m)t} + P_i(-\omega_n - \omega_m)e^{i(\omega_n+\omega_m)t}. \quad (1.5.2)$$

Since $\tilde{P}_i(\mathbf{r}, t)$ is a physically measurable quantity, it must be purely real, and hence its positive- and negative-frequency components must be related by

$$P_i(-\omega_n - \omega_m) = P_i(\omega_n + \omega_m)^*. \tag{1.5.3}$$

The electric field must also be a real quantity, and its complex frequency components must obey the analogous conditions:

$$E_j(-\omega_n) = E_j(\omega_n)^*, \tag{1.5.4a}$$

$$E_k(-\omega_m) = E_k(\omega_m)^*. \tag{1.5.4b}$$

Since the fields and polarization are related to each other through the second-order susceptibility of Eq. (1.5.1), we conclude that the positive- and negative-frequency components of the susceptibility must be related according to

$$\chi_{ijk}^{(2)}(-\omega_n - \omega_m, -\omega_n, -\omega_m) = \chi_{ijk}^{(2)}(\omega_n + \omega_m, \omega_n, \omega_m)^*. \tag{1.5.5}$$

Intrinsic Permutation Symmetry

Earlier we introduced the concept of intrinsic permutation symmetry when we rewrote the expression (1.4.51) for the nonlinear susceptibility of a classical, anharmonic oscillator in the conventional form of Eq. (1.4.52). In the present section, we treat the concept of intrinsic permutation symmetry from a more general point of view.

According to Eq. (1.5.1), one of the contributions to the nonlinear polarization $P_i(\omega_n + \omega_m)$ is the product $\chi_{ijk}^{(2)}(\omega_n + \omega_m, \omega_n, \omega_m)E_j(\omega_n)E_k(\omega_m)$. However, since j, k, n, and m are dummy indices, we could just as well have written this contribution with n interchanged with m and with j interchanged with k, that is, as $\chi_{ikj}^{(2)}(\omega_n + \omega_m, \omega_m, \omega_n)E_k(\omega_m)E_j(\omega_n)$. These two expressions are numerically equal if we require that the nonlinear susceptibility be unchanged by the simultaneous interchange of its last two frequency arguments and its last two cartesian indices:

$$\chi_{ijk}^{(2)}(\omega_n + \omega_m, \omega_n, \omega_m) = \chi_{ikj}^{(2)}(\omega_n + \omega_m, \omega_m, \omega_n). \tag{1.5.6}$$

This property is known as intrinsic permutation symmetry.

Note that this symmetry condition is introduced purely as a matter of convenience. For example, we could set one member of the pair of elements shown in Eq. (1.5.6) equal to zero and double the value of the other member. Then, when the double summation of Eq. (1.5.1) was carried out, the result for the physically meaningful quantity $P_j(\omega_n + \omega_m)$ would be left unchanged.

This symmetry condition can also be derived from a more general point of view using the concept of the nonlinear response function (Butcher, 1965; Flytzanis, 1975).

Symmetries for Lossless Media

Two additional symmetries of the nonlinear susceptibility tensor occur for the case of a lossless nonlinear medium.

The first of these conditions states that for a lossless medium all of the components of $\chi_{ijk}^{(2)}(\omega_n + \omega_m, \omega_n, \omega_m)$ are real. This result is obeyed for the classical anharmonic oscillator discussed in Section 1.4, as can be verified by evaluating the expression for $\chi^{(2)}$ in the limit in which all of the applied frequencies and their sums and differences are significantly different from the resonance frequency. The general proof that $\chi^{(2)}$ is real for a lossless medium is obtained by verifying that the quantum-mechanical expression for $\chi^{(2)}$ (which is derived in Chapter 3) is also purely real in this limit.

The second of these new symmetries is *full* permutation symmetry. This condition states that *all* of the frequency arguments of the nonlinear susceptibility can be freely interchanged, as long as the corresponding cartesian indices are interchanged simultaneously. In permuting the frequency arguments, it must be recalled that the first argument is always the sum of the latter two, and thus the signs of the frequencies must be inverted when the first frequency is interchanged with either of the latter two. Full permutation symmetry implies, for instance, that

$$\chi_{ijk}^{(2)}(\omega_3 = \omega_1 + \omega_2) = \chi_{jki}^{(2)}(-\omega_1 = \omega_2 - \omega_3). \tag{1.5.7}$$

However, according to Eq. (1.5.5) the right-hand side of this equation is equal to $\chi_{jki}^{(2)}(\omega_1 = -\omega_2 + \omega_3)^*$, which, due to the reality of $\chi^{(2)}$ for a lossless medium, is equal to $\chi_{jki}^{(2)}(\omega_1 = -\omega_2 + \omega_3)$. We hence conclude that

$$\chi_{ijk}^{(2)}(\omega_3 = \omega_1 + \omega_2) = \chi_{jki}^{(2)}(\omega_1 = -\omega_2 + \omega_3). \tag{1.5.8}$$

By an analogous procedure, one can show that

$$\chi_{ijk}^{(2)}(\omega_3 = \omega_1 + \omega_2) = \chi_{kij}^{(2)}(\omega_2 = \omega_3 - \omega_1). \tag{1.5.9}$$

A general proof of the validity of the condition of full permutation symmetry entails verifying that the quantum-mechanical expression for $\chi^{(2)}$ (which is derived in Chapter 3) obeys this condition when all of the optical frequencies are detuned many linewidths from the resonance frequencies of the optical medium. Full permutation symmetry can also be deduced from a consideration of the field energy density within a nonlinear medium, as shown below.

Field Energy Density for a Nonlinear Medium

The condition that the nonlinear susceptibility must possess full permutation symmetry for a lossless medium can be deduced from a consideration of the form of the electromagnetic field energy within a nonlinear medium. For the case of a linear medium, the energy density associated with the electric field

$$\tilde{E}_i(t) = \sum_n E_i(\omega_n)e^{-i\omega_n t} \qquad (1.5.10)$$

is given according to Poynting's theorem as

$$U = \frac{1}{8\pi}\langle \tilde{\mathbf{D}} \cdot \tilde{\mathbf{E}} \rangle = \frac{1}{8\pi}\sum_i \langle \tilde{D}_i \tilde{E}_i \rangle, \qquad (1.5.11)$$

where the angular brackets denote a time average. Since the displacement vector is given by

$$\tilde{D}_i(t) = \sum_j \epsilon_{ij}\tilde{E}_j(t) = \sum_j \sum_n \epsilon_{ij}(\omega_n)E_j(\omega_n)e^{-i\omega_n t}, \qquad (1.5.12)$$

where the dielectric tensor is given by

$$\epsilon_{ij}(\omega_n) = \delta_{ij} + 4\pi\chi_{ij}^{(1)}(\omega_n), \qquad (1.5.13)$$

we can write the energy density as

$$U = \frac{1}{8\pi}\sum_i \sum_n E_i^*(\omega_n)E_i(\omega_n) + \frac{1}{2}\sum_{ij}\sum_n E_i^*(\omega_n)\chi_{ij}^{(1)}(\omega_n)E_j(\omega_n). \qquad (1.5.14)$$

Here the first term represents the energy density associated with the electric field in vacuum and the second term represents the energy stored in the polarization of the medium.

For the case of a nonlinear medium, the expression for the electric field energy density (Kleinman, 1962; Armstrong *et al.*, 1962; Pershan, 1963) associated with the polarization of the medium takes the more general form

$$U = \frac{1}{2}\sum_{ij}\sum_n \chi_{ij}^{(1)}(\omega_n)E_i^*(\omega_n)E_j(\omega_n)$$

$$+ \frac{1}{3}\sum_{ijk}\sum_{mn}\chi_{ijk}^{(2)\prime}(-\omega_n-\omega_m,\omega_m,\omega_n)E_i^*(\omega_m+\omega_n)E_j(\omega_m)E_k(\omega_n)$$

$$+ \frac{1}{4}\sum_{ijkl}\sum_{mno}\chi_{ijkl}^{(3)\prime}(-\omega_o-\omega_n-\omega_m,\omega_m,\omega_n,\omega_o) \qquad (1.5.15)$$

$$\times E_i^*(\omega_m+\omega_n+\omega_o)E_j(\omega_m)E_k(\omega_n)E_l(\omega_o) + \cdots.$$

For the present, the quantities $\chi^{(2)\prime}$, $\chi^{(3)\prime}$, ... are to be thought of simply as coefficients in the power series expansion of U in the amplitudes of the applied field; later these quantities will be related to the nonlinear susceptibilities. Since the order in which the fields are multiplied together in determining U is immaterial, the quantities $\chi^{(n)\prime}$ clearly possess full permutation symmetry, that is, their frequency arguments can be freely permuted as long as the corresponding indices are also permuted.

In order to relate the expression (1.5.15) for the energy density to the nonlinear polarization, and subsequently to the nonlinear susceptibility, we use the result that the polarization of a medium is given (Pershan, 1963; Landau and Lifshitz, 1960) by the expression

$$P_i(\omega_n) = \frac{\partial U}{\partial E_i^*(\omega_n)}. \tag{1.5.16}$$

Thus, by differentiation of Eq. (1.5.15), we obtain an expression for the linear polarization as

$$P_i^{(1)}(\omega_m) = \sum_j \chi_{ij}^{(1)}(\omega_m) E_j(\omega_m), \tag{1.5.17a}$$

and for the nonlinear polarization as[*]

$$P_i^{(2)}(\omega_m + \omega_n) = \sum_{jk} \sum_{(mn)} \chi_{ijk}^{(2)\prime}(-\omega_m - \omega_n, \omega_m, \omega_n) E_j(\omega_m) E_k(\omega_n) \tag{1.5.17b}$$

$$P_i^{(3)}(\omega_m + \omega_n + \omega_o) = \sum_{jkl} \sum_{(mno)} \chi_{ijkl}^{(3)\prime}(-\omega_m - \omega_n - \omega_o, \omega_m, \omega_n, \omega_o) \tag{1.5.17c}$$

$$\times E_j(\omega_m) E_k(\omega_n) E_k(\omega_o).$$

We note that these last two expressions are identical to Eqs. (1.3.13) and (1.3.21), which define the nonlinear susceptibilities (except for the unimportant fact that the quantities $\chi^{(n)}$ and $\chi^{(n)\prime}$ use opposite conventions regarding the sign of the first frequency argument). Since the quantities $\chi^{(n)\prime}$ possess full permutation symmetry, we conclude that the susceptibilities $\chi^{(n)}$ do also. Note that this demonstration is valid only for the case of a lossless medium, because only in this case is the internal energy a function of state.

Kleinman's Symmetry

Quite often nonlinear optical interactions involve optical waves whose frequencies ω_i are much smaller than the lowest resonance frequency of the material system. Under these conditions, the nonlinear susceptibility is essentially independent of frequency. For example, the expression (1.4.24) for the second-order susceptibility of an anharmonic oscillator predicts a value of

---

[*] In performing the differentiation, the prefactors $\frac{1}{2}, \frac{1}{3}, \frac{1}{4}, \ldots$ of Eq. (1.5.15) disappear because 2, 3, 4, ... equivalent terms appear as the result the summations over the frequency arguments.

the susceptibility that is essentially independent of the frequencies of the applied waves whenever these frequencies are much smaller than the resonance frequency ω_0. Furthermore, under conditions of low-frequency excitation the system responds essentially instantaneously to the applied field, and we have seen in Section 1.2 that under such conditions the nonlinear polarization can be described in the time domain by the relation

$$\tilde{P}(t) = \chi^{(2)} \tilde{E}(t)^2, \tag{1.5.18}$$

where $\chi^{(2)}$ can be taken to be a constant.

Since the medium is necessarily lossless whenever the applied field frequencies ω_i are very much smaller than the resonance frequency ω_0, the condition of full permutation symmetry (1.5.7) must be valid under these circumstances. This condition states that the indices can be permuted as long as the frequencies are permuted simultaneously, and it leads to the conclusion that

$$\chi_{ijk}^{(2)}(\omega_3 = \omega_1 + \omega_2) = \chi_{jki}^{(2)}(\omega_1 = -\omega_2 + \omega_3) = \chi_{kij}^{(2)}(\omega_2 = \omega_3 - \omega_1)$$

$$= \chi_{ikj}^{(2)}(\omega_3 = \omega_2 + \omega_1) = \chi_{jik}^{(2)}(\omega_1 = \omega_3 - \omega_2)$$

$$= \chi_{kji}^{(2)}(\omega_2 = -\omega_1 + \omega_3).$$

However, under the present conditions $\chi^{(2)}$ does not actually depend on the frequencies, and we can therefore permute the indices without permuting the frequencies, leading to the result

$$\chi_{ijk}^{(2)}(\omega_3 = \omega_1 + \omega_2) = \chi_{jki}^{(2)}(\omega_3 = \omega_1 + \omega_2) = \chi_{kij}^{(2)}(\omega_3 = \omega_1 + \omega_2)$$

$$= \chi_{ikj}^{(2)}(\omega_3 = \omega_1 + \omega_2) = \chi_{jik}^{(2)}(\omega_3 = \omega_1 + \omega_2) \tag{1.5.19}$$

$$= \chi_{kji}^{(2)}(\omega_3 = \omega_1 + \omega_2).$$

This result is known as the Kleinman symmetry condition. It is valid whenever dispersion of the susceptibility can be neglected.

Contracted Notation

We now introduce a notational device that is often used when the Kleinman symmetry condition is valid. We introduce the tensor

$$d_{ijk} = \tfrac{1}{2}\chi_{ijk}^{(2)} \tag{1.5.20}$$

and for simplicity suppress the frequency arguments. The nonlinear polarization can then be written as

$$P_i(\omega_n + \omega_m) = \sum_{jk} \sum_{(nm)} 2d_{ijk} E_j(\omega_n) E_k(\omega_m). \tag{1.5.21}$$

We now assume that d_{ijk} is symmetric in its last two indices. This assumption is valid whenever Kleinman's symmetry condition is valid and in addition is valid in general for second-harmonic generation, since in this case ω_n and ω_m are equal. We then simplify the notation by introducing a contracted matrix d_{il} according to the prescription

$$
\begin{array}{ccccccc}
jk: & 11 & 22 & 33 & 23,32 & 31,13 & 12,21 \\
l: & 1 & 2 & 3 & 4 & 5 & 6
\end{array}
\tag{1.5.22}
$$

The nonlinear susceptibility tensor can then be represented as the 3×6 matrix

$$
d_{il} = \begin{bmatrix}
d_{11} & d_{12} & d_{13} & d_{14} & d_{15} & d_{16} \\
d_{21} & d_{22} & d_{23} & d_{24} & d_{25} & d_{26} \\
d_{31} & d_{32} & d_{33} & d_{34} & d_{35} & d_{36}
\end{bmatrix}.
\tag{1.5.23}
$$

If we now *explicitly* introduce the Kleinman symmetry condition, i.e., we assert that the indices d_{ijk} can be freely permuted, we find that not all of the 18 elements of d_{il} are independent. For instance, we see that

$$
d_{12} \equiv d_{122} = d_{212} \equiv d_{26}
\tag{1.5.24a}
$$

and that

$$
d_{14} \equiv d_{123} = d_{213} \equiv d_{25}.
\tag{1.5.24b}
$$

By applying this type of argument systematically, we find that d_{il} has only 10 independent elements when the Kleinman symmetry condition is valid; the form of d_{il} under these conditions is

$$
d_{il} = \begin{bmatrix}
d_{11} & d_{12} & d_{13} & d_{14} & d_{15} & d_{16} \\
d_{16} & d_{22} & d_{23} & d_{24} & d_{14} & d_{12} \\
d_{15} & d_{24} & d_{33} & d_{23} & d_{13} & d_{14}
\end{bmatrix}.
\tag{1.5.25}
$$

We can describe the nonlinear polarization leading to second-harmonic generation in terms of d_{il} by the matrix equation

$$
\begin{bmatrix}
P_x(2\omega) \\
P_y(2\omega) \\
P_z(2\omega)
\end{bmatrix}
= 2
\begin{bmatrix}
d_{11} & d_{12} & d_{13} & d_{14} & d_{15} & d_{16} \\
d_{21} & d_{22} & d_{23} & d_{24} & d_{25} & d_{26} \\
d_{31} & d_{32} & d_{33} & d_{34} & d_{35} & d_{36}
\end{bmatrix}
\begin{bmatrix}
E_x(\omega)^2 \\
E_y(\omega)^2 \\
E_z(\omega)^2 \\
2E_y(\omega)E_z(\omega) \\
2E_x(\omega)E_z(\omega) \\
2E_x(\omega)E_y(\omega)
\end{bmatrix}.
$$

$$
\tag{1.5.26}
$$

When the Kleinman symmetry condition is valid, we can describe the nonlinear polarization leading to sum-frequency generation (with $\omega_3 = \omega_1 + \omega_2$) by the equation

$$
\begin{bmatrix} P_x(\omega_3) \\ P_y(\omega_3) \\ P_z(\omega_3) \end{bmatrix} = 4 \begin{bmatrix} d_{11} & d_{12} & d_{13} & d_{14} & d_{15} & d_{16} \\ d_{21} & d_{22} & d_{23} & d_{24} & d_{25} & d_{26} \\ d_{31} & d_{32} & d_{33} & d_{34} & d_{35} & d_{36} \end{bmatrix}
$$

$$
\times \begin{bmatrix} E_x(\omega_1)E_x(\omega_2) \\ E_y(\omega_1)E_y(\omega_2) \\ E_z(\omega_1)E_z(\omega_2) \\ E_y(\omega_1)E_z(\omega_2) + E_z(\omega_1)E_y(\omega_2) \\ E_x(\omega_1)E_z(\omega_2) + E_z(\omega_1)E_x(\omega_2) \\ E_x(\omega_1)E_y(\omega_2) + E_y(\omega_1)E_x(\omega_2) \end{bmatrix}. \quad (1.5.27)
$$

As described above in relation to Eq. (1.3.17), the extra factor of 2 comes from the summation over n and m in Eq. (1.3.13).

Effective Value of d (d_{eff})

For a fixed geometry (i.e., for fixed propagation and polarization directions) it is possible to express the nonlinear polarization giving rise to sum-frequency generation by means of the scalar relationship

$$
P(\omega_3) = 4d_{\text{eff}} E(\omega_1) E(\omega_2), \quad (1.5.28)
$$

and analogously for second-harmonic generation by

$$
P(2\omega) = 2d_{\text{eff}} E(\omega)^2, \quad (1.5.29)
$$

where

$$
E(\omega) = |\mathbf{E}(\omega)| = \left[\sum_j E_j(\omega)^2 \right]^{1/2}.
$$

In each case, d_{eff} is obtained by evaluation of the summation $\sum_{jk}$ in the general equation (1.3.13).

A general prescription for calculating d_{eff} for each of the crystal classes has been presented by Midwinter and Warner (1965); see also Table 3.1 of Zernike and Midwinter (1973). They show, for example, that for a negative uniaxial crystal of crystal class $3m$ the effective value of d is given by the expression

$$d_{\text{eff}} = d_{31} \sin\theta - d_{22} \cos\theta \sin 3\phi \qquad (1.5.30\text{a})$$

under conditions (known as type I conditions) such that the two lower-frequency waves have the same polarization, and by

$$d_{\text{eff}} = d_{22} \cos^2\theta \cos 3\phi \qquad (1.5.30\text{b})$$

under conditions (known as type II conditions) such that the polarizations are orthogonal. In these equations, θ is the angle between the propagation vector and the crystalline z axis (the optic axis), and ϕ is the azimuthal angle between the propagation vector and the xz crystalline plane.

Spatial Symmetry of the Nonlinear Medium

The form of the linear and nonlinear susceptibility tensors is constrained by the symmetry properties of the optical medium. To see why this should be so, let us consider a crystal for which the x and y directions are equivalent but for which the z direction is different. By saying that the x and y directions are equivalent, we mean that if the crystal were rotated by 90 degrees about the z axis, the crystal structure would look identical after the rotation. The z axis is then said to be a fourfold axis of symmetry. For such a crystal, we would expect that the optical response would be the same for an applied optical field polarized in either the x or the y direction, and thus, for example, that the second-order susceptibility components $\chi^{(2)}_{zxx}$ and $\chi^{(2)}_{zyy}$ would be equal.

For any particular crystal, the form of the linear and nonlinear optical suscep-tibilities is determined by considering the consequences of all of the symmetry properties for that particular crystal. For this reason, it is necessary to deter-mine what types of symmetry properties can occur in a crystalline medium. By means of the mathematical method known as group theory, crystallographers have found that all crystals can be classified as belonging to one of 32 possible crystal classes depending on what is called the point group symmetry of the crystal. The details of this classification scheme lie outside of the subject matter of the present text.[*] However, by way of example, a crystal is said to belong to point group 4 if it possesses only a fourfold axis of symmetry, to point group 3 if it possesses only a threefold axis of symmetry, and to belong to point group $3m$ if it possesses a threefold axis of symmetry and in addition a plane of mirror symmetry perpendicular to this axis.

---

[*] The reader who is interested in the details should consult Buerger (1963) or any of the other books on group theory and crystal symmetry listed in the bibliography at the end of the present chapter.

Influence of Spatial Symmetry on the Linear Optical Properties of a Material Medium

As an illustration of the consequences of spatial symmetry on the optical properties of a material system, let us first consider the restrictions that this symmetry imposes on the form of the linear susceptibility tensor $\chi^{(1)}$. The results of a group theroetical analysis of this problem shows that five different cases are possible depending on the symmetry properties of the material system. These possibilities are summarized in Table 1.5.1. Each entry is labeled by the crystal system to which the material belongs. By convention, crystals are categorized in terms of seven possible crystal systems on the basis of the form of the crystal lattice. (Table 1.5.2 below gives the correspondence between crystal system and each of the 32 point groups.) For completeness, isotropic materials (such as liquids and gases) are also included in Table 1.5.1. We see from this table that cubic and isotropic materials are isotropic in their linear optical properties, because $\chi^{(1)}$ is diagonal with equal diagonal components. All of the

TABLE 1.5.1   Form of the linear susceptibility tensor $\chi^{(1)}$ as determined by the symmetry properties of the optical medium, for each of the seven crystral classes and for isotropic materials. Each nonvanishing element is denoted by its cartesian indices

Triclinic	$\begin{bmatrix} xx & xy & xz \\ yx & yy & yz \\ zx & zy & zz \end{bmatrix}.$
Monoclinic	$\begin{bmatrix} xx & 0 & xz \\ 0 & yy & 0 \\ zx & 0 & zz \end{bmatrix}.$
Orthorhombic	$\begin{bmatrix} xx & 0 & 0 \\ 0 & yy & 0 \\ 0 & 0 & zz \end{bmatrix}.$
Tetragonal Trigonal Hexagonal	$\begin{bmatrix} xx & 0 & 0 \\ 0 & xx & 0 \\ 0 & 0 & zz \end{bmatrix}.$
Cubic Isotropic	$\begin{bmatrix} xx & 0 & 0 \\ 0 & xx & 0 \\ 0 & 0 & xx \end{bmatrix}.$

other crystal systems are anisotropic in their linear optical properties (in the sense that the polarization **P** need not be parallel to the applied electric field **E**) and consequently display the property of birefringence. Tetragonal, trigonal, and hexagonal crystals are said to be uniaxial crystals because there is one particular direction (the z-axis) for which the linear optical properties display rotational symmetry. Crystals of the triclinic, monoclinic, and orthorhombic systems are said to be biaxial.

Influence of Inversion Symmetry on the Second-Order Nonlinear Response

One of the symmetry properties that some but not all crystals possess is inversion symmetry. For a material system that is centrosymmetric (i.e., possesses a center of inversion) the $\chi^{(2)}$ nonlinear susceptibility must vanish identically. Since 11 of the 32 crystal classes possess inversion symmetry, this rule is very powerful, as it immediately eliminates all crystals belonging to these classes from consideration for second-order nonlinear optical interactions.

Although the result that $\chi^{(2)}$ vanishes for a centrosymmetric medium is general in nature, we shall demonstrate this fact only for the special case of second-harmonic generation in a medium that responds instantaneously to the applied optical field. We assume that the nonlinear polarization is given by

$$\tilde{P}(t) = \chi^{(2)} \tilde{E}^2(t), \tag{1.5.31}$$

where the applied field is given by

$$\tilde{E}(t) = \mathscr{E} \cos \omega t. \tag{1.5.32}$$

If we now change the sign of the applied electric field $\tilde{E}(t)$, the sign of the induced polarization $\tilde{P}(t)$ must also change, because we have assumed that the medium possesses inversion symmetry. Hence the relation (1.5.31) must be replaced by

$$-\tilde{P}(t) = \chi^{(2)} [-\tilde{E}(t)]^2, \tag{1.5.33}$$

which shows that

$$-\tilde{P}(t) = \chi^{(2)} \tilde{E}^2(t). \tag{1.5.34}$$

By comparison of this result with Eq. (1.5.31), we see that $\tilde{P}(t)$ must equal $-\tilde{P}(t)$, which can occur only if $\tilde{P}(t)$ vanishes identically. This result shows that

$$\chi^{(2)} = 0. \tag{1.5.35}$$

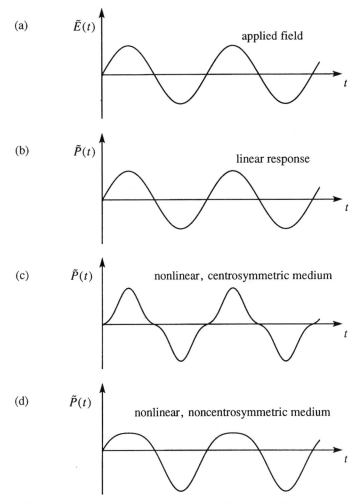

FᴵɢᴜʀE 1.5.1   Waveforms associated with the atomic response.

This result can be understood intuitively by considering the motion of an electron in a nonparabolic potential well. Because of the nonlinearity of the associated restoring force, the atomic response will show significant harmonic distortion. Part (a) of Fig. 1.5.1 shows the waveform of the incident monochromatic electromagnetic wave of frequency ω. For the case of a medium with linear response (part b), there is no distortion of the waveform associated with the polarization of the medium. Part (c) shows the induced polarization for the case of a nonlinear medium that possesses a center of symmetry and whose potential energy function has the form shown in Fig. 1.4.2. Although significant waveform distortion is evident, only odd harmonics of the fundamental frequency are present. For the case (part d) of a nonlinear, noncentrosymmetric medium having a potential energy function of the form shown in Fig. 1.4.1, both even and odd harmonics are present in the waveform associated with the atomic response. Note also the qualitative difference between the waveforms shown in parts (c) and (d). For the centrosymmetric medium (part c), the time-averaged response is zero, whereas for the noncentrosymmetric medium (part d) the time-average response is nonzero, because the medium responds differently to an electric field pointing, say, in the upward direction than to one pointing downward.*

Influence of Spatial Symmetry on the Second-Order Susceptibility

We have just seen how inversion symmetry when present requires that the second-order vanish identically. Any additional symmetry property of a nonlinear optical medium can impose additional restrictions on the form of the nonlinear susceptibility tensor. By explicit consideration of the symmetries of each of the 32 crystal classes, one can determine the allowed form of the susceptibility tensor for crystals of that class. The results of such a calculation for the second-order nonlinear optical response, which was performed originally by Butcher (1965), are presented in Table 1.5.2. Under those conditions (described following Eq. (1.5.21)) where the second-order susceptibility can be described using contracted notation, the results presented in Table 1.5.2 can usefully be displayed graphically. These results, as adapted from Zernike and Midwinter (1973), are presented in Fig. 1.5.2. Note that the influence of Kleinman symmetry is also described in the figure. As an example of how to use the table, the diagram for a crystal of class $3m$ is meant to imply that the form of the d_{il} matrix is

$$d_{il} = \begin{bmatrix} 0 & 0 & 0 & 0 & d_{31} & -d_{22} \\ -d_{22} & d_{22} & 0 & d_{31} & 0 & 0 \\ d_{31} & d_{31} & d_{33} & 0 & 0 & 0 \end{bmatrix}$$

The second-order nonlinear optical susceptibilities of a number of crystals are summarized in Table 1.5.3.

---

* Parts (a) and (b) of Fig. 1.5.1 are plots of the function $\sin \omega t$, part (c) is a plot of the function $\sin \omega t - 0.25 \sin 3\omega t$, and part (d) is a plot of $-0.2 + \sin \omega t + 0.2 \cos 2\omega t$.

TABLE 1.5.2 Form of the second-order susceptibility tensor for each of the 32 crystal classes. Each element is denoted by its cartesian indices

Crystal system	Crystal class	Nonvanishing tensor elements
Triclinic	$1 = C_1$	All elements are independent and nonzero
	$\bar{1} = S_2$	Each element vanishes
Monoclinic	$2 = C_2$	$xyz, xzy, xxy, xyx, yxx, yyy, yzz, yzx, yxz, zyz,$ zzy, zzx, zyx (twofold axis parallel to $\hat{y}$)
	$m = C_{1h}$	$xxx, xyy, xzz, xzx, xxz, yyz, yzy, yxy, yyx, zxx,$ zyy, zzz, zzx, zxz (mirror plane perpendicular to $\hat{y}$)
	$2/m = C_{2h}$	Each element vanishes
Orthorhombic	$222 = D_2$	$xyz, xzy, yzx, yxz, zxy, zyx$
	$mm2 = C_{2v}$	$xzx, xxz, yyz, yzy, zxx, zyy, zzz$
	$mmm = D_{2h}$	Each element vanishes
Tetragonal	$4 = C_4$	$xyz = -yxz, xzy = -yzx, xzx = yzy, xxz = yyz,$ $zxx = zyy, zzz, zxy = -zyx$
	$\bar{4} = S_4$	$xyz = yxz, xzy = yzx, xzx = -yzy, xxz = -yyz,$ $zxx = -zyy, zxy = zyx$
	$422 = D_4$	$xyz = -yxz, xzy = -yzx, zxy = -zyx$
	$4mm = C_{4v}$	$xzx = yzy, xxz = yyz, zxx = zyy, zzz$
	$\bar{4}2m = D_{2d}$	$xyz = yxz, xzy = yzx, zxy = zyx$
	$4/m = C_{4h}$	Each element vanishes
	$4/mmm = D_{4h}$	Each element vanishes
Cubic	$432 = O$	$xyz = -xzy = yzx = -yxz = zxy = -zyx$
	$\bar{4}3m = T_d$	$xyz = xzy = yzx = yxz = zxy = zyx$
	$23 = T$	$xyz = yzx = zxy, xzy = yxz = zyx$
	$m3 = T_h, m3m = O_h$	Each element vanishes
Trigonal	$3 = C_3$	$xxx = -xyy = -yyx = -yxy, xyz = -yxz, xzy = -yzx,$ $xzx = yzy, xxz = yyz, yyy = -yxx = -xxy = -xyx,$ $zxx = zyy, zzz, zxy = -zyx$
	$32 = D_3$	$xxx = -xyy = -yyx = -yxy, xyz = -yxz,$ $xzy = -yzx, zxy = -zyx$
	$3m = C_{3v}$	$xzx = yzy, xxz = yyz, zxx = zyy, zzz, yyy = -yxx =$ $-xxy = -xyx$ (mirror plane perpendicular to $\hat{x}$)
	$\bar{3} = S_6, \bar{3}m = D_{3d}$	Each element vanishes
Hexagonal	$6 = C_6$	$xyz = -yxz, xzy = -yzx, xzx = yzy, xxz = yyz,$ $zxx = zyy, zzz, zxy = -zyx$
	$\bar{6} = C_{3h}$	$xxx = -xyy = -yxy = -yyx,$ $yyy = -yxx = -xyx = -xxy$
	$622 = D_6$	$xyz = -yxz, xzy = -yzx, zxy = -zyx$
	$6mm = C_{6v}$	$xzx = yzy, xxz = yyz, zxx = zyy, zzz$
	$\bar{6}m2 = D_{3h}$	$yyy = -yxx = -xxy = -xyx$
	$6/m = C_{6h},$	Each element vanishes
	$6/mmm = D_{6h}$	Each element vanishes

Biaxial crystal classes

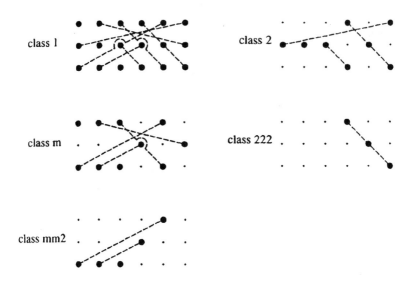

Uniaxial crystal classes

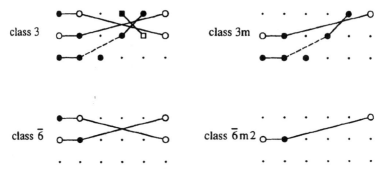

FIGURE 1.5.2   Form of the d_{il} matrix for the 21 crystal classes that lack inversion symmetry. Small dot: zero coefficient; large dot: nonzero coefficient; square: coefficient that is zero when Kleinman's symmetry condition is valid; connected symbols: numerically equal coefficients, but the open-symbol coefficient is opposite in sign to the closed symbol to which it is joined. Dashed connections are valid only under Kleinman's symmetry conditions. (After Zernike and Midwinter, 1973.)

Uniaxial crystal classes (*Continued*)

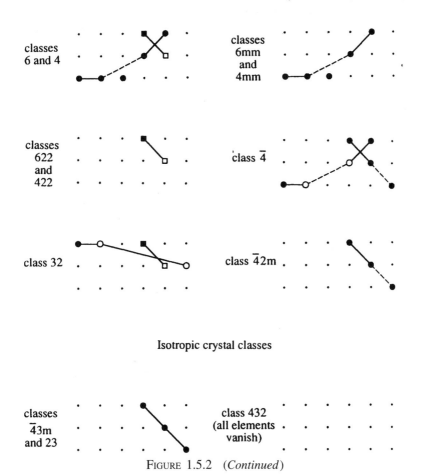

classes
6 and 4

classes
6mm
and
4mm

classes
622
and
422

class $\bar{4}$

class 32

class $\bar{4}$2m

Isotropic crystal classes

classes
$\bar{4}$3m
and 23

class 432
(all elements
vanish)

FIGURE 1.5.2   (*Continued*)

Number of Independent Elements of $\chi_{ijk}^{(2)}(\omega_3, \omega_2, \omega_1)$

We remarked above in relation to Eq. (1.5.1) that as many as 324 complex numbers must be specified in order to describe the general interaction of three optical waves. In practice, this number is often greatly reduced.

Because of the reality of the physical fields, only half of these numbers are independent (see Eq. (1.5.5)). Furthermore, the intrinsic permutation symmetry of $\chi^{(2)}$ (Eq. (1.5.6)) shows that there are only 81 independent parameters. For a lossless medium, all elements of $\chi^{(2)}$ are real and the condition of full permutation symmetry is valid, implying that only 27 of these numbers are

TABLE 1.5.3   Second-order nonlinear optical susceptibilities for several crystals

Material	Point group	d_{il} (10^{-9} cm/statvolt)
Quartz	$32 = D_3$	$d_{11} = 0.96$
		$d_{14} = 0.02$
$Ba_2NaNb_5O_{15}$	$mm2 = C_{2v}$	$d_{31} = -35$
		$d_{32} = -35$
		$d_{33} = -48$
$LiNbO_3$	$3m = C_{3v}$	$d_{22} = 7.4$
		$d_{31} = 14$
		$d_{33} = -98$
$BaTiO_3$	$4mm = C_{4v}$	$d_{15} = -41$
		$d_{31} = -43$
		$d_{33} = -16$
KH_2PO_4	$\bar{4}2m = D_{2d}$	$d_{14} = 1.2$
(KDP)		$d_{36} = 1.1$
$LiIO_3$	$6 = C_6$	$d_{35} = -13$
		$d_{36} = -10$
GaAs	$\bar{4}3m$	$d_{36} = 406$
KD_2PO_4	$\bar{4}2m = D_{2d}$	$d_{36} = 1.26$
(KD*P)		$d_{14} = 1.26$
CdS	$6mm = C_{6v}$	$d_{33} = 86$
		$d_{31} = 90$
		$d_{36} = 100$
Ag_3AsS_3	$3m = C_{3v}$	$d_{22} = 68$
(proustite)		$d_{31} = 36$
$CdGeAs_2$	$\bar{4}2m = D_{2d}$	$d_{36} = 1090$
$AgGaSe_2$	$\bar{4}2m = D_{2d}$	$d_{36} = -81$
$AgSbS_3$	$3m = C_{3v}$	$d_{31} = 30$
(pyrargyrite)		$d_{22} = 32$
beta-BaB_2O_4		$d_{11} = 4.6$
(beta barium borate)		

Notes: Values are obtained from a variety of sources. Some of the more complete tabulations are those of S. Singh in *Handbook of Lasers*, Chemical Rubber Company, Cleveland, Ohio 1971, that of A. V. Smith, available at http://www.sandia.gov/imrl/XWEB1128/xxtal.htm, and the data sheets of Cleveland Crystals, Inc, available at http://www.clevelandcrystals.com.

To convert to the MKS system using the convention that $P = dE^2$, multiply each entry by $4\pi\epsilon_0/(3 \times 10^4) = 3.71 \times 10^{-15}$ to obtain d in units of C/V^2.

To convert to the MKS system using the convention that $P = \epsilon_0 dE^2$, multiply each entry by $4\pi(3 \times 10^4) = 4.189 \times 10^{-4}$ to obtain d in units of m/V.

In any system of units, $\chi^{(2)} = 2d$ by convention.

independent. For second-harmonic generation, contracted notation can be used, and only 18 independent elements exist. When Kleinman's symmetry is valid, only 10 of these elements are independent. Furthermore, any crystalline symmetries of the nonlinear material can reduce this number further.

Distinction between Noncentrosymmetric and Cubic Crystal Classes

It is worth noting that a material can possess a cubic lattice and yet be non-centrosymmetric. In fact, gallium arsenide is an example of a material with just these properties. Gallium arsenide crystallizes in what is known as the zincblende structure (named after the well-known mineral form of zinc sulfide), which has crystal point group $\bar{4}2m$. As can be seen from Table 1.5.2 or from Fig. 1.5.2, materials of the $\bar{4}2m$ crystal class possess a nonvanishing second-order nonlinear optical response. In fact, as can be seen from Table 1.5.3, gallium arsenide has an unusually large second-order nonlinear susceptibility. However, as the zincblende crystal structure possesses a cubic lattice, gallium arsenide does not display birefringence. We shall see in Chapter 2 that it is necessary that a material possess adequate birefringence in order that the phase matching condition of nonlinear optics be satisfied. Because gallium arsenide does not possess birefringence, it cannot normally participate in phase-matched second-order interactions.

It is perhaps surprising that a material can possess the highly regular spatial arrangement of atoms characteristic of the cubic lattice and yet be noncentrosymmetric. This distinction can be appreciated by examination of Fig. 1.5.3, which shows both the diamond structure (point group $m3m$) and the zincblende

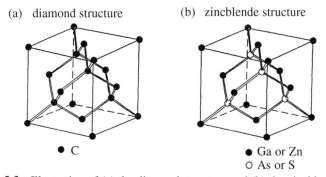

(a)   diamond structure   (b)   zincblende structure

• C

● Ga or Zn
○ As or S

FIGURE 1.5.3   Illustration of (a) the diamond structure and (b) the zincblende structure. Both possess a cubic lattice and hence cannot display birefringence, but the carbon structure is centrosymmetric whereas the zincblende structure is noncentrosymmetric.

structure (point group $\bar{4}3m$). One sees that the crystal lattice is the same in the two cases, but that the arrangement of atoms within the lattice allows carbon but not zincblende to possess a center of inversion symmetry. In detail, a point of inversion symmetry for the diamond structure is located midway between any two nearest-neighbor carbon atoms. This symmetry does not occur in the zincblende structure because the nearest neighbors are of different species.

Distinction between Noncentrosymmetric and Polar Crystal Classes

As noted above, of the 32 crystal point groups, only 21 are noncentrosymmetric and consequently can possess a nonzero second-order susceptibility $\chi^{(2)}$. A more restrictive condition is that certain crystal possess a permanent dipole moment. Crystals of this sort are known as *polar* crystals, or as *ferroelectric* crystals. This property has important technological consequences, because crystals of this sort can display the pyroelectric effect (a change of permanent dipole moment with temperature, which can be used to construct optical detectors)* or the photorefractive effect, which is described in greater detail in Chapter 11. Group theoretical arguments (see, for instance, Nye, 1985) demonstrate that the polar crystal classes are

$$1 \qquad 2 \qquad 3 \qquad 4 \qquad 6$$
$$m \quad mm2 \quad 3m \quad 4mm \quad 6mm$$

Clearly, all polar crystal classes are noncentrosymmetric, but not all noncentrosymmetric crystal classes are polar. This distinction can be seen straightforwardly by means of an example from molecular physics. Consider a molecule with tetrahedral symmetry such as CCl_4. In this molecule the four chlorine ions are arranged on the vertices of a regular tetrahedron, which is centered on the carbon ion. Clearly this arrangement cannot possess a permanent dipole moment, but this structure is nonetheless noncentrosymmetric.

Influence of Spatial Symmetry on the Third-Order Nonlinear Response

The spatial symmetry of the nonlinear optical medium also restricts the form of the third-order nonlinear optical susceptibility. The allowed form of the susceptibility has been calculated by Butcher (1965) and has been summarized by Hellwarth (1977); a minor correction to these results was later pointed out by Shang and Hsu (1987). These results are presented in Table 1.5.4. Note that for

---

* The operation of pyroelectric detectors is described for instance in Section 13.3 of *Radiometry and the Detection of Optical Radiation*, R. W. Boyd, John Wiley & Sons, New York, 1983.

TABLE 1.5.4   Form of the third-order susceptibility tensor $\chi^{(3)}$ for each of the crystal classes and for isotropic materials. Each element is denoted by its cartesian indices

---

Isotropic

There are 21 nonzero elements, of which only 3 are independent. They are:

$$yyzz = zzyy = zzxx = xxzz = xxyy = yyxx,$$
$$yzyz = zyzy = zxzx = xzxz = xyxy = yxyx,$$
$$yzzy = zyyz = zxxz = xzzx = xyyx = yxxy;$$

and

$$xxxx = yyyy = zzzz = xxyy + xyxy + xyyx.$$

Cubic

For the two classes 23 and $m3$, there are 21 nonzero elements, of which only 7 are independent. They are:

$$xxxx = yyyy = zzzz,$$
$$yyzz = zzxx = xxyy,$$
$$zzyy = xxzz = yyxx,$$
$$yzyz = zxzx = xyxy,$$
$$zyzy = xzxz = yxyx,$$
$$yzzy = zxxz = xyyx,$$
$$zyyz = xzzx = yxxy.$$

For the three classes 432, $\bar{4}3m$, and $m3m$, there 21 nonzero elements, of which only 4 are independent. They are:

$$xxxx = yyyy = zzzz,$$
$$yyzz = zzyy = zzxx = xxzz = xxyy = yyxx,$$
$$yzyz = zyzy = zxzx = xzxz = xyxy = yxyx,$$
$$yzzy = zyyz = zxxz = xzzx = xyyx = yxxy.$$

Hexagonal

For the three classes 6, $\bar{6}$, and $6/m$, there are 41 nonzero elements, of which only 19 are independent. They are:

$$zzzz,$$

$$xxxx = yyyy = xxyy + xyyx + xyxy, \quad \begin{cases} xxyy = yyxx, \\ xyyx = yxxy, \\ xyxy = yxyx, \end{cases}$$

$yyzz = xxzz,$	$xyzz = -yxzz,$
$zzyy = zzxx,$	$zzxy = -zzyx,$
$zyyz = zxxz,$	$zxyz = -zyxz,$
$yzzy = xzzx,$	$xzzy = -yzzx,$
$yzyz = xzxz,$	$xzyz = -yzxz,$
$zyzy = zxzx,$	$zxzy = -zyzx,$

$$xxxy = -yyyx = yyxy + yxyy + xyyy, \quad \begin{cases} yyxy = -xxyx, \\ yxyy = -xyxx, \\ xyyy = -yxxx. \end{cases}$$

(*continues*)

TABLE 1.5.4 *(Continued)*

For the four classes 622, $6mm$, $6/mmm$, and $\bar{6}m2$, there are 21 nonzero elements, of which only 10 are independent. They are:

$$zzzz,$$

$$xxxx = yyyy = xxyy + xyyx + xyxy, \qquad \begin{cases} xxyy = yyxx, \\ xyyx = yxxy, \\ xyxy = yxyx, \end{cases}$$

$$yyzz = xxzz,$$
$$zzyy = zzxx,$$
$$zyyz = zxxz,$$
$$yzzy = xzzx,$$
$$yzyz = xzxz,$$
$$zyzy = zxzx.$$

Trigonal

For the two classes 3 and $\bar{3}$, there are 73 nonzero elements, of which only 27 are independent. They are:

$$zzzz,$$

$$xxxx = yyyy = xxyy + xyyx + xyxy, \qquad \begin{cases} xxyy = yyxx, \\ xyyx = yxxy, \\ xyxy = yxyx, \end{cases}$$

$$yyzz = xxzz, \qquad xyzz = -yxzz,$$
$$zzyy = zzxx, \qquad zzxy = -zzyx,$$
$$zyyz = zxxz, \qquad zxyz = -zyxz,$$
$$yzzy = xzzx, \qquad xzzy = -yzzx,$$
$$yzyz = xzxz, \qquad xzyz = -yzxz,$$
$$zyzy = zxzx, \qquad zxzy = -zyzx,$$

$$xxxy = -yyyx = yyxy + yxyy + xyyy, \qquad \begin{cases} yyxy = -xxyx, \\ yxyy = -xyxx, \\ xyyy = -yxxx. \end{cases}$$

$$yyyz = -yxxz = -xyxz = -xxyz,$$
$$yyzy = -yxzx = -xyzx = -xxzy,$$
$$yzyy = -yzxx = -xzyx = -xzxy,$$
$$zyyy = -zyxx = -zxyx = -zxxy,$$
$$xxxz = -xyyz = -yxyz = -yyxz,$$
$$xxzx = -xyzy = -yxzy = -yyzx,$$
$$xzxx = -yzxy = -yzyx = -xzyy,$$
$$zxxx = -zxyy = -zyxy = -zyyx.$$

For the three classes $3m$, $\bar{3}m$, and 32, there are 37 nonzero elements, of which only 14 are independent. They are:

$$zzzz,$$

$$xxxx = yyyy = xxyy + xyyx + xyxy, \qquad \begin{cases} xxyy = yyxx, \\ xyyx = yxxy, \\ xyxy = yxyx, \end{cases}$$

TABLE 1.5.4 (*Continued*)

$$yyzz = xxzz, \quad xxxz = -xyyz = -yxyz = -yyxz,$$
$$zzyy = zzxx, \quad xxzx = -xyzy = -yxzy = -yyzx,$$
$$zyyz = zxxz, \quad xzxx = -xzyy = -yzxy = -yzyx,$$
$$yzzy = xzzx, \quad zxxx = -zxyy = -zyxy = -zyyx,$$
$$yzyz = xzxz,$$
$$zyzy = zxzx.$$

Tetragonal

For the three classes 4, $\bar{4}$, and $4/m$, there are 41 nonzero elements, of which only 21 are independent. They are:

$$xxxx = yyyy, \quad zzzz,$$

$zzxx = zzyy,$	$xyzz = -yxzz,$	$xxyy = yyxx,$	$xxxy = -yyyx,$
$xxzz = zzyy,$	$zzxy = -zzyx,$	$xyxy = yxyx,$	$xxyx = -yyxy,$
$zxzx = zyzy,$	$xzyz = -yzxz,$	$xyyx = yxxy,$	$xyxx = -yxyy,$
$xzxz = yzyz,$	$zxzy = -zyzx,$		$yxxx = -xyyy,$
$zxxz = zyyz,$	$zxyz = -zyxz,$		
$xzzx = yzzy,$	$xzzy = -yzzx.$		

For the four classes 422, $4mm$, $4/mmm$, and $\bar{4}2m$, there are 21 nonzero elements, of which only 11 are independent. They are:

$$xxxx = yyyy, \quad zzzz,$$

$yyzz = xxzz,$	$yzzy = xzzx$	$xxyy = yyxx,$
$zzyy = zzxx,$	$yzyz = xzxz$	$xyxy = yxyx,$
$zyyz = zxxz,$	$zyzy = zxzx$	$xyyx = yxxy.$

Monoclinic

For the three classes 2, m, and $2/m$, there are 41 independent nonzero elements, consisting of:

3 elements with indices all equal,
18 elements with indices equal in pairs,
12 elements with indices having two y's one x, and one z,
4 elements with indices having three x's and one z,
4 elements with indices having three z's and one x.

Orthorhombic

For all three classes, 222, $mm2$, and mmm, there are 21 independent nonzero elements, consisting of:

3 elements with indices all equal,
18 elements with indices equal in pairs,

Triclinic

For both classes, 1 and $\bar{1}$, there are 81 independent nonzero elements.

the important special case of an isotropic optical material, the results presented in Table 1.5.4 agree with the result derived explicitly in the discussion of the nonlinear refractive index in Section 4.2.

1.6. Time-Domain Description of Optical Nonlinearities

In the preceding sections, we described optical nonlinearities in terms of the response of an optical material to one or more essentially monochromatic applied fields. We found that the nonlinear polarization thereby produced consists of a discrete summation of frequency components at the harmonics and the sums and differences of the frequencies present in the applied field. In particular, we described the nonlinear response in the frequency domain by relating the frequency components $P(\omega)$ of the nonlinear polarization to those of the applied optical field, $E(\omega')$.

It is also possible to describe optical nonlinearities directly in the time domain by considering the polarization $\tilde{P}(t)$ that is produced by some arbitrary applied field $\tilde{E}(t)$. These two methods of description are entirely equivalent, although description in the time domain is more convenient for certain types of problems, such as those involving applied fields in the form of short pulses, whereas description in the frequency domain is more convenient when each input field is nearly monochromatic.

Let us first consider the special case of a material that displays purely linear response. We can describe the polarization induced in such a material by

$$\tilde{P}^{(1)}(t) = \int_0^\infty R^{(1)}(\tau)\tilde{E}(t-\tau)d\tau. \tag{1.6.1}$$

Here $R^{(1)}(\tau)$ is the linear response function, which gives the contribution to the polarization produced at time t by an electric field applied at the earlier time $t - \tau$. The total polarization is obtained by integrating these contributions over all previous times τ. In writing Eq. (1.6.1) as shown, with the lower limit of integration set equal to zero and not to $-\infty$, we have assumed that $R^{(1)}(\tau)$ obeys the causality condition $R^{(1)}(\tau) = 0$ for $\tau < 0$. This condition expresses the fact that $\tilde{P}^{(1)}(t)$ depends only on past and not on future values of $\tilde{E}(t)$.

Equation (1.6.1) can be transformed to the frequency domain by introducing the Fourier transforms of the various quantities that appear in this equation. We adopt the following definition of the Fourier transform:

$$E(\omega) = \int_{-\infty}^\infty \tilde{E}(t)e^{i\omega t}dt \tag{1.6.2a}$$

$$\tilde{E}(t) = \frac{1}{2\pi}\int_{-\infty}^\infty E(\omega)e^{-i\omega t}d\omega \tag{1.6.2b}$$

with analogous definitions for other quantities. By introducing Eq. (1.6.2b) into Eq. (1.6.1), we obtain

$$\tilde{P}^{(1)}(t) = \int_0^\infty d\tau \int_{-\infty}^\infty \frac{d\omega}{2\pi} R^{(1)}(\tau) E(\omega) e^{-i\omega(t-\tau)}$$

$$= \int_{-\infty}^\infty \frac{d\omega}{2\pi} \int_0^\infty d\tau R^{(1)}(\tau) e^{i\omega\tau} E(\omega) e^{-i\omega t} \qquad (1.6.3)$$

or

$$\tilde{P}^{(1)}(t) = \int_{-\infty}^\infty \frac{d\omega}{2\pi} \chi^{(1)}(\omega; \omega) E(\omega) e^{-i\omega t}, \qquad (1.6.4)$$

where we have introduced an explicit expression for the linear susceptibility

$$\chi^{(1)}(\omega; \omega) = \int_0^\infty d\tau R^{(1)}(\tau) e^{i\omega\tau}. \qquad (1.6.5)$$

Equation (1.6.4) gives the time-varying polarization in terms of the frequency components of the applied field and the frequency dependent susceptibility. By replacing the left-hand side of this equation with $\int P^{(1)}(\omega) \exp(-i\omega t) d\omega/2\pi$ and noting that the equality must be maintained for each frequency ω, we recover the usual frequency domain description of linear response:

$$P^{(1)}(\omega) = \chi^{(1)}(\omega; \omega) E(\omega). \qquad (1.6.6)$$

The nonlinear response can be described by analogous procedures. The contribution to the polarization second-order in the applied field strength is represented as

$$\tilde{P}^{(2)}(t) = \int_0^\infty d\tau_1 \int_0^\infty d\tau_2 R^{(2)}(\tau_1, \tau_2) E(t - \tau_1) E(t - \tau_2), \quad (1.6.7)$$

where the causality condition requires that $R^{(2)}(\tau_1, \tau_2) = 0$ if either τ_1 or τ_2 is negative. As above, we write $E(t - \tau_1)$ and $E(t - \tau_2)$ in terms of their Fourier transforms using Eq. (1.6.2b) so that the expression for the second-order polarization becomes

$$\tilde{P}^{(2)}(t) = \int_{-\infty}^\infty \frac{d\omega_1}{2\pi} \int_{-\infty}^\infty \frac{d\omega_2}{2\pi} \int_0^\infty d\tau_1 \int_0^\infty d\tau_2 R^{(2)}(\tau_1, \tau_2)$$

$$\times E(\omega_1) e^{-i\omega_1(t-\tau_1)} E(\omega_2) e^{-i\omega_2(t-\tau_2)}$$

$$= \int_{-\infty}^\infty \frac{d\omega_1}{2\pi} \int_{-\infty}^\infty \frac{d\omega_2}{2\pi} \chi^{(2)}(\omega_\sigma; \omega_1, \omega_2) E(\omega_1) E(\omega_2) e^{-i\omega_\sigma t} \quad (1.6.8)$$

where we have defined $\omega_\sigma = \omega_1 + \omega_2$ and have introduced the second-order susceptibility

$$\chi^{(2)}(\omega_\sigma; \omega_1, \omega_2) = \int_0^\infty d\tau_1 \int_0^\infty d\tau_2 R^{(2)}(\tau_1, \tau_2) e^{i(\omega_1 \tau_1 + \omega_2 \tau_2)} \quad (1.6.9)$$

This procedure can readily be generalized to higher-order susceptibilities. In particular, we can express the third-order polarization as

$$\tilde{P}^{(3)}(t) = \int_{-\infty}^\infty \frac{d\omega_1}{2\pi} \int_{-\infty}^\infty \frac{d\omega_2}{2\pi} \int_{-\infty}^\infty \frac{d\omega_3}{2\pi} \chi^{(3)}(\omega_\sigma; \omega_1, \omega_2, \omega_3)$$

$$\times E(\omega_1) E(\omega_2) E(\omega_3) e^{-i\omega_\sigma t} \quad (1.6.10)$$

where $\omega_\sigma = \omega_1 + \omega_2 + \omega_3$ and where

$$\chi^{(3)}(\omega_\sigma; \omega_1, \omega_2, \omega_3) = \int_0^\infty d\tau_1 \int_0^\infty d\tau_2 \int_0^\infty d\tau_3$$

$$\times R^{(3)}(\tau_1, \tau_2, \tau_3) e^{i(\omega_1 \tau_1 + \omega_2 \tau_2 + \omega_3 \tau_3)}. \quad (1.6.11)$$

1.7. Kramers–Kronig Relations in Linear and Nonlinear Optics

Kramers–Kronig relations are often encountered in linear optics. These conditions relate the real and imaginary parts of frequency-dependent quantities such as the linear susceptibility. They are useful because, for instance, they allow one to determine the real part of the susceptibility at some particular frequency from a knowledge of the frequency dependence of the imaginary part of the susceptibility. In this section, we review the derivation of the Kramers–Kronig relations as they are usually formulated for a system with linear response, and then show how Kramers–Kronig relations can be formulated to apply to some (but not all) nonlinear optical interactions.

Kramers–Kronig Relations in Linear Optics

We saw in the previous section that the linear susceptibility can be represented as

$$\chi^{(1)}(\omega) \equiv \chi^{(1)}(\omega; \omega) = \int_0^\infty R^{(1)}(\tau) e^{i\omega\tau} d\tau, \quad (1.7.1)$$

where the lower limit of integration has been set equal to zero to reflect the fact that $R^{(1)}(\tau)$ obeys the causality condition $R^{(1)}(\tau) = 0$ for $\tau < 0$. Note also (e.g., from Eq. (1.6.1)) that $R^{(1)}(\tau)$ is necessarily real, since it relates two inherently real quantities $\tilde{P}(t)$ and $\tilde{E}(t)$. We thus deduce from Eq. (1.7.1) that

$$\chi^{(1)}(-\omega) = \chi^{(1)}(\omega)^*. \quad (1.7.2)$$

Let us examine some of the other mathematical properties of the linear susceptibility. In doing so, it is useful, as a purely mathematical artifact, to treat the frequency ω as a complex quantity $\omega = \mathrm{Re}\,\omega + i\,\mathrm{Im}\,\omega$. An important mathematical property of $\chi(\omega)$ is the fact that it is analytic (i.e., single-valued and possessing continuous derivatives) in the upper half of the complex plane, that is, for $\mathrm{Im}\,\omega \geq 0$. In order to demonstrate that $\chi(\omega)$ is analytic in the upper half plane, it is adequate to show that the integral in Eq. (1.7.1) converges everywhere in that region. We first note that the integrand in Eq. (1.7.1) is of the form $R^{(1)}(\tau)\exp[i(\mathrm{Re}\,\omega)\tau]\exp[-(\mathrm{Im}\,\omega)\tau]$, and since $R^{(1)}(\tau)$ is everywhere finite, the presence of the factor $\exp[-(\mathrm{Im}\,\omega)\tau]$ is adequate to ensure convergence of the integral for $\mathrm{Im}\,\omega > 0$. For $\mathrm{Im}\,\omega = 0$ (that is, along the real axis) the integral can be shown to converge, either from a mathematical argument based on the fact the $R^{(1)}(\tau)$ must be square integrable or from the physical statement that $\chi(\omega)$ is a measurable quantity and hence must be finite.

To establish the Kramers–Kronig relations, we next consider the integral

$$\mathrm{Int} = \int_{-\infty}^{\infty} \frac{\chi^{(1)}(\omega')d\omega'}{\omega' - \omega}. \tag{1.7.3}$$

We adopt the convention that in expressions such as (1.7.3) we are to take the Cauchy principal value of the integral, that is,

$$\int_{-\infty}^{\infty} \frac{\chi^{(1)}(\omega')d\omega'}{\omega' - \omega} \equiv \lim_{\delta \to 0}\left[\int_{-\infty}^{\omega-\delta} \frac{\chi^{(1)}(\omega')d\omega'}{\omega' - \omega} + \int_{\omega+\delta}^{\infty} \frac{\chi^{(1)}(\omega')d\omega'}{\omega' - \omega}\right]. \tag{1.7.4}$$

We evaluate expression (1.7.3) using the techniques of contour integration, noting that the desired integral Int is given by $\mathrm{Int} = \mathrm{Int}(A) - \mathrm{Int}(B) - \mathrm{Int}(C)$ where $\mathrm{Int}(A)$, $\mathrm{Int}(B)$, and $\mathrm{Int}(C)$ are the path integrals of $\chi^{(1)}(\omega')/(\omega' - \omega)$ over the paths shown in Fig. 1.7.1. Since $\chi^{(1)}(\omega')$ is analytic in the upper half plane, the only singularity of the integrand $\chi(\omega')/(\omega' - \omega)$ in the upper half-plane is a simple pole along the real axis at $\omega' = \omega$. We thus find that $\mathrm{Int}(A) = 0$ by Cauchy's theorem since its closed path of integration contains no poles. Furthermore, $\mathrm{Int}(B) = 0$ since the integration path increases as $|\omega'|$, whereas for large $|\omega'|$ the integrand scales as $\chi(\omega')/|\omega'|$, and thus the product will tend toward zero so long as $\chi(\omega')$ approaches zero for sufficiently large ω'. Finally, by the residue theorem $\mathrm{Int}(C) = -\pi i \chi(\omega)$. By introducing these values into Eq. (1.7.3), we obtain the result

$$\chi^{(1)}(\omega) = \frac{-i}{\pi}\int_{-\infty}^{\infty} \frac{\chi^{(1)}(\omega')d\omega'}{\omega' - \omega}. \tag{1.7.5}$$

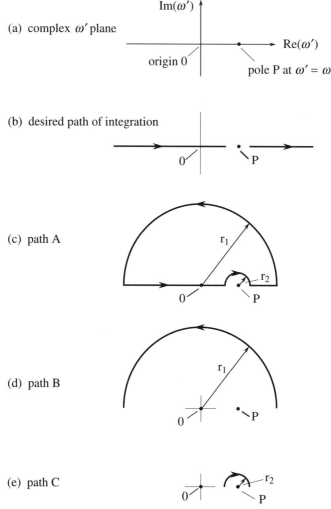

(a) complex ω' plane

origin 0

pole P at $\omega' = \omega$

(b) desired path of integration

(c) path A

(d) path B

(e) path C

FIGURE 1.7.1   Diagrams used in the contour integration of Eq. (1.7.3). (a) shows the complex ω' plane, (b) shows the desired path of integration, and (c), (d), and (e) show paths over which the integral can be evaluated using the techniques of contour integration. In performing the integration the limits $r_1 \to \infty$ and $r_2 \to 0$ are taken.

By separating $\chi^{(1)}(\omega)$ into its real and imaginary parts as $\chi^{(1)}(\omega) = \mathrm{Re}\,\chi^{(1)}(\omega) + i\,\mathrm{Im}\,\chi^{(1)}(\omega)$, we obtain one form of the Kramers–Kronig relations

$$\mathrm{Re}\,\chi^{(1)}(\omega) = \frac{1}{\pi} \int_{-\infty}^{\infty} \frac{\mathrm{Im}\,\chi^{(1)}(\omega')d\omega'}{\omega' - \omega}, \qquad (1.7.6a)$$

$$\mathrm{Im}\,\chi^{(1)}(\omega) = -\frac{1}{\pi} \int_{-\infty}^{\infty} \frac{\mathrm{Re}\,\chi^{(1)}(\omega')d\omega'}{\omega' - \omega}. \qquad (1.7.6b)$$

These integrals show how the real part of $\chi^{(1)}$ can be deduced from a knowledge of the frequency dependence of the imaginary part of $\chi^{(1)}$, and vice versa. Since it is usually easier to measure absorption spectra than the frequency dependence of the refractive index, it can be quite useful to make use of Eq. (1.7.6a) as a means of predicting the frequency dependence of the real part of $\chi^{(1)}$.

The Kramers–Kronig relations can be rewritten to involve integration over only (physically meaningful) positive frequencies. From Eq. (1.7.2), we see that

$$\operatorname{Re}\chi^{(1)}(-\omega) = \operatorname{Re}\chi^{(1)}(\omega) \qquad \operatorname{Im}\chi^{(1)}(-\omega) = -\operatorname{Im}\chi^{(1)}(\omega). \quad (1.7.7)$$

We can thus rewrite Eq. (1.7.6b) as follows:

$$\operatorname{Im}\chi^{(1)}(\omega) = -\frac{1}{\pi}\int_{-\infty}^{0}\frac{\operatorname{Re}\chi^{(1)}(\omega')d\omega'}{\omega' - \omega} - \frac{1}{\pi}\int_{0}^{\infty}\frac{\operatorname{Re}\chi^{(1)}(\omega')d\omega'}{\omega' - \omega}$$

$$= \frac{1}{\pi}\int_{0}^{\infty}\frac{\operatorname{Re}\chi^{(1)}(\omega')d\omega'}{\omega' + \omega} - \frac{1}{\pi}\int_{0}^{\infty}\frac{\operatorname{Re}\chi^{(1)}(\omega')d\omega'}{\omega' - \omega} \quad (1.7.8)$$

and hence

$$\operatorname{Im}\chi^{(1)}(\omega) = \frac{-2\omega}{\pi}\int_{0}^{\infty}\frac{\operatorname{Re}\chi^{(1)}(\omega')}{\omega'^2 - \omega^2}d\omega'. \quad (1.7.9a)$$

We similarly find that

$$\operatorname{Re}\chi^{(1)}(\omega) = \frac{2}{\pi}\int_{0}^{\infty}\frac{\omega'\operatorname{Im}\chi^{(1)}(\omega')}{\omega'^2 - \omega^2}d\omega'. \quad (1.7.9b)$$

Kramers–Kronig Relations in Nonlinear Optics

Relations analogous to the usual Kramers–Kronig relations for the linear response can be deduced for some but not all nonlinear optical interactions. Let us first consider a nonlinear susceptibility of the form $\chi^{(3)}(\omega_\sigma; \omega_1, \omega_2, \omega_3)$ with $\omega_\sigma = \omega_1 + \omega_2 + \omega_3$ and with ω_1, ω_2, and ω_3 all positive and distinct. Such a susceptibility obeys a Kramers–Kronig relation in each of the three input frequencies, for example,

$$\chi^{(3)}(\omega_\sigma; \omega_1, \omega_2, \omega_3) = \frac{1}{i\pi}\int_{-\infty}^{\infty}\frac{\chi^{(3)}(\omega_\sigma'; \omega_1, \omega_2', \omega_3)}{\omega_2' - \omega_2}d\omega_2', \quad (1.7.10)$$

where $\omega_\sigma' = \omega_1 + \omega_2' + \omega_3$ and similarly for integrals involving ω_1' and ω_3'. The proof of this result proceeds in a manner strictly analogous to that of the linear Kramers–Kronig relation. In particular, we note from Eq. (1.6.11) that $\chi^{(3)}(\omega_\sigma; \omega_1, \omega_2, \omega_3)$ is the Fourier transform of a causal response function, and hence $\chi^{(3)}(\omega_\sigma; \omega_1, \omega_2, \omega_3)$ considered as a function of its three independent variables ω_1, ω_2, and ω_3, is analytic in the region $\operatorname{Im}\omega_1 \geq 0$, $\operatorname{Im}\omega_2 \geq 0$, and Im $\omega_3 \geq 0$. We can then perform the integration indicated on the right-hand side of Eq. (1.7.10) as a contour integration closed in the upper part of the complex ω_2 plane, and obtain the indicated result. In fact, it is not at all surprising that a Kramers–Kronig-like relation should exist for the present situation; the expression $\chi^{(3)}(\omega_\sigma; \omega_1, \omega_2, \omega_3) E(\omega_1) E(\omega_2) E(\omega_3)$ is linear in the field $E(\omega_2)$ and the physical system is causal, and thus the reasoning leading to the usual linear Kramers–Kronig relation is directly relevant to the present situation.

Note that in Eq. (1.7.10), all but one of the input frequencies are held fixed. Kramers–Kronig relations can also be formulated under more general circumstances. It can be shown (see, for instance, Section 6.2 of Hutchings et al., 1992) by means of a somewhat intricate argument that

$$\chi^{(n)}(\omega_\sigma; \omega_1 + p_1\omega, \omega_2 + p_2\omega, \ldots, \omega_n + p_n\omega)$$
$$= \frac{1}{i\pi} \int_{-\infty}^{\infty} \frac{\chi^{(n)}(\omega_\sigma'; \omega_1 + p_1\omega', \omega_2 + p_2\omega', \ldots, \omega_n + p_n\omega')}{\omega' - \omega} d\omega' \quad (1.7.11)$$

where $p_i \geq 0$ for all i and where at least one p_i must be nonzero. Among the many special cases included in Eq. (1.7.11) are those involving the susceptibility for second-harmonic generation

$$\chi^{(2)}(2\omega; \omega, \omega) = \frac{1}{i\pi} \int_{-\infty}^{\infty} \frac{\chi^{(2)}(2\omega', \omega', \omega')}{\omega' - \omega} d\omega' \quad (1.7.12)$$

and for third-harmonic generation

$$\chi^{(3)}(3\omega; \omega, \omega, \omega) = \frac{1}{i\pi} \int_{-\infty}^{\infty} \frac{\chi^{(3)}(3\omega'; \omega', \omega', \omega')}{\omega' - \omega} d\omega'. \quad (1.7.13)$$

Kramers–Kronig relations can also be formulated for the induced change in refractive index, which is described by a susceptibility of the sort $\chi^{(3)}(\omega; \omega, \omega_1, -\omega_1)$. In particular, one can show (Hutchings et al., 1992) that

$$\chi^{(3)}(\omega; \omega, \omega_1, -\omega_1) = \frac{1}{i\pi} \int_{-\infty}^{\infty} \frac{\chi^{(3)}(\omega'; \omega, \omega_1, -\omega_1) d\omega'}{\omega' - \omega}. \quad (1.7.14)$$

Probably the most important process for which it is not possible to form a Kramers–Kronig relation is for the self-induced change in refractive index, that is, for processes described by the nonlinear susceptibility $\chi^{(3)}(\omega; \omega, \omega, -\omega)$. Note that this susceptibility is not of the form of Eq. (1.7.10) or of (1.7.11), because the first two applied frequencies are equal and because the third frequency is negative. Moreover, one can show by explicit calculation (see the problems at the end of this chapter) that for specific model systems the real and imaginary parts of $\chi^{(3)}$ are not related in the proper manner to satisfy the Kramers–Kronig relations.

To summarize the results of this section, we have seen that Kramers–Kronig relations, which are always valid in linear optics, are valid for some but not all nonlinear optical processes.

Problems

1. *Conversion from gaussian to SI units.* For proustite $\chi^{(2)}_{yyy}$ has the value 1.3×10^7 in gaussian units. What is its value in MKS units, assuming the convention that

$$\tilde{P}^{(2)}(t) = \epsilon_0 \chi^{(2)} \tilde{E}^2(t)? \qquad (1.7.15)$$

[Ans: 5.5×10^{-11} m/V.]

2. *Numerical estimate of nonlinear optical quantities.* A laser beam of frequency ω carrying 1 W of power is focused to a spot size of 30-μm diameter in a crystal having a refractive index of $n = 2$ and a second-order susceptibility of $\chi^{(2)} = 1 \times 10^{-7}$ esu. Calculate numerically the amplitude $P(2\omega)$ of the component of the nonlinear polarization oscillating at frequency 2ω. Estimate numerically the amplitude of the dipole moment per atom $\mu(2\omega)$ oscillating at frequency 2ω. Compare this value with the atomic unit of dipole moment (ea_0, where a_0 is the Bohr radius) and with the linear response of the atom, that is, with the component $\mu(\omega)$ of the dipole moment oscillating at frequency ω. We shall see in the next chapter that, under the conditions stated above, nearly all of the incident power can be converted to the second harmonic for a 1-cm-long crystal.

[Ans: $P(2\omega) = 1.5 \times 10^{-5}$ esu. Assuming that $N = 10^{22}$ atoms/cm^3, $\mu(2\omega) = 1.5 \times 10^{-27}$ esu $= 5.9 \times 10^{-10} ea_0$. By comparison, $P(\omega) = 2.9$ esu and $\mu(\omega) = 2.9 \times 10^{-22}$ esu $= 1.1 \times 10^{-4} ea_0$, which shows that $\mu(2\omega)/\mu(\omega) = 5.2 \times 10^{-6}$.]

3. *Perturbation expansion.* Explain why it is unnecessary to include the term $\lambda^0 \tilde{x}^{(0)}$ in the power series of Eq. (1.4.6).

4. *Tensor properties of the anharmonic oscillator model.* Starting from Eq. (1.4.52), relevant to a collection of isotropic, centrosymmetric, anharmonic oscillators, show that the nonlinear susceptibility possesses the following tensor properties:

$$\chi_{1122} = \chi_{1212} = \chi_{1221} = \chi_{1133} = \chi_{1313} = \chi_{1331} = \chi_{2233} = \chi_{2323}$$

$$= \chi_{2332} = \chi_{2211} = \chi_{2121} = \chi_{2112} = \chi_{3311} = \chi_{3131} = \chi_{3113}$$

$$= \chi_{3322} = \chi_{3232} = \chi_{3223} = \tfrac{1}{3}\chi_{1111} = \tfrac{1}{3}\chi_{2222} = \tfrac{1}{3}\chi_{3333},$$

with all other elements vanishing. Give a simple physical argument that explains why the vanishing elements do vanish. Also, give a simple physical argument that explains why χ_{ijhl} possesses off-diagonal tensor components, even though the medium is isotropic.

5. *Comparison of the centrosymmetric and noncentrosymmetric models.* For the noncentrosymmetric anharmonic oscillator described by Eq. (1.4.1), derive an expression for the third-order displacement $\tilde{x}^{(3)}$ and consequently for the third-order susceptibility $\chi_{1111}^{(3)}(\omega_q, \omega_m, \omega_n, \omega_p)$. Compare this result to that given by Eq. (1.4.52) for a purely centrosymmetric medium. Note that for a noncentrosymmetric medium both of these contributions can be present. Estimate the size of each of these contributions to see which is larger.

6. *Determination of d_{eff}.* Verify Eqs. (1.5.30a) and (1.5.30b).

7. *Formal properties of the third-order response.* Section 1.5 contains a description of some of the formal mathematical properties of the second-order susceptibility. For the present problem, you are to determine the analogous symmetry properties of the third-order susceptibility $\chi^{(3)}$. In your response, be sure to include the equations analogous to Eqs. (1.5.1), (1.5.2), (1.5.5), (1.5.6), (1.5.8), (1.5.9), and (1.5.19).

8. *Consequences of crystalline symmetry.* Through explicit consideration of the symmetry properties of each of the 32 point groups, verify the results presented in Tables 1.5.2 and 1.5.4 and in Fig. 1.5.2.

[Notes: This problem is lengthy and requires a more detailed knowledge of group theory and crystal symmetry than that presented in this text. For a list of recommended readings on these subjects, see the reference list to the present chapter. For a discussion of this problem, see also Butcher (1965).]

9. *Subtlety regarding crystal class 432.* According to Table 1.5.2, $\chi^{(2)}$ possesses nonvanishing tensor elements for crystal class 432, but according to Fig. 1.5.2 d_{il} for this crystal class vanishes identically. Justify these two statements by taking explicit account of the additional constraints that are implicit in the definition of the d_{il} matrix.

10. *Kramers–Kronig relations.* Show by explicit calculation that the linear susceptibility of an optical transition modeled in the two-level approximation obeys the Kramers–Kronig relations, but that neither the total susceptibility χ nor the third-order susceptibility $\chi^{(3)}$ obeys these relations. Explain this result by finding the location of the poles of χ and of $\chi^{(3)}$.

[Hints: $\chi^{(1)}$ and $\chi^{(3)}$ are given by Eqs. (6.3.33) and χ is given by Eq. (6.3.23).]

11. *Kramers–Kronig relations.* For the classical anharmonic oscillator model of Eq. (1.4.20) show by explicit calculation that $\chi^{(2)}(2\omega; \omega, \omega)$ obeys the

Kramers–Kronig relations in the form (1.7.12). Show also that $\chi^{(2)}(\omega_1;\omega_3,-\omega_2)$ does not satisfy Kramers–Kronig relations.

12. *Example of the third-order response.* The third-order polarization includes a term oscillating at the fundamental frequency and given by

$$P^{(3)}(\omega) = 3\chi^{(3)}|E(\omega)|^2 E(\omega).$$

Assume that the field at frequency ω includes two contributions that propagate in the directions given by wavevectors $\mathbf{k}_1$ and $\mathbf{k}_2$. Assume also that the second contribution is sufficiently weak that it can be treated linearly. Calculate the nonlinear polarization at the fundamental frequency and give the physical interpretation of its different terms.

References

General References

J. A. Armstrong, N. Bloembergen, J. Ducuing, and P. S. Pershan, *Phys. Rev.* **127**, 1918 (1962).

R. W. Boyd, *J. Mod. Opt.* **46**, 367 (1999).

R. L. Byer and S. E. Harris, *Phys. Rev.* **168**, 1064 (1968).

Cleveland Crystals, Inc, 19306 Redwood Road, Cleveland, OH 44110 USA, provides a large number of useful data sheets which may also be obtained at http://www.clevelandcrystals.com

C. Flytzanis, in *Quantum Electronics, a Treatise*, Vol. 1, Part A, edited by H. Rabin and C. L. Tang, Academic Press, New York, 1975.

P. A. Franken, A. E. Hill, C. W. Peters, and G. Weinreich, *Phys. Rev. Lett.* **7**, 118 (1961).

C. G. B. Garrett and F. N. H. Robinson, *IEEE J. Quantum Electron.* **2**, 328 (1966).

W. Kaiser and C. G. B. Garrett, *Phys. Rev. Lett.* **7**, 229 (1961).

D. A. Kleinman, *Phys. Rev.* **126**, 1977 (1962).

S. E. Harris, M. K. Oshman, and R. L. Byer, *Phys. Rev. Lett.* **18**, 732 (1967).

R. W. Hellwarth, *Prog. Quantum Electron.* **5**, 1 (1977).

L. D. Landau and E. M. Lifshitz, *Electrodynamics of Continuous Media*, Pergamon, New York, 1960, Section 10.

T. H. Maiman, *Nature* **187**, 493 (1960).

J. E. Midwinter and J. Warner, *Brit. J. Appl. Phys.* **16**, 1135 (1965).

R. C. Miller, *Appl. Phys. Lett.* **5**, 17 (1964).

A. Owyoung, *The Origins of the Nonlinear Refractive Indices of Liquids and Glasses*, Ph.D. dissertation, California Institute of Technology, 1971.

P. S. Pershan, *Phys. Rev.* **130**, 919 (1963).

C. C. Shang and H. Hsu, *IEEE J. Quantum Electron.* **23**, 177 (1987).

Y. R. Shen, *Phys. Rev.* **167**, 818 (1968).

S. Singh, in *Handbook of Lasers*, Chemical Rubber Co., Cleveland, OH, 1971.

Smith, A. V., SNLO, a public-domain nonlinear optics data base which can be obtained at
http://www.sandia.gov/imrl/XWEB1128/xxtal.htm

Books on Nonlinear Optics

G. P. Agrawal, *Nonlinear Fiber Optics*, Academic Press, Boston, 1989.

S. A. Akhmanov and R. V. Khokhlov, *Problems of Nonlinear Optics*, Gordon and Breach, New York, 1972.

G. C. Baldwin, *An Introduction to Nonlinear Optics*, Plenum Press, New York, 1969.

N. Bloembergen, *Nonlinear Optics*, Benjamin, New York, 1964.

P. N. Butcher, *Nonlinear Optical Phenomena*, Ohio State University, 1965.

P. N. Butcher and D. Cotter, *The Elements of Nonlinear Optics*, Cambridge University Press, 1990.

N. B. Delone, *Fundamentals of Nonlinear Optics of Atomic Gases*, Wiley, New York, 1988.

D. C. Hannah, M. A. Yuratich, and D. Cotter, *Nonlinear Optics of Free Atoms and Molecules*, Springer-Verlag, Berlin, 1979.

F. A. Hopf and G. I. Stegeman, *Applied Classical Electrodynamics, Vol. 1: Linear Optics*, Wiley, New York, 1985; *Vol. 2: Nonlinear Optics*, Wiley, New York, 1986.

D. N. Klyshko, *Photons and Nonlinear Optics*, Gordon and Breach, New York, 1988.

M. D. Levenson and S. Kano, *Introduction to Nonlinear Laser Spectroscopy*, Academic Press, Boston, 1988.

R. Loudon, *The Quantum Theory of Light*, Clarendon Press, Oxford, 1983.

A. C. Newell and J. V. Moloney, *Nonlinear Optics*, Addison-Wesley, Redwood City, CA, 1992.

J. F. Reintjes, *Nonlinear Optical Parametric Processes in Liquids and Gases*, Academic Press, Orlando, FL, 1984.

E. G. Sauter, *Nonlinear Optics*, Wiley, New York, 1996.

M. Schubert and B. Wilhelmi, *Nonlinear Optics and Quantum Electronics*, Wiley, New York, 1986.

Y. R. Shen, *The Principles of Nonlinear Optics*, Wiley, New York, 1984.

R. L. Sutherland, *Handbook of Nonlinear Optics*, Marcel Dekker, Inc., New York, 1996.

A. Yariv, *Quantum Electronics*, Wiley, New York, 1975.

F. Zernike and J. E. Midwinter, *Applied Nonlinear Optics*, Wiley, New York, 1973.

Books on Group Theory and Crystal Symmetry

S. Bhagavantan, *Crystal Symmetry and Physical Properties*, Academic Press, London, 1966.

W. L. Bond, *Crystal Technology*, Wiley, New York, 1976.

M. J. Buerger, *Elementary Crystallography*, Wiley, New York, 1963.

J. F. Nye, *The Physical Properties of Crystals*, Clarendon Press, Oxford, 1985.

F. C. Phillips, *An Introduction to Crystallography*, Wiley, New York, 1976.

M. Tinkham, *Group Theory and Quantum Mechanics*, McGraw-Hill, New York, 1964.

Kramers–Kronig Relations

F. Bassani and S. Scandolo, *Phys. Rev. B* **44**, 8446, 1991.

P. N. Butcher and D. Cotter, *The Elements of Nonlinear Optics*, Cambridge University Press, Cambridge, UK, 1990, Appendix 8.

W. J. Caspers, *Phys. Rev.* **133**, A1249 (1964).

D. C. Hutchings, M. Sheik-Bahae, D. J. Hagan, and E. W. Van Stryland, *Optical Quantum Electron.* **24**, 1 (1992).

S. M. Kogan, *Sov. Phys. JETP* **16**, 217 (1963).

L. D. Landau and E. M. Lifshitz, *Electrodynamics of Continuous Media*, Pergamon Press, Oxford, 1960, Section 62.

R. Loudon, *Quantum Theory of Light, Second Edition*, Clarendon Press, Oxford, 1983.

K. E. Peiponen, E. M. Vartiainen, and T. Asakura, pp. 57–94, in *Progress in Optics*, Vol. 37, edited by E. Wolf, Elsevier Science BV, 1997.

P. J. Price, *Phys. Rev.*, **130**, 1792 (1963).

F. L. Ridener, Jr. and R. H. Good, Jr., *Phys. Rev. B* **10**, 4980 (1974) and *Phys. Rev. B* **11**, 2768 (1975).

F. Smet and A van Groenendael, *Phys. Rev. A* **19**, 334 (1979).

Chapter 2

Wave-Equation Description
of Nonlinear Optical Interactions

2.1. The Wave Equation for Nonlinear Optical Media

We have seen in the last chapter how nonlinearity in the response of a material system to an intense laser field can cause the polarization of the medium to develop new frequency components not present in the incident radiation field. These new frequency components of the polarization act as sources of new frequency components of the electromagnetic field. In the present chapter, we examine how Maxwell's equations describe the generation of these new components of the field, and more generally we see how the various frequency components of the field become coupled by the nonlinear interaction.

Before developing the mathematical theory of these effects, we shall give a simple physical picture of how these frequency components are generated. For definiteness, we consider the case of sum-frequency generation as shown in part (a) of Fig. 2.1.1, where the input fields are at frequency ω_1 and ω_2. Because of nonlinearities in the atomic response, each atom develops an oscillating dipole moment which contains a component at frequency $\omega_1 + \omega_2$. An isolated atom would radiate at this frequency in the form of a dipole radiation pattern, as shown symbolically in part (b) of the figure. However, any material sample contains an enormous number N of atomic dipoles, each oscillating with a phase that is determined by the phases of the incident fields. If the relative phasing of these dipoles is correct, the field radiated by each dipole will add constructively in the forward direction, leading to radiation in the form of a well-defined beam, as illustrated in part (c) of the figure. The system

(a)

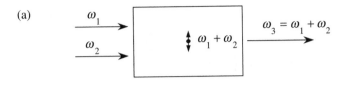

(b)

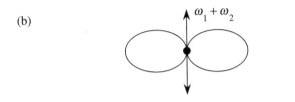

(c)

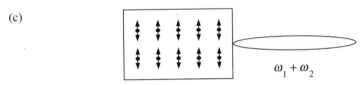

FIGURE 2.1.1   Sum-frequency generation.

will act as a phased array of dipoles when a certain condition, known as the phase-matching condition (see Eq. (2.2.15) in the next section), is satisfied. Under these conditions, the electric field strength of the radiation emitted in the forward direction will be N times larger than that due to any one atom, and consequently the intensity will be N^2 times as large.

Let us now consider the form of the wave equation for the propagation of light through a nonlinear optical medium. We begin with Maxwell's equations, which we write in gaussian units in the form*

$$\nabla \cdot \tilde{\mathbf{D}} = 4\pi \tilde{\rho}, \tag{2.1.1}$$

$$\nabla \cdot \tilde{\mathbf{B}} = 0, \tag{2.1.2}$$

$$\nabla \times \tilde{\mathbf{E}} = -\frac{1}{c}\frac{\partial \tilde{\mathbf{B}}}{\partial t}, \tag{2.1.3}$$

$$\nabla \times \tilde{\mathbf{H}} = \frac{1}{c}\frac{\partial \tilde{\mathbf{D}}}{\partial t} + \frac{4\pi}{c}\tilde{\mathbf{J}}. \tag{2.1.4}$$

* Throughout the text we use a tilde to denote a quantity that varies rapidly in time.

We are primarily interested in the solution of these equations in regions of space that contain no free charges, so that

$$\tilde{\rho} = 0, \tag{2.1.5}$$

and that contain no free currents, so that

$$\tilde{\mathbf{J}} = 0. \tag{2.1.6}$$

We assume that the material is nonmagnetic, so that

$$\tilde{\mathbf{B}} = \tilde{\mathbf{H}}. \tag{2.1.7}$$

However, we allow the material to be nonlinear in the sense that the fields $\tilde{\mathbf{D}}$ and $\tilde{\mathbf{E}}$ are related by

$$\tilde{\mathbf{D}} = \tilde{\mathbf{E}} + 4\pi \tilde{\mathbf{P}}, \tag{2.1.8}$$

where in general the polarization vector $\tilde{\mathbf{P}}$ depends nonlinearly upon the local value of the electric field strength $\tilde{\mathbf{E}}$.

We now proceed to derive the optical wave equation in the usual manner. We take the curl of the curl-$\tilde{\mathbf{E}}$ Maxwell equation (2.1.3), interchange the order of space and time derivatives on the right-hand side of the resulting equation, and use Eqs. (2.1.4), (2.1.6), and (2.1.7) to replace $\nabla \times \tilde{\mathbf{B}}$ by $(1/c)(\partial \tilde{\mathbf{D}}/\partial t)$, to obtain the equation

$$\nabla \times \nabla \times \tilde{\mathbf{E}} + \frac{1}{c^2} \frac{\partial^2}{\partial t^2} \tilde{\mathbf{D}} = 0. \tag{2.1.9a}$$

We now use Eq. (2.1.8) to eliminate $\tilde{\mathbf{D}}$ from this equation, and we thereby obtain the expression

$$\nabla \times \nabla \times \tilde{\mathbf{E}} + \frac{1}{c^2} \frac{\partial^2}{\partial t^2} \tilde{\mathbf{E}} = -\frac{4\pi}{c^2} \frac{\partial^2 \tilde{\mathbf{P}}}{\partial t^2}. \tag{2.1.9b}$$

This is the most general form of the wave equation in nonlinear optics. Under certain conditions it can be simplified. For example, by using an identity from vector calculus, we can write the first term on the left-hand side of Eq. (2.1.9b) as

$$\nabla \times \nabla \times \tilde{\mathbf{E}} = \nabla(\nabla \cdot \tilde{\mathbf{E}}) - \nabla^2 \tilde{\mathbf{E}}. \tag{2.1.10}$$

In the linear optics of isotropic source-free media, the first term on the right-hand side of this equation vanishes because the Maxwell equation $\nabla \cdot \tilde{\mathbf{D}} = 0$ implies that $\nabla \cdot \tilde{\mathbf{E}} = 0$. However, in nonlinear optics this term is generally non-vanishing even for isotropic materials, as a consequence of the more general relation (2.1.8) between $\tilde{\mathbf{D}}$ and $\tilde{\mathbf{E}}$. Fortunately, in nonlinear optics the first term on the right-hand side of Eq. (2.1.10) can usually be dropped for cases of interest.

For example, if $\tilde{\mathbf{E}}$ is of the form of a transverse, infinite plane wave, $\nabla \cdot \tilde{\mathbf{E}}$ vanishes identically. More generally, the first term can often be shown to be small, even when it does not vanish identically, especially when the slowly-varying amplitude approximation (see Section 2.2) is valid. For the remainder of this book, we shall usually assume that the contribution of $\nabla(\nabla \cdot \tilde{\mathbf{E}})$ in Eq. (2.1.10) is negligible so that the wave equation can be taken to have the form

$$-\nabla^2 \tilde{\mathbf{E}} + \frac{1}{c^2} \frac{\partial^2}{\partial t^2} \tilde{\mathbf{E}} = -\frac{4\pi}{c^2} \frac{\partial^2 \tilde{\mathbf{P}}}{\partial t^2}. \tag{2.1.11}$$

It is often convenient to split $\tilde{\mathbf{P}}$ into its linear and nonlinear parts as

$$\tilde{\mathbf{P}} = \tilde{\mathbf{P}}^{(1)} + \tilde{\mathbf{P}}^{\text{NL}}. \tag{2.1.12}$$

Here $\tilde{\mathbf{P}}^{(1)}$ is the part of $\tilde{\mathbf{P}}$ that depends linearly on the electric field strength $\tilde{\mathbf{E}}$. We can similarly decompose the displacement field $\tilde{\mathbf{D}}$ into its linear and nonlinear parts as

$$\tilde{\mathbf{D}} = \tilde{\mathbf{D}}^{(1)} + 4\pi \tilde{\mathbf{P}}^{\text{NL}}, \tag{2.1.13a}$$

where the linear part is given by

$$\tilde{\mathbf{D}}^{(1)} = \tilde{\mathbf{E}} + 4\pi \tilde{\mathbf{P}}^{(1)}. \tag{2.1.13b}$$

In terms of this quantity, the wave equation (2.1.11) becomes

$$-\nabla^2 \tilde{\mathbf{E}} + \frac{1}{c^2} \frac{\partial^2 \tilde{\mathbf{D}}^{(1)}}{\partial t^2} = -\frac{4\pi}{c^2} \frac{\partial^2 \tilde{\mathbf{P}}^{\text{NL}}}{\partial t^2}. \tag{2.1.14}$$

To see why this form of the wave equation is useful, let us first consider the case of a lossless, dispersionless medium. We can then express the relation between $\tilde{\mathbf{D}}^{(1)}$ and $\tilde{\mathbf{E}}$ in terms of a real, frequency-independent dielectric tensor $\epsilon^{(1)}$ as

$$\tilde{\mathbf{D}}^{(1)} = \epsilon^{(1)} \cdot \tilde{\mathbf{E}}. \tag{2.1.15a}$$

For the case of an isotropic material, this relation reduces to simply

$$\tilde{\mathbf{D}}^{(1)} = \epsilon^{(1)} \tilde{\mathbf{E}}, \tag{2.1.15b}$$

where $\epsilon^{(1)}$ is a scalar quantity. For this (simpler) case of an isotropic, dispersionless material, the wave equation (2.1.14) becomes

$$-\nabla^2 \tilde{\mathbf{E}} + \frac{\epsilon^{(1)}}{c^2} \frac{\partial^2 \tilde{\mathbf{E}}}{\partial t^2} = -\frac{4\pi}{c^2} \frac{\partial^2 \tilde{\mathbf{P}}^{\text{NL}}}{\partial t^2}. \tag{2.1.16}$$

This equation has the form of a driven (i.e., inhomogeneous) wave equation; the nonlinear response of the medium acts as a source term which appears on the right-hand side of this equation. In the absence of this source term,

Eq. (2.1.16) admits solutions of the form of free waves propagating with velocity c/n, where $n = [\epsilon^{(1)}]^{1/2}$ is the (linear) index of refraction.

For the case of a dispersive medium, we must consider each frequency component of the field separately. We represent the electric, linear displacement, and polarization fields as the sums of their various frequency components:

$$\tilde{\mathbf{E}}(\mathbf{r}, t) = \sum_n{}' \tilde{\mathbf{E}}_n(\mathbf{r}, t), \tag{2.1.17a}$$

$$\tilde{\mathbf{D}}^{(1)}(\mathbf{r}, t) = \sum_n{}' \tilde{\mathbf{D}}_n^{(1)}(\mathbf{r}, t), \tag{2.1.17b}$$

$$\tilde{\mathbf{P}}^{\mathrm{NL}}(\mathbf{r}, t) = \sum_n{}' \tilde{\mathbf{P}}_n^{\mathrm{NL}}(\mathbf{r}, t), \tag{2.1.17c}$$

where the summation is to be performed over positive field frequencies only, and we represent each frequency component in terms of its complex amplitude as

$$\tilde{\mathbf{E}}_n(\mathbf{r}, t) = \mathbf{E}_n(\mathbf{r})e^{-i\omega_n t} + \text{c.c.}, \tag{2.1.18a}$$

$$\tilde{\mathbf{D}}_n^{(1)}(\mathbf{r}, t) = \mathbf{D}_n^{(1)}(\mathbf{r})e^{-i\omega_n t} + \text{c.c.}, \tag{2.1.18b}$$

$$\tilde{\mathbf{P}}_n^{\mathrm{NL}}(\mathbf{r}, t) = \mathbf{P}_n^{\mathrm{NL}}(\mathbf{r})e^{-i\omega_n t} + \text{c.c.} \tag{2.1.18c}$$

If dissipation can be neglected, the relationship between $\tilde{\mathbf{D}}_n^{(1)}$ and $\tilde{\mathbf{E}}_n$ can be expressed in terms of a real, frequency-dependent dielectric tensor according to

$$\tilde{\mathbf{D}}_n^{(1)}(\mathbf{r}, t) = \epsilon^{(1)}(\omega_n) \cdot \tilde{\mathbf{E}}_n(\mathbf{r}, t). \tag{2.1.19}$$

When Eqs. (2.1.17a) through (2.1.19) are introduced into Eq. (2.1.14), we obtain a wave equation analogous to (2.1.16) that is valid for each frequency component of the field:

$$-\nabla^2 \tilde{\mathbf{E}}_n + \frac{\epsilon^{(1)}(\omega_n)}{c^2} \frac{\partial^2 \tilde{\mathbf{E}}_n}{\partial t^2} = -\frac{4\pi}{c^2} \frac{\partial^2 \tilde{\mathbf{P}}_n^{\mathrm{NL}}}{\partial t^2}. \tag{2.1.20}$$

The general case of a dissipative medium is treated by allowing the dielectric tensor to be a complex quantity that relates the complex field amplitudes according to

$$\mathbf{D}_n^{(1)}(\mathbf{r}) = \epsilon^{(1)}(\omega_n) \cdot \mathbf{E}_n(\mathbf{r}). \tag{2.1.21}$$

This expression, along with Eqs. (2.1.17) and (2.1.18), can be introduced into the wave equation (2.1.14), to obtain

$$-\nabla^2 \mathbf{E}_n(\mathbf{r}) - \frac{\omega_n^2}{c^2} \epsilon^{(1)}(\omega_n) \cdot \mathbf{E}_n(\mathbf{r}) = \frac{4\pi \omega_n^2}{c^2} \mathbf{P}_n^{\mathrm{NL}}(\mathbf{r}). \tag{2.1.22}$$

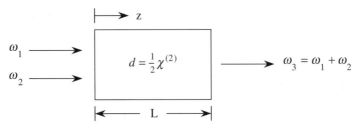

FIGURE 2.2.1   Sum-frequency generation.

2.2. The Coupled-Wave Equations for Sum-Frequency Generation

We next study how the nonlinear optical wave equation that we derived in the previous section can be used to describe specific nonlinear optical interactions. In particular, we consider sum-frequency generation in a lossless nonlinear optical medium involving collimated, monochromatic, continuous-wave input beams. We assume the configuration shown in Fig. 2.2.1, where the applied waves fall onto the nonlinear medium at normal incidence. For simplicity, we ignore double refraction effects. The treatment given here can be generalized straightforwardly to include nonnormal incidence and double refraction.[*]

The wave equation in the form (2.1.20) must hold for each frequency component of the field and in particular for the sum-frequency component at frequency ω_3. In the absence of a nonlinear source term, the solution to this equation for a plane wave at frequency ω_3 propagating in the $+z$ direction is

$$\tilde{E}_3(z, t) = A_3 e^{i(k_3 z - \omega_3 t)} + \text{c.c.}, \tag{2.2.1}$$

where[†]

$$k_3 = \frac{n_3 \omega_3}{c}, \qquad n_3 = \left[\epsilon^{(1)}(\omega_3)\right]^{1/2}, \tag{2.2.2}$$

and where the amplitude of the wave A_3 is a *constant*. We expect on physical grounds that, when the nonlinear source term is not too large, the solution to Eq. (2.1.20) will still be of the form of Eq. (2.2.1), except that A_3 will become a slowly varying function of z. We hence adopt Eq. (2.2.1) with A_3 taken to be a function of z as the form of the trial solution to the wave equation (2.1.20) in the presence of the nonlinear source term.

We represent the nonlinear source term appearing in Eq. (2.1.20) as

$$\tilde{P}_3(z, t) = P_3 e^{-i\omega_3 t} + \text{c.c.}, \tag{2.2.3}$$

---

[*] See, for example, Shen (1984), Chapter 6.

[†] For convenience, we are working in the scalar field approximation; n_3 represents the refractive index appropriate to the state of polarization of the ω_3 wave.

where according to Eq. (1.5.28)

$$P_3 = 4d_{\text{eff}}E_1E_2. \tag{2.2.4}$$

If we represent the applied fields as

$$\tilde{E}_i(z,t) = E_ie^{-i\omega_it} + \text{c.c.}, \qquad i = 1, 2, \tag{2.2.5}$$

with

$$E_i = A_ie^{ik_iz}, \qquad i = 1, 2, \tag{2.2.6}$$

the amplitude of the nonlinear polarization can be written as

$$P_3 = 4d_{\text{eff}}A_1A_2e^{i(k_1+k_2)z} \equiv p_3e^{i(k_1+k_2)z}. \tag{2.2.7}$$

We now substitute Eqs. (2.2.1), (2.2.3), and (2.2.7) into the wave equation (2.1.20). Since the fields depend only on the longitudinal coordinate z, we can replace ∇^2 by d^2/dz^2. We then obtain

$$\left[\frac{d^2A_3}{dz^2} + 2ik_3\frac{dA_3}{dz} - k_3^2A_3 + \frac{\epsilon^{(1)}(\omega_3)\omega_3^2A_3}{c^2}\right]e^{i(k_3z-\omega_3t)} + \text{c.c.}$$

$$= \frac{-16\pi d_{\text{eff}}\omega_3^2}{c^2}A_1A_2e^{i[(k_1+k_2)z-\omega_3t]} + \text{c.c.} \tag{2.2.8}$$

Since $k_3^2 = \epsilon^{(1)}(\omega_3)\omega_3^2/c^2$, the third and fourth terms on the left-hand side of this expression cancel. Note that we can drop the complex conjugate terms from each side and still maintain the equality. We can then cancel the factor $\exp(-i\omega_3t)$ on each side and write the resulting equation as

$$\frac{d^2A_3}{dz^2} + 2ik_3\frac{dA_3}{dz} = \frac{-16\pi d_{\text{eff}}\omega_3^2}{c^2}A_1A_2e^{i(k_1+k_2-k_3)z}. \tag{2.2.9}$$

It is usually permissible to neglect the first term on the left-hand side of this equation on the grounds that it is very much smaller than the second. This approximation is known as the slowly-varying amplitude approximation and is valid whenever

$$\left|\frac{d^2A_3}{dz^2}\right| \ll \left|k_3\frac{dA_3}{dz}\right|. \tag{2.2.10}$$

This condition requires that the fractional change in A_3 in a distance of the order of an optical wavelength must be much smaller than unity. When this approximation is made, Eq. (2.2.9) becomes

$$\frac{dA_3}{dz} = \frac{8\pi id_{\text{eff}}\omega_3^2}{k_3c^2}A_1A_2e^{i\Delta kz}$$

$$= \frac{2\pi i\omega_3}{n_3c}p_3e^{i\Delta kz}, \tag{2.2.11}$$

where we have introduced the quantity

$$\Delta k = k_1 + k_2 - k_3, \qquad (2.2.12)$$

which is called the wavevector (or momentum) mismatch. Equation (2.2.11) is known as a coupled-amplitude equation, because it shows how the amplitude of the ω_3 wave varies as a consequence of its coupling to the ω_1 and ω_2 waves. In general, the spatial variation of the ω_1 and ω_2 waves must also be taken into consideration, and we can derive analogous equations for the ω_1 and ω_2 fields by repeating the derivation given above for each of these frequencies. We hence find two additional coupled-amplitude equations given by

$$\frac{dA_1}{dz} = \frac{8\pi i d_{\mathrm{eff}} \omega_1^2}{k_1 c^2} A_3 A_2^* e^{-i \Delta k z} \qquad (2.2.13)$$

and

$$\frac{dA_2}{dz} = \frac{8\pi i d_{\mathrm{eff}} \omega_2^2}{k_2 c^2} A_3 A_1^* e^{-i \Delta k z}. \qquad (2.2.14)$$

Note that, in writing these equations in the forms shown, we have assumed that the medium is lossless. For a lossless medium, no explicit loss terms need be included in these equations, and furthermore we can make use of the condition of full permutation symmetry (Eq. (1.5.8)) to conclude that the coupling coefficient has the same value d_{eff} in each equation.

Phase-Matching Considerations

For simplicity, let us first assume that the amplitudes A_1 and A_2 of the input fields can be taken as constants on the right-hand side of Eq. (2.2.11). This assumption is valid whenever the conversion of the input fields into the sum-frequency field is not too large. We note that, for the special case

$$\Delta k = 0, \qquad (2.2.15)$$

the amplitude A_3 of the sum-frequency wave increases linearly with z, and consequently that its intensity increases quadratically with z. The condition (2.2.15) is known as the condition of perfect phase matching. When this condition is fulfilled, the generated wave maintains a fixed phase relation with respect to the nonlinear polarization and is able to extract energy most efficiently from the incident waves. From a microscopic point of view, when the condition (2.2.15) is fulfilled the individual atomic dipoles that constitute the material system are properly phased so that the field emitted by each dipole adds coherently in the forward direction. The total power radiated by the ensemble of atomic dipoles thus scales as the square of the number of atoms that participate.

When the condition (2.2.15) is not satisfied, the intensity of the emitted radiation is smaller than for the case of $\Delta k = 0$. The amplitude of the sum-frequency (ω_3) field at the exit plane of the nonlinear medium is given in this case by integrating Eq. (2.2.11) from $z = 0$ to $z = L$, yielding

$$A_3(L) = \frac{8\pi i d_{\text{eff}}\omega_3^2 A_1 A_2}{k_3 c^2} \int_0^L e^{i\Delta kz} dz = \frac{8\pi i d\omega_3^2 A_1 A_2}{k_3 c^2}\left(\frac{e^{i\Delta kL} - 1}{i\Delta k}\right).$$

(2.2.16)

The intensity of the ω_3 wave is given by the magnitude of the time-averaged Poynting vector, which for our definition of field amplitude is given by

$$I_i = \frac{n_i c}{2\pi}|A_i|^2, \quad i = 1, 2, 3.$$

(2.2.17)

We thus obtain

$$I_3 = \frac{32\pi d_{\text{eff}}^2 \omega_3^4 |A_1|^2 |A_2|^2 n_3}{k_3^2 c^3}\left|\frac{e^{i\Delta kL} - 1}{\Delta k}\right|^2.$$

(2.2.18)

The squared modulus that appears in this equation can be expressed as

$$\left|\frac{e^{i\Delta kL} - 1}{\Delta k}\right|^2 = L^2\left(\frac{e^{i\Delta kL} - 1}{\Delta kL}\right)\left(\frac{e^{-i\Delta kL} - 1}{\Delta kL}\right) = 2L^2\frac{(1 - \cos \Delta kL)}{(\Delta kL)^2}$$

(2.2.19)

$$= L^2\frac{\sin^2(\Delta kL/2)}{(\Delta kL/2)^2} \equiv L^2 \text{sinc}^2(\Delta kL/2).$$

Finally, our expression for I_3 can be written in terms of the intensities of the incident fields by using Eq. (2.2.17) to express $|A_i|^2$ in terms of the intensities, yielding the result

$$I_3 = \frac{512\pi^5 d_{\text{eff}}^2 I_1 I_2}{n_1 n_2 n_3 \lambda_3^2 c} L^2 \text{sinc}^2(\Delta kL/2),$$

(2.2.20)

where $\lambda_3 = 2\pi c/\omega_3$ is the vacuum wavelength of the ω_3 wave. Note that the effect of wavevector mismatch is included entirely in the factor $\text{sinc}^2(\Delta kL/2)$. This factor, which is known as the phase mismatch factor, is plotted in Fig. 2.2.2.

It should be noted that the efficiency of the three-wave mixing process decreases as $|\Delta k|L$ increases, with some oscillations occurring. The reason for this behavior is that if L is greater than approximately $1/\Delta k$, the output wave can get out of phase with its driving polarization, and power can flow from the ω_3 wave back into the ω_1 and ω_2 waves (see Eq. (2.2.11)). For this reason, one sometimes defines

$$L_c = 2/\Delta k$$

(2.2.21)

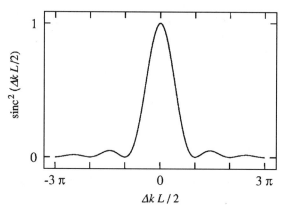

FIGURE 2.2.2   Effects of wavevector mismatch on the efficiency of sum-frequency generation.

to be the coherent buildup length of the interaction, so that the phase mismatch factor in Eq. (2.2.20) can be written as

$$\text{sinc}^2(L/L_c). \tag{2.2.22}$$

2.3.   The Manley–Rowe Relations

Let us now consider, from a general point of view, the mutual interaction of three optical waves propagating through a lossless nonlinear optical medium, as illustrated in Fig. 2.3.1.

We have just derived the coupled-amplitude equations (Eqs. (2.2.11) through (2.2.14)) that describes the spatial variation of the amplitude of each wave. Let us now consider the spatial variation of the *intensity* associated with each of these waves. Since

$$I_i = \frac{n_i c}{2\pi} A_i A_i^*, \tag{2.3.1}$$

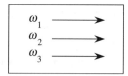

FIGURE 2.3.1   Optical waves of frequencies ω_1, ω_2, and $\omega_3 = \omega_1 + \omega_2$ interact in a lossless nonlinear optical medium.

the variation of the intensity is described by

$$\frac{dI_i}{dz} = \frac{n_i c}{2\pi}\left(A_i^* \frac{dA_i}{dz} + A_i \frac{dA_i^*}{dz}\right). \tag{2.3.2}$$

Through use of this result and Eq. (2.2.13), we find that the spatial variation of the intensity of the wave at frequency ω_1 is given by

$$\frac{dI_1}{dz} = \frac{n_i c}{2\pi}\frac{8\pi d_{\text{eff}}\omega_1^2}{k_1 c^2}(i A_1^* A_3 A_2^* e^{-i\Delta kz} + \text{c.c.})$$

$$= 4d_{\text{eff}}\omega_1(i A_3 A_1^* A_2^* e^{-i\Delta kz} + \text{c.c.})$$

or by

$$\frac{dI_1}{dz} = -8d_{\text{eff}}\omega_1 \operatorname{Im}(A_3 A_1^* A_2^* e^{-i\Delta kz}). \tag{2.3.3a}$$

We similarly find that the spatial variation of the intensities of the waves at frequencies ω_2 and ω_3 is given by

$$\frac{dI_2}{dz} = -8d_{\text{eff}}\omega_2 \operatorname{Im}(A_3 A_1^* A_2^* e^{-i\Delta kz}), \tag{2.3.3b}$$

$$\frac{dI_3}{dz} = -8d_{\text{eff}}\omega_3 \operatorname{Im}(A_3^* A_1 A_2 e^{i\Delta kz})$$

$$= 8d_{\text{eff}}\omega_3 \operatorname{Im}(A_3 A_1^* A_2^* e^{-i\Delta kz}). \tag{2.3.3c}$$

We see that the sign of dI_1/dz is the same as that of dI_2/dz but is opposite to that of dI_3/dz. We also see that the direction of energy flow depends on the relative phases of the three interacting fields.

The set of equations (2.3.3a), (2.3.3b), and (2.3.3c) shows that the total power flow is conserved, as expected for propagation through a lossless medium. To demonstrate this fact, we define the total intensity as

$$I = I_1 + I_2 + I_3. \tag{2.3.4}$$

We then find that the spatial variation of the total intensity is given by

$$\frac{dI}{dz} = \frac{dI_1}{dz} + \frac{dI_2}{dz} + \frac{dI_3}{dz}$$

$$= -8d_{\text{eff}}(\omega_1 + \omega_2 - \omega_3)\operatorname{Im}(A_3 A_1^* A_2^* e^{i\Delta kz}) = 0, \tag{2.3.5}$$

where we have made use of Eqs. (2.3.3a), (2.3.3b), and (2.3.3c) and where the last equality follows from the fact that $\omega_3 = \omega_1 + \omega_2$.

The set of equations (2.3.3a), (2.3.3b), and (2.3.3c) also implies that

$$\frac{d}{dz}\left(\frac{I_1}{\omega_1}\right) = \frac{d}{dz}\left(\frac{I_2}{\omega_2}\right) = -\frac{d}{dz}\left(\frac{I_3}{\omega_3}\right), \tag{2.3.6}$$

as can be verified by inspection. These equalities are known as the Manley–Rowe relations (Manley and Rowe, 1959). Since the energy of a photon of frequency ω_i is $\hbar\omega_i$, the quantity I_i/ω_i that appears in these relations is proportional to the intensity of the wave measured in photons per unit area per unit time. The Manley–Rowe relations can alternatively be expressed as

$$\frac{d}{dz}\left(\frac{I_2}{\omega_2}+\frac{I_3}{\omega_3}\right)=0, \quad \frac{d}{dz}\left(\frac{I_1}{\omega_1}+\frac{I_3}{\omega_3}\right)=0, \quad \frac{d}{dz}\left(\frac{I_1}{\omega_1}-\frac{I_2}{\omega_2}\right)=0. \quad (2.3.7)$$

These equations can be formally integrated to obtain the three conserved quantities (conserved in the sense that they are spatially invariant) M_1, M_2, and M_3, which are given by

$$M_1=\frac{I_2}{\omega_2}+\frac{I_3}{\omega_3}, \quad M_2=\frac{I_1}{\omega_1}+\frac{I_3}{\omega_3}, \quad M_3=\frac{I_1}{\omega_1}-\frac{I_2}{\omega_2}. \quad (2.3.8)$$

These relations tell us that the rate at which photons at frequency ω_1 are created is equal to the rate at which photons at frequency ω_2 are created and is equal to the rate at which photons at frequency ω_3 are destroyed. This result can be understood intuitively by means of the energy level description of a three-wave mixing process, which is shown in Figure 2.3.2. This diagram shows that, for a lossless medium, the creation of an ω_1 photon must be accompanied by the creation of an ω_2 photon and the annihilation of an ω_3 photon. It seems at first sight surprising that the Manley–Rowe relations should be consistent with this quantum-mechanical interpretation, when our derivation of these relations appears to be entirely classical. Note, however, that our derivation implicitly assumes that the nonlinear susceptibility possesses full permutation symmetry in that we have taken the coupling constant d_{eff} to have the same value in each of the coupled-amplitude equations (2.2.11), (2.2.13), and (2.2.14). We remarked earlier (following Eq. (1.5.9)) that in a sense the condition of full permutation symmetry is a consequence of the laws of quantum mechanics.

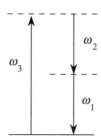

FIGURE 2.3.2   Photon description of the interaction of three optical waves.

2.4. Sum-Frequency Generation

In Section 2.2, we treated the process of sum-frequency generation in the simple limit in which the two input fields are undepleted by the nonlinear interaction. In the present section, we treat this process more generally. We assume the configuration shown in Fig. 2.4.1.

The coupled-amplitude equations describing this interaction were derived above and appear as Eqs. (2.2.11) through (2.2.14). These equations can be solved exactly in terms of the Jacobi elliptic functions. We shall not present the details of this solution, because the method is very similar to the one that we use in Section 2.6 to treat second-harmonic generation. Details can be found in Armstrong *et al.* (1962); see also Problem 2 at the end of this chapter.

Instead, we treat the somewhat simpler (but more illustrative) case in which one of the applied fields (taken to be at frequency ω_2) is strong, but the other field (at frequency ω_1) is weak. This situation would apply to the conversion of a weak infrared signal of frequency ω_1 to a visible frequency ω_3 by mixing with an intense laser beam of frequency ω_2 (see, for example, Boyd and Townes, 1977). This process is known as upconversion, because in this process the information-bearing beam is converted to a higher frequency. Usually optical-frequency waves are easier to detect with good sensitivity than are infrared waves. Since we can assume that the amplitude A_2 of the field at frequency ω_2 is unaffected by the interaction, we can take A_2 as a constant in the coupled-amplitude equations (Eqs. (2.2.11) through (2.2.14)), which then reduce to the simpler set

$$\frac{dA_1}{dz} = K_1 A_3 e^{-i\,\Delta k z}, \tag{2.4.1a}$$

$$\frac{dA_3}{dz} = K_3 A_1 e^{+i\,\Delta k z}, \tag{2.4.1b}$$

where we have introduced the quantities

$$K_1 = \frac{8\pi i \omega_1^2 d}{k_1 c^2} A_2^*, \qquad K_3 = \frac{8\pi i \omega_3^2 d}{k_3 c^2} A_2, \tag{2.4.2a}$$

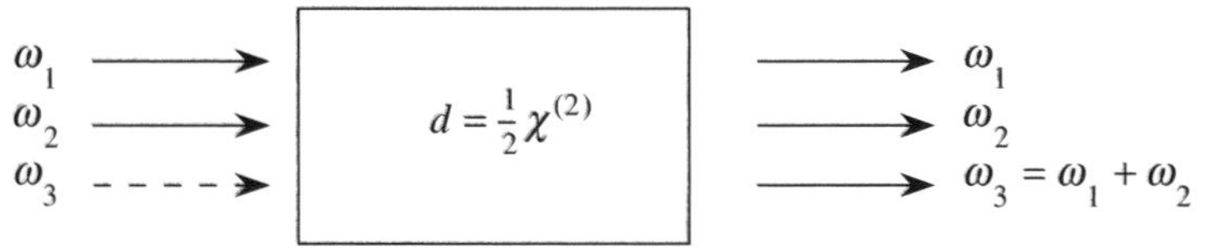

FIGURE 2.4.1 Sum-frequency generation. Typically, no input field is applied at frequency ω_3.

and

$$\Delta k = k_1 + k_2 - k_3. \tag{2.4.2b}$$

The solution to Eqs. (2.4.1) is particularly simple if we set $\Delta k = 0$, and we first treat this case. We take the derivative of Eq. (2.4.1a) to obtain

$$\frac{d^2 A_1}{dz^2} = K_1 \frac{dA_3}{dz}. \tag{2.4.3}$$

We now use Eq. (2.4.1b) to eliminate dA_3/dz from the right-hand side of this equation to obtain an equation involving only $A_1(z)$:

$$\frac{d^2 A_1}{dz^2} = -\kappa^2 A_1, \tag{2.4.4}$$

where we have introduced the *positive* coupling coefficient κ^2 defined by

$$\kappa^2 \equiv -K_1 K_3 = \frac{64\pi^2 \omega_1^2 \omega_3^2 d_{\text{eff}}^2 |A_2|^2}{k_1 k_3 c^4}. \tag{2.4.5}$$

The general solution to Eq. (2.4.4) is

$$A_1(z) = B \cos \kappa z + C \sin \kappa z. \tag{2.4.6a}$$

We now obtain the form of $A_3(z)$ through use of Eq. (2.4.1a), which shows that $A_3(z) = (dA_1/dz)/K_1$, or

$$A_3(z) = \frac{-B\kappa}{K_1} \sin \kappa z + \frac{C\kappa}{K_1} \cos \kappa z. \tag{2.4.6b}$$

We next find the solution that satisfies the appropriate boundary conditions. We assume that the ω_3 field is not present at the input, so that the boundary conditions become $A_3(0) = 0$ with $A_1(0)$ specified. We find from Eq. (2.4.6b) that the boundary condition $A_3(0) = 0$ implies that $C = 0$, and from Eq. (2.4.6a) that $B = A_1(0)$. The solution for the ω_1 field is thus given by

$$A_1(z) = A_1(0) \cos \kappa z \tag{2.4.7}$$

and for the ω_3 field by

$$A_3(z) = -A_1(0) \frac{\kappa}{K_1} \sin \kappa z. \tag{2.4.8}$$

To simplify the form of this equation we express the ratio κ/K_1 as follows:

$$\frac{\kappa}{K_1} = \frac{8\pi \omega_1 \omega_3 d_{\text{eff}} |A_2|}{(k_1 k_3)^{1/2} c^2} \frac{k_1 c^2}{8\pi i \omega_1^2 d A_2^*} = -i \left(\frac{n_1 \omega_3}{n_3 \omega_1} \right)^{1/2} \frac{|A_2|}{A_2^*}.$$

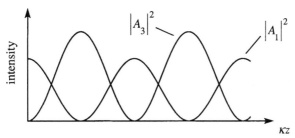

FIGURE 2.4.2   Variation of $|A_1|^2$ and $|A_3|^2$ for the case of perfect phase matching in the undepleted-pump approximation.

The ratio $|A_2|/A_2^*$ can be represented as

$$\frac{|A_2|}{A_2^*} = \frac{A_2}{A_2}\frac{|A_2|}{A_2^*} = \frac{A_2|A_2|}{|A_2|^2} = \frac{A_2}{|A_2|} = e^{i\phi_2},$$

where ϕ_2 denotes the phase of A_2. We hence find that

$$A_3(z) = i\left(\frac{n_1\omega_3}{n_3\omega_1}\right)^{1/2} A_1(0)\sin\kappa z e^{i\phi_2}. \qquad (2.4.9)$$

The nature of the solution given by Eqs. (2.4.7) and (2.4.9) is illustrated in Fig. 2.4.2.

Let us next solve Eqs. (2.4.1) for the general case of arbitrary wave vector mismatch. We seek a solution to these equations of the form

$$A_1(z) = (Fe^{igz} + Ge^{-igz})e^{-i\Delta kz/2}, \qquad (2.4.10)$$

$$A_3(z) = (Ce^{igz} + De^{-igz})e^{i\Delta kz/2}, \qquad (2.4.11)$$

where g gives the rate of spatial variation of the fields and where C, D, F, and G are constants whose values depend on the boundary conditions. We take this form for the trial solution because we expect the ω_1 and ω_3 waves to display the same spatial variation, since they are coupled together. We separate out the factors $e^{\pm i\Delta kz/2}$ because doing so simplifies the final form of the solution. Equations (2.4.10) and (2.4.11) are now substituted into Eq. (2.4.1a), to obtain

$$(igFe^{igz} - igGe^{-igz})e^{-(1/2)i\Delta kz} - \frac{1}{2}i\Delta k(Fe^{igz} + Ge^{-igz})e^{-(1/2)i\Delta kz}$$
$$= (K_1Ce^{igz} + K_1De^{-igz})e^{-(1/2)i\Delta kz}. \qquad (2.4.12)$$

Since this equation must hold for all values of z, the terms that vary as e^{igz} and e^{-igz} must each maintain the equality separately; the coefficients of these terms thus must be related by

$$F\left(ig - \tfrac{1}{2}i\Delta k\right) = K_1C, \qquad (2.4.13)$$

$$-G\left(ig + \tfrac{1}{2}i\Delta k\right) = K_1D. \qquad (2.4.14)$$

In a similar fashion, we find by substituting the trial solution into Eq. (2.4.1b) that

$$(igCe^{igz} - igDe^{-igz})e^{(1/2)i\Delta kz} + \tfrac{1}{2}i\,\Delta k(Ce^{igz} + De^{-igz})e^{(1/2)i\Delta kz}$$
$$= (K_3 Fe^{igz} + K_3 Ge^{-igz})e^{(1/2)i\Delta kz}, \tag{2.4.15}$$

and in order for this equation to hold for all values of z, the coefficients must satisfy

$$C\left(ig + \tfrac{1}{2}i\,\Delta k\right) = K_3 F, \tag{2.4.16}$$

$$-D\left(ig - \tfrac{1}{2}i\,\Delta k\right) = K_3 G. \tag{2.4.17}$$

Equations (2.4.13) and (2.4.16) constitute simultaneous equations for F and C. We write these equations in matrix form as

$$\begin{bmatrix} i\left(g - \tfrac{1}{2}\Delta k\right) & -K_1 \\ -K_3 & i\left(g + \tfrac{1}{2}\Delta k\right) \end{bmatrix} \begin{bmatrix} F \\ C \end{bmatrix} = 0.$$

A solution to this set of equations exists only if the determinant of the matrix of coefficients vanishes, i.e., if

$$g^2 = -K_1 K_3 + \tfrac{1}{4}\Delta k^2. \tag{2.4.18}$$

As before (cf. Eq. (2.4.5)), we introduce the positive quantity $\kappa^2 = -K_1 K_3$, so that we can express the solution to Eq. (2.4.18) as

$$g = \sqrt{\kappa^2 + \tfrac{1}{4}\Delta k^2}. \tag{2.4.19}$$

In determining g we take only the positive square root in the foregoing expression, since our trial solution (2.4.10) and (2.4.11) explicitly contains both the e^{+gz} and e^{-gz} spatial variations.

The general solution to our original set of equations (2.4.1) is given by Eqs. (2.4.10) and (2.4.11) with g given by Eq. (2.4.19). We evaluate the arbitrary constants C, D, F, and G appearing in the general solution by applying appropriate boundary conditions. We assume that the fields A_1 and A_3 are specified at the input plane $z = 0$ of the nonlinear medium, so that $A_1(0)$ and $A_3(0)$ are known. Then, by evaluating Eqs. (2.4.10) and (2.4.11) at $z = 0$, we find that

$$A_1(0) = F + G, \tag{2.4.20}$$

$$A_3(0) = C + D. \tag{2.4.21}$$

Equations (2.4.13) and (2.4.14) give two additional relations among the quantities C, D, F, and G. Consequently there are four independent linear equations relating the four quantities C, D, F, and G, and their simultaneous solution specifies these four quantities. The values of C, D, F, and G thereby obtained are introduced into the trial solution (2.4.10) and (2.4.11) to obtain the solution that meets the boundary conditions. This solution is given by

$$A_1(z) = \left[A_1(0) \cos gz + \left(\frac{K_1}{g} A_3(0) + \frac{i\Delta k}{2g} A_1(0) \right) \sin gz \right] e^{-(1/2)i\Delta kz},$$

(2.4.22)

$$A_3(z) = \left[A_3(0) \cos gz + \left(\frac{-i\Delta k}{2g} A_3(0) + \frac{K_3}{g} A_1(0) \right) \sin gz \right] e^{(1/2)i\Delta kz}.$$

(2.4.23)

In order to interpret this result, let us consider the special case in which no sum-frequency field is incident on the medium, so that $A_3(0) = 0$. Equation (2.4.23) then reduces to

$$A_3(z) = \frac{K_3}{g} A_1(0) \sin gz \, e^{(1/2)i\Delta kz}$$

(2.4.24)

and the intensity of the generated wave is proportional to

$$|A_3(z)|^2 = |A_1(0)|^2 \frac{|K_3|^2}{g^2} \sin^2 gz,$$

(2.4.25)

where g is given as before by Eq. (2.4.19). We note that the characteristic scale length g^{-1} of the interaction becomes shorter as Δk increases. However, as Δk increases the maximum intensity of the generated wave decreases. Since, according to Eq. (2.4.25), the intensity of the generated wave is inversely proportional to g^2, we see that as Δk is increased the maximum intensity of the generated wave is decreased by the factor $\kappa^2/(\kappa^2 + \frac{1}{4}\Delta k^2)$. This sort of behavior is illustrated in Fig. 2.4.3, in which the predictions of Eq. (2.4.25) are displayed graphically.

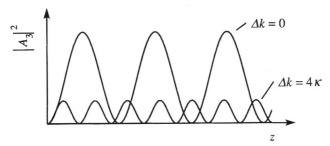

FIGURE 2.4.3 Spatial variation of the sum-frequency wave in the undepleted-pump approximation.

2.5. Difference-Frequency Generation and Parametric Amplification

Let us now consider the situation shown in Fig. 2.5.1, in which optical waves at frequencies ω_3 and ω_1 interact in a lossless nonlinear optical medium to produce an output wave at the difference frequency $\omega_2 = \omega_3 - \omega_1$. For simplicity, we assume that the ω_3 wave is a strong wave (i.e., is undepleted by the nonlinear interaction, so that we can treat A_3 as being essentially constant), and for the present we assume that no field at frequency ω_2 is incident on the medium.

The coupled-amplitude equations describing this interaction are obtained by a method analogous to that used in Section 2.2 to obtain the equations describing sum-frequency generation and have the form

$$\frac{dA_1}{dz} = \frac{8\pi i \omega_1^2 d_{\text{eff}}}{k_1 c^2} A_3 A_2^* e^{i\Delta kz}, \tag{2.5.1a}$$

$$\frac{dA_2}{dz} = \frac{8\pi i \omega_2^2 d_{\text{eff}}}{k_2 c^2} A_3 A_1^* e^{i\Delta kz}, \tag{2.5.1b}$$

where

$$\Delta k = k_3 - k_1 - k_2. \tag{2.5.2}$$

We first solve these equations for the case of perfect phase matching, that is, $\Delta k = 0$. We differentiate Eq. (2.5.1b) with respect to z and introduce the complex conjugate of Eq. (2.5.1a) to eliminate dA_1^*/dz from the right-hand side. We thereby obtain the equation

$$\frac{d^2 A_2}{dz^2} = \frac{64\pi^2 \omega_1^2 \omega_2^2 d^2}{k_1 k_2 c^4} A_3 A_3^* A_2 \equiv \kappa^2 A_2, \tag{2.5.3}$$

where we have introduced the coupling constant

$$\kappa^2 = \frac{64\pi^2 d^2 \omega_1^2 \omega_2^2}{k_1 k_2 c^4} |A_3|^2. \tag{2.5.4}$$

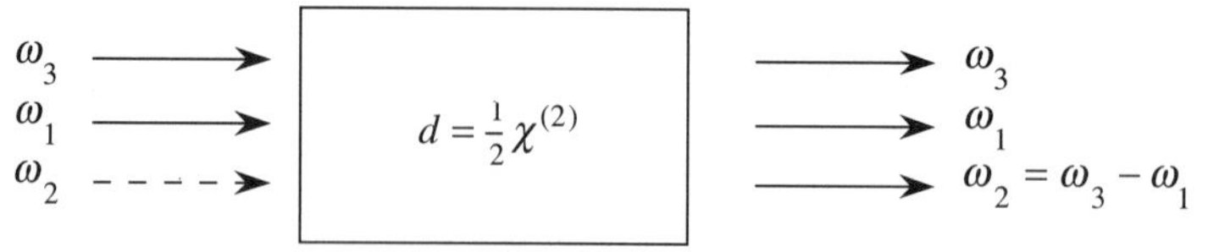

$\omega_3 \longrightarrow$ | $d = \frac{1}{2}\chi^{(2)}$ | $\longrightarrow \omega_3$
$\omega_1 \longrightarrow$ | | $\longrightarrow \omega_1$
$\omega_2 \dashrightarrow$ | | $\longrightarrow \omega_2 = \omega_3 - \omega_1$

FIGURE 2.5.1   Difference-frequency generation. Typically, no input field is applied at frequency ω_2.

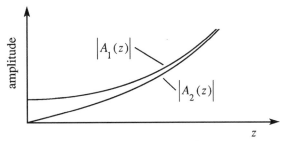

FIGURE 2.5.2   Spatial evolution of A_1 and A_2 for difference-frequency generation for the case $\Delta k = 0$ in the constant-pump approximation.

The general solution to this equation is

$$A_2(z) = C \sinh \kappa z + D \cosh \kappa z, \tag{2.5.5}$$

where C and D are integration constants whose values depend on the boundary conditions.

We now assume the boundary conditions

$$A_2(0) = 0, \qquad A_1(0) \text{ arbitrary.} \tag{2.5.6}$$

The solution to Eqs. (2.5.1a) and (2.5.1b) that meets these boundary conditions is readily found to be

$$A_1(z) = A_1(0) \cosh \kappa z, \tag{2.5.7}$$

$$A_2(z) = i \left(\frac{n_1 \omega_2}{n_2 \omega_1} \right)^{1/2} \frac{A_3}{|A_3|} A_1^*(0) \sinh \kappa z. \tag{2.5.8}$$

The nature of this solution is shown in Fig. 2.5.2. Note that both the ω_1 and the ω_2 fields experience monotonic growth and that each grows asymptotically (i.e., for $\kappa z \gg 1$) as $e^{\kappa z}$. We see from the form of the solution that the ω_1 field retains its initial phase and is simply amplified by the interaction, while the generated wave at frequency ω_2 has a phase that depends both on that of the pump wave and on that of the ω_1 wave. This behavior of monotonic growth of both waves is qualitatively dissimilar from that of sum-frequency generation, where oscillatory behavior occurs.

The reason for the different behavior in this case can be understood intuitively in terms of the energy level diagram shown in Fig. 2.5.3. We can think of diagram (a) as showing how the presence of a field at frequency ω_1 stimulates the downward transition that leads to the generation of the ω_2 field. Likewise, diagram (b) shows that the ω_2 field stimulates the generation of the ω_1 field. Hence the generation of the ω_1 field reinforces the generation of the ω_2 field, and vice versa, leading to the exponential growth of each wave.

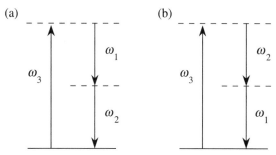

FIGURE 2.5.3   Difference-frequency generation.

Since the ω_1 field is amplified by the process of difference-frequency generation, which is a parametric process, this process is also known as parametric
amplification. In this language, one says that the signal wave (the ω_1 wave) is
amplified by the nonlinear mixing process, and an idler wave (at $\omega_2 = \omega_3 - \omega_1$)
is generated by the process. If mirrors that are highly reflecting at frequencies
ω_1 and/or ω_2 are placed on either side of the nonlinear medium to form an
optical resonator, oscillation can occur as a consequence of the gain of the
parametric amplification process. Such a device is known as a parametric
oscillator.

The first cw optical parametric oscillator was built by Giordmaine and Miller
in 1965. The theory of parametric amplification and parametric oscillators has
been reviewed by Byer and Herbst (1977).

The solution to the coupled-amplitude equations (2.5.1) for the general case
of arbitrary $\Delta k \neq 0$ makes a good exercise for the reader (see Problem 4 at the
end of this chapter). The solution for the case of arbitrary boundary conditions
(i.e., both $A_1(0)$ and $A_2(0)$ specified) is given by

$$A_1(z) = \left[A_1(0)\left(\cosh gz - \frac{i\Delta k}{2g}\sinh gz \right) + \frac{\kappa_1}{g} A_2^*(0)\sinh gz \right] e^{i\Delta kz/2},$$

(2.5.9a)

$$A_2(z) = \left[A_2(0)\left(\cosh gz - \frac{i\Delta k}{2g}\sinh gz \right) + \frac{\kappa_2}{g} A_1^*(0)\sinh gz \right] e^{i\Delta kz/2},$$

(2.5.9b)

where the coefficient g (which is not the same as that of Eq. (2.4.19)) is given
by

$$g = \left[\kappa_1\kappa_2^* - \left(\frac{\Delta k}{2} \right)^2 \right]^{1/2}$$

(2.5.10a)

with

$$\kappa_j = \frac{8\pi i \omega_j^2 d_{\text{eff}} A_3}{k_j c^2}. \tag{2.5.10b}$$

2.6. Second-Harmonic Generation

In this section we present a mathematical description of the process of second-harmonic generation, shown symbolically in Fig. 2.6.1. We assume that the medium is lossless both at the fundamental frequency ω_1 and at the second-harmonic frequency $\omega_2 = 2\omega_1$, so that the nonlinear susceptibility obeys the condition of full permutation symmetry. Our discussion closely follows that of one of the first theoretical treatments of second-harmonic generation (Armstrong *et al.*, 1962).

We take the total electric field within the nonlinear medium to be given by

$$\tilde{E}(z, t) = \tilde{E}_1(z, t) + \tilde{E}_2(z, t), \tag{2.6.1}$$

where each component is expressed in terms of a complex amplitude $E_j(z)$ and slowly varying amplitude $A_j(z)$ according to

$$\tilde{E}_j(z, t) = E_j(z)e^{-i\omega_j t} + \text{c.c.}, \tag{2.6.2}$$

where

$$E_j(z) = A_j(z)e^{ik_j z}, \tag{2.6.3}$$

and where the wave number and refractive index are given by

$$k_j = n_j \omega_j/c, \qquad n_j = \left[\epsilon^{(1)}(\omega_j)\right]^{1/2}. \tag{2.6.4}$$

We assume that each frequency component of the electric field obeys the driven wave equation (see also Eq. (2.1.20))

$$\frac{\partial^2 \tilde{E}_j}{\partial z^2} - \frac{\epsilon^{(1)}(\omega_j)}{c^2}\frac{\partial^2 \tilde{E}_j}{\partial t^2} = \frac{4\pi}{c^2}\frac{\partial^2}{\partial t^2}\tilde{P}_j. \tag{2.6.5}$$

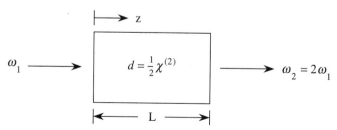

FIGURE 2.6.1 Second-harmonic generation.

The nonlinear polarization is represented as

$$\tilde{P}^{\mathrm{NL}}(z,t) = \tilde{P}_1(z,t) + \tilde{P}_2(z,t) \tag{2.6.6}$$

with

$$\tilde{P}_j(z,t) = P_j(z)e^{-i\omega_j t} + \text{c.c.}, \quad j = 1, 2. \tag{2.6.7}$$

The expressions for P_j are given according to Eqs. (1.5.28) and (1.5.29) by

$$P_1(z) = 4d_{\mathrm{eff}} E_2 E_1^* = 4d_{\mathrm{eff}} A_2 A_1^* e^{i(k_2 - k_1)z} \tag{2.6.8}$$

and

$$P_2(z) = 2d_{\mathrm{eff}} E_1^2 = 2d_{\mathrm{eff}} A_1^2 e^{2ik_1 z}. \tag{2.6.9}$$

Note that the degeneracy factor appearing in these two expressions are different. We obtain coupled-amplitude equations for the two frequency components by methods analogous to those used in Section 2.2 in deriving the coupled-amplitude equations for sum-frequency generation. We find that

$$\frac{dA_1}{dz} = \frac{8\pi i \omega_1^2 d_{\mathrm{eff}}}{k_1 c^2} A_2 A_1^* e^{-i\Delta k z} \tag{2.6.10}$$

and

$$\frac{dA_2}{dz} = \frac{4\pi i \omega_2^2 d_{\mathrm{eff}}}{k_2 c^2} A_1^2 e^{i\Delta k z}, \tag{2.6.11}$$

where

$$\Delta k = 2k_1 - k_2. \tag{2.6.12}$$

In the undepleted-pump approximation (i.e., A_1 constant), Eq. (2.6.11) can be integrated immediately to obtain an expression for the spatial dependence of the second-harmonic field amplitude. More generally, the pair of coupled equations must be solved simultaneously. To do so, it is convenient to work with the modulus and phase of each of the field amplitudes rather than with the complex quantities themselves. It is also convenient to express these amplitudes in dimensionless form. We thus write the complex, slowly varying field amplitudes as

$$A_1 = \left(\frac{2\pi I}{n_1 c}\right)^{1/2} u_1 e^{i\phi_1}, \tag{2.6.13}$$

$$A_2 = \left(\frac{2\pi I}{n_2 c}\right)^{1/2} u_2 e^{i\phi_2}. \tag{2.6.14}$$

Here we have introduced the total intensity of the two waves,

$$I = I_1 + I_2, \tag{2.6.15}$$

where the intensity of each wave is given by

$$I_j = \frac{n_j c}{2\pi} |A_j|^2. \tag{2.6.16}$$

As a consequence of the Manley–Rowe relations, the total intensity I is a constant. The new field amplitudes u_1 and u_2 are defined in such a manner that $u_1^2 + u_2^2$ is also a conserved (i.e., spatially invariant) quantity normalized such that

$$u_1(z)^2 + u_2(z)^2 = 1. \tag{2.6.17}$$

We next introduce a normalized distance parameter

$$\zeta = z/l, \tag{2.6.18}$$

where

$$l = \left(\frac{n_1^2 n_2 c^3}{2\pi I} \right)^{1/2} \frac{1}{8\pi \omega_1 d_{\text{eff}}} \tag{2.6.19}$$

is the characteristic distance over which the fields exchange energy. We also introduce the relative phase of the interacting fields,

$$\theta = 2\phi_1 - \phi_2 + \Delta k z, \tag{2.6.20}$$

and a normalized phase mismatch parameter

$$\Delta s = \Delta k l. \tag{2.6.21}$$

The quantities u_j, ϕ_j, ζ, and Δs defined in Eqs. (2.6.13) through (2.6.21) are now introduced into the coupled-amplitude equations (2.6.10) and (2.6.11), which reduce after straightforward (but lengthy) algebra to the set of coupled equations for the three real quantities u_1, u_2, and θ:

$$\frac{du_1}{d\zeta} = u_1 u_2 \sin \theta, \tag{2.6.22}$$

$$\frac{du_2}{d\zeta} = -u_1^2 \sin \theta, \tag{2.6.23}$$

$$\frac{d\theta}{d\zeta} = \Delta s + \frac{\cos \theta}{\sin \theta} \frac{d}{d\zeta} \left(\ln u_1^2 u_2 \right). \tag{2.6.24}$$

This set of equations has been solved under general conditions by Armstrong *et al.* We shall return later to a discussion of the general solution, but for now we assume the case of perfect phase matching so that Δk and hence Δs vanish.

It is easy to verify by direct differentiation that, for $\Delta s = 0$, Eq. (2.6.24) can be rewritten as

$$\frac{d}{d\zeta} \ln \left(\cos \theta u_1^2 u_2 \right) = 0. \tag{2.6.25}$$

Hence the quantity $\ln(\cos \theta u_1^2 u_2)$ is a constant, which we call $\ln \Gamma$, so that the solution to Eq. (2.6.25) can be expressed as

$$u_1^2 u_2 \cos \theta = \Gamma, \tag{2.6.26}$$

where the constant Γ is independent of the normalized propagation distance ζ. The value of Γ can be determined from the known values of u_1, u_2, and θ at the entrance face to the nonlinear medium $\zeta = 0$.

We have thus found two conserved quantities: $u_1^2 + u_2^2$ (according to Eq. (2.6.17)) and $u_1^2 u_2 \cos \theta$ (according to Eq. (2.6.26)). These conserved quantities can be used to decouple the set of equations (2.6.22)–(2.6.24). Equation (2.6.23), for instance, can be written using Eq. (2.6.17) and the identity $\sin^2 \theta + \cos^2 \theta = 1$ as

$$\frac{du_2}{d\zeta} = \pm (1 - u_2^2)(1 - \cos^2 \theta)^{1/2}. \tag{2.6.27}$$

Equations (2.6.26) and (2.6.17) are next used to express $\cos^2 \theta$ in terms of the conserved quantity Γ and the unknown function u_2; the resulting expression is substituted into Eq. (2.6.27), which becomes

$$\frac{du_2}{d\zeta} = \pm (1 - u_2^2)\left(1 - \frac{\Gamma^2}{u_1^4 u_2^2}\right)^{1/2} = \pm (1 - u_2^2)\left(1 - \frac{\Gamma^2}{(1 - u_2^2)^2 u_2^2}\right)^{1/2}. \tag{2.6.28}$$

This result is simplified algebraically to give

$$u_2 \frac{du_2}{d\zeta} = \pm \left[(1 - u_2^2)^2 u_2^2 - \Gamma^2 \right]^{1/2},$$

or

$$\frac{du_2^2}{d\zeta} = \pm 2 \left[(1 - u_2^2)^2 u_2^2 - \Gamma^2 \right]^{1/2}. \tag{2.6.29}$$

This equation is of a standard form, whose solution can be expressed in terms of the Jacobi elliptic functions. An example of the solution for one particular choice of initial conditions is illustrated in Fig. 2.6.2. Note that, in general, the fundamental and second-harmonic fields interchange energy periodically.

The solution of Eq. (2.6.29) becomes particularly simple for the special case in which the constant Γ is equal to zero. The condition $\Gamma = 0$ occurs whenever the amplitude of either of the two input fields is equal to zero or whenever the

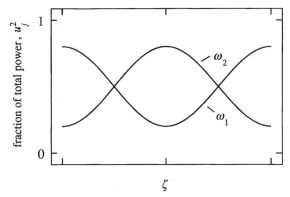

FIGURE 2.6.2   Typical solution to Eq. (2.6.29), after Armstrong *et al.* (1962).

fields are initially phased so that $\cos\theta = 0$. We note that since Γ is a conserved quantity, it is then equal to zero for all values of ζ, which in general requires (see Eq. (2.6.26)) that

$$\cos\theta = 0. \tag{2.6.30a}$$

For definiteness, we assume that

$$\sin\theta = -1 \tag{2.6.30b}$$

(rather than $+1$). We hence see that the relative phase of the interacting fields is spatially invariant for the case of $\Gamma = 0$. In addition, when $\Gamma = 0$ the coupled-amplitude equations take on the relatively simple forms

$$\frac{du_1}{d\zeta} = -u_1 u_2 \tag{2.6.31}$$

$$\frac{du_2}{d\zeta} = u_1^2. \tag{2.6.32}$$

This second equation can be transformed through use of Eq. (2.6.17) to obtain

$$\frac{du_2}{d\zeta} = 1 - u_2^2, \tag{2.6.33}$$

whose solution is

$$u_2 = \tanh(\zeta + \zeta_0), \tag{2.6.34}$$

where ζ_0 is a constant of integration.

We now assume that the initial conditions are

$$u_1(0) = 1, \qquad u_2(0) = 0. \tag{2.6.35}$$

These conditions imply that no second-harmonic light is incident on the nonlinear crystal, as is the case in most experiments. Then, since $\tanh 0 = 0$, we
see that the integration constant ζ_0 is equal to 0 and hence that

$$u_2(\zeta) = \tanh \zeta. \tag{2.6.36}$$

The amplitude u_1 of the fundamental wave is found through use of Eq. (2.6.32)
(or through use of Eq. (2.6.17)) to be given by

$$u_1(\zeta) = \mathrm{sech}\ \zeta. \tag{2.6.37}$$

Recall that $\zeta = z/l$. For the case in which only the fundamental field is present
at $z = 0$, the length parameter of Eq. (2.6.19) is given by

$$l = \frac{(n_1 n_2)^{1/2} c}{8\pi \omega_1 d_{\mathrm{eff}} |A_1(0)|}. \tag{2.6.38}$$

The solution given by Eqs. (2.6.36) and (2.6.37) is shown graphically in Fig.
2.6.3. We see that all of the incident radiation is converted into the second
harmonic in the limit $\zeta \to \infty$. In addition, we note that $\tanh(\zeta + \zeta_0)$ has the
same asymptotic behavior for any finite value of ζ_0. Thus, whenever Γ is equal
to zero, all of the radiation at the fundamental frequency will eventually be
converted to the second harmonic, for any initial ratio of u_1 to u_2.

As mentioned above, Armstrong *et al.* have also solved the coupled-
amplitude equations describing second-harmonic generation for arbitrary Δk.
They find that the solution can be expressed in terms of elliptic integrals. We
shall not reproduce their derivation here; instead we summarize their results

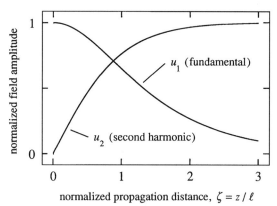

FIGURE 2.6.3   Spatial variations of the fundamental and second-harmonic field
amplitudes for the case of perfect phase matching and the boundary condition $u_2(0) = 0$.

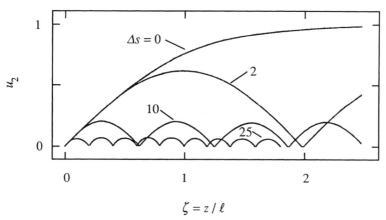

$$\zeta = z / \ell$$

FIGURE 2.6.4   Effect of wavevector mismatch on the efficiency of second-harmonic generation.

graphically in Fig. 2.6.4 for the case in which no radiation is incident at the second-harmonic frequency. We see from the figure that the effect of a nonzero propagation-vector mismatch is to lower the conversion efficiency.

As an illustration of how to apply the formulas derived in this section, we estimate the conversion efficiency for second-harmonic generation attainable using typical cw lasers. We first estimate the numerical value of the parameter ζ given by Eqs. (2.6.18) and (2.6.38) at the plane $z = L$, where L is the length of the nonlinear crystal. We assume that the incident laser beam carries power P and is focused to a spot size w_0 at the center of the crystal. The field strength A_1 can then be estimated by the expression

$$I_1 = \frac{P}{\pi w_0^2} = \frac{n_1 c}{2\pi} A_1^2. \qquad (2.6.39)$$

We assume that the beam is optimally focused in the sense that the focal spot size w_0 is chosen so that the depth b of the focal region is equal to the length L of the crystal, that is,*

$$b \equiv \frac{2\pi w_0^2}{\lambda_1 / n_1} = L, \qquad (2.6.40)$$

where λ_1 denotes the wavelength of the incident wave in vacuum. From Eqs. (2.6.39) and (2.6.40), the laser field amplitude under conditions of optimum

---

* See also the discussion of nonlinear interactions involving focused gaussian beams presented in Section 2.10.

focusing is seen to be given by

$$A_1 = \left(\frac{4\pi P}{c\lambda_1 L} \right)^{1/2},$$ (2.6.41)

and hence the parameter $\zeta = L/l$ is given through use of Eq. (2.6.38) by

$$\zeta = \left(\frac{1024\pi^5 d_{\mathrm{eff}}^2 L P}{n_1 n_2 c \lambda_1^3} \right)^{1/2}.$$ (2.6.42)

Typical values of the parameters appearing in this equation are $d_{\mathrm{eff}} = 1 \times 10^{-8}$ esu, $L = 1$ cm, $P = 1$ W $= 1 \times 10^7$ erg/sec, $\lambda = 0.5 \times 10^{-4}$ cm, and $n = 2$, which lead to the value $\zeta = 0.14$. The efficiency η for conversion of power from the ω_1 wave to the ω_2 wave can be defined by

$$\eta = \frac{u_2^2(L)}{u_1^2(0)},$$ (2.6.43)

and from Eq. (2.6.36), we see that for the values given above, η is of the order of 2%.

2.7. Phase-Matching Considerations

We saw in Section 2.2 that for sum-frequency generation involving undepleted input beams, the intensity of the generated field at frequency $\omega_3 = \omega_1 + \omega_2$ varies with the wavevector mismatch

$$\Delta k = k_1 + k_2 - k_3$$ (2.7.1)

according to

$$I_3 = I_3(\mathrm{max}) \frac{\sin^2(\Delta k L/2)}{(\Delta k L/2)^2}.$$ (2.7.2)

This expression predicts a dramatic decrease in the efficiency of the sum-frequency generation when the condition of perfect phase matching, $\Delta k = 0$, is not satisfied.

Behavior of the sort predicted by Eq. (2.7.2) was first observed experimentally by Maker et al. (1962) and is illustrated in Fig. 2.7.1. Their experiment involved focusing the output of a pulsed ruby laser into a single crystal of quartz and measuring how the intensity of the second-harmonic signal varied as the crystal was rotated, thus varying the effective path length L through the crystal. The wavevector mismatch Δk was nonzero and approximately the same for all orientations used in their experiment.

(a)

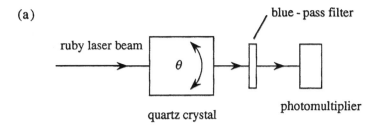

quartz crystal

photomultiplier

(b)

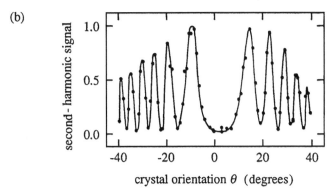

crystal orientation θ (degrees)

FIGURE 2.7.1   (a) Experimental setup of Maker *et al.* (b) Their experimental results.

For nonlinear mixing processes that are sufficiently efficient to lead to depletion of the input beams, the functional dependence of the efficiency of the process on the phase mismatch is no longer given by Eq. (2.7.2). However, even in this case the efficient generation of the output field requires that the condition $\Delta k = 0$ be maintained.

The phase-matching condition $\Delta k = 0$ is often difficult to achieve because the refractive index of materials that are lossless in the range ω_1 to ω_3 (we assume that $\omega_1 \leq \omega_2 \leq \omega_3$) shows an effect known as normal dispersion: the refractive index is an increasing function of frequency. As a result, the condition for perfect phase matching with collinear beams,

$$n_1\omega_1 + n_2\omega_2 = n_3\omega_3, \tag{2.7.3}$$

where

$$\omega_1 + \omega_2 = \omega_3, \tag{2.7.4}$$

cannot be achieved. For the case of second-harmonic generation, with $\omega_1 = \omega_2$, $\omega_3 = 2\omega_1$, these conditions require that

$$n(\omega_1) = n(2\omega_1), \tag{2.7.5}$$

which is clearly not possible when $n(\omega)$ increases monotonically with ω. For the case of sum-frequency generation, the argument is slightly more complicated, but the conclusion is the same. To show that phase matching is not possible in this case, we first rewrite Eq. (2.7.3) as

$$n_3 = \frac{n_1\omega_1 + n_2\omega_2}{\omega_3}. \tag{2.7.6}$$

This result is now used to express the refractive index difference $n_3 - n_2$ as

$$n_3 - n_2 = \frac{n_1\omega_1 + n_2\omega_2 - n_2\omega_3}{\omega_3} = \frac{n_1\omega_1 - n_2(\omega_3 - \omega_2)}{\omega_3} = \frac{n_1\omega_1 - n_2\omega_1}{\omega_3},$$

or finally as

$$n_3 - n_2 = (n_1 - n_2)\frac{\omega_1}{\omega_3}. \tag{2.7.7}$$

For normal dispersion, n_3 must be greater than n_2, and hence the left-hand side of this equation must be positive. However, n_2 must also be greater than n_1, showing that the right-hand side must be negative, which demonstrates that Eq. (2.7.7) cannot possess a solution.

In principle, it is possible to achieve the phase-matching condition by making use of anomalous dispersion, that is, the decrease in refractive index with increasing frequency that occurs near an absorption feature. However, the most common procedure for achieving phase matching is to make use of the birefringence displayed by many crystals. Birefringence is the dependence of the refractive index on the direction of polarization of the optical radiation. Not all crystals display birefringence; in particular, crystals belonging to the cubic crystal system are optically isotropic (i.e., show no birefringence) and thus are not phase-matchable.

The linear optical properties of the various crystal systems are summarized in Table 2.7.1.

TABLE 2.7.1   Linear optical classification of the various crystal systems

System	Linear optical classification
Triclinic, monoclinic, orthorhombic	Biaxial
Trigonal, tetragonal, hexagonal	Uniaxial
Cubic	Isotropic

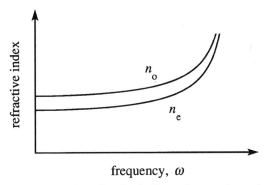

frequency, ω

FIGURE 2.7.2   Dispersion of the refractive indices of a negative uniaxial crystal. For the opposite case of a positive uniaxial crystal, the extraordinary index n_e is greater than the ordinary index n_o.

In order to achieve phase matching through the use of birefringent crystals, the highest-frequency wave $\omega_3 = \omega_1 + \omega_2$ is polarized in the direction that gives it the lower of the two possible refractive indices. For the case of a negative uniaxial crystal, as in the example shown in Fig. 2.7.2, this choice corresponds to the extraordinary polarization. There are two choices for the polarizations of the lower-frequency waves. Midwinter and Warner (1965) define type I phase matching to be the case in which the two lower-frequency waves have the same polarization, and type II to be the case where the polarizations are orthogonal. The possibilities are summarized in Table 2.7.2. No assumptions regarding the relative magnitudes of ω_1 and ω_2 are implied by the classification scheme. However, for type II phase matching it is easier to achieve the phase-matching condition (i.e., less birefringence is required) if $\omega_2 > \omega_1$ for the choice of ω_1 and ω_2 used in writing the table. Also, independent of the relative magnitudes of ω_1 and ω_2, type I phase matching is easier to achieve than type II.

Careful control of the refractive indices at each of the three optical frequencies is required in order to establish the phase-matching condition ($\Delta k = 0$). Typically phase matching is achieved by one of two methods: angle tuning and temperature tuning.

TABLE 2.7.2   Phase-matching methods for uniaxial crystals

	Positive uniaxial $(n_e > n_o)$	Negative uniaxial $(n_e < n_o)$
Type I	$n_3^o \omega_3 = n_1^e \omega_1 + n_2^e \omega_2$	$n_3^e \omega_3 = n_1^o \omega_1 + n_2^o \omega_2$
Type II	$n_3^o \omega_3 = n_1^o \omega_1 + n_2^e \omega_2$	$n_3^e \omega_3 = n_1^e \omega_1 + n_2^o \omega_2$

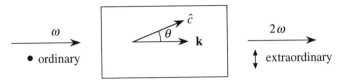

FIGURE 2.7.3   Geometry of angle-tuned phase matching of second-harmonic generation.

Angle Tuning

This method involves precise angular orientation of the crystal with respect to the propagation direction of the incident light. It is most simply described for the case of a uniaxial crystal, and the following discussion will be restricted to this case. Uniaxial crystals are characterized by a particular direction known as the optic axis (or c axis or z axis). Light polarized perpendicular to the plane containing the propagation vector $\mathbf{k}$ and the optic axis is the ordinary polarization. Such light experiences a refractive index n_o called the ordinary refractive index. Light polarized in the plane containing $\mathbf{k}$ and the optic axis is the extraordinary polarization and experiences a refractive index $n_e(\theta)$ that depends on the angle θ between the optic axis and $\mathbf{k}$ according to the relation[*]

$$\frac{1}{n_e(\theta)^2} = \frac{\sin^2 \theta}{\bar{n}_e^2} + \frac{\cos^2 \theta}{n_o^2}. \tag{2.7.8}$$

Here $\bar{n}_e$ is the principal value of the extraordinary refractive index. Note that $n_e(\theta)$ is equal to the principal value $\bar{n}_e$ for $\theta = 90$ degrees and is equal to n_o for $\theta = 0$. Phase matching is achieved by adjusting the angle θ to achieve the value of $n_e(\theta)$ for which the condition $\Delta k = 0$ is satisfied.

As an illustration of angle phase matching, we consider the case of type I second-harmonic generation in a negative uniaxial crystal, as shown in Fig. 2.7.3. Since n_e is less than n_o for a negative uniaxial crystal, one chooses the fundamental to be an ordinary wave and the second harmonic to be an extraordinary wave, in order that the birefringence of the material can compensate for the dispersion. The phase-matching condition (2.7.5) then becomes

$$n_e(2\omega, \theta) = n_o(\omega), \tag{2.7.9}$$

or

$$\frac{\sin^2 \theta}{\bar{n}_e(2\omega)^2} + \frac{\cos^2 \theta}{n_o(2\omega)^2} = \frac{1}{n_o(\omega)^2}. \tag{2.7.10}$$

[*] For a derivation of this relation, see, for example, Born and Wolf (1975), Section 14.3; Klein (1970), Eq. (11.160a); or Zernike and Midwinter (1973), Eq. (1.26).

In order to simplify this equation, we replace $\cos^2 \theta$ by $1 - \sin^2 \theta$ and solve for $\sin^2 \theta$ to obtain

$$\sin^2 \theta = \frac{\dfrac{1}{n_o(\omega)^2} - \dfrac{1}{n_o(2\omega)^2}}{\dfrac{1}{\bar{n}_e(2\omega)^2} - \dfrac{1}{n_o(2\omega)^2}}. \tag{2.7.11}$$

This equation shows how the crystal should be oriented in order to achieve the phase-matching condition. Note that under arbitrary conditions this equation does not necessarily possess a solution for a physically meaningful orientation angle θ. For example, if for some material the dispersion in the linear refractive index is too large or the birefringence is too small, the right-hand side of this equation can have a magnitude larger than unity and consequently the equation will have no solution.

Temperature Tuning

There is one serious drawback to the use of angle tuning. Whenever the angle θ between the propagation direction and the optic axis has a value other than 0 or 90 degrees, the Poynting vector **S** and the propagation vector **k** are not parallel for extraordinary rays. As a result, ordinary and extraordinary rays with parallel propagation vectors quickly diverge from one another as they propagate through the crystal. This walkoff effect limits the spatial overlap of the two waves and decreases the efficiency of any nonlinear mixing process involving such waves.

For some crystals, notably lithium niobate, the amount of birefringence is strongly temperature-dependent. As a result, it is possible to phase-match the mixing process by holding θ fixed at 90 degrees and varying the temperature of the crystal. The temperature dependence of the refractive indices of lithium niobate has been given by Hobden and Warner (1966).

2.8. Optical Parametric Oscillators

We noted earlier in Section 2.5 that the process of difference-frequency generation necessarily leads to the amplification of the lower-frequency input field. This amplification process is known as optical parametric amplification, and the gain resulting from this process can be used to construct a device known as an optical parametric oscillator (OPO). Some of these issues are summarized in Fig. 2.8.1. Part (a) of the figure shows that in generating the difference frequency $\omega_i = \omega_p - \omega_s$, the lower-frequency input wave ω_s is amplified. Conventionally, ω_p is known as the pump frequency, ω_s the signal frequency,

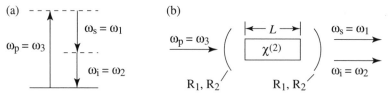

FIGURE 2.8.1   (a) Relationship between difference-frequency generation and optical parametric amplification. (b) The gain associated with the process of optical parametric amplification can be used to construct the device shown, which is known as an optical parametric oscillator.

and ω_i the idler frequency. The gain associated with the process of optical parametric amplification can in the presence of feedback produce oscillation, as shown in part (b) of the figure. If the end mirrors of this device are highly reflecting at both frequencies ω_s and ω_i, the device is known as a doubly resonant oscillator; if they are highly reflecting at ω_s or ω_i but not at both, the device is known as a singly resonant oscillator. Note that when an OPO is operated near the point of degeneracy ($\omega_s = \omega_i$) it tends to operate as a doubly resonant oscillator.* The optical parametric oscillator has proven to be a versatile source of frequency-tunable radiation throughout the infrared, visible, and ultraviolet spectral regions. It can produce either a continuous-wave output or pulses of nanosecond, picosecond, or femtosecond duration.

Let us recall how to calculate the gain of the process of optical parametric amplification. For convenience, we label the pump, signal, and idler frequencies as $\omega_p = \omega_3$, $\omega_s = \omega_1$, and $\omega_i = \omega_2$. We take the coupled amplitude equations to have the form (see also Eqs. (2.5.1))

$$\frac{dA_1}{dz} = \frac{8\pi i \omega_1^2 d}{k_1 c^2} A_3 A_2^* e^{i\Delta k z},$$
(2.8.1a)

$$\frac{dA_2}{dz} = \frac{8\pi i \omega_2^2 d}{k_2 c^2} A_3 A_1^* e^{i\Delta k z}$$
(2.8.1b)

where $\Delta k \equiv k_3 - k_1 - k_2$. These equations possess the solution

$$A_1(z) = \left[A_1(0) \left(\cosh gz - \frac{i\Delta k}{2g} \sinh gz \right) + \frac{\kappa_1}{g} A_2^*(0) \sinh gz \right] e^{i\Delta k z/2},$$
(2.8.2a)

$$A_2(z) = \left[A_2(0) \left(\cosh gz - \frac{i\Delta k}{2g} \sinh gz \right) + \frac{\kappa_2}{g} A_1^*(0) \sinh gz \right] e^{i\Delta k z/2}$$
(2.8.2b)

---

* In principle, polarization effects can be used to suppress cavity feedback for either the signal or idler wave for the case of type-II phase matching.

where we have introduced the quantities

$$g = [\kappa_1 \kappa_2^* - (\Delta k/2)^2]^{1/2} \quad \text{and} \quad \kappa_i = \frac{8\pi i \omega_1^2 d A_3}{k_j c^2}. \tag{2.8.3}$$

For the special case of perfect phase matching ($\Delta k = 0$) and under the assumption that the input amplitude of field A_2 vanishes ($A_2(0) = 0$), the solution reduces to

$$A_1(z) = A_1(0) \cosh gz \Rightarrow \tfrac{1}{2} A_1(0) \exp(gz) \tag{2.8.4a}$$

$$A_2(z) = i \left(\frac{n_1 \omega_2}{n_2 \omega_1}\right)^{1/2} \frac{A_3}{|A_3|} A_1^*(0) \sinh gz \Rightarrow O(1) A_1^*(0) \exp(gz). \tag{2.8.4b}$$

In each expression, the last form gives the asymptotic value for large z, and the symbol $O(1)$ means of the order of unity. One sees that asymptotically both waves experience exponential growth, with an amplitude gain coefficient of g.

Threshold for Parametric Oscillation

We next consider the threshold condition for the establishment of parametric oscillation. We treat the device shown in Fig. 2.8.1b, in which the two end mirrors are assumed to be indentical but are allowed to have different (intensity) reflectivities R_1 and R_2 at the signal and idler frequencies.

As a first approximation, we express the threshold condition as a statement that the fractional energy gain per pass must equal the fractional energy loss per pass. Under the assumptions of exact cavity resonance, of perfect phase matching ($\Delta k = 0$), and that the cavity is doubly resonant with the same reflectivity at the signal and idler frequencies (that is, $R_1 = R_2 \equiv R$, $(1 - R) \ll 1$), this condition can be expressed as

$$(e^{2gL} - 1) = 2(1 - R). \tag{2.8.5}$$

Under the realistic condition that the single-pass exponential gain $2gL$ is not large compared to unity, this condition becomes

$$gL = 1 - R. \tag{2.8.6}$$

This is the threshold condition first formulated by Giordmaine and Miller (1965).

The threshold condition for optical parametric oscillation can be formulated more generally as a statement that the fields within the resonator must replicate themselves each round trip. For arbitrary end-mirror reflectivities at the signal and idler frequencies, this condition can be expressed, again assuming perfect phase matching, as

$$A_1(0) = \left[A_1(0) \cosh gL + \frac{\kappa_1}{g} A_2^*(0) \sinh gL \right](1 - l_1), \quad (2.8.7a)$$

$$A_2^*(0) = \left[A_2^*(0) \cosh gL + \frac{\kappa_2^*}{g} A_1(0) \sinh gL \right](1 - l_2), \quad (2.8.7b)$$

where $l_i = 1 - R_i e^{-\alpha_i L}$ is the fractional amplitude loss per pass, α_i being the absorption coefficient of the crystal at frequency ω_i. By requiring that Eqs. (2.8.7) be satisfied simultaneously, we find the threshold condition to be

$$\cosh gL = 1 + \frac{l_1 l_2}{2 - l_1 - l_2}. \quad (2.8.8)$$

The threshold conditions for both doubly resonant oscillators and singly resonant oscillators are contained in this result. The doubly resonant oscillator is described by taking the limit of low loss for both the signal and idler waves ($l_1, l_2 \ll 1$). In this limit, $\cosh gL$ can be approximated by $1 + \frac{1}{2} g^2 L^2$, leading to the conclusion that the threshold condition for a doubly resonant oscillator is

$$g^2 L^2 = l_1 l_2, \quad (2.8.9)$$

in consistency with Eq. (2.8.6).

The threshold condition for a singly resonant oscillator can be obtained by assuming that there is no feedback for the idler frequency, that is, that $l_2 = 1$. If we assume low loss for the signal frequency (that is, $l_1 \ll 1$), the threshold condition becomes

$$g^2 L^2 = 2 l_1. \quad (2.8.10)$$

Note that the threshold value of gL for a singly resonant oscillator is larger than that of the doubly resonant oscillator by a factor of $(2/l_2)^{1/2}$. Despite this fact, it is usually desirable to configure optical parametric oscillators to be singly resonant because of their increased stability, for reasons that are explained below.

For simplicity, the treatment of this subsection has assumed the case of perfect phase matching. It is easy to show that the threshold condition for the case $\Delta k \neq 0$ can be obtained by replacing g^2 by $g^2 \operatorname{sinc}^2 \Delta k L/2$ in Eqs. (2.8.9) and (2.8.10).

Wavelength Tuning of an OPO

The condition of energy conservation $\omega_s + \omega_i = \omega_p$ allows any frequency ω_s smaller than ω_p to be generated by an optical parametric oscillator. The output frequency ω_s can be controlled through the phase-matching condition

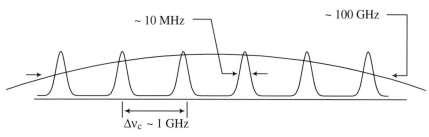

FIGURE 2.8.2   Schematic representation of the gain spectrum (the broad curve) and cavity mode structure of an OPO. Note that typically many cavity modes lie beneath the gain profile of the OPO.

$\Delta k = 0$, which invariably can be satisfied for at most one pair of frequencies ω_s and ω_i. The output frequency bandwidth can often be narrowed by placing wavelength-selective items (such as etalons) inside the OPO cavity.

The principles of phase matching were described earlier in Section 2.7. Recall that phase matching can be achieved either by varying the orientation of the nonlinear crystal (angle phase matching) or by varying the temperature of the crystal.

Influence of Cavity Mode Structure on OPO Tuning

Let us next take a more detailed look at the tuning characteristics of an OPO. We shall see that both the tuning and stability characteristics of an OPO are very different for the singly resonant and doubly resonant cases.

Note first that under typical conditions the cavity mode spacing and cavity resonance width tend to be much smaller than the width of the gain curve of the optical parametric amplification process. This circumstance is illustrated in Fig. 2.8.2.* Let us next consider which of these cavity modes will actually undergo oscillation.

For the case of a singly resonant oscillator (displayed in part a of Fig. 2.8.3), the situation is relatively simple. Oscillation occurs on the cavity mode closest to the peak of the gain curve. Note also that (barring mechanical instabilities, etc.) oscillation will occur on only one cavity mode. The reason for this

---

* This example assumes that the cavity length L_c is 15 cm so that the cavity mode spacing $\Delta v_c = c/2L_c$ is 1 GHz, that the cavity finesse $\mathscr{F}$ is 100 so that the linewidth associated with each mode is 1 GHz/$\mathscr{F}$ = 10 MHz, and that the width of the gain curve is 100 GHz. This gain linewidth is estimated by assuming that $\Delta k L$ (which is zero at the center of the gain line and where L is the crystal length) drops to the value π at the edge of the gain line. If we then assume that Δk changes with signal frequency because of material dispersion, and that dn/dv is of the order of 10^{-15} sec, we obtain 100 GHz as the gain bandwidth.

(a)

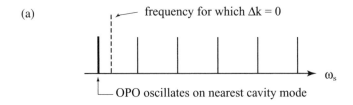

(b)

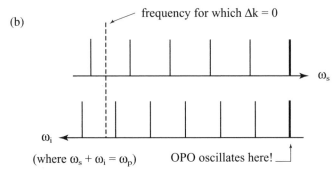

(where $\omega_s + \omega_i = \omega_p$)   OPO oscillates here! ⌐

FIGURE 2.8.3   (a) Symbolic representation of the mode structure of a singly resonant OPO. (b) Symbolic representation of the mode structure of a doubly resonant OPO. The signal-frequency and idler-frequency axes increase in opposite directions, such that at each horizontal point $\omega_s + \omega_i$ has the fixed value ω_p. Thus any point on the axis represents a point where the energy conservation relation $\omega_s + \omega_i = \omega_p$ is satisfied, although only at points where signal and idler modes occur at the same horizontal point is the double-resonance condition satisfied.

behavior is that once oscillation commences on the cavity mode closest to the peak of the gain curve, the pump power is depleted, thus lowering the gain to the value of the loss for this mode. By assumption, the gain will be smaller at the frequencies of the other cavity modes, and thus these modes will be below threshold for oscillation. This behavior is very much analogous to that of a homogeneously broadened laser, which tends to oscillate on a single-cavity mode.

Consider now the different situation of a doubly resonant oscillator (Fig. 2.8.3b). For a doubly resonant oscillator, oscillation is very much favored under conditions such that a signal and its corresponding idler mode can simultaneously support oscillation. Note from the figure that neither of these modes is necessarily the mode closest to the peak of the gain curve (which occurs at $\Delta k = 0$). As a consequence doubly resonant oscillators tend not to tune smoothly. Moreover, such devices tend not to run stably, because, for example, small fluctuations in the pump frequency or the cavity length L can lead to disproportionately large variations in the signal frequency.

The argument just presented, based on the structure of Fig. 2.8.3b, presupposes that the cavity modes are not equally spaced. In fact, it is easy to show that the cavity mode spacing for a cavity of length L_c filled with a dispersive medium is given by

$$\Delta \nu_c = \frac{1}{n^{(g)}} \frac{c}{2L_c} \quad \text{where} \quad n^{(g)} = n + \nu \frac{dn}{d\nu} \quad (2.8.11)$$

(see problems 7 and 8 at the end of this chapter), which clearly is not a constant as a function of frequency. Here $n^{(g)}$ is known as the group index.

Let us next examine more quantitatively the effect described pictorially in Fig. 2.8.3b. We first estimate the characteristic frequency separation $\delta\omega$ between the peak of the gain curve and the frequency of actual oscillation. To do so, it is convenient to introduce the quantity

$$\Delta\omega \equiv \omega_p - \omega_s^{(m)} - \omega_i^{(m)} \quad (2.8.12)$$

where $\omega_s^{(m)}$ is one of the signal cavity mode frequencies and similarly for $\omega_i^{(m)}$. Clearly, oscillation can occur only for a pair of modes such that $\Delta\omega \approx 0$ (or more precisely where $\Delta\omega \lesssim \delta\omega_c$ where $\delta\omega_c$ is the spectral width of the cavity resonance). Note next that in jumping by one cavity mode for both ω_s and ω_i, the quantity $\Delta\omega$ will change by the amount

$$\delta(\Delta\omega) = 2\pi \left(\frac{c}{2n_s^{(g)} L_c} - \frac{c}{2n_i^{(g)} L_c} \right) = \frac{\pi c}{L_c} \left(\frac{n_i^{(g)} - n_s^{(g)}}{n_s^{(g)} n_i^{(g)}} \right). \quad (2.8.13)$$

We next estimate the value of the frequncy separation $\delta\omega$ by noting that it corresponds to a change in $\Delta\omega$ from its value near the point $\Delta k = 0$ to its value (≈ 0) at the oscillation point. Unless the length of the OPO cavity is actively controlled, the value of $\Delta\omega$ near $\Delta k = 0$ can be as large as one-half of a typical mode spacing or

$$\Delta\omega_0 \simeq \frac{1}{2} \left(\frac{2\pi c}{2n^{(g)} L_c} \right) = \frac{\pi c}{2n^{(g)} L_c}, \quad (2.8.14)$$

where $n^{(g)}$ is some typical value of the group index. The number of modes between the peak of the gain curve and the actual operating point under this situation is thus of the order of

$$N = \frac{\Delta\omega_0}{\delta(\Delta\omega)} = \frac{n^{(g)}}{2\left(n_s^{(g)} - n_i^{(g)}\right)} \quad (2.8.15)$$

and the characteristic frequency separation $\delta\omega$ is thus given by

$$\delta\omega = \Delta\omega_c N \simeq \frac{2\pi c}{2n^{(g)} L_c} N = \frac{\pi c}{2L_c} N \frac{1}{\left(n_s^{(g)} - n_i^{(g)}\right)}. \quad (2.8.16)$$

Note that this shift can be very large for $n_s^{(g)} \approx n_i^{(g)}$.

The model just presented can be used to estimate an important quantity, the operational linewidth $\delta\omega^{(\mathrm{OPO})}$ of the oscillator. We noted above that in principle an OPO should oscillate on a single cavity mode. However, because of unavoidable technical noise, an OPO might be expected to oscillate (simultaneously or sequentially) on many different cavity modes. The technical noise might be in the form of mechanical vibrations of the OPO cavity, leading to a jitter of amount $\delta\omega_c$ in the resonance frequency of each cavity mode. Alternatively, the technical noise might be in the form of the spectral breadth $\delta\omega_p$ of the pump radiation. Whichever effect is larger might be expected to dominate, and thus that the effective value of the technical noise is given by $\delta\omega_{\mathrm{eff}} = \max(\delta\omega_c, \delta\omega_p)$. Analogously to Eq. (2.8.15), one then expects that the number of modes that undergo oscillation is given by

$$N_{\mathrm{OPO}} = \frac{\delta\omega_{\mathrm{eff}}}{\delta(\Delta\omega)} = \frac{\max(\delta\omega_p, \delta\omega_c)}{\delta(\Delta\omega)}. \qquad (2.8.17)$$

Consequently, the the OPO linewidth is expected to be

$$\delta\omega^{(\mathrm{OPO})} = N_{\mathrm{OPO}}\Delta\omega_c = \frac{n_g}{n_g^{(s)} - n_g^{(i)}} \max(\delta\omega_p, \delta\omega_c). \qquad (2.8.18)$$

Note that (unless extreme care, such as active stabilization, is employed) the linewidth of an OPO tends to be much greater than that of the pump field or that of the bare OPO cavity.

Equation (2.8.18) has important implications in the design of OPOs. Note that this expression formally diverges at the point of degeneracy for a type-I (but not a type-II) OPO. The narrower linewidth of a type-II OPO than for a type-I OPO constructed of the same material has been observed in practice by Bosenberg and Tang (1990).

We conclude this section with a brief historical summary of progress in the development of OPOs. The first operating OPO was demonstrated by Giordmaine and Miller (1965); it utilized the nonlinear optical response of lithium niobate and worked in the pulsed regime. Continuous-wave operation of an OPO was demonstrated by Smith et al. (1968) and utilized a $Ba_2NaNb_5O_{15}$ nonlinear crystal. Interest in the development of OPOs was renewed in the 1980s as a consequence of the availability of new nonlinear materials such as β-BaB_2O_4 (beta-barium borate or BBO), LiB_3O_5 (lithium borate or LBO), and $KTiOPO_4$ (KTP), which possessed high nonlinearity, high resistance to laser damage, and large birefringence. These materials led to the rapid development of new OPO capabilities, such as continuous tunability from 0.42 to 2.3 μm in a BBO OPO with conversion efficiencies as large as 32% (Bosenberg et al., 1989), and OPOs that can produce tunable femtosecond pulses in KTP (Edelstein *et al.*, 1989). The use of quasi-phase-matching in periodically poled lithium niobate has also been utilized to produce novel OPOs.

2.9.   Quasi-Phase-Matching

The previous section describes techniques that utilize the birefringence of an optical material to achieve the phase-matching condition of nonlinear optics. This condition must be maintained for the efficient generation of new frequency components in any nonlinear optical interaction. However, there are circumstances under which these techniques are not suitable. For instance, a particular material may possess no birefringence (an example is gallium arsenide) or may possess insufficient birefringence to compensate for the dispersion of the linear refractive indices over the wavelength range of interest. The problem of insufficient birefringence becomes increasingly acute at shorter wavelengths, because (as illustrated very schematically in Fig. 2.7.2) the refractive index of a given material tends to increase rapidly with frequency at high frequencies, whereas the birefringence (that is, the difference between the ordinary and extraordinary refractive indices) tends to be more nearly constant. Another circumstance under which birefringence phase matching cannot be used is when a particular application require the use of the d_{33} nonlinear coefficient, which tends to be much larger than the off-diagonal coefficients. However, the d_{33} nonlinear coefficient can be accessed only if all the interacting waves are polarized in the same direction. Under this circumstance, even if birefringence is present it cannot be used to compensate for dispersion.

There is a technique known as quasi-phase-matching that can be used when normal phase matching cannot be implemented. The idea of quasi-phase-matching is illustrated in Fig. 2.9.1, which shows both a single crystal of

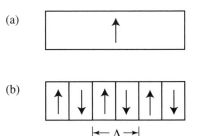

FIGURE 2.9.1   Schematic representations of a second-order nonlinear optical material in the form of (a) a homogeneous single crystal and (b) a periodically poled material in which the positive c axis alternates in orientation with period Λ.

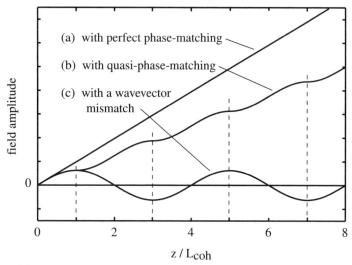

FIGURE 2.9.2   Comparison of the spatial variation of the field amplitude of the generated wave in a nonlinear optical interaction for three different phase matching conditions: Curve (a) assumes that the phase matching condition is perfectly satisfied, and consequently the field amplitude grows linearly with propagation distance. Curve (c) assumes that the wavevector mismatch Δk is nonzero, and consequently the field amplitude of the generated wave oscillates periodically with distance. Curve (b) assumes the case of a quasi-phase-matched interaction, in which the orientation of the positive c axis is periodically modulated with a period of twice the coherent buildup length L_{coh}, in order to compensate for the influence of wavevector mismatch. In this case the field amplitude grows monotonically with propagation distance, although less rapidly than in the case of a perfectly phase-matched interaction.

nonlinear optical material (part a) and a periodically poled material (part b). A periodically poled material is a structure that has been fabricated in such a manner that the orientation of one of the crystalline axes, often the c axis of a ferroelectric material, is inverted periodically as a function of postion within the material. An inversion in the direction of the c axis has the consequence of inverting the sign of the nonlinear coupling coefficient d_{eff}. This periodic alternation of the sign of d_{eff} can compensate for a nonzero wavevector mismatch Δk. The nature of this effect is illustrated in Fig. 2.9.2. Part (a) of this figure shows that, in a perfectly phase matched interaction in an ordinary single-crystal nonlinear optical material, the field strength of the generated wave grows linearly with propagation distance. In the presence of a wavevector mismatch (part c), the field amplitude of the generated wave oscillates with propagation distance. The nature of quasi-phase-matching is shown in part

(b) of this figure. Here it is assumed that the period Λ of the alternation of the crystalline axis has been set equal to twice the coherent buildup length L_{coh} of the nonlinear interaction. Then, each time the field amplitude of the generated wave is about to begin to decrease as a consequence of the wavevector mismatch, a reversal of the sign of d_{eff} occurs which allows the field amplitude to continue to grow monotonically.

A mathematical description of quasi-phase-matching can be formulated as follows. We let $d(z)$ denote the spatial dependence of the nonlinear coupling coefficient. In the example shown in part (b) of Fig. 2.9.1, $d(z)$ is simply the square-wave function which can be represented as

$$d(z) = d_{eff}\,\text{sign}[\cos(2\pi z/\Lambda)]; \tag{2.9.1}$$

more complicated spatial variations are also possible. In this equation, d_{eff} denotes the nonlinear coefficient of the homogeneous material. The spatial variation of the nonlinear coefficient leads to a modification of the coupled amplitude equations describing the nonlinear optical interaction. The nature of the modification can be deduced by noting that in the derivation of the coupled amplitude equations, the constant quantity d_{eff} appearing in Eq. (2.2.7) must be replaced by the spatially varying quantity $d(z)$. It is useful to describe this spatial variation of $d(z)$ in terms of a Fourier series as

$$d(z) = d_{eff} \sum_{m=-\infty}^{m=\infty} G_m \exp(ik_m z), \tag{2.9.2}$$

where $k_m = 2\pi m/\Lambda$ is the grating vector associated with the mth Fourier component of $d(z)$. For the form of modulation given in the example of Eq. (2.9.1), the coefficients G_m are readily shown to be given by

$$G_m = (2/m\pi)\sin(m\pi/2), \tag{2.9.3}$$

from which it follows that the fundamental amplitude G_1 is given by $G_1 = 2/\pi$. Coupled amplitude equations are now derived as in Section 2.2. In performing this derivation, one assumes that one particular Fourier component of $d(z)$ provides the dominant coupling among the interacting waves. After making the slowly varying amplitude approximation, one obtains the set of equations

$$\frac{dA_1}{dz} = \frac{8\pi i\omega_1 d_Q}{n_1 c} A_3 A_2^* e^{-i\Delta k_Q z} \tag{2.9.4a}$$

$$\frac{dA_2}{dz} = \frac{8\pi i\omega_2 d_Q}{n_2 c} A_3 A_1^* e^{-i\Delta k_Q z} \tag{2.9.4b}$$

$$\frac{dA_3}{dz} = \frac{8\pi i\omega_3 d_Q}{n_3 c} A_1 A_2 e^{i\Delta k_Q z} \tag{2.9.4c}$$

where d_Q is the nonlinear coupling coefficient which depends on the Fourier order m according to

$$d_Q = d_{\text{eff}} G_m \tag{2.9.5}$$

and where the wavevector mismatch for order m is given by

$$\Delta k_Q = k_1 + k_2 - k_3 - k_m. \tag{2.9.6}$$

Note that these coupled amplitude equations are formally identical to those derived above (that is, Eqs. (2.2.11), (2.2.13), and (2.2.14)) for a homogeneous material, but they involve modified values of the nonlinear coupling coefficient d_{eff} and wavevector mismatch Δk. Because of the tendency for d_Q to decrease with increasing values of m (see Eq. (2.9.3)), it is most desirable to achive quasi-phase-matching through use of a first-order ($m = 1$) interaction for which

$$\Delta k_Q = k_1 + k_2 - k_3 - 2\pi/\Lambda, \qquad d_Q = (2/\pi)d_{\text{eff}}. \tag{2.9.7}$$

From the first of these relations, we see that the optimum period for the quasi-phase-matched structure is given by

$$\Lambda = 2L_{\text{coh}} = 2\pi/(k_1 + k_2 - k_3). \tag{2.9.8}$$

As a numerical example, L_{coh} is equal to 3.4 μm for second-harmonic generation of radiation at a wavelength of 1.06 μm in lithium niobate.

A number of different approaches have been proposed for the fabrication of quasi-phase-matched structures. The idea of quasi-phase-matching originates in a very early paper by Armstrong *et al.* (1962), who suggest slicing a nonlinear optical medium into thin segments and rotating alternating segments by 180 degrees. This approach while feasible is hampered by the required thinness of the individual layers. More recent work has involved the study of techniques that lead to the growth of crystals with a periodic alternation in the orientation of the crystalline c axis or techniques that allow the orientation of the c axis to be inverted locally in an existing crystal. Progress in this field is reviewed by Byer (1997). A particularly promising approach, which originated with Yamada *et al.* (1993), is the use of a static electric field to invert the orientation of the ferroelectric domains (and consequently of the crystalline c axis) in a thin sample of lithium niobate. In this approach, a metallic electrode pattern in the form of long stripes is deposited on the top surface a lithium niobate crystal, whereas the bottom surface is uniformly coated to act as a ground plane. A static electric field of the order of 21 kV/mm is then applied to the material, which leads to domain reversal only of the material directly under the top electrode. Khanarian *et al.* (1990) have demonstrated that polymeric materials can be periodically poled by the application of a static electric field. Quasi-phase-matched materials offer promise for many applications of nonlinear optics, some of which are outlined in the review of Byer (1997).

2.10. Nonlinear Optical Interactions with Focused Gaussian Beams

In the past several sections we have treated nonlinear optical interactions in the approximation in which all of the interacting waves are taken to be infinite plane waves. However, in practice, the incident radiation is usually focused into the nonlinear optical medium in order to increase its intensity and hence to increase the efficiency of the nonlinear optical process. This section explores the nature of nonlinear optical interactions that are excited by focused laser beams.

Paraxial Wave Equation

We assume that each frequency component of the beam obeys a wave equation of the form of Eq. (2.1.20), and thus is given by

$$\nabla^2 \tilde{\mathbf{E}}_n - \frac{1}{(c/n)^2} \frac{\partial^2 \tilde{\mathbf{E}}_n}{\partial t^2} = \frac{4\pi}{c^2} \frac{\partial^2 \tilde{\mathbf{P}}_n}{\partial t^2}. \tag{2.10.1}$$

We next represent the electric field $\tilde{\mathbf{E}}_n$ and polarization $\tilde{\mathbf{P}}_n$ as

$$\tilde{\mathbf{E}}_n(\mathbf{r}, t) = \mathbf{A}_n(\mathbf{r})e^{i(k_n z - \omega_n t)} + \text{c.c.}, \tag{2.10.2a}$$

$$\tilde{\mathbf{P}}_n(\mathbf{r}, t) = \mathbf{p}_n(\mathbf{r})e^{i(k_n' z - \omega_n t)} + \text{c.c.} \tag{2.10.2b}$$

Here we allow $\tilde{\mathbf{E}}_n$ and $\tilde{\mathbf{P}}_n$ to represent non-plane waves by allowing the complex amplitudes $\mathbf{A}_n$ and $\mathbf{p}_n$ to be spatially varying quantities. In addition, we allow the possibility of a wavevector mismatch by allowing the wavevector of $\tilde{\mathbf{P}}_n$ to be different from that of $\tilde{\mathbf{E}}_n$. We next substitute Eqs. (2.10.2) into (2.10.1). Since we have specified the z direction as the dominant direction of propagation of the wave $\tilde{\mathbf{E}}_n$, it is useful to express the laplacian operator as $\nabla^2 = \partial^2/\partial z^2 + \nabla_T^2$, where the transverse laplacian is given by $\nabla_T^2 = \partial^2/\partial x^2 + \partial^2/\partial y^2$ in rectangular coordinates and is given by $\nabla_T^2 = (1/r)(\partial/\partial r)(r\partial/\partial r) + (1/r)^2\partial^2/\partial\phi^2$ in cylindrical coordinates. As in the derivation of Eq. (2.2.11), we now make the slowly-varying amplitude approximation, that is, we assume that the variation of $\mathbf{A}_n$ with z occurs only over distances much larger than an optical wavelength. We hence find that Eq. (2.10.1) becomes

$$2ik_n \frac{\partial \mathbf{A}_n}{\partial z} + \nabla_T^2 \mathbf{A}_n = -\frac{4\pi \omega_n^2}{c^2} \mathbf{p}_n e^{i\Delta k z}, \tag{2.10.3}$$

where $\Delta k = k'_n - k_n$. This result is known as the paraxial wave equation, because the approximation of neglecting the contribution $\partial^2 A/\partial z^2$ on the left-hand side is justifiable insofar as the wave $\mathbf{E}_n$ is propagating primarily along the z axis.

Gaussian Beams

Let us first study the nature of the solution to Eq. (2.10.3) for the case of the free propagation of an optical wave, that is, for the case in which the source term containing $\mathbf{p}_n$ vanishes. The paraxial wave equation is solved in such a case by a beam having a transverse intensity distribution that is everywhere a gaussian and that can be represented as (Kogelnik and Li, 1966)

$$A(r, z) = \mathscr{A}\frac{w_0}{w(z)}e^{-r^2/w(z)^2}e^{ikr^2/2R(z)}e^{i\Phi(z)}, \qquad (2.10.4a)$$

where

$$w(z) = w_0\left[1 + \left(\lambda z/\pi w_0^2\right)^2\right]^{1/2} \qquad (2.10.4b)$$

represents the $1/e$ radius of the field distribution, where

$$R(z) = z\left[1 + \left(\pi w_0^2/\lambda z\right)^2\right] \qquad (2.10.4c)$$

represents the radius of curvature of the optical wavefront, and where

$$\Phi(z) = -\arctan\left(\lambda z/\pi w_0^2\right) \qquad (2.10.4d)$$

represents the spatial variation of the phase of the wave (measured with respect to that of an infinite plane wave). In these formulas, w_0 represents the beam waist radius (that is, the value of w at the plane $z = 0$), and $\lambda = 2\pi c/n\omega$ represents the wavelength of the radiation in the medium. The angular divergence of the beam in the far field is given by $\theta_{\rm ff} = \lambda/\pi w_0$. The nature of this solution is illustrated in Fig. 2.10.1.

For theoretical work it is often convenient to represent the gaussian beam in the more compact (but less intuitive) form (see Problem 10 at the end of the chapter)

$$A(r, z) = \frac{\mathscr{A}}{1 + i\zeta}e^{-r^2/w_0^2(1+i\zeta)}. \qquad (2.10.5a)$$

Here[*]

$$\zeta = 2z/b \qquad (2.10.5b)$$

---

[*] Note that the quantity ζ defined here bears no relation to the quantity ζ introduced in Eq. (2.6.18) in our discussion of second-harmonic generation.

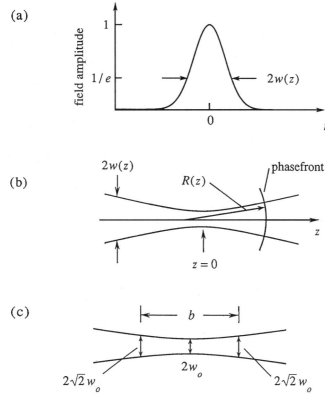

FIGURE 2.10.1   (a) Field amplitude distribution of a gaussian laser beam. (b) Variation of the beam radius w and wavefront radius of curvature R with position z. (c) Relation between the beam waist radius and the confocal parameter b.

is a dimensionless longitudinal coordinate defined in terms of the confocal parameter

$$b = 2\pi w_0^2/\lambda = k w_0^2, \qquad (2.10.5c)$$

which, as illustrated in part (c) of Fig. 2.10.1, is a measure of the longitudinal extent of the focal region of the gaussian beam. The total power $\mathscr{P}$ carried by a gaussian laser beam can be calculated by integrating over the transverse intensity distribution of the beam. Since $\mathscr{P} = \int I \, 2\pi r \, dr$, where the intensity is given by $I = (nc/2\pi)|A|^2$, we find that

$$\mathscr{P} = \tfrac{1}{4} n c w_0^2 |\mathscr{A}|^2. \qquad (2.10.6)$$

Harmonic Generation Using Focused Gaussian Beams

Let us now treat harmonic generation excited by a gaussian fundamental beam. For generality, we consider the generation of the qth harmonic. According to Eq. (2.10.3), the amplitude A_q of the $\omega_q = q\omega$ frequency component of the optical field must obey the equation

$$2ik_q\frac{\partial A_q}{\partial z} + \nabla_T^2 A_q = -\frac{4\pi\omega_q^2}{c^2}\chi^{(q)}A_1^q e^{i\Delta kz} \tag{2.10.7}$$

where $\Delta k = qk_1 - k_q$ and where we have set the complex amplitude p_q of the nonlinear polarization equal to $p_q = \chi^{(q)}A_1^q$. Here $\chi^{(q)}$ is the nonlinear susceptibility describing qth-harmonic generation, i.e., $\chi^{(q)} = \chi^{(q)}(q\omega = \omega + \omega + \cdots + \omega)$, and A_1 is the complex amplitude of the fundamental wave, which according to Eq. (2.10.5a) can be represented as

$$A_1(r, z) = \frac{\mathscr{A}_1}{1 + i\zeta}e^{-r^2/w_0^2(1+i\zeta)}. \tag{2.10.8}$$

We work in the constant-pump approximation. We solve Eq. (2.10.7) by adopting the trial solution

$$A_q(r, z) = \frac{\mathscr{A}_q(z)}{1 + i\zeta}e^{-qr^2/w_0^2(1+i\zeta)}, \tag{2.10.9}$$

where $\mathscr{A}_q(z)$ is a function of z. One might guess this form for the trial solution because its radial dependence is identical to that of the source term in Eq. (2.10.7). Note also that (ignoring the spatial variation of $\mathscr{A}_q(z)$) the trial solution corresponds to a beam with the same confocal parameter as the fundamental beam (2.10.8); this behavior makes sense in that the harmonic wave is generated coherently over a region whose longitudinal extent is equal to that of the fundamental wave. If the trial solution (2.10.9) is substituted into Eq. (2.10.7), we find that it satisfies this equation so long as $\mathscr{A}_q(z)$ obeys the (ordinary) differential equation

$$\frac{d\mathscr{A}_q}{dz} = \frac{i2\pi q\omega}{nc}\chi^{(q)}\mathscr{A}_1^q\frac{e^{i\Delta kz}}{(1+i\zeta)^{q-1}}. \tag{2.10.10}$$

This equation can be integrated directly to obtain

$$\mathscr{A}_q(z) = \frac{i2\pi q\omega}{nc}\chi^{(q)}\mathscr{A}_1^q J_q(\Delta k, z_0, z), \tag{2.10.11a}$$

where

$$J_q(\Delta k, z_0, z) = \int_{z_0}^{z}\frac{e^{i\Delta kz'}dz'}{(1 + 2iz'/b)^{q-1}}, \tag{2.10.11b}$$

and where z_0 represents the value of z at the entrance to the nonlinear medium. We see that the harmonic radiation is generated with a confocal parameter equal to that of the incident laser beam. Hence the beam waist radius of the qth harmonic radiation is $q^{1/2}$ times smaller than that of the incident beam, and the far-field diffraction angle $\theta_{ff} = \lambda/\pi w_0$ is $q^{1/2}$ times smaller than that of the incident laser beam. We have solved Eq. (2.10.7) by guessing the correct form (Eq. (2.10.9)) for the trial solution; a constructive solution to Eq. (2.10.7) has been presented by Kleinman et al. (1966) for second-harmonic generation and by Ward and New (1969) for the general case of qth-harmonic generation.

The integral appearing in Eq. (2.10.11b) can be evaluated analytically for certain special cases. One such case in the plane-wave limit, where $b \gg |z_0|, |z|$. In this limit the integral reduces to

$$J_q(\Delta k, z_0, z) = \int_{z_0}^{z} e^{i\Delta k z'} dz' = \frac{e^{i\Delta kz} - e^{i\Delta kz_0}}{i\,\Delta k}, \qquad (2.10.12a)$$

which implies that

$$|J_q(\Delta k, z_0, z)|^2 = L^2 \text{sinc}^2\left(\frac{\Delta k L}{2}\right) \qquad (2.10.12b)$$

where $L = z - z_0$ is the length of the interaction region.

The opposite limiting case is that in which the fundamental wave is focused tightly within the interior of the nonlinear medium; this condition implies that $z_0 = -|z_0|$, $z = |z|$, and $b \ll |z_0|, |z|$. In this limit the integral in Eq. (2.10.11b) can be approximated by replacing the upper limit of integration by infinity, that is,

$$J_q(\Delta k, z_0, z) = \int_{-\infty}^{\infty} \frac{e^{i\Delta k z'} dz'}{(1 + 2iz'/b)^{q-1}}. \qquad (2.10.13a)$$

This integral can be evaluated by means of a straightforward contour integration. One finds that

$$J_q(\Delta k, z_0, z) = \begin{cases} 0, & \Delta k \le 0, \\[2mm] \dfrac{b}{2} \dfrac{2\pi}{(q-2)!}\left(\dfrac{b\Delta k}{2}\right)^{q-2} e^{-b\Delta k/2}, & \Delta k > 0. \end{cases} \qquad (2.10.13b)$$

This functional form is illustrated for the case of third-harmonic generation ($q = 3$) in Fig. 2.10.2. We find the somewhat surprising result that the efficiency of third-harmonic generation in the tight-focusing limit vanishes identically for the case of perfect phase matching ($\Delta k = 0$) and is maximized through the use of a positive wavevector mismatch. This behavior can be understood in terms of the phase shift of π radians that any beam of light experiences in passing through its focus. This effect is known as the phase anomaly and

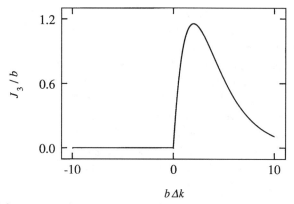

FIGURE 2.10.2   Dependence of the phase-matching factor J_3 for third-harmonic generation on the normalized confocal parameter $b\Delta k$, in the tight-focusing limit.

was first studied systematically by Gouy (1890). For the case of nonlinear optics, this effect has important consequences over and above the phase shift imparted to the transmitted light beam, because in general the nonlinear polarization $p = \chi^{(q)} A_1^q$ will experience a phase shift that is q times larger than that experienced by the incident wave of amplitude A_1. Consequently the nonlinear polarization will be unable to couple efficiently to the generated wave of amplitude A_q unless a wavevector mismatch Δk is introduced to compensate for the phase shift due to the passage of the incident wave through its focus. The reason why Δk should be positive in order for this compensation to occur can be understood intuitively in terms of the argument presented in Fig. 2.10.3.

Boyd and Kleinman (1968) have considered how to adjust the focus of the incident laser beam to optimize the efficiency of second-harmonic generation. They find that the highest efficiency is obtained when beam walkoff effects (mentioned in Section 2.7) are negligible, when the incident laser beam is focused so that the beam waist is located at the center of the crystal and the ratio L/b is equal to 2.84, and when the wavevector mismatch is set equal to $\Delta k = 3.2/L$. In this case, the power generated at the second-harmonic frequency is equal to

$$\mathscr{P}_{2\omega} = 1.068 \left[\frac{128\pi^2 \omega_1^3 d_{\text{eff}}^2 L}{c^4 n_1 n_2} \right] \mathscr{P}_{\omega}^2. \qquad (2.10.14)$$

In addition, Boyd and Kleinman show heuristically that other parametric processes, such as sum- and difference-frequency generation, are optimized by choosing the same confocal parameter for both input waves and applying the same criteria used to optimize second-harmonic generation.

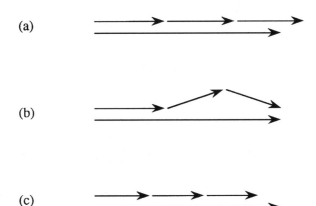

FIGURE 2.10.3 Illustration of why a positive value of Δk is desirable in harmonic generation with focused laser beams. (a) Wavevector diagram for third-harmonic generation with Δk positive. Even though the process is phase mismatched, the fundamental beam contains an angular spread of wavevectors and the phase-matched process illustrated in (b) can occur with high efficiency. (c) Conversely, for Δk negative, efficient harmonic generation cannot occur.

2.11. Nonlinear Optics at an Interface

There are certain nonlinear optical processes associated with the interface between two dissimilar optical materials. Two such examples are shown schematically in Fig. 2.11.1. Part (a) shows an optical wave falling onto a second-order nonlinear optical material. We saw earlier (in Section 2.6) how to predict the amplitude of the second-harmonic wave generated in the forward direction. But in fact a weaker second-harmonic wave is generated in reflection at the interface separating the two materials. We shall see in the present section how to predict the intensity of this reflected harmonic wave. Part (b) of the figure shows a wave falling onto a centrosymmetric nonlinear optical material. Such a material cannot possess a bulk second-order nonlinear optical susceptibility, but the presence of the interface breaks the inversion symmetry for a thin region (of the order of one molecular diameter in thickness) near the interface, and this thin layer can emit a second-harmonic wave. The intensity of the light emitted by this surface layer depends sensitively on the structural properties of the surface and especially upon the presence of molecules absorbed onto the surface. For this reason surface second-harmonic generation is an important diagnostic method in surface science.

Let us now consider in greater detail the situation considered in part (a) of Fig. 2.11.1. We assume that the wave at the fundamental frequency incident

(a)

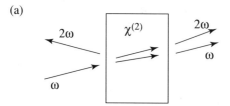

(b)

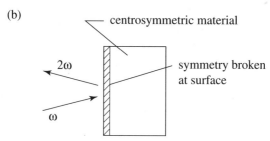

FIGURE 2.11.1   Illustration of second-harmonic generation in reflection at the surface of (a) a second-order nonlinear optical material and (b) a centrosymmetric nonlinear optical material.

on the interface can be described by

$$\tilde{\mathbf{E}}_i(\mathbf{r}, t) = \mathbf{E}_i(\omega_i)e^{-i\omega_i t} + \text{c.c.} \quad \text{where} \quad \mathbf{E}_i(\omega_i) = \mathbf{A}_i(\omega_i)e^{i\mathbf{k}_i(\omega_i)\cdot\mathbf{r}}. \quad (2.11.1)$$

This wave will be partially reflected and partially transmitted into the nonlinear optical material. Let us represent the transmitted component as

$$\mathbf{E}_T(\omega_i) = \mathbf{A}_T(\omega_i)e^{i\mathbf{k}_T(\omega_i)\cdot\mathbf{r}}, \quad (2.11.2)$$

where the amplitude $\mathbf{A}_T(\omega_i)$ and propagation direction $\mathbf{k}_T(\omega_i)$ can be determined from the standard Fresnel equations of linear optics. For simplicity, in the present discussion we ignore the effects of birefringence; we note that birefringence vanishes identically in crytals (such as GaAs) that are noncentrosymmetric yet possess a cubic lattice. The transmitted fundamental wave will create a nonlinear polarization at frequency $\omega_s = 2\omega_i$ within the medium which we represent as

$$\tilde{\mathbf{P}}(\mathbf{r}, t) = \mathbf{P}\,e^{-i\omega_s t} + \text{c.c.} \quad \text{where} \quad \mathbf{P} = \mathbf{p}\,e^{i\mathbf{k}_s(\omega_s)\cdot\mathbf{r}} \quad \text{and} \quad \mathbf{p} = \chi_{\text{eff}}^{(2)}A_T^2(\omega_i) \quad (2.11.3)$$

and where $\mathbf{k}_s(\omega_s) = 2\mathbf{k}_T(\omega_i)$.

The details of the ensuing analysis differ depending upon whether $\mathbf{p}$ is parallel or perpendicular to the plane of incidence. Here we treat only the case of $\mathbf{p}$ perpendicular to the plane of incidence; a treatment of the other case can be found for instance in Bloembergen and Pershan (1962) or in

Shen (1984). As described by Eq. (2.1.22), this nonlinear polarization will give rise to radiation at the second-harmonic frequency ω_s. The generation of this radiation is governed by the equation

$$\nabla^2 \mathbf{E}\left(\omega_s\right) + \left[\epsilon\left(\omega_s\right) \omega_s^2/c^2\right] \mathbf{E}\left(\omega_s\right) = -4\pi \left(\omega_s^2/c^2\right) \mathbf{p}_\perp \, e^{i\mathbf{k}_s \cdot \mathbf{r}} \qquad (2.11.4)$$

where $\mathbf{p}_\perp$ is the component of $\mathbf{p}$ perpendicular to the plane of incidence. The formal solution to this equation consists of any particular solution plus a general solution to the homogeneous version of this equation obtained by setting its right-hand side equal to zero. It turns out that we can meet all of the appropriate boundary conditions by assuming that the homogeneous solution is an infinite plane wave of as yet unspecified amplitude $\mathbf{A}\left(\omega_s\right)$ and wavevector $\mathbf{k}_T\left(\omega_s\right)$. We thus represent the solution to Eq. (2.11.4) as

$$\mathbf{E}_T\left(\omega_s\right) = \mathbf{A}\left(\omega_s\right) e^{i\mathbf{k}_T(\omega_s)\cdot\mathbf{r}} + \frac{4\pi \omega_s^2/c^2}{|k_s|^2 - |k_T(\omega_s)|^2} \, \mathbf{p}_\perp \, e^{i\mathbf{k}_s\cdot\mathbf{r}}, \qquad (2.11.5)$$

where $|k_T(\omega_s)|^2 = \epsilon_T\left(\omega_s\right) \omega_s^2/c^2$; as mentioned above, the direction of $\mathbf{k}_T(\omega_s)$ is at present undetermined. The electromagnetic boundary conditions at the interface require that the components of $\mathbf{E}$ and of $\mathbf{H}$ tangential to the plane of the interface be continuous. These boundary conditions can be satisfied only if we postulate the existence of a reflected, second-harmonic wave which we represent as

$$\mathbf{E}_R(\omega_s) = \mathbf{A}_R(\omega_s)e^{i\mathbf{k}_R(\omega_s)\cdot\mathbf{r}}. \qquad (2.11.6)$$

In order that the boundary conditions be met at each point along the interface, it is necessary that the nonlinear polarization of wavevector $\mathbf{k}_s = 2\mathbf{k}_T(\omega_i)$, the transmitted second-harmonic wave of wavevector $\mathbf{k}_T(\omega_s)$, and the reflected second-harmonic wave of wavevector $\mathbf{k}_R(\omega_s)$ have identical wavevector components along the plane of the interface. This situation is illustrated in Fig. 2.11.2, where we let x be a coordinate measured along the interface in the plane of incidence and let z denote a coordinate measured perpendicular to the plane of incidence. We thus require that

$$k_x^s = k_x^R(\omega_s) = k_x^T(\omega_s) \qquad (2.11.7)$$

(note that $k_x^s \equiv 2k_x^T(\omega_i)$). Furthermore, we can express the magnitude of each of the propagation vectors in terms of the dielectric constant of each medium as

$$k_T(\omega_s) = \epsilon_T^{1/2}(\omega_s)\omega_s/c, \qquad (2.11.8a)$$

$$k_R(\omega_s) = \epsilon_R^{1/2}(\omega_s)\omega_s/c, \qquad (2.11.8b)$$

$$k_i(\omega_i) = \epsilon_R^{1/2}(\omega_i)\omega_i/c, \qquad (2.11.8c)$$

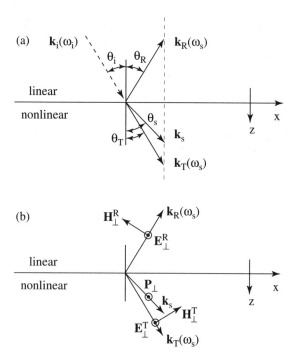

FIGURE 2.11.2   (a) Geometry showing the creation of a transmitted and reflected second-harmonic wave at the surface of a second-order nonlinear optical material. (b) Definition of the electric and magnetic field vectors for the case in which **P** is perpendicular to the plane of incidence.

where ϵ_R denotes the dielectric constant of the linear, incident medium and ϵ_T denotes the linear dielectric constant of the nonlinear medium. For mathematical convenience, we also introduce a fictitious dielectric constant ϵ_s associated with the nonlinear polarization defined such that

$$k_s = \epsilon_s^{1/2}\, \omega_s/c. \qquad (2.11.9)$$

From Eqs. (2.11.7) through (2.11.9) we can readily determine expressions relating the angles θ_i, θ_R, θ_s, and θ_T (see Fig. 2.11.2), which are given by

$$\epsilon_R^{1/2}(\omega_i) \sin\theta_i = \epsilon_R^{1/2}(\omega_s) \sin\theta_R = \epsilon_T^{1/2}(\omega_s) \sin\theta_s = \epsilon_s^{1/2} \sin\theta_s. \qquad (2.11.10)$$

This equation can be considered to be the nonlinear optical generalization of Snell's law.

We next apply explicitly the boundary conditions at the interface between the linear and nonlinear medium. According to Eq. (2.11.5), this component will lead to the generation of an electric field in the $E_\perp = E_y$ direction, and in accordance with Maxwell's equations the associated magnetic field will lie in the xz plane (see part b of Fig. 2.11.2). The continuity of the tangential components of $\mathbf{E}$ and $\mathbf{H}$ then leads to the equations

$$E_y: A_\perp^R = A_\perp^T + 4\pi p_\perp / [\epsilon_s - \epsilon_T(\omega_s)],$$

$$H_x: -\epsilon_R^{1/2}(\omega_s) A_\perp^R \cos\theta_R = \epsilon_T^{1/2}(\omega_s) A_\perp^T \cos\theta_T \qquad (2.11.11)$$

$$+ 4\pi p_\perp \cos\theta_s \epsilon_s^{1/2} / [\epsilon_s - \epsilon_T(\omega_s)].$$

These equations are readily solved simultaneously to obtain expression for $A_\perp^R$ and $A_\perp^T$. These expressions are then introduced into Eqs. (2.11.5) and (2.11.6) to find that the transmitted and reflected fields are given by

$$E_\perp^R = \frac{-4\pi p_\perp e^{i\mathbf{k}_R(\omega_s)\cdot\mathbf{r}}}{\left[\epsilon_T^{1/2}(\omega_s)\cos\theta_T + \epsilon_R^{1/2}(\omega_s)\cos\theta_R\right]\left[\epsilon_T^{1/2}(\omega_s)\cos\theta_T + \epsilon_s^{1/2}\cos\theta_s\right]},$$

$$\equiv A_\perp^R e^{i\mathbf{k}_R(\omega_s)\cdot\mathbf{r}} \qquad (2.11.12a)$$

$$E_\perp^T = \frac{-4\pi p_\perp}{\epsilon_T(\omega_s) - \epsilon_s} \left[e^{i\mathbf{k}_s\cdot\mathbf{r}} - \frac{\epsilon_s^{1/2}\cos\theta_s + \epsilon_R^{1/2}(\omega_s)\cos\theta_R}{\epsilon_T^{1/2}(\omega_s)\cos\theta_T + \epsilon_R^{1/2}(\omega_s)\cos\theta_R} e^{i\mathbf{k}_T(\omega_s)\cdot\mathbf{r}} \right].$$

$$(2.11.12b)$$

The transmitted second-harmonic wave is thus composed of a homogeneous wave with propagation vector $\mathbf{k}_T$ and an inhomogeneous wave with propagation vector $\mathbf{k}_s$. We see from Fig. 2.11.2 that $\mathbf{k}_s - \mathbf{k}_T$ must lie in the z direction and is given by

$$\mathbf{k}_s - \mathbf{k}_T = \Delta k\hat{z} = (\omega_s/c)\left[\epsilon_s^{1/2}\cos\theta_s - \epsilon_T^{1/2}(\omega_s)\cos\theta_T\right]\hat{z}. \qquad (2.11.13)$$

If this result is introduced into Eq. (2.11.12b), we can express the transmitted field in the form

$$E_\perp^T = \left[A_\perp^R + \frac{4\pi(\omega_s/c)^2 p_\perp}{2k_T(\omega_s)} \left(\frac{e^{i\Delta kz} - 1}{\Delta k} \right) \right] e^{i\mathbf{k}_T(\omega_s)\cdot\mathbf{r}}$$

$$(2.11.14)$$

$$\equiv A_\perp^T e^{i\mathbf{k}_T(\omega_s)\cdot\mathbf{r}}.$$

This equation has the form of a plane wave with a spatially varying amplitude; the spatial variation is a manifestation of imperfect phase matching of the non-linear optical interaction. The present formalism demonstrates that the origin of the spatial variation is the interference of the homogeneous and inhomogeneous solutions of the driven wave equation.

Let us interpret further the result given by Eq. (2.11.14). Let us assume that Δkz is much smaller than unity for all propagation distances z of interest. We then find that, correct to first order in Δk, the amplitude of the transmitted wave is given by

$$A_\perp^T = A_\perp^R + \frac{4\pi(\omega/c)^2 p_\perp(iz)}{2k_T(\omega_s)} = A_\perp^R + \frac{2\pi i(\omega/c)p_\perp z}{\epsilon^{1/2}(\omega_s)}. \qquad (2.11.15)$$

We see that the amplitude of the generated wave thus grows linearly from its boundary value $A_\perp^R$. We also see from Eq. (2.11.12a) that $A_\perp^R$ will be given to order of magnitude by

$$A_\perp^R \simeq -\frac{\pi p_\perp}{\epsilon} \qquad (2.11.16)$$

where ϵ is some characteristic value of the dielectric constant of the region near the interface. On the basis of this result, Eq. (2.11.15) can be approximated as

$$A_\perp^T \simeq -\frac{\pi p_\perp}{\epsilon}[1 - 2ik_T(\omega_s)z]. \qquad (2.11.17)$$

This result shows that the surface term makes a contribution comparable to that of the bulk term for a thickness t given by

$$t = \lambda/4\pi. \qquad (2.11.18)$$

Let us next examine the situation of Fig. 2.11.1b, which considers harmonic generation at the interface between two centrosymmetric media. An accurate treatment of such a situation would require that we know the nonlinear optical properties of the region near the interface at a molecular level, which is not possible at the present level of description (because we can rigorously deduce macroscopic properties from microscopic properties, but not vice versa). Nonetheless, we can make an order-of-magnitude estimate of the amplitude of the reflected wave for typical materials. Let us model the interface between two centrosymmetric materials as possessing a second-order susceptibility $\chi^{(2)}$ confined to a thickness of the order of a molecular dimension a_0. Here $\chi^{(2)}$ is a typical value of the second-order susceptibility of a noncentrosymmetric material. This assumption taken in conjunction with Eq. (2.11.18) leads to the prediction

$$A_\perp^T(\text{centrosymmetric}) = \frac{4\pi a_0}{\lambda} A_\perp^T(\text{noncentrosymmetric})$$

$$(2.11.19)$$

$$\simeq 10^{-3} A_\perp^T(\text{noncentrosymmetric}).$$

This result is in agreement with the predictions of more detailed models (see, for instance, Mizrahi and Sipe, 1988).

Problems

1. *Infrared upconversion.* One means of detecting infrared radiation is to first convert the infrared radiation to the visible by the process of sum-frequency generation. Assume that infrared radiation of frequency ω_1 is mixed with an intense laser beam of frequency ω_2 to form the upconverted signal at frequency $\omega_3 = \omega_1 + \omega_2$. Derive a formula that shows how the quantum efficiency for converting infrared photons to visible photons depends on the length L and nonlinear coefficient d_{eff} of the mixing crystal, and on the phase mismatch Δk. Estimate numerically the value of the quantum efficiency for upconversion of 10-μm infrared radiation using a 1-cm-long proustite crystal, 1 W of laser power at a wavelength of 0.65 μm, and the case of perfect phase matching and optimum focusing.

[Ans.: $\eta_Q = 2\%$.]

2. *Sum-frequency generation.* Solve the coupled-wave equations describing sum-frequency generation (Eqs. (2.2.11) through (2.2.14)) for the case of perfect phase matching ($\Delta k = 0$) but without making the approximation of Section 2.4 that the amplitude of the ω_2 wave can be taken to be constant.

[Hint: This problem is very challenging. For help, see Armstrong *et al.* (1962).]

3. *Systems of units.* Rewrite each of the displayed equations in Sections 2.1 through 2.3 in the SI system of units.

4. *Difference-frequency generation.* Solve the coupled-amplitude equations describing difference-frequency generation in the constant-pump limit, and thereby verify Eqs. (2.5.9) of the text. Assume that $\omega_1 + \omega_2 = \omega_3$, where the amplitude A_3 of the ω_3 pump wave is constant, that the medium is lossless at each of the optical frequencies, that the momentum mismatch Δk is arbitrary, and that in general there can be an input signal at each of the frequencies ω_1 and ω_2. Interpret your results by sketching representative cases of the solution and by taking special limiting cases such as that of perfect phase matching and of only two input fields.

5. *Second-harmonic generation.* Verify that Eq. (2.6.29) possesses solutions of the sort shown in Fig. 2.6.2.

6. *Second-harmonic generation.* Solve the coupled amplitude equations for the case of second-harmonic generation with the initial conditions $A_2 = 0$ but A_1 arbitrary at $z = 0$. Assume that Δk is arbitrary. Sketch how $|A(2\omega)|^2$ varies with z for several values of Δk, and hence verify the results shown in Fig. 2.6.4.

7. *Mode structure of an optical cavity.* Verify Eq. (2.8.11).

[Ans.: Assume that the refractive index n is a function of ν and require that an integral number m of half wavelengths fit within the cavity of length L_c. Thus $m\lambda/2 = L_c$ or, since

$\lambda = c/n\nu$, we obtain $n\nu = cm/2L_c$. We want to determine the frequency separation of adjacent modes. Thus $\Delta(n\nu) = \Delta(cm/2L_c)$ where Δ refers to the change in the indicated quantity between adjacent modes. Note that $\Delta(n\nu) = n\Delta\nu + \nu\Delta n = n\Delta\nu + \nu(dn/d\nu)\Delta\nu = [n + \nu(dn/d\nu)]\Delta\nu$ and that $\Delta(cm/2L_c) = c/2L_c\Delta(m) = c/2L_c$. Thus

$$\Delta\nu = \frac{c}{2L_c\,(n + \nu\,dn/d\nu)} = \frac{v_g}{2L_c} = \frac{c}{2n_gL_c}$$

where $v_g = c/[n + \nu(dn/d\nu)]$ is the usual expression for the group velocity and where $n_g = n + \nu(dn/d\nu)$ is the group index.]

8. *Mode structure of an optical cavity.* Generalize the result of the previous problem to the situation in which the cavity length is L but the material medium has length $L_c < L$.

9. *Quasi-phase-matching.* Generalize the discussion of the text leading from Eq. (2.9.1) to Eq. (2.9.6) by allowing the lengths of the inverted and noninverted sections of nonlinear optical material to be different. Let Λ be the period of the structure and l be the length of the inverted region. Show how each of the equations in this range is modified by this different assumption, and comment explicitly on the resulting modification to the value of d_Q and to the condition for the establishment of quasi-phase-matching.

10. *Gaussian laser beams.* Verify that Eqs. (2.10.4a) and (2.10.5a) are equivalent descriptions of a gaussian laser beam, and verify that they satisfy the paraxial wave Eq. (2.10.3).

11. *Gaussian laser beams.* Verify the statement made in the text that the trial solution given by Eq. (2.10.9) satisfies the paraxial wave equation in the form of Eq. (2.10.7) if the amplitude $\mathscr{A}_q(z)$ satisfies the ordinary differential equation (2.10.10).

12. *Phase matching with focused beams.* Evaluate the integral appearing in Eq. (2.10.13a) and thereby verify Eq. (2.10.13b).

13. *Third-harmonic generation.* Assuming the condition of perfect phase matching, derive and solve exactly the coupled-amplitude equations describing third-harmonic generation. You may assume that the nonlinear optical material is lossless. Include in your analysis the processes described by the two susceptibility elements $\chi^{(3)}(3\omega; \omega, \omega, \omega)$ and $\chi^{(3)}(\omega; 3\omega, -\omega, -\omega)$. Calculate the intensity of the third harmonic wave as a function of the length of the interaction region for the following two situations: (a) In the limit in which the undepleted pump approximation is valid. (b) For the general case in which the pump intensity cannot be assumed to remain constant.

14. *Poynting's theorem.* Derive the form of Poynting's theorem valid for a nonlinear optical material for which $\tilde{D} = \tilde{E} + 4\pi\tilde{P}$ with $\tilde{P} = \chi^{(1)}\tilde{E} + \chi^{(2)}\tilde{E}^2 + \chi^{(3)}\tilde{E}^3$. Assume that the material is nonmagnetic in the sense that $\tilde{B} = \tilde{H}$.

15. *Backward second-harmonic generation.* Part (c) of Fig. 2.1.1 implies that second-harmonic generation is radiated in the forward but is not appreciable radiated in the backward direction. Verify that this conclusion is in fact correct by deriving the coupled amplitude equation for a second-harmonic field propagating in the backward direction, and show that the amplitude of this wave can never become appreciable. (Note that a more rigorous calculation that reaches the same conclusion is presented in Section 2.11.)

16. *Second-harmonic generation.* Consider the process of second-harmonic generation both with $\Delta k = 0$ and $\Delta k \neq 0$ in a lossless material. State the conditions under which the following types of behavior occur: (i) The fundamental and second-harmonic fields periodically exchange energy. (ii) The second harmonic field asymptotically acquires all of the energy. (iii) The fundamental field asymptotically acquires all of the energy. (iv) Part of the energy resides in each component, and this fraction does not vary with z.

17. *Manley–Rowe relations.* Derive the Manley–Rowe relations for the process of second-harmonic generation. The derivation is analogous to that presented in Section 2.3 for the process of sum-frequency generation.

18. *Phase-matching requirements.* Explain why processes such as second-harmonic generation can be efficient only if the phase-matching relation $\Delta k = 0$ is satisfied, whereas no such requirement occurs for the case of two-photon absorption.

19. *Cascaded optical nonlinearities.* The intent of this problem is to develop an understanding of the phenomenon known as cascaded optical nonlinearities. By cascaded optical nonlinearities, one means that, through propagation, a second-order nonlinearity can mimic a third-order nonlinearity. In particular, in this problem you are to calculate the phase shift acquired by an optical wave in propagating through a second-order nonlinear optical material under conditions of nearly phase-matched second-harmonic generation, and to determine the conditions under which the phase shift acquired by the fundamental wave is approximately proportional to the product of the path length and the intensity.

To proceed, start for example with Eqs. (2.6.10) and (2.6.11), and show that one can eliminate A_2 to obtain the equation

$$\frac{d^2 A_1}{dz^2} + i \Delta k \frac{d A_1}{dz} - \Gamma^2 (1 - 2|A_1/A_0|^2) A_1 = 0$$

where Γ is a constant (give an expression for it) and A_0 is the incident value of the fundamental field. Show that under proper conditions (give specifics) the solution to this equation corresponds to a wave whose phase increases linearly with the length L of the nonlinear material and with the intensity I of the incident wave.

References

Sections 2.1 through 2.7

J. A. Armstrong, N. Bloembergen, J. Ducuing, and P. S. Pershan, *Phys. Rev.* **127**, 1918 (1962).

M. Born and E. Wolf, *Principles of Optics*, Pergamon Press, Oxford, 1975.

R. W. Boyd and C. H. Townes, *Appl. Phys. Lett.* **31**, 440 (1977).

R. L. Byer and R. L. Herbst, in *Tunable Infrared Generation*, Y. R. Shen, ed., Springer-Verlag, Berlin, 1977.

J. A. Giordmaine and R. C. Miller, *Phys. Rev. Lett.* **14**, 973 (1965); *Appl. Phys. Lett.* **9**, 298 (1966).

M. V. Hobden and J. Warner, *Phys. Lett.* **22**, 243 (1966).

M. V. Klein, *Optics*, Wiley, New York, 1970.

P. D. Maker, R. W. Terhune, M. Nisenoff, and C. M. Savage, *Phys. Rev. Lett.* **8**, 21 (1962).

J. M. Manley and H. E. Rowe, *Proc. IRE* **47**, 2115 (1959).

J. E. Midwinter and J. Warner, *Brit. J. Appl. Phys.* **16**, 1135 (1965).

Y. R. Shen, *The Principles of Nonlinear Optics*, Wiley, New York, 1984.

F. Zernike and J. E. Midwinter, *Applied Nonlinear Optics*, Wiley, New York, 1973.

Nonlinear Optical Interactions with Focused Gaussian Beams

G. C. Bjorklund, *IEEE J. Quantum Electron*, **QE-11**, 287 (1975).

G. D. Boyd and D. A. Kleinman, *J. Appl. Phys.* **39**, 3597 (1968).

C. R. Gouy, *Acad. Sci. Paris* **110**, 1251 (1890).

D. A. Kleinman, A. Ashkin, and G. D. Boyd, *Phys. Rev.* **145**, 338 (1966).

H. Kogelnik and T. Li, *Appl. Opt.* **5**, 1550 (1966).

R. B. Miles and S. E. Harris, *IEEE J. Quantum Electron.* **QE-9**, 470 (1973).

J. F. Ward and G. H. C. New, *Phys. Rev.* **185**, 57 (1969).

Optical Parametric Oscillators

W. R. Bosenberg, W. S. Pelouch, and C. L. Tang, *Appl. Phys. Lett.* **55**, 1952 (1989).

W. R. Bosenberg and C. L. Tang, *Appl. Phys. Lett.* **56**, 1819 (1990).

R. L. Byer, H. Rabin, and C. L. Tang, eds., *Treatise in Quantum Electronics*, Academic Press, New York, 1973.

M. Ebrahimzadeh and M. H. Dunn, in *Handbook of Optics IV*, 2nd ed., McGraw-Hill, New York, 2001.

D. C. Edelstein, E. S. Wachman, and C. L. Tang, *Appl. Phys. Lett.* **54**, 1728 (1989).

J. A. Giordmaine and R. C. Miller, *Phys. Rev. Lett.* **14**, 973 (1965).

J. A. Giordmaine and R. C. Miller, *Appl. Phys. Lett.* **9**, 298 (1966).

L. E. Myers, R. C. Eckardt, M. M. Fejer, R. L. Byer, W. R. Bosenberg, and J. W. Pierce, *J. Opt. Soc. Am. B* **12**, 2102 (1995).

R. G. Smith, J. E. Geusic, J. H. Levinstein, J. J. Rubin, S. Singh, and L. G. van Uitent, *Appl. Phys. Lett.* **12**, 308 (1968).

U. Simon and F. K. Tittel, in *Methods of Experimental Physics*, Vol. III, R. G. Hulet and F. B. Dunning, eds., Academic Press, San Diego, 1994.

Quasi-Phase-Matching

R. L. Byer, *J. Nonlinear Opt. Phys. Mater.* **6**, 549, 1997.

M. M. Fejer, G. A. Magel, D. H. Jundt, and R. L. Byer, *IEEE J. Quantum Electron.* **28**, 2631 (1992).

G. Khanarian, R. A. Norwood, D. Haas, B. Feuer, and D. Karim, *Appl. Phys. Lett.* **57**, 977 (1990).

M. Yamada, N. Nada, M. Saitoh, and K. Watanabe, *Appl. Phys. Lett.* **62**, 435 (1993).

Nonlinear Optics at an Interface

N. Bloembergen and P. S. Pershan, *Phys. Rev.* **128**, 602 (1962).

Y. R. Shen, *The Principles of Nonlinear Optics*, Wiley-Interscience, New York, 1984. See especially Section 6.4.

V. Mizrahi and J. E. Sipe, *J. Opt. Soc. Am. B* **5**, 660 (1988).

Chapter 3

Quantum-Mechanical Theory
of the Nonlinear Optical
Susceptibility

3.1. Introduction

In this chapter, we use the laws of quantum mechanics to derive explicit expressions for the nonlinear optical susceptibility. The motivation for obtaining these expressions is threefold: (1) these expressions display the functional form of the nonlinear optical susceptibility and hence show how the susceptibility depends on material parameters such as dipole transition moments and atomic energy levels, (2) these expressions display the internal symmetries of the susceptibility, and (3) these expressions can be used to obtain predictions of the numerical values of the nonlinear susceptibilities. These numerical predictions are particularly reliable for the case of atomic vapors, because the atomic parameters (such as atomic energy levels and dipole transition moments) that appear in the quantum-mechanical expressions are often known with high accuracy. In addition, since the energy levels of free atoms are very sharp (as opposed to the case of most solids, where allowed energies have the form of broad bands), it is possible to obtain very large values of the nonlinear susceptibility through the technique of resonance enhancement. The idea behind resonance enhancement of the nonlinear optical susceptibility is shown schematically in Fig. 3.1.1 for the case of third-harmonic generation. In part (a) of this figure, we show the process of third-harmonic generation in terms of the virtual levels introduced in Chapter 1. In part (b) we also show real atomic levels, indicated by solid horizontal lines. If one of the real atomic levels is nearly coincident with one of the virtual levels of the indicated process, the

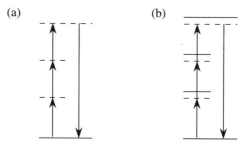

FIGURE 3.1.1   Third-harmonic generation described in terms of virtual levels (a) and with real atomic levels indicated (b).

coupling between the radiation and the atom is particularly strong and the nonlinear optical susceptibility becomes large.

Three possible strategies for enhancing the efficiency of third-harmonic generation through the technique of resonance enhancement are illustrated in Fig. 3.1.2. In part (a), the one-photon transition is nearly resonant, in part (b) the two-photon transition is nearly resonant, and in part (c) the three-photon transition is nearly resonant. The formulas derived later in this chapter demonstrate that all three procedures are equally effective at increasing the value of the third-order nonlinear susceptibility. However, the method shown in part (b) is usually the preferred way in which to generate the third-harmonic field with high efficiency, for the following reason: For the case of a one-photon resonance (part a), the incident field experiences linear absorption and is rapidly attenuated as it propagates through the medium. Similarly, for the case of the three-photon resonance (part c), the generated field experiences linear absorption. However, for the case of a two-photon resonance (part b), there is no linear absorption to limit the efficiency of the process.

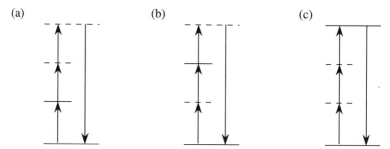

FIGURE 3.1.2   Three strategies for enhancing the process of third-harmonic generation.

3.2. Schrödinger Equation Calculation of the Nonlinear Optical Susceptibility

In this section, we present a derivation of the nonlinear optical susceptibility based on quantum-mechanical perturbation theory of the atomic wave function. The expressions that we derive using this formalism can be used to make accurate predictions of the *nonresonant* response of atomic and molecular systems. Relaxation processes, which are important for the case of near-resonant excitation, cannot be adequately described by this formalism. Relaxation processes are discussed later in this chapter in connection with the density matrix formulation of the theory of the nonlinear optical susceptibility. Even though the density matrix formalism provides results that are more generally valid, the calculation of the nonlinear susceptibility is much more complicated when performed using this method. For this reason, we first present a calculation of the nonlinear susceptibility based on the properties of the atomic wavefunction, since this method is somewhat simpler and for this reason gives a clearer picture of the underlying physics of the nonlinear interaction.

One of the fundamental assumption of quantum mechanics is that all of the properties of the atomic system can be described in terms of the atomic wavefunction $\psi(\mathbf{r}, t)$, which is the solution to the time-dependent Schrödinger equation

$$i\hbar \frac{\partial \psi}{\partial t} = \hat{H}\psi. \tag{3.2.1}$$

Here $\hat{H}$ is the Hamiltonian operator

$$\hat{H} = \hat{H}_0 + \hat{V}(t), \tag{3.2.2}$$

which is written as the sum of the Hamiltonian $\hat{H}_0$ for a free atom and an interaction Hamiltonian, $\hat{V}(t)$, which describes the interaction of the atom with the electromagnetic field. We usually take the interaction Hamiltonian to be of the form

$$\hat{V}(t) = -\hat{\mu} \cdot \tilde{\mathbf{E}}(t), \tag{3.2.3}$$

where $\hat{\mu} = -e\hat{\mathbf{r}}$ is the electric dipole moment operator and $-e$ is the charge of the electron, and where we assume that $\tilde{\mathbf{E}}(t)$ can be represented as a discrete sum of (positive and negative) frequency components as

$$\tilde{\mathbf{E}}(t) = \sum_p \mathbf{E}(\omega_p)e^{-i\omega_p t}. \tag{3.2.4}$$

Energy Eigenstates

For the case in which no external field is applied to the atom, the Hamiltonian $\hat{H}$ is simply equal to $\hat{H}_0$, and Schrödinger's equation (3.2.1) possesses solutions in the form of energy eigenstates. These states are also known as stationary states, because the time of evolution of these states is given by a simple exponential phase factor. These states have the form

$$\psi_n(\mathbf{r}, t) = u_n(\mathbf{r})e^{-i\omega_n t}. \tag{3.2.5a}$$

By substituting this form into the Schrödinger equation (3.2.1), we find that the spatially varying part of the wavefunction $u_n(\mathbf{r})$ must satisfy the eigenvalue equation (known as the time-independent Schrödinger equation)

$$\hat{H}_0 u_n(\mathbf{r}) = E_n u_n(\mathbf{r}), \tag{3.2.5b}$$

where $E_n = \hbar\omega_n$. For future convenience, we assume that these solutions are chosen in such a manner that they constitute a complete, orthonormal set satisfying the condition

$$\int u_m^* u_n d^3 r = \delta_{mn}. \tag{3.2.6}$$

Perturbation Solution to Schrödinger's Equation

For the general case in which the atom is exposed to an electromagnetic field, Schrödinger's equation (3.2.1) usually cannot be solved exactly. In such cases, it is often adequate to solve Schrödinger's equation through the use of perturbation theory. In order to solve Eq. (3.2.1) systematically in terms of a perturbation expansion, we replace the Hamiltonian (3.2.2) by

$$\hat{H} = \hat{H}_0 + \lambda\hat{V}(t), \tag{3.2.7}$$

where λ is a continuously varying parameter ranging from zero to unity that characterizes the strength of the interaction; the value $\lambda = 1$ corresponds to the actual physical situation. We now seek a solution to Schrödinger's equation in the form of a power series in λ:

$$\psi(\mathbf{r}, t) = \psi^{(0)}(\mathbf{r}, t) + \lambda\psi^{(1)}(\mathbf{r}, t) + \lambda^2\psi^{(2)}(\mathbf{r}, t) + \cdots. \tag{3.2.8}$$

By requiring that the solution be of this form for any value of λ, we assure that $\psi^{(N)}$ will be that part of the solution which is of order N in the interaction energy V. We now introduce Eq. (3.2.8) into Eq. (3.2.1) and require that all terms proportional to $\lambda^{(N)}$ satisfy the equality separately. We thereby obtain the set of equations

$$ i\hbar \frac{\partial \psi^{(0)}}{\partial t} = \hat{H}_0 \psi^{(0)}, \tag{3.2.9a}$$

$$ i\hbar \frac{\partial \psi^{(N)}}{\partial t} = \hat{H}_0 \psi^{(N)} + \hat{V} \psi^{(N-1)}, \quad N = 1, 2, 3 \ldots . \tag{3.2.9b}$$

Equation (3.2.9a) is simply Schrödinger's equation for the atom in the absence of its interaction with the applied field; we assume for definiteness that initially the atom is in state g (typically the ground state) so that the solution to this equation is

$$ \psi^{(0)}(\mathbf{r}, t) = u_g(\mathbf{r}) e^{-iE_g t/\hbar}. \tag{3.2.10}$$

The remaining equations in the perturbation expansion (Eq. (3.2.9b)) are solved by making use of the fact that the energy eigenfunctions for the free atom constitute a complete set of basis functions, in terms of which any function can be expanded. In particular, we represent the Nth-order contribution to the wavefunction $\psi^{(N)}(\mathbf{r}, t)$ as the sum

$$ \psi^{(N)}(\mathbf{r}, t) = \sum_l a_l^{(N)}(t) u_l(\mathbf{r}) e^{-i\omega_l t}. \tag{3.2.11}$$

Here $a_l^{(N)}(t)$ gives the probability amplitude that, to Nth order in the perturbation, the atom is in energy eigenstate l at time t. If Eq. (3.2.11) is substituted into Eq. (3.2.9b), we find that the probability amplitudes obey the system of equations

$$ i\hbar \sum_l \dot{a}_l^{(N)} u_l(\mathbf{r}) e^{-i\omega_l t} = \sum_l a_l^{(N-1)} \hat{V} u_l(\mathbf{r}) e^{-i\omega_l t}, \tag{3.2.12}$$

where the dot denotes a total time derivative. This equation relates all of the probability amplitudes of order N to all of the amplitudes of order $N - 1$. To simplify this equation, we multiply each side from the left by u_m^* and we integrate the resulting equation over all space. Then through use of the orthonormality condition (3.2.6), we obtain the equation

$$ \dot{a}_m^{(N)} = (i\hbar)^{-1} \sum_l a_l^{(N-1)} V_{ml} e^{i\omega_{ml} t}, \tag{3.2.13}$$

where $\omega_{ml} \equiv \omega_m - \omega_l$ and where we have introduced the matrix elements of the perturbing Hamiltonian, which are defined by

$$ V_{ml} \equiv \langle u_m | \hat{V} | u_l \rangle = \int u_m^* \hat{V} u_l \, d^3 r. \tag{3.2.14}$$

The form of Eq. (3.2.13) demonstrates the usefulness of the perturbation technique; once the probability amplitudes of order $N - 1$ are determined, the amplitudes of the next higher order (N) can be obtained by straightforward time integration. In particular, we find that

$$a_m^{(N)}(t) = (i\hbar)^{-1} \sum_l \int_{-\infty}^t dt' V_{ml}(t') a_l^{(N-1)}(t') e^{i\omega_{ml}t'} \qquad (3.2.15)$$

We shall eventually be interested in determining the linear, second-order, and third-order optical susceptibilities. To do so, we shall require explicit expressions for the probability amplitudes up to third order in the perturbation expansion. We now determine the form of these amplitudes.

To determine the first-order amplitudes $a_m^{(1)}(t)$, we set $a_l^{(0)}$ in Eq. (3.2.15) equal to δ_{lg} (corresponding to an atom known to be in state g in zeroth order) and, through use of Eqs. (3.2.3) and (3.2.4), replace $V_{ml}(t')$ by $-\sum_p \boldsymbol{\mu}_{ml} \cdot \mathbf{E}(\omega_p) \exp(-i\omega_p t')$, where $\boldsymbol{\mu}_{ml} = \int u_m^* \hat{\boldsymbol{\mu}} u_l d^3 r$ is known as the electric dipole transition moment. We next evaluate the integral appearing in Eq. (3.2.15) and assume that the contribution from the lower limit of integration vanishes; we thereby find that

$$a_m^{(1)}(t) = \frac{1}{\hbar} \sum_p \frac{\boldsymbol{\mu}_{mg} \cdot \mathbf{E}(\omega_p)}{\omega_{mg} - \omega_p} e^{i(\omega_{mg} - \omega_p)t}. \qquad (3.2.16)$$

We next determine the second-order correction to the probability amplitude by using Eq. (3.2.15) once again, but with N set equal to 2. We introduce Eq. (3.2.16) for $a_m^{(1)}$ into the right-hand side of this equation and perform the integration to find that

$$a_n^{(2)}(t) = \frac{1}{\hbar^2} \sum_{pq} \sum_m \frac{[\boldsymbol{\mu}_{nm} \cdot \mathbf{E}(\omega_q)][\boldsymbol{\mu}_{mg} \cdot \mathbf{E}(\omega_p)]}{(\omega_{ng} - \omega_p - \omega_q)(\omega_{mg} - \omega_p)} e^{i(\omega_{ng} - \omega_p - \omega_q)t}. \qquad (3.2.17)$$

Analogously, through an additional use of Eq. (3.2.15), we find that the third-order correction to the probability amplitude is given by

$$a_\nu^{(3)}(t) = \frac{1}{\hbar^3} \sum_{pqr} \sum_{mn} \frac{[\boldsymbol{\mu}_{\nu n} \cdot \mathbf{E}(\omega_r)][\boldsymbol{\mu}_{nm} \cdot \mathbf{E}(\omega_q)][\boldsymbol{\mu}_{mg} \cdot \mathbf{E}(\omega_p)]}{(\omega_{\nu g} - \omega_p - \omega_q - \omega_r)(\omega_{ng} - \omega_p - \omega_q)(\omega_{mg} - \omega_p)}$$

$$\times e^{i(\omega_{\nu g} - \omega_p - \omega_q - \omega_r)t}. \qquad (3.2.18)$$

Linear Susceptibility

Let us now use the results just obtained to determine the linear optical properties of a material system. The expectation value of the electric dipole moment is given by

$$\langle \tilde{\mathbf{p}} \rangle = \langle \psi | \hat{\boldsymbol{\mu}} | \psi \rangle, \qquad (3.2.19)$$

We find that the lowest-order contribution to $\langle \tilde{\mathbf{p}} \rangle$ (i.e., the contribution linear in the applied field amplitude) is given by

$$\langle \tilde{\mathbf{p}}^{(1)} \rangle = \langle \psi^{(0)} | \hat{\boldsymbol{\mu}} | \psi^{(1)} \rangle + \langle \psi^{(1)} | \hat{\boldsymbol{\mu}} | \psi^{(0)} \rangle, \qquad (3.2.20)$$

where $\psi^{(0)}$ is given by Eq. (3.2.10) and $\psi^{(1)}$ is given by Eqs. (3.2.11) and (3.2.16). By substituting these forms into Eq. (3.2.20) we find that

$$\langle \tilde{\mathbf{p}}^{(1)} \rangle = \frac{1}{\hbar} \sum_p \sum_m \left(\frac{\boldsymbol{\mu}_{gm}[\boldsymbol{\mu}_{mg} \cdot \mathbf{E}(\omega_p)]}{\omega_{mg} - \omega_p} e^{-i\omega_p t} + \frac{[\boldsymbol{\mu}_{mg} \cdot \mathbf{E}(\omega_p)]^* \boldsymbol{\mu}_{mg}}{\omega_{mg}^* - \omega_p} e^{i\omega_p t} \right).$$
$$(3.2.21)$$

In writing Eq. (3.2.21) in the form shown, we have formally allowed the possibility that the transition frequency ω_{mg} is a complex quantity. We have done this because a crude way of incorporating damping phenomena into the theory is to take ω_{mg} to be the complex quantity $\omega_{mg} = \omega_{mg}^0 - i\Gamma_m/2$, where ω_{mg}^0 is the (real) transition frequency and Γ_m is the population decay rate of the upper level m. This procedure is not totally acceptable, because it cannot describe the cascade of population among the excited states nor can it describe dephasing processes that are not accompanied by the transfer of population. Nonetheless, for the remainder of the present section, we shall allow the transition frequency to be a complex quantity in order to provide an indication of how damping effects could be incorporated into the present theory.

Equation (3.2.21) is written as a summation over all positive and negative field frequencies ω_p. This result is easier to interpret if we formally replace ω_p by $-\omega_p$ in the second term, in which case the expression becomes

$$\langle \tilde{\mathbf{p}}^{(1)} \rangle = \frac{1}{\hbar} \sum_p \sum_m \left(\frac{\boldsymbol{\mu}_{gm}[\boldsymbol{\mu}_{mg} \cdot \mathbf{E}(\omega_p)]}{\omega_{mg} - \omega_p} + \frac{[\boldsymbol{\mu}_{gm} \cdot \mathbf{E}(\omega_p)]\boldsymbol{\mu}_{mg}}{\omega_{mg}^* + \omega_p} \right) e^{-i\omega_p t}.$$
$$(3.2.22)$$

We next use this result to calculate the form of the linear susceptibilty. We take the linear polarization to be $\tilde{\mathbf{P}}^{(1)} = N \langle \tilde{\mathbf{p}}^{(1)} \rangle$, where N is the number density of atoms. We next express the polarization in terms of its complex amplitude as $\tilde{\mathbf{P}}^{(1)} = \sum_p \mathbf{P}^{(1)}(\omega_p) \exp(-i\omega_p t)$. Finally, we introduce the linear susceptibility defined through the relation $P_i^{(1)}(\omega_p) = \sum_j \chi_{ij}^{(1)} E_j(\omega_p)$. We thereby find that

$$\chi_{ij}^{(1)}(\omega_p) = \frac{N}{\hbar} \sum_m \left(\frac{\mu_{gm}^i \mu_{mg}^j}{\omega_{mg} - \omega_p} + \frac{\mu_{gm}^j \mu_{mg}^i}{\omega_{mg}^* + \omega_p} \right). \qquad (3.2.23)$$

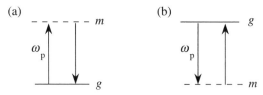

FIGURE 3.2.1   The resonant (a) and antiresonant (b) contributions to the linear susceptibility of Eq. (3.2.23).

The first and second terms in Eq. (3.2.23) can be interpreted as the resonant and antiresonant contributions to the susceptibility, as illustrated in Fig. 3.2.1. In this figure we have indicated where level m would have to be located in order for the corresponding term to become resonant. Note that if g denotes the ground state, it is impossible for the second term to become resonant, which is why it is called the antiresonant contribution.

Second-Order Susceptibility

The expression for the second-order susceptibility is derived in a manner analogous to that used for the linear susceptibility. The second-order contribution (i.e., the contribution second order in $\hat{V}$) to the induced dipole moment per atom is given by

$$\langle \tilde{\mathbf{p}}^{(2)} \rangle = \langle \psi^{(0)}|\hat{\boldsymbol{\mu}}|\psi^{(2)} \rangle + \langle \psi^{(1)}|\hat{\boldsymbol{\mu}}|\psi^{(1)} \rangle + \langle \psi^{(2)}|\hat{\boldsymbol{\mu}}|\psi^{(0)} \rangle, \qquad (3.2.24)$$

where $\psi^{(0)}$ is given by Eq. (3.2.10), and $\psi^{(1)}$ and $\psi^{(2)}$ are given by Eqs. (3.2.11), (3.2.16), and (3.2.17). We find that $\langle \tilde{\mathbf{p}}^{(2)} \rangle$ is given explicitly by

$$\langle \tilde{\mathbf{p}}^{(2)} \rangle = \frac{1}{\hbar^2} \sum_{pq} \sum_{mn} \left(\frac{\mu_{gn}[\mu_{nm} \cdot \mathbf{E}(\omega_q)][\mu_{mg} \cdot \mathbf{E}(\omega_p)]}{(\omega_{ng} - \omega_p - \omega_q)(\omega_{mg} - \omega_p)} e^{-i(\omega_p + \omega_q)t} \right.$$

$$+ \frac{[\mu_{ng} \cdot \mathbf{E}(\omega_q)]^* \mu_{nm}[\mu_{mg} \cdot \mathbf{E}(\omega_q)]}{(\omega_{ng}^* - \omega_q)(\omega_{mg} - \omega_p)} e^{-i(\omega_p - \omega_q)t} \qquad (3.2.25)$$

$$+ \left. \frac{[\mu_{ng} \cdot \mathbf{E}(\omega_q)]^*[\mu_{nm} \cdot \mathbf{E}(\omega_p)]^* \mu_{mg}}{(\omega_{ng}^* - \omega_q)(\omega_{mg}^* - \omega_q - \omega_q)} e^{i(\omega_p + \omega_q)t} \right).$$

As in the case of the linear susceptibility, this equation can be rendered more transparent by replacing ω_q by $-\omega_q$ in the second term and by replacing ω_q by $-\omega_q$ and ω_p by $-\omega_p$ in the third term; these substitutions are permissible because the expression is to be summed over frequencies ω_p and ω_q.

We thereby obtain the result

$$
\langle \tilde{\mathbf{p}}^{(2)} \rangle = \frac{1}{\hbar^2} \sum_{pq} \sum_{mn} \left(\frac{\mu_{gn}[\mu_{nm} \cdot \mathbf{E}(\omega_q)][\mu_{mg} \cdot \mathbf{E}(\omega_p)]}{(\omega_{ng} - \omega_p - \omega_q)(\omega_{mg} - \omega_p)} \right.
$$
$$
+ \frac{[\mu_{gn} \cdot \mathbf{E}(\omega_q)]\mu_{nm}[\mu_{mg} \cdot \mathbf{E}(\omega_p)]}{(\omega_{ng}^* + \omega_q)(\omega_{mg} - \omega_p)} \tag{3.2.26}
$$
$$
\left. + \frac{[\mu_{gn} \cdot \mathbf{E}(\omega_q)][\mu_{nm} \cdot \mathbf{E}(\omega_p)]\mu_{mg}}{(\omega_{ng}^* + \omega_q)(\omega_{mg}^* + \omega_q + \omega_q)} \right) e^{-i(\omega_p + \omega_q t)}.
$$

We next take the second-order polarization to be $\tilde{\mathbf{P}}^{(2)} = N\langle \tilde{\mathbf{p}}^{(2)} \rangle$ and represent it in terms of its frequency components as $\tilde{\mathbf{P}}^{(2)} = \sum_r \mathbf{P}^{(2)}(\omega_r)\exp(-i\omega_r t)$. We also introduce the standard definition of the second-order susceptibility (see also Eq. (1.3.13)):

$$
P_i^{(2)} = \sum_{jk} \sum_{(pq)} \chi_{ijk}^{(2)}(\omega_p + \omega_q, \omega_q, \omega_p) E_j(\omega_q) E_k(\omega_p)
$$

and find that the second-order susceptibility is given by

$$
\chi_{ijk}^{(2)}(\omega_p + \omega_q, \omega_q, \omega_p) = \frac{N}{\hbar^2} \mathscr{P}_I \sum_{mn} \left(\frac{\mu_{gn}^i \mu_{nm}^j \mu_{mg}^k}{(\omega_{ng} - \omega_p - \omega_q)(\omega_{mg} - \omega_p)} \right.
$$
$$
+ \frac{\mu_{gn}^j \mu_{nm}^i \mu_{mg}^k}{(\omega_{ng}^* + \omega_q)(\omega_{mg} - \omega_p)} \tag{3.2.27}
$$
$$
\left. + \frac{\mu_{gn}^j \mu_{nm}^k \mu_{mg}^i}{(\omega_{ng}^* + \omega_q)(\omega_{mg}^* + \omega_p + \omega_q)} \right).
$$

In this expression, the symbol $\mathscr{P}_I$ denotes the intrinsic permutation operator. This operator tells us to average the expression that follows it over both permutations of the frequencies ω_p and ω_q of the applied fields. The cartesian indices j and k are to be permuted simultaneously. We introduce the intrinsic permutation operator into Eq. (3.2.27) to ensure that the resulting expression obeys the condition of intrinsic permutation symmetry, as described in the discussion of Eqs. (1.4.52) and (1.5.6). The nature of the expression (3.2.27) for the second-order susceptibility can be understood in terms of the energy level diagrams shown in Fig. 3.2.2, which show where the levels m and n would have to be located in order for each term in the expression to become resonant.

The quantum-mechanical expression for the second-order susceptibility actually comprises six terms; through use of the intrinsic permutation operator $\mathscr{P}_I$, we have been able to express the susceptibility in the form (3.2.27), in which only three terms are displayed explicitly. For the case of highly

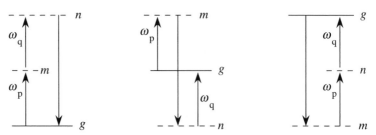

FIGURE 3.2.2   Resonant structure of the three terms of the second-order susceptibility of Eq. (3.2.27).

nonresonant excitation, such that the resonance frequencies ω_{mg} and ω_{ng} can be taken to be real quantities, the expression for $\chi^{(2)}$ can be simplified still further. In particular, under such circumstances Eq. (3.2.27) can be expressed as

$$\chi^{(2)}_{ijk}(\omega_\sigma, \omega_q, \omega_p) = \frac{N}{\hbar^2} \mathscr{P}_F \sum_{mn} \frac{\mu^i_{gn} \mu^j_{nm} \mu^k_{mg}}{(\omega_{ng} - \omega_\sigma)(\omega_{mg} - \omega_p)} \qquad (3.2.28)$$

where $\omega_\sigma = \omega_p + \omega_q$. Here we have introduced the full permutation operator, $\mathscr{P}_F$ defined such that the expression that follows it is to be summed over all permutations of the frequencies ω_p, ω_q, and $-\omega_\sigma$, that is, over all input and output frequencies. The cartesian indices are to be permuted along with the frequencies. The final result is then to be divided by the number of permutations of the input frequencies. The equivalence of Eqs. (3.2.27) and (3.2.28) can be verified by explicitly expanding the right-hand side of each equation into all six terms. The six permutations denoted by the operator $\mathscr{P}_F$ are

$$(-\omega_\sigma, \omega_q, \omega_p) \rightarrow (-\omega_\sigma, \omega_p, \omega_q), (\omega_q, -\omega_\sigma, \omega_p), (\omega_q, \omega_p, -\omega_\sigma),$$

$$(\omega_p, -\omega_\sigma, \omega_q), (\omega_p, \omega_q, -\omega_\sigma).$$

Since we can express the nonlinear susceptibility in the form of Eq. (3.2.28), we have proven the statement made in Section 1.5 that the nonlinear susceptibility of a lossless medium possesses full permutation symmetry.

Third-Order Susceptibility

We now calculate the third-order susceptibility. The dipole moment per atom, correct to third order in perturbation theory, is given by

$$\langle \tilde{\mathbf{p}}^{(3)} \rangle = \langle \psi^{(0)} | \hat{\boldsymbol{\mu}} | \psi^{(3)} \rangle + \langle \psi^{(1)} | \hat{\boldsymbol{\mu}} | \psi^{(2)} \rangle + \langle \psi^{(2)} | \hat{\boldsymbol{\mu}} | \psi^{(1)} \rangle + \langle \psi^{(3)} | \hat{\boldsymbol{\mu}} | \psi^{(0)} \rangle.$$

$$(3.2.29)$$

Formulas for $\psi^{(0)}, \psi^{(1)}, \psi^{(2)}, \psi^{(3)}$, are given by Eqs. (3.2.10), (3.2.11), (3.2.16), (3.2.17), and (3.2.18). We thus find that

$$\langle \tilde{\mathbf{p}}^{(3)} \rangle = \frac{1}{\hbar^3} \sum_{pqr} \sum_{mnv}$$

$$\times \left(\frac{\boldsymbol{\mu}_{gv}[\boldsymbol{\mu}_{vn} \cdot \mathbf{E}(\omega_r)][\boldsymbol{\mu}_{nm} \cdot \mathbf{E}(\omega_q)][\boldsymbol{\mu}_{mg} \cdot \mathbf{E}(\omega_p)]}{(\omega_{vg} - \omega_r - \omega_q - \omega_p)(\omega_{ng} - \omega_q - \omega_p)(\omega_{mg} - \omega_p)} \right.$$

$$\times e^{-i(\omega_p + \omega_q + \omega_r)t}$$

$$+ \frac{[\boldsymbol{\mu}_{vg} \cdot \mathbf{E}(\omega_r)]^* \boldsymbol{\mu}_{vn}[\boldsymbol{\mu}_{nm} \cdot \mathbf{E}(\omega_q)][\boldsymbol{\mu}_{mg} \cdot \mathbf{E}(\omega_p)]}{(\omega_{vg}^* - \omega_r)(\omega_{ng} - \omega_q - \omega_p)(\omega_{mg} - \omega_p)}$$

$$\times e^{-i(\omega_p + \omega_q - \omega_r)t} \tag{3.2.30}$$

$$+ \frac{[\boldsymbol{\mu}_{vg} \cdot \mathbf{E}(\omega_r)]^*[\boldsymbol{\mu}_{nv} \cdot \mathbf{E}(\omega_q)]^* \boldsymbol{\mu}_{nm}[\boldsymbol{\mu}_{mg} \cdot \mathbf{E}(\omega_p)]}{(\omega_{vg}^* - \omega_r)(\omega_{ng}^* - \omega_r - \omega_q)(\omega_{mg} - \omega_p)}$$

$$\times e^{-i(\omega_p - \omega_q - \omega_r)t}$$

$$+ \frac{[\boldsymbol{\mu}_{vg} \cdot \mathbf{E}(\omega_r)]^*[\boldsymbol{\mu}_{nv} \cdot \mathbf{E}(\omega_q)]^*[\boldsymbol{\mu}_{mn} \cdot \mathbf{E}(\omega_p)]^* \boldsymbol{\mu}_{mg}}{(\omega_{vg}^* - \omega_r)(\omega_{ng}^* - \omega_r - \omega_q)(\omega_{mg}^* - \omega_r - \omega_q - \omega_p)}$$

$$\left. \times e^{+i(\omega_p + \omega_q + \omega_r)t} \right).$$

Since the expression is summed over all positive and negative values of ω_p, ω_q, and ω_r, we can replace these quantities by their negatives in those expressions where the complex conjugate of a field amplitude appears. We thereby obtain the expression

$$\langle \tilde{\mathbf{p}}^{(3)} \rangle = \frac{1}{\hbar^3} \sum_{pqr} \sum_{mnv}$$

$$\times \left(\frac{\boldsymbol{\mu}_{gv}[\boldsymbol{\mu}_{vn} \cdot \mathbf{E}(\omega_r)][\boldsymbol{\mu}_{nm} \cdot \mathbf{E}(\omega_q)][\boldsymbol{\mu}_{mg} \cdot \mathbf{E}(\omega_p)]}{(\omega_{vg} - \omega_r - \omega_q - \omega_p)(\omega_{ng} - \omega_q - \omega_p)(\omega_{mg} - \omega_p)} \right.$$

$$+ \frac{[\boldsymbol{\mu}_{gv} \cdot \mathbf{E}(\omega_r)]\boldsymbol{\mu}_{vn}[\boldsymbol{\mu}_{nm} \cdot \mathbf{E}(\omega_q)][\boldsymbol{\mu}_{mg} \cdot \mathbf{E}(\omega_p)]}{(\omega_{vg}^* + \omega_r)(\omega_{ng} - \omega_q - \omega_p)(\omega_{mg} - \omega_p)}$$

$$+ \frac{[\boldsymbol{\mu}_{gv} \cdot \mathbf{E}(\omega_r)][\boldsymbol{\mu}_{vn} \cdot \mathbf{E}(\omega_q)]\boldsymbol{\mu}_{nm}[\boldsymbol{\mu}_{mg} \cdot \mathbf{E}(\omega_p)]}{(\omega_{vg}^* + \omega_r)(\omega_{ng}^* + \omega_r + \omega_q)(\omega_{mg} - \omega_p)} \tag{3.2.31}$$

$$+ \frac{[\boldsymbol{\mu}_{gv} \cdot \mathbf{E}(\omega_r)][\boldsymbol{\mu}_{vn} \cdot \mathbf{E}(\omega_q)][\boldsymbol{\mu}_{nm} \cdot \mathbf{E}(\omega_p)]\boldsymbol{\mu}_{mg}}{(\omega_{vg}^* + \omega_r)(\omega_{ng}^* + \omega_r + \omega_q)(\omega_{mg}^* + \omega_r + \omega_q + \omega_p)} \right)$$

$$\times e^{-i(\omega_p + \omega_q + \omega_r)t}.$$

We now use this result to calculate the third-order susceptibility: We let $\tilde{\mathbf{P}}^{(3)} = N\langle\tilde{\mathbf{p}}^{(3)}\rangle = \sum_s \mathbf{P}^{(3)}(\omega_s)\exp(-i\omega_s t)$ and introduce the definition (1.3.21) of the third-order susceptibility:

$$P_k(\omega_p + \omega_q + \omega_r) = \sum_{hij}\sum_{(pqr)} \chi^{(3)}_{kjih}(\omega_\sigma, \omega_r, \omega_q, \omega_p) E_j(\omega_r) E_i(\omega_q) E_h(\omega_p).$$

We thereby obtain the result

$$\chi^{(3)}_{kjih}(\omega_\sigma, \omega_r, \omega_q, \omega_p)$$

$$= \frac{N}{\hbar^3}\mathscr{P}_I \sum_{mnv}\left[\frac{\mu^k_{gv}\mu^j_{vn}\mu^i_{nm}\mu^h_{mg}}{(\omega_{vg} - \omega_r - \omega_q - \omega_p)(\omega_{ng} - \omega_q - \omega_p)(\omega_{mg} - \omega_p)} \right.$$

$$+ \frac{\mu^j_{gv}\mu^k_{vn}\mu^i_{nm}\mu^h_{mg}}{(\omega^*_{vg} + \omega_r)(\omega_{ng} - \omega_q - \omega_p)(\omega_{mg} - \omega_p)} \qquad (3.2.32)$$

$$+ \frac{\mu^j_{gv}\mu^i_{vn}\mu^k_{nm}\mu^h_{mg}}{(\omega^*_{vg} + \omega_r)(\omega^*_{ng} + \omega_r + \omega_q)(\omega_{mg} - \omega_p)}$$

$$\left. + \frac{\mu^j_{gv}\mu^i_{vn}\mu^h_{nm}\mu^k_{mg}}{(\omega^*_{vg} + \omega_r)(\omega^*_{ng} + \omega_r + \omega_q)(\omega^*_{mg} + \omega_r + \omega_q + \omega_p)} \right].$$

Here we have again made use of the intrinsic permutation operator $\mathscr{P}_I$ defined following Eq. (3.2.27). The complete expression for the third-order susceptibility actually contains 24 terms, of which only four are displayed explicitly in Eq. (3.2.33); the others can be obtained through permutations of the frequencies (and cartesian indices) of the applied fields. The locations of the resonances in the displayed terms of this expression are illustrated in Fig. 3.2.3.

As in the case of the second-order susceptibility, the expression for $\chi^{(3)}$ can be written very compactly for the case of highly nonresonant excitation such that the imaginary parts of the resonant frequencies (recall that $\omega_{lg} =$

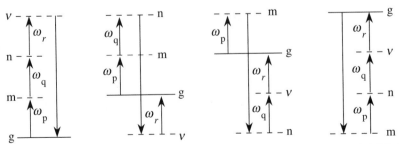

FIGURE 3.2.3   Locations of the resonances of each term in the expression (3.2.32) for the third-order susceptibility.

$\omega_{lg}^0 - i\,\Gamma_l/2)$ can be ignored. In this case, the expression for $\chi^{(3)}$ can be written as

$$\chi_{kjih}^{(3)}(\omega_\sigma, \omega_r, \omega_q, \omega_p)$$

$$= \frac{N}{\hbar^3}\mathscr{P}_F \sum_{mnv} \frac{\mu_{gv}^k \mu_{vn}^j \mu_{nm}^i \mu_{mg}^h}{(\omega_{vg} - \omega_\sigma)(\omega_{ng} - \omega_q - \omega_p)(\omega_{mg} - \omega_p)},$$

(3.2.33)

where $\omega_\sigma = \omega_p + \omega_q + \omega_r$ and where we have made use of the full permutation operator $\mathscr{P}_F$ defined following Eq. (3.2.28).

Third-Harmonic Generation in Alkali Metal Vapors

As an example of the use of Eq. (3.2.33), we next calculate the nonlinear optical susceptibility describing third-harmonic generation in a vapor of sodium atoms. Except for minor changes in notation, our treatment is similar to the original treatment of Miles and Harris (1973). We assume that the incident radiation is linearly polarized in the z direction. Consequently, the nonlinear polarization will have only a z component, and we can suppress the tensor nature of the nonlinear interaction. If we represent the applied field as

$$\tilde{E}(\mathbf{r}, t) = E_1(\mathbf{r})e^{-i\omega t} + \text{c.c.},$$

(3.2.34)

we find that the nonlinear polarization can be represented as

$$\tilde{P}(\mathbf{r}, t) = P_3(\mathbf{r})e^{-i3\omega t} + \text{c.c.},$$

(3.2.35)

where

$$P_3(\mathbf{r}) = \chi^{(3)}(3\omega)E_1^3$$

(3.2.36)

(with $\chi^{(3)}(3\omega) \equiv \chi^{(3)}(3\omega = \omega + \omega + \omega)$) and where the nonlinear susceptibility describing third-harmonic generation is given, ignoring damping effects, by

$$\chi^{(3)}(3\omega) = \frac{N}{\hbar^3}\sum_{mnv}\mu_{gv}\mu_{vn}\mu_{nm}\mu_{mg}$$

$$\times \left[\frac{1}{(\omega_{vg} - 3\omega)(\omega_{ng} - 2\omega)(\omega_{mg} - \omega)} \right.$$

$$+ \frac{1}{(\omega_{vg} + \omega)(\omega_{ng} - 2\omega)(\omega_{mg} - \omega)}$$

(3.2.37)

$$+ \frac{1}{(\omega_{vg} + \omega)(\omega_{ng} + 2\omega)(\omega_{mg} - \omega)}$$

$$\left. + \frac{1}{(\omega_{vg} + \omega)(\omega_{ng} + 2\omega)(\omega_{mg} + 3\omega)} \right].$$

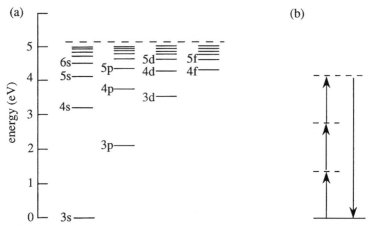

FIGURE 3.2.4   (a) Energy-level diagram of the sodium atom. (b) The third-harmonic generation process.

Equation (3.2.37) can be readily evaluated through use of the known energy level structure and dipole transition moments of the sodium atom. Figure 3.2.4 shows an energy level diagram of the low-lying states of the sodium atom and a photon energy level diagram describing the process of third-harmonic generation. We see that only the first contribution to Eq. (3.2.37) can become fully resonant. This term becomes fully resonant when ω is nearly equal to ω_{mg}, 2ω is nearly equal to ω_{ng}, and 3ω is nearly equal to ω_{vg}. In performing the summation over excited levels m, n, and v, the only levels that contribute are those that obey the selection rules for electric dipole transitions. In particular, since the ground state is an s state, the matrix element μ_{mg} will be nonzero only if m denotes a p state. Similarly, since m denotes a p state, the matrix element μ_{nm} will be nonzero only if n denotes an s state. In either case, v must denote a p state, since only in this case can both μ_{vn} and μ_{gv} be nonzero. The two types of coupling schemes that contribute to $\chi^{(3)}$ are shown in Fig. 3.2.5.

Through use of tabulated values of the matrix elements for the sodium atom, Miles and Harris have calculated numerically the value of $\chi^{(3)}$ as a function of the frequency ω of the incident laser field. The results of this calculation are shown in Fig. 3.2.6. A number of strong resonances in the nonlinear suscepti-bility are evident. Each such resonance is labeled by the quantum number of the level and the type of resonance that leads to the resonance enhancement. The peak labeled $3p(3\omega)$, for example, is due to a three-photon resonance with the $3p$ level of sodium. Miles and Harris also presented experimental results that confirm predictions of their theory. Because atomic vapors are

(a)   (b)

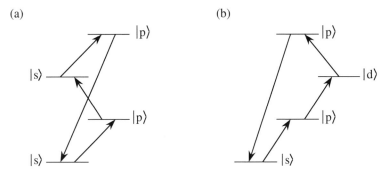

FIGURE 3.2.5   Two coupling schemes that contribute to the third-order susceptibility.

centrosymmetric, they cannot produce a second-order response. Nonetheless, the present of a static electric field can break the inversion symmetry of the material medium, allowing processes such as sum-frequency generation to occur. These effects can be particularly large if the optical fields excite the high-lying Rydberg levels of an atomic system. The details of this process have

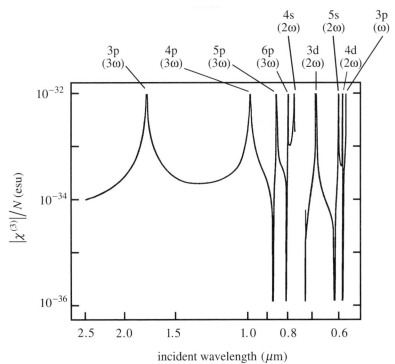

FIGURE 3.2.6   The nonlinear susceptibility describing third-harmonic generation in atomic sodium vapor plotted versus the wavelength of the fundamental radiation (after Miles and Harris, 1973).

been described theoretically by Boyd and Xiang (1982), with experimental confirmation presented by Gauthier *et al.* (1983) and Boyd *et al.* (1984).

3.3. Density Matrix Formalism of Quantum Mechanics

In the present section through Section 3.7, we calculate the nonlinear optical susceptibility through use of the density matrix formulation of quantum mechanics. We use this formalism because it is capable of treating effects, such as collisional broadening of the atomic resonances, that cannot be treated by the simple theoretical formalism based on the atomic wave function. It is important that the formalism be capable of treating collisional effects for the following reasons: We saw in the last section that the nonlinear response will be particularly large when the incident laser frequency or one of its harmonics (or, more generally, when the sum or difference of the applied field frequencies) is tuned close to one of the resonance frequencies of the atomic system. For those cases in which the detuning from resonance is comparable to or less than the width of the resonance, it is necessary that the theory include a treatment of the physical processes that produce the line broadening.

Let us begin by reviewing how the density matrix formalism follows from the basic laws of quantum mechanics.[*] If a quantum-mechanical system (such as an atom) is known to be in a particular quantum-mechanical state which we designate s, we can describe all of the physical properties of the system in terms of the wavefunction $\psi_s(\mathbf{r}, t)$ appropriate to this state. This wavefunction obeys the Schrödinger equation

$$i\hbar \frac{\partial \psi_s(\mathbf{r}, t)}{\partial t} = \hat{H}\psi_s(\mathbf{r}, t). \tag{3.3.1}$$

where $\hat{H}$ denotes the Hamiltonian operator of the system. We assume that $\hat{H}$ can be represented as

$$\hat{H} = \hat{H}_0 + \hat{V}(t), \tag{3.3.2}$$

where $\hat{H}_0$ is the Hamiltonian for a free atom and $\hat{V}(t)$ represents the interaction energy. In order to determine how the wavefunction evolves in time, it is often helpful to make explicit use of the fact that the energy eigenstates of the free-atom Hamiltonian $\hat{H}_0$ form a complete set of basis functions. We can hence

---

[*] The reader who is already familiar with the density matrix formalism can skip directly to Section 3.4.

represent the wavefunction of state s as

$$\psi_s(\mathbf{r}, t) = \sum_n C_n^s(t) u_n(\mathbf{r}), \tag{3.3.3}$$

where, as in Section 3.2, the functions $u_n(\mathbf{r})$ are the energy eigensolutions to the time-independent Schrödinger equation

$$\hat{H}_0 u_n(\mathbf{r}) = E_n u_n(\mathbf{r}) \tag{3.3.4}$$

and are assumed to be orthonormal in that they obey the relation

$$\int u_m^*(\mathbf{r}) u_n(\mathbf{r}) d^3 r = \delta_{mn}. \tag{3.3.5}$$

The expansion coefficient $C_n^s(t)$ gives the probability amplitude that the atom, which is known to be in state s, is in energy eigenstate n at time t. The time evolution of $\psi_s(\mathbf{r}, t)$ can be specified in terms of the time evolution of each of the expansion coefficient $C_n^s(t)$. To determine how these coefficients evolve in time, we introduce the expansion equation (3.3.3) into Schrödinger's equation (3.3.1) to obtain

$$i\hbar \sum_n \frac{dC_n^s(t)}{dt} u_n(\mathbf{r}) = \sum_n C_n^s(t) \hat{H} u_n(\mathbf{r}). \tag{3.3.6}$$

Each side of this equation involves a summation over all of the energy eigenstates of the system. In order to simplify this equation, we multiply each side from the left by $u_m^*(\mathbf{r})$ and integrate over all space. The summation on the left-hand side of the resulting equation reduces to a single term through use of the orthogonality condition of Eq. (3.3.5). The right-hand side is simplified by introducing the matrix elements of the Hamiltonian operator $\hat{H}$, defined through

$$H_{mn} = \int u_m^*(\mathbf{r}) \hat{H} u_n(\mathbf{r}) d^3 r. \tag{3.3.7}$$

We thereby obtain the result

$$i\hbar \frac{d}{dt} C_m^s(t) = \sum_n H_{mn} C_n^s(t). \tag{3.3.8}$$

This equation is entirely equivalent to the Schrödinger equation (3.3.1), but is written in terms of the probability amplitudes $C_n^s(t)$.

The expectation value of any observable quantity can be calculated in terms of the wavefunction of the system. A basic postulate of quantum mechanics states that any observable quantity A is associated with a Hermitian operator $\hat{A}$. The expectation value of A is then obtained according to the prescription

$$\langle A \rangle = \int \psi_s^* \hat{A} \psi_s d^3 r. \qquad (3.3.9)$$

Here the angular brackets denote a quantum-mechanical average. This relationship is conveniently written in Dirac notation as

$$\langle A \rangle = \langle \psi_s | \hat{A} | \psi_s \rangle = \langle s | \hat{A} | s \rangle. \qquad (3.3.10)$$

where we shall alternatively use $| \psi_s \rangle$ or $| s \rangle$ to denote the state s. The expectation value $\langle A \rangle$ can also be expressed in terms of the probability amplitudes $C_n^s(t)$ by introducing Eq. (3.3.3) into Eq. (3.3.9) to obtain

$$\langle A \rangle = \sum_{mn} C_m^{s*} C_n^s A_{mn}, \qquad (3.3.11)$$

where we have introduced the matrix elements A_{mn} of the operator $\hat{A}$, defined through

$$A_{mn} = \langle u_m | \hat{A} | u_n \rangle = \int u_m^* \hat{A} u_n d^3 r. \qquad (3.3.12)$$

As long as the initial state of the system and the Hamiltonian operator $\hat{H}$ for the system are known, the formalism described above by Eqs. (3.3.1) through (3.3.12) is capable of providing a complete description of the time evolution of the system and of all of its observable properties. However, there are circumstances under which the state of the system is not known in a precise manner. An example is a collection of atoms in an atomic vapor, where the atoms can interact with one another by means of collisions. Each time a collision occurs, the wave function of each interacting atom is modified. If the collisions are sufficiently weak, the modification may involve only an overall change in the phase of the wave function. However, since it is computationally infeasible to keep track of the phase of each atom within the atomic vapor, from a practical point of view the state of each atom is not known.

Under such circumstances, where the precise state of the system is unknown, the density matrix formalism can be used to describe the system in a statistical sense. Let us denote by $p(s)$ the probability that the system is in the state s. The quantity $p(s)$ is to be understood as a classical rather than a quantum-mechanical probability. Hence $p(s)$ simply reflects our lack of knowledge of the actual quantum-mechanical state of the system; it is not a consequence of any sort of quantum-mechanical uncertainty relation. In terms of $p(s)$, we define the elements of the density matrix of the system by

$$\rho_{nm} = \sum_s p(s) C_m^{s*} C_n^s. \tag{3.3.13}$$

This relation can also be written symbolically as

$$\rho_{nm} = \overline{C_m^* C_n}, \tag{3.3.14}$$

where the overbar denotes an ensemble average, that is, an average over all of the possible states of the system. In either form, the indices n and m are understood to run over all of the energy eigenstates of the system.

The elements of the density matrix have the following physical interpretation: The diagonal elements ρ_{nn} give the probability that the system is in energy eigenstate n. The off-diagonal elements have a somewhat more abstract interpretation: ρ_{nm} gives the "coherence" between levels n and m, in the sense that ρ_{nm} will be nonzero only if the system is in a coherent superposition of energy eigenstate n and m. We show below that the off-diagonal elements of the density matrix are, in certain circumstances, proportional to the induced electric dipole moment of the atom.

The density matrix is useful because it can be used to calculate the expectation value of any observable quantity. Since the expectation value of an observable quantity A for a system known to be in the quantum state s is given according to Eq. (3.3.11) by $\langle A \rangle = \sum_{mn} C_m^{s*} C_n^s A_{mn}$, the expectation value for the case in which the exact state of the system is not known is obtained by averaging Eq. (3.3.11) over all possible states of the system, to yield

$$\overline{\langle A \rangle} = \sum_s p(s) \sum_{nm} C_m^{s*} C_n^s A_{mn}. \tag{3.3.15}$$

The notation used on the left-hand side of this equation means that we are calculating the ensemble average of the quantum-mechanical expectation value of the observable quantity A.[*] Through use of Eq. (3.3.13), this quantity can alternatively be expressed as

$$\overline{\langle A \rangle} = \sum_{nm} \rho_{nm} A_{mn}. \tag{3.3.16}$$

The double summation in the equation can be simplified as follows:

$$\sum_{nm} \rho_{nm} A_{mn} = \sum_n \left(\sum_m \rho_{nm} A_{mn} \right) = \sum_n (\hat{\rho}\hat{A})_{nn} \equiv \mathrm{tr}(\hat{\rho}\hat{A}),$$

---

[*] In later sections of this chapter, we shall follow conventional notation and omit the overbar from expressions such as $\overline{\langle A \rangle}$, allowing the angular brackets to denote both a quantum and a classical average.

where we have introduced the trace operation, which is defined for any operator $\hat{M}$ by tr $\hat{M} = \sum_n M_{nn}$. The expectation value of A is hence given by

$$\overline{\langle A \rangle} = \text{tr}(\hat{\rho}\hat{A}). \tag{3.3.17}$$

The notation used in these equations is that $\hat{\rho}$ denotes the density operator, whose n, m matrix component is denoted ρ_{nm}; $\hat{\rho}\hat{A}$ denotes the product of $\hat{\rho}$ with the operator $\hat{A}$; and $(\hat{\rho}\hat{A})_{nn}$ denotes the n, n component of the matrix representation of this product.

We have just seen that the expectation value of any observable quantity can be determined straightforwardly in terms of the density matrix. In order to determine how any expectation value evolves in time, it is thus necessary only to determine how the density matrix itself evolves in time. By direct time differentiation of Eq. (3.3.13), we obtain

$$\dot{\rho}_{nm} = \sum_s \frac{dp(s)}{dt} C_m^{s*} C_n^s + \sum_s p(s) \left(C_m^{s*} \frac{dC_n^s}{dt} + \frac{dC_m^{s*}}{dt} C_n^s \right). \tag{3.3.18}$$

For the present, let us assume that $p(s)$ does not vary in time, so that the first term in this expression vanishes. We can then evaluate the second term straightforwardly by using Schrödinger's equation for the time evolution of the probability amplitudes equation (3.3.8). From this equation we obtain the expressions

$$C_m^{s*} \frac{dC_n^s}{dt} = \frac{-i}{\hbar} C_m^{s*} \sum_\nu H_{n\nu} C_\nu^s,$$

$$C_n^s \frac{dC_m^{s*}}{dt} = \frac{i}{\hbar} C_n^s \sum_\nu H_{m\nu}^* C_\nu^{s*} = \frac{i}{\hbar} C_n^s \sum_\nu H_{\nu m} C_\nu^{s*}.$$

These results are now substituted into Eq. (3.3.18) (with the first term on the right-hand side omitted) to obtain

$$\dot{\rho}_{nm} = \sum_s p(s) \frac{i}{\hbar} \sum_\nu \left(C_n^s C_\nu^{s*} H_{\nu m} - C_m^{s*} C_\nu^s H_{n\nu} \right). \tag{3.3.19}$$

The right-hand side of this equation can be written more compactly by introducing the form (3.3.13) for the density matrix to obtain

$$\dot{\rho}_{nm} = \frac{i}{\hbar} \sum_\nu (\rho_{n\nu} H_{\nu m} - H_{n\nu} \rho_{\nu m}). \tag{3.3.20}$$

Finally, the summation over ν can be performed formally to write this result as

$$\dot{\rho}_{nm} = \frac{i}{\hbar} (\hat{\rho}\hat{H} - \hat{H}\hat{\rho})_{nm} = \frac{-i}{\hbar} [\hat{H}, \hat{\rho}]_{nm}. \tag{3.3.21}$$

We have written the last form in terms of the commutator, defined for any two operators $\hat{A}$ and $\hat{B}$ by $[\hat{A}, \hat{B}] = \hat{A}\hat{B} - \hat{B}\hat{A}$.

Equation (3.3.21) describes how the density matrix evolves in time as the result of interactions that are included in the Hamiltonian $\hat{H}$. However, as mentioned above, there are certain interactions (such as those resulting from collisions between atoms) that cannot conveniently be included in a Hamiltonian description. Such interactions can lead to a change in the state of the system, and hence to a nonvanishing value of $dp(s)/dt$. We include such effects in the formalism by adding phenomenological damping terms to the equation of motion (3.3.21). There is more than one way to model such decay processes. For the most part, we shall model such processes by taking the density matrix equations to have the form

$$\dot{\rho}_{nm} = \frac{-i}{\hbar}[\hat{H}, \hat{\rho}]_{nm} - \gamma_{nm}\left(\rho_{nm} - \rho_{nm}^{\mathrm{eq}}\right). \tag{3.3.22}$$

Here the second term on the right-hand side is a phenomenological damping term, which indicates that ρ_{nm} relaxes to its equilibrium value ρ_{nm}^{eq} at rate γ_{nm}. Since γ_{nm} is a decay rate, we assume that $\gamma_{nm} = \gamma_{mn}$. In addition, we make the physical assumption that

$$\rho_{nm}^{\mathrm{eq}} = 0 \quad \text{for} \quad n \neq m. \tag{3.3.23}$$

We are hence assuming that in thermal equilibrium the excited states of the system may contain population (i.e., ρ_{nn}^{eq} can be nonzero) but that thermal excitation, which is expected to be an incoherent process, cannot produce any coherent superpositions of atomic states ($\rho_{nm}^{\mathrm{eq}} = 0$ for $n \neq m$).

An alternative method of describing decay phenomena is to assume that the off-diagonal elements of the density matrix are damped in the manner described above, but to describe the damping of the diagonal elements by allowing population to decay from higher-lying levels to lower-lying levels. In such a case, the density matrix equations of motion are given by

$$\dot{\rho}_{nm} = -i\hbar^{-1}[\hat{H}, \hat{\rho}]_{nm} - \gamma_{nm}\rho_{nm}, \quad n \neq m, \tag{3.3.24a}$$

$$\dot{\rho}_{nn} = -i\hbar^{-1}[\hat{H}, \hat{\rho}]_{nn} + \sum_{E_m > E_n} \Gamma_{nm}\rho_{mm} - \sum_{E_m < E_n} \Gamma_{mn}\rho_{nn}. \tag{3.3.24b}$$

Here Γ_{nm} gives the rate per atom at which population decays from level m to level n, and, as above, γ_{nm} gives the damping rate of the ρ_{nm} coherence.

The damping rates γ_{nm} for the off-diagonal elements of the density matrix are not entirely independent of the damping rates of the diagonal elements. In fact, under quite general conditions the off-diagonal elements can be

represented as

$$\gamma_{nm} = \tfrac{1}{2}(\Gamma_n + \Gamma_m) + \gamma_{nm}^{\text{col}}. \tag{3.3.25}$$

Here, Γ_n and Γ_m denote the total decay rates of population out of levels n and m, respectively. In the notation of Eq. (3.3.24b), for example, Γ_n is given by the expression

$$\Gamma_n = \sum_{n'\ (E_{n'} < E_n)} \Gamma_{n'n}. \tag{3.3.26}$$

The quantity γ_{nm}^{col} in Eq. (3.3.25) is the dipole dephasing rate due to processes (such as elastic collisions) that are not associated with the transfer of population; γ_{nm}^{col} is sometimes called the proper dephasing rate. To see why Eq. (3.3.25) depends upon the population decay rates in the manner indicated, we note that if level n has lifetime $\tau_n = 1/\Gamma_n$, the probability to be in level n must decay as

$$|C_n(t)|^2 = |C_n(0)|^2 e^{-\Gamma_n t}, \tag{3.3.27}$$

and hence the probability amplitude must vary in time as

$$C_n(t) = C_n(0)e^{-i\omega_n t}e^{-\Gamma_n t/2}. \tag{3.3.28}$$

Likewise, the probability amplitude of being in level m must vary as

$$C_m(t) = C_m(0)e^{-i\omega_m t}e^{-\Gamma_m t/2}. \tag{3.3.29}$$

Thus the coherence between the two levels must vary as

$$C_n^*(t)C_m(t) = C_n^*(0)C_m(0)e^{-i\omega_{mn} t}e^{-(\Gamma_n + \Gamma_m)t/2}. \tag{3.3.30}$$

But since the ensemble average of $C_n^*C_m$ is just ρ_{mn}, whose damping rate is denoted γ_{mn}, it follows that

$$\gamma_{mn} = \tfrac{1}{2}(\Gamma_n + \Gamma_m). \tag{3.3.31}$$

Example: Two-Level Atom

As an example of the use of the density matrix formalism, we apply it to the simple case illustrated in Fig. 3.3.1, in which only the two atomic states a and b interact appreciably with the incident optical field. The wavefunction

FIGURE 3.3.1   A two-level atom.

describing state s of such an atom is given by

$$\psi_s(\mathbf{r}, t) = C_a^s(t) u_a(\mathbf{r}) + C_b^s(t) u_b(\mathbf{r}), \qquad (3.3.32)$$

and hence the density matrix describing the atom is the two-by-two matrix given explicitly by

$$\begin{bmatrix} \rho_{aa} & \rho_{ab} \\ \rho_{ba} & \rho_{bb} \end{bmatrix} = \begin{bmatrix} \overline{C_a C_a^*} & \overline{C_a C_b^*} \\ \overline{C_b C_a^*} & \overline{C_b C_b^*} \end{bmatrix}. \qquad (3.3.33)$$

The matrix representation of the dipole moment operator is

$$\hat{\mu} \Rightarrow \begin{bmatrix} 0 & \mu_{ab} \\ \mu_{ba} & 0 \end{bmatrix}, \qquad (3.3.34)$$

where $\mu_{ij} = \mu_{ji}^* = -e\langle i|\hat{z}|j\rangle$, $-e$ is the electron charge, and $\hat{z}$ is the position operator for the electron. We have set the diagonal elements of the dipole moment operator equal to zero on the basis of the implicit assumption that states a and b have definite parity, in which case $\langle a|\hat{\mathbf{r}}|a\rangle$ and $\langle b|\hat{\mathbf{r}}|b\rangle$ vanish identically as a consequence of symmetry considerations. The expectation value of the dipole moment is given according to Eq. (3.3.17) by $\langle \hat{\mu} \rangle = \text{tr}(\hat{\rho}\hat{\mu})$. Explicitly, $\hat{\rho}\hat{\mu}$ is represented as

$$\hat{\rho}\hat{\mu} \Rightarrow \begin{bmatrix} \rho_{aa} & \rho_{ab} \\ \rho_{ba} & \rho_{bb} \end{bmatrix} \begin{bmatrix} 0 & \mu_{ab} \\ \mu_{ba} & 0 \end{bmatrix} = \begin{bmatrix} \rho_{ab}\mu_{ba} & \rho_{aa}\mu_{ab} \\ \rho_{bb}\mu_{ba} & \rho_{ba}\mu_{ab} \end{bmatrix} \qquad (3.3.35)$$

and hence the expectation value of the induced dipole moment is given by

$$\overline{\langle \mu \rangle} = \text{tr}(\hat{\rho}\hat{\mu}) = \rho_{ab}\mu_{ba} + \rho_{ba}\mu_{ab}. \qquad (3.3.36)$$

As stated in connection with Eq. (3.3.14), the expectation value of the dipole moment is seen to depend upon the off-diagonal elements of the density matrix.

The density matrix treatment of the two-level atom is developed more fully in Chapter 6.

3.4. Perturbation Solution of the Density Matrix Equation of Motion

In the last section, we saw that the density matrix equation of motion with the phenomenological inclusion of damping is given by

$$\dot{\rho}_{nm} = \frac{-i}{\hbar}[\hat{H}, \hat{\rho}]_{nm} - \gamma_{nm}(\rho_{nm} - \rho_{nm}^{\text{eq}}). \qquad (3.4.1)$$

In general, this equation cannot be solved exactly for physical systems of interest, and for this reason we develop a perturbative technique for solving it.

This technique presupposes that, as in Eq. (3.3.2) in the preceding section, the Hamiltonian can be split into two parts as

$$\hat{H} = \hat{H}_0 + \hat{V}(t), \tag{3.4.2}$$

where $\hat{H}_0$ represents the Hamiltonian of the free atom and $\hat{V}(t)$ represents the energy of interaction of the atom with the externally applied radiation field. This interaction is assumed to be weak in the sense that the expectation value and matrix elements of $\hat{V}$ are much smaller than the expectation value of $\hat{H}_0$. We usually assume that this interaction energy is given adequately by the electric dipole approximation as

$$\hat{V} = -\hat{\mu} \cdot \tilde{\mathbf{E}}(t), \tag{3.4.3}$$

where $\hat{\mu} = -e\hat{\mathbf{r}}$ denotes the electric dipole moment operator of the atom. However, for generality and for compactness of notation, we shall introduce Eq. (3.4.3) only when necessary.

When Eq. (3.4.2) is introduced into Eq. (3.4.1), the commutator $[\hat{H}, \hat{\rho}]$ splits into two terms. We examine first the commutator of $\hat{H}_0$ with $\hat{\rho}$. We assume that the states n represent the energy eigenfunctions u_n of the unperturbed Hamiltonian $\hat{H}_0$ and hence satisfy the equation $\hat{H}_0 u_n = E_n u_n$. (See also Eq. (3.3.4).) As a consequence, the matrix representation of $\hat{H}_0$ is diagonal, that is,

$$H_{0,nm} = E_n \delta_{nm}. \tag{3.4.4}$$

The commutator can thus be expanded as

$$[\hat{H}_0, \hat{\rho}]_{nm} = (\hat{H}_0 \hat{\rho} - \hat{\rho} \hat{H}_0)_{nm} = \sum_\nu (H_{0,n\nu} \rho_{\nu m} - \rho_{n\nu} H_{0,\nu m})$$

$$= \sum_\nu (E_n \delta_{n\nu} \rho_{\nu m} - \rho_{n\nu} \delta_{\nu m} E_m) \tag{3.4.5}$$

$$= E_n \rho_{nm} - E_m \rho_{nm} = (E_n - E_m) \rho_{nm}.$$

For future convenience, we define the transition frequency (in angular frequency units) as

$$\omega_{nm} = \frac{E_n - E_m}{\hbar}. \tag{3.4.6}$$

Through use of Eqs. (3.4.2), (3.4.5), and (3.4.6), the density matrix equation of motion (3.4.1) thus becomes

$$\dot{\rho}_{nm} = -i\omega_{nm}\rho_{nm} - \frac{i}{\hbar}[\hat{V}, \hat{\rho}]_{nm} - \gamma_{nm}\left(\rho_{nm} - \rho_{nm}^{\text{eq}}\right). \tag{3.4.7}$$

We can also expand the commutator of $\hat{V}$ with $\hat{\rho}$ to obtain the density matrix equation of motion in the form*

$$\dot{\rho}_{nm} = -i\omega_{nm}\rho_{nm} - \frac{i}{\hbar}\sum_{\nu}(V_{n\nu}\rho_{\nu m} - \rho_{n\nu}V_{\nu m}) - \gamma_{nm}(\rho_{nm} - \rho_{nm}^{\text{eq}}). \qquad (3.4.8)$$

For most problems of physical interest, Eq. (3.4.8) cannot be solved analytically. We therefore seek a solution in the form of a perturbation expansion. In order to carry out this procedure, we replace V_{ij} in Eq. (3.4.8) by λV_{ij}, where λ is a parameter ranging between zero and one that characterizes the strength of the perturbation. The value $\lambda = 1$ is taken to represent the actual physical situation. We now seek a solution to Eq. (3.4.8) in the form of a power series in λ, that is,

$$\rho_{nm} = \rho_{nm}^{(0)} + \lambda\rho_{nm}^{(1)} + \lambda^2\rho_{nm}^{(2)} + \cdots. \qquad (3.4.9)$$

We require that Eq. (3.4.9) be a solution of Eq. (3.4.8) for any value of the parameter λ. In order for this condition to hold, the coefficients of each power of λ must satisfy Eq. (3.4.8) separately. We thereby obtain the set of equations

$$\dot{\rho}_{nm}^{(0)} = -i\omega_{nm}\rho_{nm}^{(0)} - \gamma_{nm}(\rho_{nm}^{(0)} - \rho_{nm}^{\text{eq}}), \qquad (3.4.10a)$$

$$\dot{\rho}_{nm}^{(1)} = -(i\omega_{nm} + \gamma_{nm})\rho^{(1)} - i\hbar^{-1}[\hat{V}, \hat{\rho}^{(0)}]_{nm}, \qquad (3.4.10b)$$

$$\dot{\rho}_{nm}^{(2)} = -(i\omega_{nm} + \gamma_{nm})\rho^{(2)} - i\hbar^{-1}[\hat{V}, \hat{\rho}^{(1)}]_{nm}, \qquad (3.4.10c)$$

etc. This system of equations can now be integrated directly, since, if the set of equations is solved in the order shown, each equation contains only linear homogeneous terms and inhomogeneous terms that are already known.

Equation (3.4.10a) describes the time evolution of the system in the absence of any external field. We take the steady-state solution to this equation to be

$$\rho_{nm}^{(0)} = \rho_{nm}^{\text{eq}} \qquad (3.4.11a)$$

where (for reasons given earlier; see Eq. (3.3.23))

$$\rho_{nm}^{\text{eq}} = 0 \quad \text{for} \quad n \neq m. \qquad (3.4.11b)$$

---

* In this section, we are describing the time evolution of the system in the Schrödinger picture. It is sometimes convenient to describe the time evolution instead in the interaction picture. To find the analogous equation of motion in the interaction picture, we define new quantities σ_{nm} and σ_{nm}^{eq} through

$$\rho_{nm} = \sigma_{nm}e^{-i\omega_{nm}t} \qquad \rho_{nm}^{\text{eq}} = \sigma_{nm}^{\text{eq}}e^{-i\omega_{nm}t}.$$

In terms of these new quantities, Eq. (3.4.8) becomes

$$\dot{\sigma}_{nm} = -\frac{i}{\hbar}\sum_{\nu}[V_{n\nu}\sigma_{\nu m}e^{i\omega_{n\nu}t} - \sigma_{n\nu}e^{i\omega_{\nu m}t}V_{\nu m}] - \gamma_{nm}(\sigma_{nm} - \sigma_{nm}^{\text{eq}}).$$

Now that $\rho_{nm}^{(0)}$ is known, Eq. (3.4.10b) can be integrated. In order to do so, we make a change of variables by representing $\rho_{nm}^{(1)}$ as

$$\rho_{nm}^{(1)}(t) = S_{nm}^{(1)}(t)e^{-(i\omega_{nm}+\gamma_{nm})t}. \tag{3.4.12}$$

The derivative $\dot{\rho}_{nm}^{(1)}$ can be represented in terms of $S_{nm}^{(1)}$ as

$$\dot{\rho}_{nm}^{(1)} = -(i\omega_{nm} + \gamma_{nm})S_{nm}^{(1)}e^{-(i\omega_{nm}+\gamma_{nm})t} + \dot{S}_{nm}^{(1)}e^{-(i\omega_{nm}+\gamma_{nm})t}. \tag{3.4.13}$$

These forms are substituted into Eq. (3.4.10b), which then becomes

$$\dot{S}_{nm}^{(1)} = \frac{-i}{\hbar}[\hat{V}, \hat{\rho}^{(0)}]_{nm}e^{(i\omega_{nm}+\gamma_{nm})t} \tag{3.4.14}$$

and which can be integrated to give

$$S_{nm}^{(1)} = \int_{-\infty}^{t} \frac{-i}{\hbar}[\hat{V}(t'), \hat{\rho}^{(0)}]_{nm}e^{(i\omega_{nm}+\gamma_{nm})t'}dt'. \tag{3.4.15}$$

This expression is now substituted back into Eq. (3.4.12) to obtain

$$\rho_{nm}^{(1)}(t) = \int_{-\infty}^{t} \frac{-i}{\hbar}[\hat{V}(t'), \hat{\rho}^{(0)}]_{nm}e^{(i\omega_{nm}+\gamma_{nm})(t'-t)}dt'. \tag{3.4.16}$$

In similar fashion, all of the higher-order corrections to the density matrix can be obtained. These expressions are formally identical to Eq. (3.4.16). The expression for $\rho_{nm}^{(q)}$, for example, is obtained by replacing $\hat{\rho}^{(0)}$ with $\hat{\rho}^{(q-1)}$ on the right-hand side of Eq. (3.4.16).

3.5. Density Matrix Calculation of the Linear Susceptibility

As a first application of the perturbation solution to the density matrix equations of motion, we calculate the linear susceptibility of an atomic system. The relevant starting equation for this calculation is Eq. (3.4.16), which we write in the form

$$\rho_{nm}^{(1)}(t) = e^{-(i\omega_{nm}+\gamma_{nm})t} \int_{-\infty}^{t} dt' \frac{-i}{\hbar}[\hat{V}(t'), \hat{\rho}^{(0)}]_{nm}e^{(i\omega_{nm}+\gamma_{nm})t'}. \tag{3.5.1}$$

As before, the interaction Hamiltonian is given by Eq. (3.4.3) as

$$\hat{V}(t') = -\hat{\mu} \cdot \tilde{\mathbf{E}}(t'), \tag{3.5.2}$$

and we assume that the unperturbed density matrix is given by (see also Eqs. (3.4.11))

$$\rho_{nm}^{(0)} = 0 \quad \text{for} \quad n \neq m. \tag{3.5.3}$$

We represent the applied field as

$$\tilde{\mathbf{E}}(t) = \sum_{p} \mathbf{E}(\omega_p)e^{-i\omega_p t}. \tag{3.5.4}$$

The first step is to obtain an explicit expression for the commutator appearing in Eq. (3.5.1):

$$\left[\hat{V}(t),\hat{\rho}^{(0)}\right]_{nm} = \sum_{v}\left[V(t)_{nv}\rho^{(0)}_{vm} - \rho^{(0)}_{nv}V(t)_{vm}\right]$$

$$= -\sum_{v}\left[\boldsymbol{\mu}_{nv}\rho^{(0)}_{vm} - \rho^{(0)}_{nv}\boldsymbol{\mu}_{vm}\right]\cdot\tilde{\mathbf{E}}(t) \qquad (3.5.5)$$

$$= -\left(\rho^{(0)}_{mm} - \rho^{(0)}_{nn}\right)\boldsymbol{\mu}_{nm}\cdot\tilde{\mathbf{E}}(t).$$

Here the second form is obtained by introducing $\hat{V}(t)$ explicitly from Eq. (3.5.2), and the third form is obtained by performing the summation over all v and utilizing the condition (3.5.3). This expression for the commutator is introduced into Eq. (3.5.1) to obtain

$$\rho^{(1)}_{nm}(t) = \frac{i}{\hbar}\left(\rho^{(0)}_{mm} - \rho^{(0)}_{nn}\right)\boldsymbol{\mu}_{nm}\cdot e^{-(i\omega_{nm}+\gamma_{nm})t}\int_{-\infty}^{t}\tilde{\mathbf{E}}(t')e^{(i\omega_{nm}+\gamma_{nm})t'}dt'. \qquad (3.5.6)$$

We next introduce Eq. (3.5.4) for $\tilde{\mathbf{E}}(t)$ to obtain

$$\rho^{(1)}_{nm}(t) = \frac{i}{\hbar}\left(\rho^{(0)}_{mm} - \rho^{(0)}_{nn}\right)\boldsymbol{\mu}_{nm}\cdot\sum_{p}\mathbf{E}(\omega_p)$$

$$\times e^{-(i\omega_{nm}+\gamma_{nm})t}\int_{-\infty}^{t}e^{[i\omega_{nm}-\omega_p)+\gamma_{nm}]t'}dt'. \qquad (3.5.7)$$

The second line of this expression can be evaluated explicitly as

$$e^{-(i\omega_{nm}+\gamma_{nm})t}\left(\frac{e^{[i(\omega_{nm}-\omega_p)+\gamma_{nm}]t'}}{i(\omega_{nm}-\omega_p)+\gamma_{nm}}\right)\Bigg|_{-\infty}^{t} = \frac{e^{i\omega_p t}}{i(\omega_{nm}-\omega_p)+\gamma_{nm}}, \qquad (3.5.8)$$

and $\rho^{(1)}_{nm}$ is hence seen to be given by

$$\rho^{(1)}_{nm} = \hbar^{-1}\left(\rho^{(0)}_{mm} - \rho^{(0)}_{nn}\right)\sum_{p}\frac{\boldsymbol{\mu}_{nm}\cdot\mathbf{E}(\omega_p)e^{-i\omega_p t}}{(\omega_{nm}-\omega_p)-i\gamma_{nm}}. \qquad (3.5.9)$$

We next use this result to calculate the expectation value of the induced dipole moment*:

$$\langle\tilde{\boldsymbol{\mu}}(t)\rangle = \mathrm{tr}\left(\hat{\rho}^{(1)}\hat{\boldsymbol{\mu}}\right) = \sum_{nm}\rho^{(1)}_{nm}\boldsymbol{\mu}_{mn}$$

$$= \sum_{nm}\hbar^{-1}\left(\rho^{(0)}_{mm} - \rho^{(0)}_{nn}\right)\sum_{p}\frac{\boldsymbol{\mu}_{mn}[\boldsymbol{\mu}_{nm}\cdot\mathbf{E}(\omega_p)]e^{-i\omega_p t}}{(\omega_{nm}-\omega_p)-i\gamma_{nm}}. \qquad (3.5.10)$$

* Here and throughout the remainder of this chapter we are omitting the bar over quantities such as $\langle\mu\rangle$ for simplicity of notation. Hence the angular brackets are meant to imply both a quantum and an ensemble average.

We decompose $\langle \tilde{\boldsymbol{\mu}}(t) \rangle$ into its frequency components according to

$$\langle \tilde{\boldsymbol{\mu}}(t) \rangle = \sum_p \langle \boldsymbol{\mu}(\omega_p) \rangle e^{-i\omega_p t} \qquad (3.5.11)$$

and define the linear susceptibility tensor $\chi^{(1)}(\omega)$ by the equation

$$\mathbf{P}(\omega_p) = N \langle \boldsymbol{\mu}(\omega_p) \rangle = \chi^{(1)}(\omega_p) \cdot \mathbf{E}(\omega_p), \qquad (3.5.12)$$

where N denotes the atomic number density. By comparing this equation with Eq. (3.5.10), we find that the linear susceptibility is given by

$$\chi^{(1)}(\omega_p) = \frac{N}{\hbar} \sum_{nm} \left(\rho_{mm}^{(0)} - \rho_{nn}^{(0)} \right) \frac{\mu_{mn}\mu_{nm}}{(\omega_{nm} - \omega_p) - i\gamma_{nm}}. \qquad (3.5.13)$$

The result given by Eqs. (3.5.12) and (3.5.13) can be written in cartesian component form as

$$P_i(\omega_p) = N \langle \mu_i(\omega_p) \rangle = \sum_j \chi_{ij}^{(1)}(\omega_p) E_j(\omega_p) \qquad (3.5.14)$$

with

$$\chi_{ij}^{(1)}(\omega_p) = \frac{N}{\hbar} \sum_{nm} \left(\rho_{mm}^{(0)} - \rho_{nn}^{(0)} \right) \frac{\mu_{mn}^i \mu_{nm}^j}{(\omega_{nm} - \omega_p) - i\gamma_{nm}}. \qquad (3.5.15)$$

We see that the linear susceptibility is proportional to the population difference $\rho_{mm}^{(0)} - \rho_{nn}^{(0)}$; hence if levels m and n contain equal populations, the $m \to n$ transition does not contribute to the linear susceptibility.

Equation (3.5.15) is an extremely compact way of representing the linear susceptibility. At times it is more intuitive to express the susceptibility in an expanded form. We first rewrite Eq. (3.5.15) as

$$\chi_{ij}^{(1)}(\omega_p) = \frac{N}{\hbar} \sum_{nm} \rho_{mm}^{(0)} \frac{\mu_{mn}^i \mu_{nm}^j}{(\omega_{nm} - \omega_p) - i\gamma_{nm}} - \frac{N}{\hbar} \sum_{nm} \rho_{nn}^{(0)} \frac{\mu_{mn}^i \mu_{nm}^j}{(\omega_{nm} - \omega_p) - i\gamma_{nm}}. \qquad (3.5.16)$$

We next interchange the dummy indices n and m in the second summation, so that the two summations can be recombined as

$$\chi_{ij}^{(1)}(\omega_p) = \frac{N}{\hbar} \sum_{nm} \rho_{mm}^{(0)} \left[\frac{\mu_{mn}^i \mu_{nm}^j}{(\omega_{nm} - \omega_p) - i\gamma_{nm}} - \frac{\mu_{nm}^i \mu_{mn}^j}{(\omega_{mn} - \omega_p) - i\gamma_{mn}} \right]. \qquad (3.5.17)$$

We now use the fact that $\omega_{mn} = -\omega_{nm}$ and $\gamma_{nm} = \gamma_{mn}$ to write this result as

$$\chi_{ij}^{(1)}(\omega_p) = \frac{N}{\hbar} \sum_{nm} \rho_{mm}^{(0)} \left[\frac{\mu_{mn}^i \mu_{nm}^j}{(\omega_{nm} - \omega_p) - i\gamma_{nm}} + \frac{\mu_{nm}^i \mu_{mn}^j}{(\omega_{nm} + \omega_p) + i\gamma_{nm}} \right]. \qquad (3.5.18)$$

In order to interpret this result, let us first make the simplifying assumption that all of the population is in one level (typically the ground state), which we denote as level a. Mathematically, this assumption can be stated as

$$\rho_{aa}^{(0)} = 1, \qquad \rho_{mm}^{(0)} = 0 \quad \text{for} \quad m \neq a. \tag{3.5.19}$$

We now perform the summation over m in Eq. (3.5.18) to obtain

$$\chi_{ij}^{(1)}(\omega_p) = \frac{N}{\hbar} \sum_n \left[\frac{\mu_{an}^i \mu_{na}^j}{(\omega_{na} - \omega_p) - i\gamma_{na}} + \frac{\mu_{na}^i \mu_{an}^j}{(\omega_{na} + \omega_p) + i\gamma_{na}} \right]. \tag{3.5.20}$$

We see that for positive frequencies (i.e., for $\omega_p > 0$), only the first term can become resonant. The second term is known as the antiresonant or counter-rotating term. We can often drop the second term, especially when ω_p is close to one of the resonance frequencies of the atom. Let us assume that ω_p is nearly resonant with the transition frequency ω_{na}. Then to good approximation the linear susceptibility is given by

$$\chi_{ij}^{(1)}(\omega_p) = \frac{N}{\hbar} \frac{\mu_{an}^i \mu_{na}^j}{(\omega_{na} - \omega_p) - i\gamma_{na}} = \frac{N}{\hbar} \mu_{an}^i \mu_{na}^j \frac{(\omega_{na} - \omega_p) + i\gamma_{na}}{(\omega_{na} - \omega_p)^2 + \gamma_{na}^2}. \tag{3.5.21}$$

The real and imaginary parts of this expression are shown in Fig. 3.5.1. We see that the imaginary part of χ_{ij} has the form of a Lorentzian line shape with a linewidth (full width at half maximum) equal to $2\gamma_{na}$.

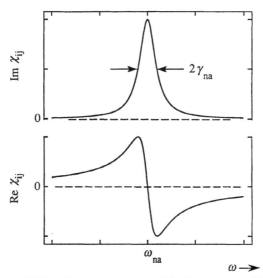

FIGURE 3.5.1   Resonance nature of the linear susceptibility.

Linear Dispersion Theory

Since linear dispersion theory plays a key role in our understanding of optical phenomena, the remainder of this section is devoted to the interpretation of the results derived above. Let us first specialize our results to the case of an isotropic material. As a consequence of symmetry considerations, $\mathbf{P}$ must then be parallel to $\mathbf{E}$ in such a medium, and we can therefore express the linear susceptibility as the scalar quantity $\chi^{(1)}(\omega)$ defined through $\mathbf{P}(\omega) = \chi^{(1)}(\omega)\mathbf{E}(\omega)$, which is given by

$$\chi^{(1)}(\omega) = N\hbar^{-1} \sum_n \tfrac{1}{3}|\boldsymbol{\mu}_{na}|^2 \left[\frac{1}{(\omega_{na} - \omega) - i\gamma_{na}} + \frac{1}{(\omega_{na} + \omega) + i\gamma_{na}} \right].$$

(3.5.22)

For simplicity we are assuming the case of a nondegenerate ground state (e.g., $J = 0$). We have included the factor of $\frac{1}{3}$ in the numerator of this expression for the following reason: The summation over n includes all of the magnetic sublevels of the atomic excited states. However, on average only one-third of the $a \to n$ transitions will have their dipole transition moments parallel to the polarization vector of the incident field, and hence only one-third of these transitions contribute effectively to the susceptibility.

It is useful to introduce the *oscillator strength* of the $a \to n$ transition. This quantity is defined by

$$f_{na} = \frac{2m\omega_{na}|\boldsymbol{\mu}_{na}|^2}{3\hbar e^2}.$$

(3.5.23)

Standard books on quantum mechanics (see, for example, Bethe and Salpeter, 1977) show that this quantity obeys the *oscillator strength sum rule*, that is,

$$\sum_n f_{na} = 1.$$

(3.5.24)

If a is the atomic ground state, the frequency ω_{na} is necessarily positive, and the sum rule hence shows that the oscillator strength is a positive quantity bounded by unity, that is, $0 \le f_{na} \le 1$. The expression (3.5.22) for the linear susceptibility can be written in terms of the oscillator strength as

$$\chi^{(1)}(\omega) = \sum_n \frac{Nf_{na}e^2}{2m\omega_{na}} \left[\frac{1}{(\omega_{na} - \omega) - i\gamma_{na}} + \frac{1}{(\omega_{na} + \omega) + i\gamma_{na}} \right]$$

(3.5.25)

$$\simeq \sum_n f_{na} \left[\frac{Ne^2/m}{\omega_{na}^2 - \omega^2 - 2i\omega\gamma_{na}} \right].$$

In the latter form, the expression in square brackets is formally identical to the expression for the linear susceptibility predicted by the classical Lorentz model of the atom (see also Eq. (1.4.17)). We see that the quantum-mechanical prediction differs from that of the Lorentz model only in that in the quantum-mechanical theory there can be more than one resonance frequency ω_{na}. The strength of each such transition is given by the value of the oscillator strength.

Let us next see how to calculate the refractive index and absorption coefficient. The refractive index $n(\omega)$ is related to the linear dielectric constant $\epsilon(\omega)$ and linear susceptibility $\chi^{(1)}(\omega)$ through

$$n(\omega) = \sqrt{\epsilon(\omega)} = \sqrt{1 + 4\pi \chi^{(1)}(\omega)} \simeq 1 + 2\pi \chi^{(1)}(\omega). \quad (3.5.26)$$

In obtaining the last expression, we have assumed that the medium is sufficiently dilute (i.e., N sufficiently small) that $4\pi \chi^{(1)} \ll 1$. For the remainder of the present section, we shall assume that this assumption is valid, both so that we can use Eq. (3.5.26) as written and also so that we can ignore local-field corrections (cf. Section 3.8). The significance of the refractive index $n(\omega)$ is that the propagation of a plane wave through the material system is described by

$$\tilde{E}(z, t) = E_0 e^{i(kz - \omega t)} + \text{c.c.}, \quad (3.5.27)$$

where the propagation constant k is given by

$$k = n(\omega)\omega/c. \quad (3.5.28)$$

Hence the intensity $I = (nc/4\pi)\langle \tilde{E}(z, t)^2 \rangle$ of this wave varies with position in the medium according to

$$I(z) = I_0 e^{-\alpha z}, \quad (3.5.29)$$

where the *absorption coefficient* α is given by

$$\alpha = 2n''\omega/c, \quad (3.5.30)$$

and where we have defined the real and imaginary parts of the refractive index as $n(\omega) = n' + in''$. Alternatively, through use of Eq. (3.5.26), we can represent the absorption coefficient in terms of the susceptibility as

$$\alpha = 4\pi \chi^{(1)''}\omega/c, \quad (3.5.31a)$$

where $\chi^{(1)}(\omega) = \chi^{(1)'} + i\chi^{(1)''}$. Through use of Eq. (3.5.25), we find that the absorption coefficient of the material system is given by

$$\alpha = \sum_n \frac{2\pi f_{na} N e^2}{mc\gamma_{na}} \left[\frac{\gamma_{na}^2}{(\omega_{na} - \omega)^2 + \gamma_{na}^2} \right]. \quad (3.5.31b)$$

It is often useful to describe the response of a material system to an applied field in terms of microscopic rather than macroscopic quantities. We define the atomic polarizability $\alpha^{(1)}(\omega)$ as the coefficient relating the induced dipole moment $\langle \boldsymbol{\mu}(\omega) \rangle$ and the applied field $\mathbf{E}(\omega)^*$:

$$\langle \boldsymbol{\mu}(\omega) \rangle = \alpha^{(1)}(\omega)\mathbf{E}(\omega). \qquad (3.5.32)$$

The susceptibility and polarizability are related (when local-field corrections can be ignored) through

$$\chi^{(1)}(\omega) = N\alpha^{(1)}(\omega), \qquad (3.5.33)$$

and we hence find from Eq. (3.5.22) that the polarizability is given by

$$\alpha^{(1)}(\omega) = \hbar^{-1} \sum_{n} \tfrac{1}{3}|\mu_{na}|^2 \left[\frac{1}{(\omega_{na} - \omega) - i\gamma_{na}} + \frac{1}{(\omega_{na} + \omega) + i\gamma_{na}} \right].$$

$$(3.5.34)$$

Another microscopic quantity that is often encountered is the absorption cross section σ, which is defined through the relation

$$\sigma = \alpha/N. \qquad (3.5.35)$$

The cross section can hence be interpreted as the effective area of an atom for removing radiation from an incident beam of light. By comparison with Eqs. (3.5.31a) and (3.5.33), we see that the absorption cross section is related to the atomic polarizability $\alpha^{(1)} = \alpha^{(1)\prime} + i\alpha^{(1)\prime\prime}$ through

$$\sigma = 4\pi\alpha^{(1)\prime\prime}\omega/c. \qquad (3.5.36)$$

Equation (3.5.34) shows how the polarizability can be calculated in terms of the transition frequencies ω_{na}, the dipole transition moments μ_{na}, and the dipole dephasing rates γ_{na}. The transition frequencies and dipole moments are inherent properties of any atomic system and can be obtained either by solving Schrödinger's equation for the atom or through laboratory measurement. The dipole dephasing rate, however, depends not only on the inherent atomic properties but also upon the local environment. We saw in Eq. (3.3.25) that the dipole dephasing rate γ_{mn} can be represented as

$$\gamma_{nm} = \tfrac{1}{2}(\Gamma_n + \Gamma_m) + \gamma_{nm}^{\text{col}}. \qquad (3.5.37)$$

Next we calculate the maximum values that the polarizability and absorption cross section can attain. We consider the case of resonant excitation ($\omega = \omega_{na}$) of some excited level n. We find, through use of Eq. (3.5.34) and dropping the nonresonant contribution, that the polarizability is purely imaginary and is

---

* Note that α denotes the absorption coefficient and $\alpha^{(1)}$ denotes the polarizability.

given by

$$\alpha_{\text{res}}^{(1)} = \frac{i|\mu_{n'a}|^2}{\hbar\gamma_{n'a}}. \tag{3.5.38}$$

We have let n' designate the state associated with level n that is excited by the incident light. Note that the factor of $\frac{1}{3}$ no longer appears in Eq. (3.5.38), because we have performed the summation over all states (i.e., magnetic sublevels) of level n. The polarizability will take on its maximum possible value if $\gamma_{n'a}$ is as small as possible, which according to Eq. (3.5.37) occurs when $\gamma_{n'a}^{\text{col}} = 0$. If a is the atomic ground state, as we have been assuming, its decay rate Γ_a must vanish, and hence the minimum possible value of $\gamma_{n'a}$ is $\frac{1}{2}\Gamma_{n'}$.

The population decay rate out of state n' is usually dominated by spontaneous emission. If state n' can decay only to the ground state, this decay rate is equal to the Einstein A coefficient and is given by

$$\Gamma_{n'} = \frac{4\omega_{na}^3|\mu_{n'a}|^2}{3\hbar c^3}. \tag{3.5.39}$$

If $\gamma_{n'a} = \frac{1}{2}\Gamma_{n'}$ is inserted into Eq. (3.5.38), we find that the maximum possible value that the polarizability can possess is

$$\alpha_{\max}^{(1)} = i\frac{3}{2}\left(\frac{\lambda}{2\pi}\right)^3. \tag{3.5.40}$$

We find the value of the absorption cross section associated with this value of the polarizability through use of Eq. (3.5.36):

$$\sigma_{\max} = \frac{3\lambda^2}{2\pi}. \tag{3.5.41}$$

These results show that under resonant excitation an atomic system possesses an effective linear dimension approximately equal to an optical wavelength.

3.6.   Density Matrix Calculation of the Second-Order Susceptibility

In this section we calculate the second-order (i.e., $\chi^{(2)}$) susceptibility of an atomic system. We present the calculation in considerable detail, for the following two reasons: (1) the second-order susceptibility is intrinsically important for many applications; and (2) the calculation of the third-order susceptibility proceeds along lines that are analogous to those followed by the present derivation. However, the expression for the third-order susceptibility $\chi^{(3)}$ is so complicated (it contains 48 terms) that it is infeasible to show all of the steps in the calculation of $\chi^{(3)}$. Thus the present development serves as a template for the calculation of higher-order susceptibilities.

From the perturbation expansion (3.4.16), the general result for the second-order correction to $\hat{\rho}$ is given by

$$\rho_{nm}^{(2)} = e^{-(i\omega_{nm}+\gamma_{nm})t} \int_{-\infty}^{t} \frac{-i}{\hbar} [\hat{V}, \hat{\rho}^{(1)}]_{nm} e^{(i\omega_{nm}+\gamma_{nm})t'} dt', \qquad (3.6.1)$$

where the commutator can be expressed (by analogy with Eq. (3.5.5)) as

$$[\hat{V}, \hat{\rho}^{(1)}]_{nm} = -\sum_{\nu} \left(\boldsymbol{\mu}_{n\nu}\rho_{\nu m}^{(1)} - \rho_{n\nu}^{(1)}\boldsymbol{\mu}_{\nu m}\right) \cdot \tilde{\mathbf{E}}(t). \qquad (3.6.2)$$

In order to evaluate this commutator, the first-order solution given by Eq. (3.5.9) is written with changes in the dummy indices as

$$\rho_{\nu m}^{(1)} = \hbar^{-1}\left(\rho_{mm}^{(0)} - \rho_{\nu\nu}^{(0)}\right) \sum_{p} \frac{\boldsymbol{\mu}_{\nu m} \cdot \mathbf{E}(\omega_p)}{(\omega_{\nu m} - \omega_p) - i\gamma_{\nu m}} e^{-i\omega_p t} \qquad (3.6.3)$$

and as

$$\rho_{n\nu}^{(1)} = \hbar^{-1}\left(\rho_{\nu\nu}^{(0)} - \rho_{nn}^{(0)}\right) \sum_{p} \frac{\boldsymbol{\mu}_{n\nu} \cdot E(\omega_p)}{(\omega_{n\nu} - \omega_p) - i\gamma_{n\nu}} e^{-i\omega_p t}. \qquad (3.6.4)$$

The applied optical field $\tilde{\mathbf{E}}(t)$ is expressed as

$$\tilde{\mathbf{E}}(t) = \sum_{q} \mathbf{E}(\omega_q)e^{-i\omega_q t}. \qquad (3.6.5)$$

The commutator of Eq. (3.6.2) thus becomes

$$[\hat{V}, \hat{\rho}^{(1)}]_{nm} = -\hbar^{-1} \sum_{\nu} \left(\rho_{mm}^{(0)} - \rho_{\nu\nu}^{(0)}\right)$$

$$\times \sum_{pq} \frac{[\boldsymbol{\mu}_{n\nu} \cdot \mathbf{E}(\omega_q)][\boldsymbol{\mu}_{\nu m} \cdot \mathbf{E}(\omega_p)]}{(\omega_{\nu m} - \omega_q) - i\gamma_{\nu m}} e^{-i(\omega_p+\omega_q)t}$$

$$+\hbar^{-1} \sum_{\nu} \left(\rho_{\nu\nu}^{(0)} - \rho_{nn}^{(0)}\right)$$

$$\times \sum_{pq} \frac{[\boldsymbol{\mu}_{n\nu} \cdot \mathbf{E}(\omega_p)][\boldsymbol{\mu}_{\nu m} \cdot \mathbf{E}(\omega_q)]}{(\omega_{n\nu} - \omega_p) - i\gamma_{n\nu}} e^{-i(\omega_p+\omega_q)t}. \qquad (3.6.6)$$

This expression is now inserted into Eq. (3.6.1), and the integration is performed to obtain

$$\rho_{nm}^{(2)} = \sum_{\nu} \sum_{pq} e^{-i(\omega_p+\omega_q)t}$$

$$\times \left\{ \frac{\rho_{mm}^{(0)} - \rho_{\nu\nu}^{(0)}}{\hbar^2} \frac{[\boldsymbol{\mu}_{n\nu} \cdot \mathbf{E}(\omega_q)][\boldsymbol{\mu}_{\nu m} \cdot \mathbf{E}(\omega_p)]}{[(\omega_{nm} - \omega_p - \omega_q) - i\gamma_{nm}][(\omega_{\nu m} - \omega_p) - i\gamma_{\nu m}]} \right.$$

$$\left. - \frac{\rho_{\nu\nu}^{(0)} - \rho_{nn}^{(0)}}{\hbar^2} \frac{[\boldsymbol{\mu}_{n\nu} \cdot \mathbf{E}(\omega_q)][\boldsymbol{\mu}_{\nu m} \cdot \mathbf{E}(\omega_q)]}{[(\omega_{nm} - \omega_p - \omega_q) - i\gamma_{nm}][(\omega_{n\nu} - \omega_p) - i\gamma_{n\nu}]} \right\}$$

$$\equiv \sum_{\nu} \sum_{pq} K_{nm\nu} e^{-i(\omega_p+\omega_q)t}. \qquad (3.6.7)$$

We have given the complicated expression in curly braces the label K_{nmv} because it appears in many subsequent equations.

We next calculate the expectation value of the atomic dipole moment, which (according to Eq. (3.3.16)) is given by

$$\langle \tilde{\mu} \rangle = \sum_{nm} \rho_{nm} \mu_{mn}. \tag{3.6.8}$$

We are interested in the various frequency components of $\langle \tilde{\mu} \rangle$, whose complex amplitudes $\langle \mu(\omega_r) \rangle$ are defined through

$$\langle \tilde{\mu} \rangle = \sum_{r} \langle \mu(\omega_r) \rangle e^{-i\omega_r t}. \tag{3.6.9}$$

Then, in particular, the complex amplitude of the component of the atomic dipole moment oscillating at frequency $\omega_p + \omega_q$ is given by

$$\langle \mu(\omega_p + \omega_q) \rangle = \sum_{nmv} \sum_{(pq)} K_{nmv} \mu_{mn}, \tag{3.6.10}$$

and consequently the complex amplitude of the component of the nonlinear polarization oscillating at frequency $\omega_p + \omega_q$ is given by

$$\mathbf{P}^{(2)}(\omega_p + \omega_q) = N \langle \mu(\omega_p + \omega_q) \rangle = N \sum_{nmv} \sum_{(pq)} K_{nmv} \mu_{mn}. \tag{3.6.11}$$

We define the nonlinear susceptibility through the equation

$$P_i^{(2)}(\omega_p + \omega_q) = \sum_{jk} \sum_{(pq)} \chi_{ijk}^{(2)}(\omega_p + \omega_q, \omega_q, \omega_p) E_j(\omega_q) E_k(\omega_p), \tag{3.6.12}$$

using the same notation as that used earlier (see also Eq. (1.3.13)). By comparison of Eqs. (3.6.7), (3.6.11), and (3.6.12), we obtain a tentative expression for the susceptibility tensor given by

$$\chi_{ijk}^{(2)\prime}(\omega_p + \omega_q, \omega_q, \omega_p) = \frac{N}{\hbar^2}$$

$$\times \sum_{mnv} \left\{ \left(\rho_{mm}^{(0)} - \rho_{vv}^{(0)} \right) \frac{\mu_{mn}^i \mu_{nv}^j \mu_{vm}^k}{[(\omega_{nm} - \omega_p - \omega_q) - i\gamma_{nm}][(\omega_{vm} - \omega_p) - i\gamma_{vm}]} \tag{a}$$

$$- \left(\rho_{vv}^{(0)} - \rho_{nn}^{(0)} \right) \frac{\mu_{mn}^i \mu_{vm}^j \mu_{nv}^k}{[(\omega_{nm} - \omega_p - \omega_q) - i\gamma_{nm}][(\omega_{nv} - \omega_p) - i\gamma_{nv}]} \right\}. \tag{b}$$

$$\tag{3.6.13}$$

We have labeled the two terms that appear in this expression (a) and (b) so that we can keep track of how these terms contribute to our final expression for the second-order susceptibility.

Equation (3.6.13) can be used in conjunction with Eq. (3.6.12) to make proper predictions of the nonlinear polarization, which is a physically meaningful quantity. However, Eq. (3.6.13) does not possess intrinsic permutation symmetry (cf. Section 1.5), which we require the susceptibility to possess. We therefore define the nonlinear susceptibility to be one-half the sum of the right-hand side of Eq. (3.6.13) with an analogous expression obtained by simultaneously interchanging ω_p with ω_q and j with k. We thereby obtain the result

$$\chi_{ijk}^{(2)}(\omega_p + \omega_q, \omega_q, \omega_p) = \frac{N}{2\hbar^2}$$

$$\times \sum_{mnv} \left\{ \left(\rho_{mm}^{(0)} - \rho_{vv}^{(0)}\right) \left[\frac{\mu_{mn}^i \mu_{nv}^j \mu_{vm}^k}{[(\omega_{nm} - \omega_p - \omega_q) - i\gamma_{nm}][(\omega_{vm} - \omega_p) - i\gamma_{vm}]} \right. \right. \tag{a_1}$$

$$\left. + \frac{\mu_{mn}^i \mu_{nv}^k \mu_{vm}^j}{[(\omega_{nm} - \omega_p - \omega_q) - i\gamma_{nm}][(\omega_{vm} - \omega_q) - i\gamma_{vm}]} \right] \tag{a_2}$$

$$- \left(\rho_{vv}^{(0)} - \rho_{nn}^{(0)}\right) \left[\frac{\mu_{mn}^i \mu_{vm}^j \mu_{nv}^k}{[(\omega_{nm} - \omega_p - \omega_q) - i\gamma_{nm}][(\omega_{nv} - \omega_p) - i\gamma_{nv}]} \right. \tag{b_1}$$

$$\left. \left. + \frac{\mu_{mn}^i \mu_{vm}^k \mu_{nv}^j}{[(\omega_{nm} - \omega_p - \omega_q) - i\gamma_{nm}][(\omega_{nv} - \omega_q) - i\gamma_{nv}]} \right] \right\}. \tag{b_2}$$

$$\tag{3.6.14}$$

This expression displays intrinsic permutation symmetry and gives the nonlinear susceptibility in a reasonably compact fashion. It is clear from its form that certain contributions to the susceptibility vanish when two of the levels associated with the contribution contain equal populations. We shall examine the nature of this cancellation in greater detail later (see Eq. (3.6.17)). Note that the population differences that appear in this expression are always associated with the two levels separated by a one-photon resonance, as we can see by inspection of the detuning factors that appear in the denominator.

The expression for the second-order nonlinear susceptibility can be rewritten in several different forms, all of which are equivalent, but which provide different insights in to the resonant nature of the nonlinear coupling. Since the indices m, n, and v are summed over, they constitute dummy indices. We can therefore replace the indices v, n, and m in the last two terms of Eq. (3.6.15) by m, v, and n, respectively, so that the population difference term is the same as that of the first two terms. We thereby recast the second-order susceptibility into the form

$$\chi_{ijk}^{(2)}(\omega_p + \omega_q, \omega_q, \omega_p) = \frac{N}{2\hbar^2} \sum_{mn\nu} \left(\rho_{mm}^{(0)} - \rho_{\nu\nu}^{(0)}\right)$$

$$\times \left\{ \frac{\mu_{mn}^i \mu_{n\nu}^j \mu_{\nu m}^k}{[(\omega_{nm} - \omega_p - \omega_q) - i\gamma_{nm}][(\omega_{\nu m} - \omega_p) - i\gamma_{\nu m}]} \right. \qquad (a_1)$$

$$+ \frac{\mu_{mn}^i \mu_{n\nu}^k \mu_{\nu m}^j}{[(\omega_{nm} - \omega_p - \omega_q) - i\gamma_{nm}][(\omega_{\nu m} - \omega_q) - i\gamma_{\nu m}]} \qquad (a_2) \quad (3.6.15)$$

$$- \frac{\mu_{n\nu}^i \mu_{mn}^j \mu_{\nu m}^k}{[(\omega_{\nu n} - \omega_p - \omega_q) - i\gamma_{\nu n}][(\omega_{\nu m} - \omega_p) - i\gamma_{\nu m}]} \qquad (b_1)$$

$$\left. - \frac{\mu_{n\nu}^i \mu_{mn}^k \mu_{\nu m}^j}{[(\omega_{\nu n} - \omega_p - \omega_q) - i\gamma_{\nu n}][(\omega_{\nu m} - \omega_q) - i\gamma_{\nu m}]} \right\}. \qquad (b_2)$$

We can make this result more transparent by making another change in dummy indices: we replace indices m, ν, and n by l, m, and n, respectively. In addition, we replace ω_{lm}, ω_{ln}, and ω_{mn} by $-\omega_{ml}$, $-\omega_{nl}$, and $-\omega_{nm}$, respectively, whenever one of them appears. Also, we reorder the product of matrix elements in the numerator so that the subscripts n, m, and l are "chained" in the sense shown, and thereby obtain the result

$$\chi_{ijk}^{(2)}(\omega_p + \omega_q, \omega_q, \omega_p) = \frac{N}{2\hbar^2} \sum_{lmn} \left(\rho_{ll}^{(0)} - \rho_{mm}^{(0)}\right)$$

$$\times \left\{ \frac{\mu_{ln}^i \mu_{nm}^j \mu_{ml}^k}{[(\omega_{nl} - \omega_p - \omega_q) - i\gamma_{nl}][(\omega_{ml} - \omega_p) - i\gamma_{ml}]} \right. \qquad (a_1)$$

$$+ \frac{\mu_{ln}^i \mu_{nm}^k \mu_{ml}^j}{[(\omega_{nl} - \omega_p - \omega_q) - i\gamma_{nl}][(\omega_{ml} - \omega_q) - i\gamma_{ml}]} \qquad (a_2) \quad (3.6.16)$$

$$+ \frac{\mu_{ln}^j \mu_{nm}^i \mu_{ml}^k}{[(\omega_{nm} + \omega_p + \omega_q) + i\gamma_{nm}][(\omega_{ml} - \omega_p) - i\gamma_{ml}]} \qquad (b_1)$$

$$\left. + \frac{\mu_{ln}^k \mu_{nm}^i \mu_{ml}^j}{[(\omega_{nm} + \omega_p + \omega_q) + i\gamma_{nm}][(\omega_{ml} - \omega_q) - i\gamma_{ml}]} \right\}. \qquad (b_2)$$

One way of interpreting this result is to consider where levels l, m, and n would have to be located in order for each of the terms to become resonant. The positions of these energies are illustrated in Fig. 3.6.1. For definiteness, we have drawn the figure with ω_p and ω_q positive. In each case the magnitude of the contribution to the nonlinear susceptibility is proportional to the population difference between levels l and m.

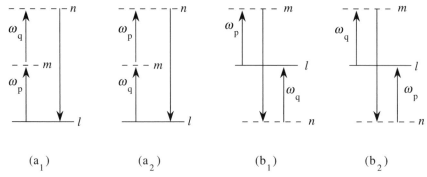

FIGURE 3.6.1   The resonance structure of Eq. (3.6.16).

In order to illustrate how to apply Eq. (3.6.16) and in order to examine the nature of the cancellation that can occur when more than one of the atomic levels contains population, we consider the simple three-level atomic system illustrated in Fig. 3.6.2. We assume that only levels a, b, and c interact appreciably with the optical fields, and that the applied field at frequency ω_1 is nearly resonant with the $a \to b$ transition, the applied field at frequency ω_2 is nearly resonant with the $b \to c$ transition, and the generated field frequency $\omega_3 = \omega_1 + \omega_2$ is nearly resonant with the $c \to a$ transition. If we now perform the summation over the dummy indices l, m, and n in Eq. (3.6.16) and retain only those terms in which both factors in the denominator are resonant, we find that the nonlinear susceptibility is given by

$$\chi_{ijk}^{(2)}(\omega_3, \omega_2, \omega_1)$$

$$= \frac{N}{2\hbar^2} \left\{ \left(\rho_{aa}^{(0)} - \rho_{bb}^{(0)} \right) \left[\frac{\mu_{ac}^i \mu_{cb}^j \mu_{ba}^k}{[(\omega_{ca} - \omega_3) - i\gamma_{ca}][(\omega_{ba} - \omega_1) - i\gamma_{ba}]} \right] \right.$$

$$\left. + \left(\rho_{cc}^{(0)} - \rho_{bb}^{(0)} \right) \left[\frac{\mu_{ac}^i \mu_{cb}^j \mu_{ba}^k}{[(\omega_{ca} - \omega_3) - i\gamma_{ca}][(\omega_{cb} - \omega_2) - i\gamma_{cb}]} \right] \right\}. \quad (3.6.17)$$

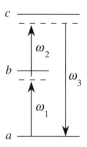

FIGURE 3.6.2   Three-level atomic system.

Here the first term comes from the first term in Eq. (3.6.16), and the second term comes from the last (fourth) term in Eq. (3.6.16). Note that the first term vanishes if $\rho_{aa}^{(0)} = \rho_{bb}^{(0)}$ and that the second term vanishes if $\rho_{bb}^{(0)} = \rho_{cc}^{(0)}$. If all three populations are equal, the resonant contribution vanishes identically.

For some purposes it is useful to express the general result (3.6.16) for the second-order susceptibility in terms of a summation over populations rather than a summation over population differences. In order to cast the susceptibility in such a form, we change the dummy indices l, m, and n to n, l, and m in the summation containing $\rho_{mm}^{(0)}$, but leave them unchanged in the summation containing $\rho_{ll}^{(0)}$. We thereby obtain the result

$$\chi_{ijk}^{(2)}(\omega_p + \omega_q, \omega_q, \omega_p)$$

$$= \frac{N}{2\hbar^2} \sum_{lmn} \rho_{ll}^{(0)} \left\{ \frac{\mu_{ln}^i \mu_{nm}^j \mu_{ml}^k}{[(\omega_{nl} - \omega_p - \omega_q) - i\gamma_{nl}][(\omega_{ml} - \omega_p) - i\gamma_{ml}]} \right. \tag{a_1}$$

$$+ \frac{\mu_{ln}^i \mu_{nm}^k \mu_{ml}^j}{[(\omega_{nl} - \omega_p - \omega_q) - i\gamma_{nl}][(\omega_{ml} - \omega_q) - i\gamma_{ml}]} \tag{a_2}$$

$$+ \frac{\mu_{ln}^k \mu_{nm}^i \mu_{ml}^j}{[(\omega_{mn} - \omega_p - \omega_q) - i\gamma_{mn}][(\omega_{nl} + \omega_p) + i\gamma_{nl}]} \tag{a_1'}$$

$$+ \frac{\mu_{ln}^j \mu_{nm}^i \mu_{ml}^k}{[(\omega_{mn} - \omega_p - \omega_q) - i\gamma_{mn}][(\omega_{nl} + \omega_q) + i\gamma_{nl}]} \tag{a_2'}$$

$$+ \frac{\mu_{ln}^j \mu_{nm}^i \mu_{ml}^k}{[(\omega_{nm} + \omega_p + \omega_q) + i\gamma_{nm}][(\omega_{ml} - \omega_p) - i\gamma_{ml}]} \tag{b_1}$$

$$+ \frac{\mu_{ln}^k \mu_{nm}^i \mu_{ml}^j}{[(\omega_{nm} + \omega_p + \omega_q) + i\gamma_{nm}][(\omega_{ml} - \omega_q) - i\gamma_{ml}]} \tag{b_2}$$

$$+ \frac{\mu_{ln}^k \mu_{nm}^j \mu_{ml}^i}{[(\omega_{ml} + \omega_p + \omega_q) + i\gamma_{ml}][(\omega_{nl} + \omega_p) + i\gamma_{nl}]} \tag{b_1'}$$

$$\left. + \frac{\mu_{ln}^j \mu_{nm}^k \mu_{ml}^i}{[(\omega_{ml} + \omega_p + \omega_q) + i\gamma_{ml}][(\omega_{nl} + \omega_q) + i\gamma_{nl}]} \right\}. \tag{b_2'}$$

$$\tag{3.6.18}$$

As before, we can interpret this result by considering the conditions under which each term of the equation can become resonant. Figure 3.6.3 shows where the energy levels l, m, and n would have to be located in order for each term to become resonant, under the assumption the ω_p and ω_q are both positive. Note that the unprimed diagrams are the same as those of Fig. 3.6.1 (which

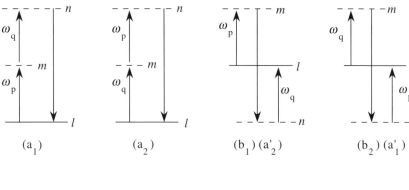

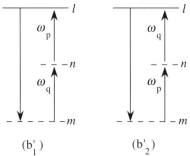

FIGURE 3.6.3   The resonances of Eq. (3.6.18).

represents Eq. (3.6.16)), but that diagrams b_1' and b_2' represent new resonances not present in Fig. 3.6.1.

Another way of making sense of the general eight-term expression for $\chi^{(2)}$ Eq. (3.6.18) is to keep track of how the density matrix is modified in each order of perturbation theory. Through examination of Eqs. (3.6.1) through (3.6.7), we find that the terms of type a, a', b, b' occur as the result of the following perturbation expansion:

$$
\begin{align}
&\text{(a): } \rho_{mm}^{(0)} \to \rho_{vm}^{(1)} \to \rho_{nm}^{(2)}, &&\text{(a'): } \rho_{vv}^{(0)} \to \rho_{vm}^{(1)} \to \rho_{nm}^{(2)}, \\
&\text{(b): } \rho_{vv}^{(0)} \to \rho_{nv}^{(1)} \to \rho_{nm}^{(2)}, &&\text{(b'): } \rho_{nn}^{(0)} \to \rho_{nv}^{(1)} \to \rho_{nm}^{(2)}.
\end{align}
\tag{3.6.19}
$$

However, in writing Eq. (3.6.18) in the displayed form, we have changed the dummy indices appearing in it. In terms of these new indices, the perturbation expansion is

$$
\begin{align}
&\text{(a): } \rho_{ll}^{(0)} \to \rho_{ml}^{(1)} \to \rho_{nl}^{(2)}, &&\text{(a'): } \rho_{ll}^{(0)} \to \rho_{ln}^{(1)} \to \rho_{mn}^{(2)}, \\
&\text{(b): } \rho_{ll}^{(0)} \to \rho_{ml}^{(1)} \to \rho_{mn}^{(2)}, &&\text{(b'): } \rho_{ll}^{(0)} \to \rho_{ln}^{(1)} \to \rho_{lm}^{(2)}.
\end{align}
\tag{3.6.20}
$$

Note that the various terms differ in whether it is the left or right index that is changed by each elementary interaction and by the order in which such a modification occurs.

A convenient way of keeping track of the order in which the elementary interactions occur is by means of double-sided Feynman diagrams. These diagrams represent the way in which the *density operator* is modified by the interaction of the atom with the laser field. We represent the density operator as

$$\hat{\rho} = \overline{|\psi\rangle\langle\psi|}, \tag{3.6.21}$$

where $|\psi\rangle$ represents the ket vector for some state of the system, $\langle\psi|$ (the bra vector) represents the Hermitian adjoint of $\langle\psi|$, and the overbar represents an ensemble average. The elements of the density matrix are related to the density operator $\hat{\rho}$ through the equation

$$\rho_{nm} = \langle n|\hat{\rho}|m\rangle. \tag{3.6.22}$$

Figure 3.6.4 gives a pictorial description of the modification of the density matrix as indicated by the expressions (3.6.20). The left-hand side of each diagram indicates the time evolution of $|\psi\rangle$, and the right-hand side indicates

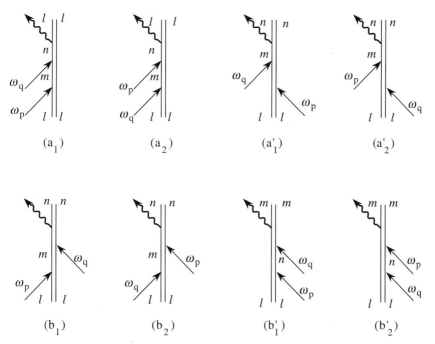

FIGURE 3.6.4   Double-sided Feynman diagrams.

the time evolution of $\langle\psi|$, with time increasing vertically upward. Each interaction with the applied field is indicated by a solid arrow labeled by the field frequency. The trace operation, which corresponds to calculating the output field, is indicated by the wavy arrow.* It should be noted that there are several different conventions concerning the rules for drawing double-sided Feynman diagrams (Yee and Gustafson, 1978; Prior, 1984; Boyd and Mukamel, 1984).

$\chi^{(2)}$ in the Limit of Nonresonant Excitation

When all of the frequencies ω_p, ω_q, and $\omega_p + \omega_q$ differ significantly from any resonance frequency of the atomic system, the imaginary contributions to the denominators in Eq. (3.6.18) can be ignored. In this case, the expression for $\chi^{(2)}$ can be simplified. In particular, terms (a_2') and (b_1) can be combined into a single term, and similarly for terms (a_1') and (b_2). We note that the numerators of terms (a_2') and (b_1) are identical, and that their denominators can be combined as follows:

$$\frac{1}{(\omega_{mn} - \omega_p - \omega_q)(\omega_{nl} + \omega_q)} + \frac{1}{(-\omega_{mn} + \omega_p + \omega_q)(\omega_{ml} - \omega_p)}$$

$$= \frac{1}{(\omega_{mn} - \omega_p - \omega_q)}\left[\frac{1}{\omega_{nl} + \omega_q} - \frac{1}{\omega_{ml} - \omega_p}\right]$$

$$= \frac{1}{(\omega_{mn} - \omega_p - \omega_q)}\left[\frac{\omega_{ml} - \omega_p - \omega_{nl} - \omega_q}{(\omega_{nl} + \omega_q)(\omega_{ml} - \omega_p)}\right] \qquad (3.6.23)$$

$$= \frac{1}{(\omega_{mn} - \omega_p - \omega_q)}\left[\frac{\omega_{mn} - \omega_p - \omega_q}{(\omega_{nl} + \omega_q)(\omega_{ml} - \omega_p)}\right]$$

$$= \frac{1}{(\omega_{nl} + \omega_q)(\omega_{ml} - \omega_p)}.$$

The same procedure can be performed on terms (a_1') and (b_2); the only difference between this case and the one treated in Eq. (3.6.23) is that ω_p and ω_q have switched roles. The frequency dependence is thus

$$\frac{1}{(\omega_{nl} + \omega_p)(\omega_{ml} - \omega_q)}. \qquad (3.6.24)$$

---

* In drawing Fig. 3.6.4, we have implicitly assumed that all of the applied field frequencies are positive, which corresponds to the absorption of an incident photon. The interaction with a negative field frequency which corresponds to the emission of a photon, is sometimes indicated by a solid arrow pointing diagonally upward and away from (rather than towards) the central double line.

The expression for $\chi^{(2)}$ in the off-resonance case thus becomes

$$\chi^{(2)}_{ijk}(\omega_p + \omega_q, \omega_q, \omega_p)$$

$$= \frac{N}{2\hbar^2} \sum_{lmn} \rho^{(0)}_{ll} \left\{ \frac{\mu^i_{ln}\mu^j_{nm}\mu^k_{ml}}{[(\omega_{nl} - \omega_p - \omega_q)(\omega_{ml} - \omega_p)} \right. \tag{a_1}$$

$$+ \frac{\mu^i_{ln}\mu^k_{nm}\mu^j_{ml}}{(\omega_{nl} - \omega_p - \omega_q)(\omega_{ml} - \omega_q)} \tag{a_2}$$

$$+ \frac{\mu^j_{ln}\mu^i_{nm}\mu^k_{ml}}{(\omega_{nl} + \omega_q)(\omega_{ml} - \omega_p)} \tag{b_1), (a'_2}$$

$$+ \frac{\mu^k_{ln}\mu^i_{nm}\mu^j_{ml}}{(\omega_{nl} + \omega_p)(\omega_{ml} - \omega_q)} \tag{b_2), (a'_1}$$

$$+ \frac{\mu^k_{ln}\mu^j_{nm}\mu^i_{ml}}{(\omega_{ml} + \omega_p + \omega_q)(\omega_{nl} + \omega_p)} \tag{b'_1}$$

$$+ \left. \frac{\mu^j_{ln}\mu^k_{nm}\mu^i_{ml}}{(\omega_{ml} + \omega_p + \omega_q)(\omega_{nl} + \omega_q)} \right\}. \tag{b'_2}$$

$$\tag{3.6.25}$$

Note that only six terms appear in this expression for the off-resonance susceptibility, whereas eight terms appear in the general expression of Eq. (3.6.18). One can verify by explicit calculation that Eq. (3.6.25) satisfies the condition of full permutation symmetry (see also Eq. (1.5.7)). In addition, one can see by inspection that Eq. (3.6.25) is identical to the result obtained above (Eq. (3.2.27)) based on perturbation theory of the atomic wavefunction.

There are several diagrammatic methods that can be used to interpret this expression. One of the simplest is to plot the photon energies on an atomic energy-level diagram. This method displays the conditions under which each contribution can become resonant. The results of such an analysis gives exactly the same diagrams displayed in Fig. 3.6.3. Equation (3.6.25) can also be understood in terms of a diagrammatic approach introduced by Ward (1965).

3.7. Density Matrix Calculation of the Third-Order Susceptibility

The third-order correction to the density matrix is given by the perturbation expansion of Eq. (3.4.16) as

$$\rho^{(3)}_{nm} = e^{-(i\omega_{nm} + \gamma_{nm})t} \int_{-\infty}^{t} \frac{-i}{\hbar} [\hat{V}, \hat{\rho}^{(2)}]_{nm} e^{(i\omega_{nm} + \gamma_{nm})t'} dt', \tag{3.7.1}$$

where the commutator can be represented explicitly as

$$\left[\hat{V}, \hat{\rho}^{(2)}\right]_{nm} = -\sum_{v}\left(\boldsymbol{\mu}_{nv}\rho_{vm}^{(2)} - \rho_{nv}^{(2)}\boldsymbol{\mu}_{vm}\right)\cdot\tilde{\mathbf{E}}(t). \qquad (3.7.2)$$

Expressions for $\rho_{vm}^{(2)}$ and $\rho_{nv}^{(2)}$ are available from Eq. (3.6.7). Since these expressions are very complicated, we use the abbreviated notation introduced there:

$$\rho_{vm}^{(2)} = \sum_{l}\sum_{pq} K_{vml}e^{-i(\omega_p+\omega_q)t}, \qquad (3.7.3)$$

where

$$K_{vml} = \frac{\rho_{mm}^{(0)} - \rho_{ll}^{(0)}}{\hbar^2}\frac{[\boldsymbol{\mu}_{vl}\cdot\mathbf{E}(\omega_q)][\boldsymbol{\mu}_{lm}\cdot\mathbf{E}(\omega_p)]}{[(\omega_{vm}-\omega_p-\omega_q)-i\gamma_{vm}][(\omega_{lm}-\omega_p)-i\gamma_{lm}]}$$
$$-\frac{\rho_{ll}^{(0)} - \rho_{vv}^{(0)}}{\hbar^2}\frac{[\boldsymbol{\mu}_{vl}\cdot\mathbf{E}(\omega_p)][\boldsymbol{\mu}_{lm}\cdot\mathbf{E}(\omega_q)]}{[(\omega_{vm}-\omega_p-\omega_q)-i\gamma_{vm}][(\omega_{vl}-\omega_p)-i\gamma_{vl}]} \qquad (3.7.4)$$

and

$$\rho_{nv}^{(2)} = \sum_{l}\sum_{pq} K_{nvl}e^{-i(\omega_p+\omega_q)t}, \qquad (3.7.5)$$

where

$$K_{nvl} = \frac{\rho_{vv}^{(0)} - \rho_{ll}^{(0)}}{\hbar^2}\frac{[\boldsymbol{\mu}_{nl}\cdot\mathbf{E}(\omega_q)][\boldsymbol{\mu}_{lv}\cdot\mathbf{E}(\omega_p)]}{[(\omega_{nv}-\omega_p-\omega_q)-i\gamma_{nv}][(\omega_{lv}-\omega_p)-i\gamma_{lv}]}$$
$$-\frac{\rho_{ll}^{(0)} - \rho_{nn}^{(0)}}{\hbar^2}\frac{[\boldsymbol{\mu}_{vl}\cdot\mathbf{E}(\omega_p)][\boldsymbol{\mu}_{lm}\cdot\mathbf{E}(\omega_q)]}{[(\omega_{nv}-\omega_p-\omega_q)-i\gamma_{nv}][(\omega_{nl}-\omega_p)-i\gamma_{nl}]}. \qquad (3.7.6)$$

We also represent the electric field as

$$\tilde{\mathbf{E}}(t) = \sum_{r}\mathbf{E}(\omega_r)e^{-i\omega_r t}. \qquad (3.7.7)$$

The commutator thus becomes

$$\left[\hat{V}, \hat{\rho}^{(2)}\right]_{nm} = -\sum_{vl}\sum_{pqr}[\boldsymbol{\mu}_{nv}\cdot\mathbf{E}(\omega_r)]K_{vml}e^{-i(\omega_p+\omega_q+\omega_r)t}$$
$$+\sum_{vl}\sum_{pqr}[\boldsymbol{\mu}_{vm}\cdot\mathbf{E}(\omega_r)]K_{nvl}e^{-i(\omega_p+\omega_q+\omega_r)t}. \qquad (3.7.8)$$

The integration of Eq. (3.7.1) with the commutator given by Eq. (3.7.8) can now be performed. We obtain

$$
\rho_{nm}^{(3)} = \frac{1}{\hbar} \sum_{vl} \sum_{pqr} \left\{ \frac{[\boldsymbol{\mu}_{nv} \cdot \mathbf{E}(\omega_r)] K_{vml}}{(\omega_{nm} - \omega_p - \omega_q - \omega_r) - i\gamma_{nm}} \right.
$$

$$
\left. - \frac{[\boldsymbol{\mu}_{vm} \cdot \mathbf{E}(\omega_r)] K_{nvl}}{(\omega_{nm} - \omega_p - \omega_q - \omega_r) - i\gamma_{nm}} \right\} e^{-i(\omega_p + \omega_q + \omega_r)t}.
$$

$$(3.7.9)$$

The nonlinear polarization oscillating at frequency $\omega_p + \omega_q + \omega_r$ is given by

$$
\mathbf{P}(\omega_p + \omega_q + \omega_r) = N \langle \boldsymbol{\mu}(\omega_p + \omega_q + \omega_r) \rangle, \qquad (3.7.10)
$$

where

$$
\langle \tilde{\boldsymbol{\mu}} \rangle = \sum_{nm} \rho_{nm} \boldsymbol{\mu}_{\mathbf{mn}} \equiv \sum_s \langle \boldsymbol{\mu}(\omega_s) \rangle e^{-i\omega_s t}. \qquad (3.7.11)
$$

We express the nonlinear polarization in terms of the third-order susceptibility defined by (see also Eq. (1.3.21))

$$
P_k(\omega_p + \omega_q + \omega_r) = \sum_{hij} \sum_{pqr} \chi_{kjih}^{(3)}(\omega_p + \omega_q + \omega_r, \omega_r, \omega_q, \omega_p)
$$

$$
\times E_j(\omega_r) E_i(\omega_q) E_h(\omega_p). \qquad (3.7.12)
$$

By combining Eqs. (3.7.9) through (3.7.12), we find that the third-order susceptibility is given by

$$
\chi_{kjih}^{(3)}(\omega_p + \omega_q + \omega_r, \omega_r, \omega_q, \omega_p) = \frac{N}{\hbar^3} \mathscr{P}_I \sum_{nmvl}
$$

$$
\left\{ \frac{\left(\rho_{mm}^{(0)} - \rho_{ll}^{(0)}\right) \mu_{mn}^k \mu_{nv}^j \mu_{vl}^i \mu_{lm}^h}{[(\omega_{nm} - \omega_p - \omega_q - \omega_r) - i\gamma_{nm}][(\omega_{vm} - \omega_p - \omega_q) - i\gamma_{vm}][(\omega_{lm} - \omega_p) - i\gamma_{lm}]} \right. \quad (a)
$$

$$
- \frac{\left(\rho_{ll}^{(0)} - \rho_{vv}^{(0)}\right) \mu_{mn}^k \mu_{nv}^j \mu_{lm}^i \mu_{vl}^h}{[(\omega_{nm} - \omega_p - \omega_q - \omega_r) - i\gamma_{nm}][(\omega_{vm} - \omega_p - \omega_q) - i\gamma_{vm}][(\omega_{vl} - \omega_p) - i\gamma_{vl}]} \quad (b)
$$

$$
- \frac{\left(\rho_{vv}^{(0)} - \rho_{ll}^{(0)}\right) \mu_{mn}^k \mu_{vm}^j \mu_{nl}^i \mu_{lv}^h}{[(\omega_{nm} - \omega_p - \omega_q - \omega_r) - i\gamma_{nm}][(\omega_{nv} - \omega_p - \omega_q) - i\gamma_{nv}][(\omega_{lv} - \omega_p) - i\gamma_{lv}]} \quad (c)
$$

$$
\left. + \frac{\left(\rho_{ll}^{(0)} - \rho_{nn}^{(0)}\right) \mu_{mn}^k \mu_{vm}^j \mu_{lv}^i \mu_{nl}^h}{[(\omega_{nm} - \omega_p - \omega_q - \omega_r) - i\gamma_{nm}][(\omega_{nv} - \omega_p - \omega_q) - i\gamma_{nv}][(\omega_{nl} - \omega_p) - i\gamma_{nl}]} \right\}. \quad (d)
$$

$$(3.7.13)$$

Here we have again made use of the intrinsic permutation operator $\mathscr{P}_I$, whose meaning is that everything to the right of it is to be averaged over all possible permutations of the input frequencies ω_p, ω_q, and ω_r, with the cartesian indices h, i, j permuted simultaneously. Next, we rewrite this equation as eight separate terms by changing the dummy indices so that l is always the index of $\rho_{ii}^{(0)}$. We also require that only positive resonance frequencies appear if the energies are ordered so that $E_\nu > E_n > E_m > E_l$, and we arrange the matrix elements so that they appear in "natural" order, $l \to m \to n \to \nu$ (reading right to left). We obtain

$$\chi_{kjih}^{(3)}(\omega_p + \omega_q + \omega_r, \omega_r, \omega_q, \omega_p) = \frac{N}{\hbar^3} \mathscr{P}_I \sum_{\nu nml} \rho_{ll}^{(0)}$$

$$\times \left\{ \frac{\mu_{l\nu}^k \mu_{\nu n}^j \mu_{nm}^i \mu_{ml}^h}{[(\omega_{\nu l} - \omega_p - \omega_q - \omega_r) - i\gamma_{\nu l}][(\omega_{nl} - \omega_p - \omega_q) - i\gamma_{nl}][(\omega_{ml} - \omega_p) - i\gamma_{ml}]} \right. \quad (a_1)$$

$$+ \frac{\mu_{l\nu}^h \mu_{\nu n}^k \mu_{nm}^j \mu_{ml}^i}{[(\omega_{n\nu} - \omega_p - \omega_q - \omega_r) - i\gamma_{n\nu}][(\omega_{m\nu} - \omega_p - \omega_q) - i\gamma_{m\nu}][(\omega_{\nu l} + \omega_p) + i\gamma_{\nu l}]} \quad (a_2)$$

$$+ \frac{\mu_{l\nu}^j \mu_{\nu n}^k \mu_{nm}^i \mu_{ml}^h}{[(\omega_{n\nu} - \omega_p - \omega_q - \omega_r) - i\gamma_{n\nu}][(\omega_{\nu m} + \omega_p + \omega_q) + i\gamma_{\nu m}][(\omega_{ml} - \omega_p) - i\gamma_{ml}]} \quad (b_1)$$

$$+ \frac{\mu_{l\nu}^h \mu_{\nu n}^i \mu_{nm}^k \mu_{ml}^j}{[(\omega_{m\nu} - \omega_p - \omega_q - \omega_r) - i\gamma_{m\nu}][(\omega_{nl} + \omega_p + \omega_q) + i\gamma_{nl}][(\omega_{\nu l} + \omega_p) + i\gamma_{\nu l}]} \quad (b_2)$$

$$+ \frac{\mu_{l\nu}^j \mu_{\nu n}^k \mu_{nm}^i \mu_{ml}^h}{[(\omega_{\nu n} + \omega_p + \omega_q + \omega_r) + i\gamma_{\nu n}][(\omega_{nl} - \omega_p - \omega_q) - i\gamma_{nl}][(\omega_{ml} - \omega_p) - i\gamma_{ml}]} \quad (c_1)$$

$$+ \frac{\mu_{l\nu}^h \mu_{\nu n}^j \mu_{nm}^k \mu_{ml}^i}{[(\omega_{nm} + \omega_p + \omega_q + \omega_r) + i\gamma_{nm}][(\omega_{m\nu} - \omega_p - \omega_q) - i\gamma_{m\nu}][(\omega_{\nu l} + \omega_p) + i\gamma_{\nu l}]} \quad (c_2)$$

$$+ \frac{\mu_{l\nu}^i \mu_{\nu n}^j \mu_{nm}^k \mu_{ml}^h}{[(\omega_{nm} + \omega_p + \omega_q + \omega_r) + i\gamma_{nm}][(\omega_{\nu m} + \omega_p + \omega_q) + i\gamma_{\nu m}][(\omega_{ml} - \omega_p) - i\gamma_{ml}]} \quad (d_1)$$

$$\left. + \frac{\mu_{l\nu}^h \mu_{\nu n}^i \mu_{nm}^j \mu_{ml}^k}{[(\omega_{ml} + \omega_p + \omega_q + \omega_r) + i\gamma_{ml}][(\omega_{nl} + \omega_p + \omega_q) + i\gamma_{nl}][(\omega_{\nu l} + \omega_p) + i\gamma_{\nu l}]} \right\}. \quad (d_2)$$

$$(3.7.14)$$

For the general case in which ω_p, ω_q, and ω_r are distinct, six permutations of the field frequencies occur, and hence the expression for $\chi^{(3)}$ consists of 48 different terms once the permutation operator $\mathscr{P}_I$ is expanded. The resonance structure of this expression can be understood in terms of the energy level diagrams shown in Fig. 3.7.1. Furthermore, the nature of the perturbation expansion leading to Eq. (3.7.14) can be understood in terms of the double-sided Feynman diagrams shown in Fig. 3.7.2.

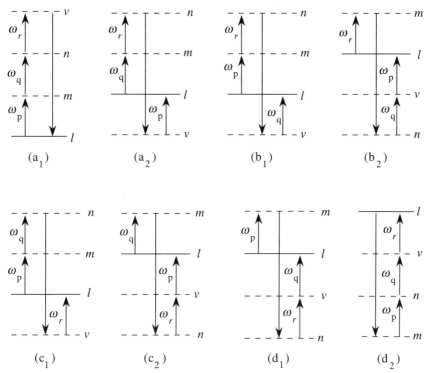

FIGURE 3.7.1   The resonance structure of the expression (3.7.14) for the third-order nonlinear susceptibility.

We saw in Section 3.2 that the general expression for the third-order susceptibility calculated using perturbation theory applied to the atomic wavefunction contained 24 terms. Equation (3.2.33) shows four of these terms explicitly; the other terms are obtained from the six permutations of the frequencies of the applied field. It can be shown that Eq. (3.7.14) reduces to Eq. (3.2.33) in the limit of nonresonant excitation, where the imaginary contributions ($i\gamma_{\alpha\beta}$) appearing in Eq. (3.7.14) can be ignored. One can demonstrate this fact by means of a calculation similar to that used to derive Eq. (3.6.24), which applies to the case of the second-order susceptibility (see Problem 5 at the end of this chapter).

In fact, even in the general case in which the imaginary contributions $i\gamma_{\alpha\beta}$ appearing in Eq. (3.7.14) are retained, it is possible to rewrite the 48-term expression (3.7.14) in the form of the 24-term expression (3.2.33), by allowing the coefficient of each of the 24 terms to be weakly frequency-dependent. These frequency-dependent coefficients usually display resonances at frequencies other than those that appear in Fig. 3.7.1, and these new resonances occur only if

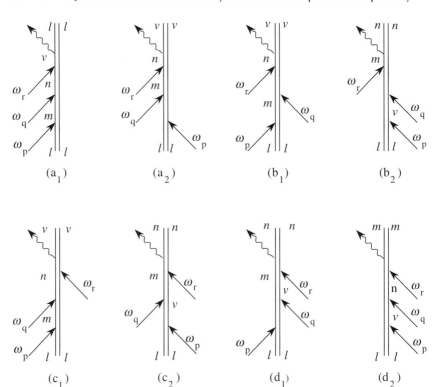

FIGURE 3.7.2   Double-sided Feynman diagrams associated with the various terms in Eq. (3.7.14).

the line-broadening mechanism is collisional (rather than radiative). The nature of these *collision-induced resonances* has been discussed by Bloembergen *et al.* (1978), Prior (1984), and Rothberg (1987).

3.8.   Local-Field Corrections to the Nonlinear Optical Susceptibility

The treatment of the nonlinear optical susceptibility presented thus far has made the implicit assumption that the electric field acting on each atom or molecule is the macroscopic electric field that appears in Maxwell's equations. In general, one has to distinguish between the macroscopic electric field and the effective electric field that each atom experiences, which is also known as the Lorentz local field. The distinction between these two fields is important except for the case of a medium that is so dilute that its linear dielectric constant is nearly equal to unity.

Local-Field Effects in Linear Optics

Let us first review the theory of local field effects in linear optics. The electric field $\tilde{E}$ that appears in Maxwell's equations in the form of Eqs. (2.1.1) through (2.1.8) is known as the macroscopic or Maxwell field. This field is obtained by performing a spatial average of the actual (or microscopic) electric field over a region of space whose linear dimensions are of the order of at least several atomic diameters. It is useful to perform such an average to smooth out the wild variations in electric field that occur in the immediate vicinity of the atomic nuclei and electrons. The macroscopic electric field thus has contributions from sources external to the material system and from the dipole moments of all of the dipoles that constitute the system.

Let us now see how to calculate the dipole moment induced in a representative molecule contained within the material system. We assume for simplicity that the medium is lossless and dispersionless, so that we can conveniently represent the fields as time-varying quantities rather than having to introduce complex field amplitudes. We let $\tilde{E}$ represent the macroscopic field and $\tilde{P}$ the polarization within the bulk of the material. Furthermore, we represent the dipole moment induced in a typical molecule as

$$\tilde{p} = \alpha \tilde{E}_{loc}, \tag{3.8.1}$$

where α is the usual linear polarizability and where $\tilde{E}_{loc}$ is the local field, that is, the effective electric field that acts on the molecule. The local field is the field resulting from all external sources and from all molecules within the sample *except* the one under consideration.

We calculate this field through use of a procedure described by Lorentz (1952). We imagine drawing a small sphere centered on the molecule under consideration, as shown in Fig. 3.8.1. This sphere is assumed to be sufficiently large that it contains many atoms. The electric field produced at the center of the sphere by molecules contained within the sphere (not including the molecule at the center) will tend to cancel, and for the case of a liquid, gas, or cubic crystal, this cancellation can be shown to be exact. We can then imagine removing these molecules from the sphere, leaving only the molecule under

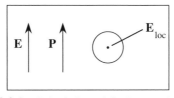

FIGURE 3.8.1   Calculation of the Lorentz local field.

consideration, which is then located at the center of an evacuated sphere within an otherwise uniformly polarized medium. It is then a simple problem in electrostatics to calculate the value of the field at the center of the sphere. The field, which we identify as the Lorentz local field, is given by (see also Born and Wolf, 1975, Section 2.3, or Jackson, 1975, Section 4.5)

$$\tilde{\mathbf{E}}_{loc} = \tilde{\mathbf{E}} + \tfrac{4}{3}\pi\tilde{\mathbf{P}}. \tag{3.8.2}$$

By definition, the polarization of the material is given by

$$\tilde{\mathbf{P}} = N\tilde{\mathbf{p}}, \tag{3.8.3}$$

where N is the number density of atoms and $\tilde{\mathbf{p}}$ is the dipole moment per atom, which under the present circumstances is given by Eq. (3.8.1). By combining Eqs. (3.8.1) through (3.8.3), we find that the polarization and macroscopic field are related by

$$\tilde{\mathbf{P}} = N\alpha\big(\tilde{\mathbf{E}} + \tfrac{4}{3}\pi\tilde{\mathbf{P}}\big). \tag{3.8.4}$$

It is useful to express this result in terms of the linear susceptibility $\chi^{(1)}$, defined by

$$\tilde{\mathbf{P}} = \chi^{(1)}\tilde{\mathbf{E}}. \tag{3.8.5}$$

If we substitute this expression for $\tilde{\mathbf{P}}$ into Eq. (3.8.4) and solve the resulting equation for $\chi^{(1)}$, we find that

$$\chi^{(1)} = \frac{N\alpha}{1 - \tfrac{4}{3}\pi N\alpha}. \tag{3.8.6}$$

For the usual case in which the polarizability α is positive, we see that the susceptibility is larger than the value $N\alpha$ predicted by theories that ignore local-field corrections. We also see that the susceptibility increases with N more rapidly than linearly.

Alternatively, we can express the result given by Eq. (3.8.6) in terms of the linear dielectric constant

$$\epsilon^{(1)} = 1 + 4\pi\chi^{(1)}. \tag{3.8.7}$$

If the left-hand side of Eq. (3.8.6) is replaced by $\chi^{(1)} = (\epsilon^{(1)} - 1)/4\pi$ and the resulting equation is rearranged so that its right-hand side is linear in α, we find that the dielectric constant is given by the expression

$$\frac{\epsilon^{(1)} - 1}{\epsilon^{(1)} + 2} = \frac{4}{3}\pi N\alpha. \tag{3.8.8a}$$

This equation (often with $\epsilon^{(1)}$ replaced by n^2) is known as the Lorentz–Lorenz law. Note that, through rearrangement, Eq. (3.8.8a) can be written as

$$\frac{\epsilon^{(1)} + 2}{3} = \frac{1}{1 - \frac{4}{3}\pi N\alpha}. \tag{3.8.8b}$$

Equation (3.8.6) can thus be expressed as

$$\chi^{(1)} = \frac{\epsilon^{(1)} + 2}{3} N\alpha. \tag{3.8.8c}$$

This result shows that $\chi^{(1)}$ is larger than $N\alpha$ by the factor $(\epsilon^{(1)} + 2)/3$. The factor $(\epsilon^{(1)} + 2)/3$ can thus be interpreted as the local-field correction factor for the linear susceptibility.

Local-Field Effects in Nonlinear Optics

In the nonlinear-optical case, the Lorentz local field is still given by Eq. (3.8.2), but the polarization now has both linear and nonlinear contributions:

$$\tilde{\mathbf{P}} = \tilde{\mathbf{P}}^{\mathrm{L}} + \tilde{\mathbf{P}}^{\mathrm{NL}}. \tag{3.8.9}$$

We represent the linear contribution as

$$\tilde{\mathbf{P}}^{\mathrm{L}} = N\alpha\tilde{\mathbf{E}}_{\mathrm{loc}}. \tag{3.8.10}$$

Note that this contribution is linear in the sense that it is linear in the strength of the local field. In general it is not linear in the strength of the macroscopic field. We next introduce Eqs. (3.8.2) and (3.8.9) into this equation to obtain

$$\tilde{\mathbf{P}}^{\mathrm{L}} = N\alpha\left(\tilde{\mathbf{E}} + \tfrac{4}{3}\pi\tilde{\mathbf{P}}^{\mathrm{L}} + \tfrac{4}{3}\pi\tilde{\mathbf{P}}^{\mathrm{NL}}\right). \tag{3.8.11}$$

We now solve this equation for $\tilde{\mathbf{P}}^{\mathrm{L}}$ and use Eqs. (3.8.6) and (3.8.7) to express the factor $N\alpha$ that appears in the resulting expression in terms of the linear dielectric constant. We thereby obtain

$$\tilde{\mathbf{P}}^{\mathrm{L}} = \frac{\epsilon^{(1)} - 1}{4\pi}\left(\tilde{\mathbf{E}} + \tfrac{4}{3}\pi\tilde{\mathbf{P}}^{\mathrm{NL}}\right). \tag{3.8.12}$$

Next we consider the displacement vector

$$\tilde{\mathbf{D}} = \tilde{\mathbf{E}} + 4\pi\tilde{\mathbf{P}} = \tilde{\mathbf{E}} + 4\pi\tilde{\mathbf{P}}^{\mathrm{L}} + 4\pi\tilde{\mathbf{P}}^{\mathrm{NL}}. \tag{3.8.13}$$

If the expression (3.8.12) for the linear polarization is substituted into this expression, we obtain

$$\tilde{\mathbf{D}} = \epsilon^{(1)}\tilde{\mathbf{E}} + 4\pi\left(\frac{\epsilon^{(1)} + 2}{3}\right)\tilde{\mathbf{P}}^{\mathrm{NL}}. \tag{3.8.14}$$

We see that the second term is not simply $4\pi \tilde{\mathbf{P}}^{\mathrm{NL}}$, as might have been expected, but that the nonlinear polarization appears multiplied by the factor $(\epsilon^{(1)}+2)/3$. We recall that in the derivation of the polarization-driven wave equation of nonlinear optics, a nonlinear source term appears when the second time derivative of $\tilde{\mathbf{D}}$ is calculated (see, for example, Eq. (2.1.9a)). As a consequence of Eq. (3.8.14), we see that the nonlinear source term is actually the nonlinear polarization $\tilde{\mathbf{P}}^{\mathrm{NL}}$ multiplied by the factor $(\epsilon^{(1)}+2)/3$. To emphasize this point, Bloembergen (1965) introduces the *nonlinear source polarization* defined by

$$\tilde{\mathbf{P}}^{\mathrm{NLS}} = \left(\frac{\epsilon^{(1)}+2}{3}\right)\tilde{\mathbf{P}}^{\mathrm{NL}} \tag{3.8.15}$$

so that Eq. (3.8.14) can be expressed as

$$\tilde{\mathbf{D}} = \epsilon^{(1)}\tilde{\mathbf{E}} + 4\pi\tilde{\mathbf{P}}^{\mathrm{NLS}}. \tag{3.8.16}$$

When the derivation of the wave equation is carried out as in Section 2.1 using this expression for $\tilde{\mathbf{D}}$, we obtain the result

$$\nabla \times \nabla \times \tilde{\mathbf{E}} + \frac{\epsilon^{(1)}}{c^2}\frac{\partial^2\tilde{\mathbf{E}}}{\partial t^2} = -\frac{4\pi}{c^2}\frac{\partial^2\tilde{\mathbf{P}}^{\mathrm{NLS}}}{\partial t^2}. \tag{3.8.17}$$

This result shows how local-field corrections are incorporated into the wave equation.

The distinction between the local and macroscopic fields also arises in that the field that induces a dipole moment in each atom is the local field, whereas by definition the nonlinear susceptibility relates the nonlinear source polarization to the macroscopic field. To good approximation, we can relate the local and macroscopic fields by replacing $\tilde{\mathbf{P}}$ by $\tilde{\mathbf{P}}^L$ in Eq. (3.8.2) to obtain

$$\tilde{\mathbf{E}}_{\mathrm{loc}} = \tilde{\mathbf{E}} + \tfrac{4}{3}\pi\chi^{(1)}\tilde{\mathbf{E}} = \left(1 + \frac{4\pi}{3}\frac{\epsilon^{(1)}-1}{4\pi}\right)\tilde{\mathbf{E}},$$

or

$$\tilde{\mathbf{E}}_{\mathrm{loc}} = \left(\frac{\epsilon^{(1)}+2}{3}\right)\tilde{\mathbf{E}}. \tag{3.8.18}$$

We now apply the results of Eqs. (3.8.17) and (3.8.18) to the case of second-order nonlinear interactions. We define the nonlinear susceptibility by means of the equation (see also Eq. (1.3.13))

$$P_i^{\mathrm{NLS}}(\omega_m + \omega_n) = \sum_{jk}\sum_{(mn)}\chi_{ijk}^{(2)}(\omega_m + \omega_n, \omega_m, \omega_n)E_j(\omega_m)E_k(\omega_n), \tag{3.8.19}$$

where

$$P_i^{\text{NLS}}(\omega_m + \omega_n) = \left(\frac{\epsilon^{(1)}(\omega_m + \omega_n) + 2}{3}\right) P_i^{\text{NL}}(\omega_m + \omega_n) \quad (3.8.20)$$

and where the quantities $E_j(\omega_m)$ represent macroscopic fields. The nonlinear polarization (i.e., the second-order contribution to the dipole moment per unit volume) can be represented as

$$P_i^{\text{NL}}(\omega_m + \omega_n) = N \sum_{jk} \sum_{(mn)} \beta_{ijk}(\omega_m + \omega_n, \omega_m, \omega_n) E_j^{\text{loc}}(\omega_m) E_k^{\text{loc}}(\omega_n),$$

$$(3.8.21)$$

where the proportionality constant β_{ijk} is known as the second-order hyperpolarizability. The local fields appearing in this expression are related to the macroscopic fields according to Eq. (3.8.18), which we now rewrite as

$$E_j^{\text{loc}}(\omega_m) = \left(\frac{\epsilon^{(1)}(\omega_m) + 2}{3}\right) E_j(\omega_m). \quad (3.8.22)$$

By combining Eqs. (3.8.19) through (3.8.22), we find that the nonlinear susceptibility can be represented as

$$\chi_{ijk}^{(2)}(\omega_m + \omega_n, \omega_m, \omega_n)$$

$$= \mathscr{L}^{(2)}(\omega_m + \omega_n, \omega_m, \omega_n) N\beta_{ijk}(\omega_m + \omega_n, \omega_m, \omega_n), \quad (3.8.23)$$

where

$$\mathscr{L}^{(2)}(\omega_m + \omega_n, \omega_m, \omega_n)$$

$$= \left(\frac{\epsilon^{(1)}(\omega_m + \omega_n) + 2}{3}\right)\left(\frac{\epsilon^{(1)}(\omega_m) + 2}{3}\right)\left(\frac{\epsilon^{(1)}(\omega_n) + 2}{3}\right) \quad (3.8.24)$$

gives the local-field correction factor for the second-order susceptibility. For example, Eq. (3.6.18) for $\chi^{(2)}$ should be multiplied by this factor to obtain the correct expression including local-field effects.

This result is readily generalized to higher-order nonlinear interaction. For example, the expression for $\chi^{(3)}$ obtained ignoring local-field corrections should be multiplied by the factor

$$\mathscr{L}^{(3)}(\omega_l + \omega_m + \omega_n, \omega_l, \omega_m, \omega_n) =$$

$$\left(\frac{\epsilon^{(1)}(\omega_l + \omega_m + \omega_n) + 2}{3}\right)\left(\frac{\epsilon^{(1)}(\omega_l) + 2}{3}\right)\left(\frac{\epsilon^{(1)}(\omega_m) + 2}{3}\right)\left(\frac{\epsilon^{(1)}(\omega_n) + 2}{3}\right).$$

$$(3.8.25)$$

Our derivation of the form of the local-field correction factor has essentially followed the procedure of Bloembergen (1965). The nature of local-field corrections in nonlinear optics can be understood from a very different point of view introduced by Mizrahi and Sipe (1986). This method has the desirable feature that, unlike the procedure described above, it does not require that we maintain the somewhat arbitrary distinction between the nonlinear polarization and the nonlinear source polarization. For simplicity, we describe this procedure only for the case of third-harmonic generation in the scalar field approximation. We assume that the total polarization (including both linear and nonlinear contributions) at the third-harmonic frequency is given by

$$P(3\omega) = N\alpha(3\omega)E_{\text{loc}}(3\omega) + N\gamma(3\omega, \omega, \omega, \omega)E_{\text{loc}}^3(\omega), \quad (3.8.26)$$

where $\alpha(3\omega)$ is the linear polarizability for radiation at frequency 3ω and where $\gamma(3\omega, \omega, \omega, \omega)$ is the hyperpolarizability leading to third-harmonic generation. We next use Eqs. (3.8.2) and (3.8.18) to rewrite Eq. (3.8.26) as

$$P(3\omega) = N\alpha(3\omega)\left[E(3\omega) + \tfrac{4}{3}\pi P(3\omega)\right]$$
$$+ N\gamma(3\omega, \omega, \omega, \omega)\left(\frac{\epsilon^{(1)}(\omega) + 2}{3}\right)^3 E^3(\omega). \quad (3.8.27)$$

This equation is now solved algebraically for $P(3\omega)$ to obtain

$$P(3\omega) = \frac{N\alpha(3\omega)E(3\omega)}{1 - \tfrac{4}{3}\pi N\alpha(3\omega)} + \frac{N\gamma(3\omega, \omega, \omega, \omega)}{1 - \tfrac{4}{3}\pi N\alpha(3\omega)}\left(\frac{\epsilon^{(1)}(\omega) + 2}{3}\right)^3 E(\omega)^3. \quad (3.8.28)$$

We can identify the first and second terms of this expression as the linear and third-order polarizations, which we represent as

$$P(3\omega) = \chi^{(1)}(3\omega)E(3\omega) + \chi^{(3)}(3\omega, \omega, \omega, \omega)E(\omega)^3, \quad (3.8.29)$$

where (in agreement with the unusual Lorentz–Lorenz law) the linear susceptibility is given by

$$\chi^{(1)}(3\omega) = \frac{N\alpha(3\omega)}{1 - \tfrac{4}{3}\pi N\alpha(3\omega)}, \quad (3.8.30)$$

and where the third-order susceptibility is given by

$$\chi^{(3)}(3\omega, \omega, \omega, \omega) = \left(\frac{\epsilon^{(1)}(\omega) + 2}{3}\right)^3 \left(\frac{\epsilon^{(1)}(3\omega) + 2}{3}\right) N\gamma(3\omega, \omega, \omega, \omega). \quad (3.8.31)$$

We have made use of Eq. (3.8.8b) in writing Eq. (3.8.31) in the form shown. Note that the result (3.8.31) agrees with the previous result described by Eq. (3.8.25).

Experimental results demonstrating the influence of local-field effects on the linear and nonlinear optical response have been presented by Maki *et al.* (1991). The analysis given above has assumed that the material is homogenous in its structural properties. The analysis of local-field effects in composite materials comprised of two or more consituents is an area of active current research. In a composite material, the local electric field can vary considerably in space, and this effect can lead to an overall enhancement of the nonlineaar optical response. These effects have been described by Fischer *et al.* (1995) and by Nelson and Boyd (1999).

Problems

1. *Estimate of the refractive index of an atomic vapor.* Starting (for instance) from Eq. (3.5.20), perform an estimate of the magnitude of the on-resonance absorption coefficient of a dense atomic vapor assuming that the atomic number density is $N = 10^{17}$ cm^{-3}, that $\mu = 2.5 e a_0$, that the transition vacuum wavelength is 0.6 μm, and that the transition is homogeneously broadened with a linewidth (FWHM) of 10 GHz. Under the same conditions, calculate the maximum value of the real part of the refractive index near the peak of the absorption line. (These values are realistic under laboratory conditions. See for instance J. J. Maki *et al.*, 1991.)

[Ans.: $\alpha = 8 \times 10^4$ cm^{-1}, n(max) = 1.2]

2. *Estimate of the refractive index of glass.* Starting (for instance) from Eq. (3.5.20), perform an estimate of the magnitude of the real part of the refractive index of glass at visible wavelengths. Choose realistic values for the atomic number density, dipole transition moment, and detuning from resonance.

3. *Permutation symmetry of the nonlinear susceptibility.* Show that Eq. (3.6.25) possesses full permutation symmetry.

4. *Resonant nonlinear optical response.* Derive, using the density matrix formalism, an expression for the resonant contribution to the third-order susceptibility $\chi^{(3)}$ describing third-harmonic generation as illustrated below. Assume that, in thermal equilibrium, all of the population resides in the ground state. Note that since all input frequencies are equal and since only the resonant contribution is required, the answer will consist of one term and not 48 terms, which occur for the most general case of $\chi^{(3)}$. Work this problem by starting with the perturbation expansion (3.4.16) derived in the text and specializing the ensuing derivation to the interaction shown in the figure.

[Ans.: $\chi^{(3)}_{kjih}(3\omega,\omega,\omega,\omega) = \dfrac{N}{\hbar^3} \dfrac{\mu^k_{ad}\mu^j_{dc}\mu^i_{cb}\mu^h_{ba}}{[(\omega_{da}-3\omega)-i\gamma_{da}][(\omega_{ca}-2\omega)-i\gamma_{ca}][(\omega_{ba}-\omega)-i\gamma_{ba}]}$].

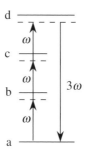

5. *Model calculation of the nonlinear susceptibility.* Consider the mutual interaction of four optical fields as illustrated in the figure below. Assume that all of the fields have the same linear polarization and that in thermal equilibrium all of the population is contained in level a. Assume that the waves are tuned sufficiently closely to the indicated resonances that only these contributions to the nonlinear interaction need be taken into account.

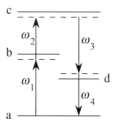

You may work this problem either by specializing the general result of Eq. (3.7.14) to the interaction shown in the figure or by repeating the derivation given in the text and specializing at each step to this interaction.

(a) Calculate the four nonlinear susceptibilities

$$\chi^{(3)}(\omega_4 = \omega_1 + \omega_2 - \omega_3), \qquad \chi^{(3)}(\omega_3 = \omega_1 + \omega_2 - \omega_4),$$

$$\chi^{(3)}(\omega_1 = \omega_3 + \omega_4 - \omega_2), \qquad \chi^{(3)}(\omega_2 = \omega_3 + \omega_4 - \omega_1)$$

that describe the four-wave mixing process, and determine the conditions under which these quantities are equal.

(b) In addition, calculate the nonlinear susceptibilities

$$\chi^{(3)}(\omega_1 = \omega_1 + \omega_2 - \omega_2), \qquad \chi^{(3)}(\omega_2 = \omega_2 + \omega_{-}\omega_1)$$

that describe two-photon absorption of the ω_1 and ω_2 fields, and determine the conditions under which they are equal. [Experimental investigation of some of the effects described by these quantities is reported by Malcuit *et al.*, *Phys. Rev. Lett.* **55**, 1086 (1985).]

6. *Generalization of Problems 4 and 5.* Repeat the calculation of the resonant contributions to $\chi^{(3)}$ for the cases studied in Problems 4 and 5 for the more general situation in which each of the levels can contain population in thermal equilibrium. Interpret your results.

[Note: The solution to this problem is very lengthy.]

7. *Pressure-induced resonances in nonlinear optics.* Verify the statement made in the text that Eq. (3.7.14) reduces to Eq. (3.2.33) in the limit in which damping effects are negligible. Show also that, even when damping is not negligible, the general 48-term expression for $\chi^{(3)}$ can be cast into an expression containing 24 terms, 12 of which contain "pressure-induced" resonances.

8. *Electromagnetically induced transparency.* The goal of this problem is to determine how the linear susceptibility $\chi^{(1)}(2\omega_1 + \omega_c)$ and the nonlinear optical susceptibility $\chi^{(3)}(\omega_{\text{sum}} = \omega_1 + \omega_1 + \omega_c)$ are modified when the field at frequency ω_c is a strong saturating field. What we shall find is that under appropriate circumstances the presence of the strong field can significantly decrease the (unwanted) linear absorption experienced by the sum frequency field while leaving the magnitude of the nonlinear response relatively unaffected. Work this problem by determining how the atomic wavefunction is modified by the presence of the applied laser fields, using the coupling scheme shown in the accompanying figure. Note that level b is appreciably detuned from a one-photon resonance, but that all other excited states are excited at a near resonance frequency.

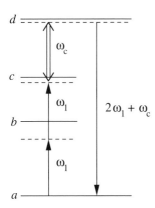

For the determination of $\chi^{(1)}(2\omega_1 + \omega_c)$, assume that the applied laser field has the form

$$\tilde{E}(t) = E_{\text{sum}}e^{-i\omega_{\text{sum}}t} + E_c e^{-i\omega_c t} + \text{c.c.} \tag{3.8.32}$$

and determine how the wavefunction is modified correct to first order in the perturbation induced by the field E_{sum} but correct to all orders in E_c. For the determination of $\chi^{(3)}(\omega_{\text{sum}} = \omega_1 + \omega_1 + \omega_c)$, assume that the applied laser field has the form

$$\tilde{E}(t) = E_1 e^{-i\omega_1 t} + E_c e^{-i\omega_c t} + \text{c.c.} \tag{3.8.33}$$

and determine how the wavefunction is modified correct to second order in the perturbation induced by the field E_1 but correct to all orders in E_c. Include the effects of damping into your calculation by assuming that population can decay out of level j with a decay rate of $2\gamma_j$.

Note that for appropriate values of γ_c and γ_d, linear absorption of the sum-frequency radiation can be dramatically reduced by the application of the saturating field, while leaving the nonlinear susceptibility relatively unchanged (S. E. Harris, J. E. Field, and A. Imamoğlu, *Phys. Rev. Lett.* **64**, 1107 (1990).)

[Alternatively, this problem can be solved by means of the density matrix formalism, in which case the effects of damping can be included by introducing separate dipole decay rates and population decay rates.]

References

Quantum Mechanics

H. A. Bethe and E. E. Salpeter, *Quantum Mechanics of One- and Two-Electron Atoms*, Plenum, New York, 1977.

E. Merzbacher, *Quantum Mechanics*, Wiley, New York, 1970.

M. Sargent III, M. O. Scully, and W. E. Lamb, Jr., *Laser Physics*, Addison-Wesley, Reading, MA, 1974.

Quantum-Mechanical Theories of the Nonlinear Optical Susceptibility

J. A. Armstrong, N. Bloembergen, J. Ducuing, and P. S. Pershan, *Phys. Rev.* **127**, 1918 (1962).

N. Bloembergen, *Nonlinear Optics*, Benjamin, New York, 1965.

N. Bloembergen and Y. R. Shen, *Phys. Rev.* **133**, A37 (1964).

N. Bloembergen, H. Lotem, and R. T. Lynch, Jr., *Indian J. Pure Appl. Phys.* **16**, 151 (1978).

R. W. Boyd and S. Mukamel, *Phys. Rev.* **A29**, 1973 (1984).

R. W. Boyd and L.-Q. Xiang, *IEEE J. Quantum Electron.* **18**, 1242 (1982).

R. W. Boyd, D. J. Gauthier, J. Krasinski, and M. S. Malcuit, *IEEE J. Quantum Electron.* **20**, 1074 (1984).

P. N. Butcher, *Nonlinear Optical Phenomena*, Ohio State University, 1965.

J. Ducuing, in *Quantum Optics* (R. J. Glauber, ed.), Academic Press, New York, 1969.

C. Flytzanis, in *Quantum Electronics, a Treatise*, Vol. 1, Part A (H. Rabin and C. L. Tang, eds.), Academic Press, New York, 1975.

D. J. Gauthier, J. Krasinski, and R. W. Boyd, *Opt. Lett.* **8**, 211 (1983).

D. C. Hanna, M. A. Yuratich, and D. Cotter, *Nonlinear Optics of Free Atoms and Molecules*, Springer-Verlag, Berlin, 1979.

D. Marcuse, *Principles of Quantum Electronics*, Academic Press, New York, 1980.

R. B. Miles and S. E. Harris, *IEEE J. Quantum Electron.* **9**, 470 (1973).

B. J. Orr and J. F. Ward, *Mol. Phys.* **20**, 513 (1971).

Y. Prior, *IEEE J. Quantum Electron.* **20**, 37 (1984).

L. Rothberg, in *Progress in Optics XXIV* (E. Wolf, ed.), Elsevier, 1987.

Y. R. Shen, *Principles of Nonlinear Optics*, Wiley, New York, 1984.

J. F. Ward, *Rev. Mod. Phys.* **37**, 1 (1965).

T. K. Yee and T. K. Gustafson, *Phys. Rev.* **A18**, 1597 (1978).

Local-Field Effects

M. Born and E. Wolf, *Principles of Optics*, Pergamon Press, Oxford, 1975.

G. L. Fischer, R. W. Boyd, R. J. Gehr, S. A. Jenekhe, J. A. Osaheni, J. E. Sipe, and L. A. Weller-Brophy, *Phys. Rev. Lett.* **74**, 1871 (1995).

J. D. Jackson, *Classical Electrodynamics*, Wiley, New York, 1975.

H. A. Lorentz, *The Theory of Electrons*, Dover, New York, 1952.

J. J. Maki, M. S. Malcuit, J. E. Sipe, and R. W. Boyd, *Phys. Rev. Lett.* **68**, 972 (1991).

V. Mizrahi and J. E. Sipe, *Phys. Rev.* **B34**, 3700 (1986).

R. L. Nelson and R. W. Boyd, *Appl. Phys. Lett.* **74**, 2417 (1999).

Chapter 4

The Intensity-Dependent Refractive Index

The refractive index of many optical materials depends upon the intensity of the light used to measure the refractive index. In this chapter, we examine some of the mathematical descriptions of the nonlinear refractive index and examine some of the physical processes that give rise to this effect. In the following chapter, we study the intensity-dependent refractive index resulting from the resonant response of an atomic system, and in Chapter 6 we study some physical processes that result from the nonlinear refractive index.

4.1. Descriptions of the Intensity-Dependent Refractive Index

The refractive index of many materials can be described by the relation

$$n = n_0 + \bar{n}_2 \langle \tilde{E}^2 \rangle, \tag{4.1.1}$$

where n_0 represents the usual, weak-field refractive index and $\bar{n}_2$ is a new optical constant (sometimes called the second-order index of refraction) that gives the rate at which the refractive index increases with increasing optical intensity.* The angular brackets surrounding the quantity $\tilde{E}^2$ represent a

---

* We place a bar over the symbol n_2 to prevent confusion with a different definition of n_2, introduced in Eq. (4.1.15) below. In accordance with conventional usage, the bar will be omitted in cases where little chance of confusion is likely.

189

time average. Thus, if the optical field is of the form

$$\tilde{E}(t) = E(\omega)e^{-i\omega t} + \text{c.c.}, \tag{4.1.2}$$

so that

$$\langle \tilde{E}(t)^2 \rangle = 2E(\omega)E(\omega)^* = 2|E(\omega)|^2, \tag{4.1.3}$$

we find that

$$n = n_0 + 2\bar{n}_2 |E(\omega)|^2. \tag{4.1.4}$$

The change in refractive index described by Eq. (4.1.1) or (4.1.4) is sometimes called the optical Kerr effect, by analogy with the traditional Kerr electrooptic effect, in which the refractive index of a material changes by an amount that is proportional to the square of the strength of an applied static electric field.

Of course, the interaction of a beam of light with a nonlinear optical medium can also be described in terms of the nonlinear polarization. The part of the nonlinear polarization that influences the propagation of a beam of frequency ω is

$$P^{\text{NL}}(\omega) = 3\chi^{(3)}(\omega = \omega + \omega - \omega)|E(\omega)|^2 E(\omega). \tag{4.1.5}$$

For simplicity we are assuming here that the light is linearly polarized and are suppressing the tensor indices of $\chi^{(3)}$; the tensor nature of $\chi^{(3)}$ is addressed explicitly in the following section. The total polarization of the material system is then described by

$$P^{\text{TOT}}(\omega) = \chi^{(1)}E(\omega) + 3\chi^{(3)}|E(\omega)|^2 E(\omega) \equiv \chi_{\text{eff}}E(\omega), \tag{4.1.6}$$

where we have introduced the effective susceptibility

$$\chi_{\text{eff}} = \chi^{(1)} + 3\chi^{(3)}|E(\omega)|^2. \tag{4.1.7}$$

In order to relate the nonlinear susceptibility $\chi^{(3)}$ to the nonlinear refractive index n_2, we note that it is generally true that

$$n^2 = 1 + 4\pi\chi_{\text{eff}}, \tag{4.1.8}$$

and by introducing Eq. (4.1.4) on the left-hand side and Eq. (4.1.7) on the right-hand side of this equation we find that

$$[n_0 + 2\bar{n}_2|E(\omega)|^2]^2 = 1 + 4\pi\chi^{(1)} + 12\pi\chi^{(3)}|E(\omega)|^2. \tag{4.1.9}$$

Correct to terms of order $|E(\omega)|^2$, this expression when expanded becomes $n_0^2 + 4n_0\bar{n}_2|E(\omega)|^2 = (1 + 4\pi\chi^{(1)}) + [12\pi\chi^{(3)}|E(\omega)|^2]$, which shows that the

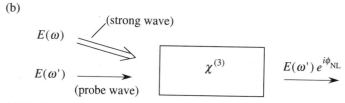

FIGURE 4.1.1 Two ways of measuring the intensity-dependent refractive index. In part (a), a strong beam of light modifies its own propagation, whereas in part (b), a strong beam of light influences the propagation of a weak beam.

linear and nonlinear refractive indices are related to the linear and nonlinear susceptibilities by

$$n_0 = \left(1 + 4\pi \chi^{(1)}\right)^{1/2} \qquad (4.1.10)$$

and

$$\bar{n}_2 = \frac{3\pi \chi^{(3)}}{n_0}. \qquad (4.1.11)$$

The discussion just given has implicitly assumed that the refractive index is measured using a single laser beam, as shown in part (a) of Fig. 4.1.1. Another way of measuring the intensity-dependent refractive index is to use two separate beams, as illustrated in part (b) of the figure. Here the presence of the strong beam of amplitude $E(\omega)$ leads to a modification of the refractive index experienced by a weak probe wave of amplitude $E(\omega')$. The nonlinear polarization affecting the probe wave is given by

$$P^{\mathrm{NL}}(\omega') = 6\chi^{(3)}(\omega' = \omega' + \omega - \omega)|E(\omega)|^2 E(\omega'). \qquad (4.1.12)$$

Note that the degeneracy factor (6) for this case is twice as large as that for the single-beam case of Eq. (4.1.5). In fact, for the two-beam case the degeneracy factor is equal to 6 even if ω' is equal to ω, because the probe beam is physically distinguishable from the strong pump beam owing to its different direction of propagation. The probe wave hence experiences a refractive index given by

$$n = n_0 + 2\bar{n}_2^{(\mathrm{cross})}|E(\omega)|^2, \qquad (4.1.13)$$

where

$$\bar{n}_2^{(\text{cross})} = \frac{6\pi \chi^{(3)}}{n_0}. \tag{4.1.14}$$

Note that the nonlinear coefficient $\bar{n}_2^{(\text{cross})}$ describing cross-coupling effects is twice as large as the coefficient $\bar{n}_2$ of Eq. (4.1.11) which describes self-action effects. Hence a strong wave affects the refractive index of a weak wave of the same frequency twice as much as it affects its own refractive index. This effect (for the case in which n_2 is positive) is known as weak-wave retardation (Chiao *et al.*, 1966).

An alternative way of defining the intensity-dependent refractive index* is by means of the equation

$$n = n_0 + n_2 I, \tag{4.1.15}$$

where I denotes the time-averaged intensity of the optical field, given by

$$I = \frac{n_0 c}{2\pi} |E(\omega)|^2. \tag{4.1.16}$$

Since the total refractive index n must be the same using either description of the nonlinear contribution, we see by comparing Eqs. (4.1.4) and (4.1.15) that

$$2\bar{n}_2 |E(\omega)|^2 = n_2 I, \tag{4.1.17}$$

and hence that $\bar{n}_2$ and n_2 are related by

$$n_2 = \frac{4\pi}{n_0 c} \bar{n}_2, \tag{4.1.18}$$

where we have made use of Eq. (4.1.16). If Eq. (4.1.11) is introduced into this expression, we find that n_2 is related to $\chi^{(3)}$ by

$$n_2 = \frac{12\pi^2}{n_0^2 c} \chi^{(3)}. \tag{4.1.19}$$

It is often convenient to measure I in units of W/cm^2, in which case n_2 is measured in units of cm^2/W. We then find that numerically

$$n_2 \left(\frac{\text{cm}^2}{\text{W}} \right) = \frac{12\pi^2}{n_0^2 c} 10^7 \chi^{(3)}(\text{esu}) = \frac{0.0395}{n_0^2} \chi^{(3)}(\text{esu}). \tag{4.1.20}$$

Some of the physical processes that can produce a nonlinear change in the refractive index are listed in Table 4.1.1, along with typical values of n_2, of

---

* For definiteness, we are treating the single-beam case of part (a) of Fig. 4.1.1. The extension to the two-beam case is straightforward.

TABLE 4.1.1 Typical values of the nonlinear refractive index[a]

Mechanism	n_2 (cm²/W)	$\chi_{1111}^{(3)}$ (esu)	Response time (sec)
Electronic polarization	10^{-16}	10^{-14}	10^{-15}
Molecular orientation	10^{-14}	10^{-12}	10^{-12}
Electrostriction	10^{-14}	10^{-12}	10^{-9}
Saturated atomic absorption	10^{-10}	10^{-8}	10^{-8}
Thermal effects	10^{-6}	10^{-4}	10^{-3}
Photorefractive effect [b]	(large)	(large)	(intensity-dependent)

[a] For linearly polarized light.

[b] The photorefractive effect often leads to a very strong nonlinear response. This response usually cannot be described in terms of a $\chi^{(3)}$ (or an n_2) nonlinear susceptibility, because the nonlinear polarization does not depend on the applied field strength in the same manner as the other mechanisms listed.

$\chi^{(3)}$, and of the characteristic time scale for the nonlinear response to develop. Electronic polarization, molecular orientation, and thermal effects are discussed in the present chapter, saturated absorption is discussed in Chapter 7, electrostriction is discussed in Chapter 9, and the photorefractive effect is described in Chapter 11.

In Table 4.1.2 the experimentally measured values of the nonlinear susceptibility are presented for several materials. Some of the methods that are used to measure the nonlinear susceptibility have been reviewed by Hellwarth (1977). As an example of the use of Table 4.1.2, note that for carbon disulfide the value of n_2 is approximately 3×10^{-14} cm²/W. Thus, a laser beam of intensity $I = 1$ MW/cm² can produce a refractive index change of 3×10^{-8}. Even though this change is rather small, refractive index changes of this order of magnitude can lead to dramatic nonlinear optical effects (some of which are described in Chapter 7) for the case of phase-matched nonlinear optical interactions.

4.2. Tensor Nature of the Third-Order Susceptibility

The third-order susceptibility $\chi_{ijkl}^{(3)}$ is a fourth-rank tensor, and thus is described in terms of 81 separate elements. For crystalline solids with low symmetry, all 81 of these elements are independent and can be nonzero (Butcher, 1965). However, for materials possessing a higher degree of spatial symmetry, the number of independent elements is very much reduced; as we show below, there are only three independent elements for an isotropic material.

Let us see how to determine the tensor nature of the third-order susceptibility for the case of an isotropic material such as a glass, a liquid, or a vapor. We begin

TABLE 4.1.2   Third-order nonlinear optical coefficients of various materials[a]

Material	n_0	$\chi^{(3)}$ (esu)	n_2 (cm²/W)	Comments and references[b]
Crystals				
Al_2O_3	1.8	2.2×10^{-14}	2.9×10^{-16}	1
CdS	2.34	7.0×10^{-12}	5.1×10^{-14}	1,1.06 μm
Diamond	2.42	1.8×10^{-13}	1.3×10^{-15}	1
GaAs	3.47	1.0×10^{-10}	3.3×10^{-13}	1,1.06 μm
Ge	4.0	4.0×10^{-11}	9.9×10^{-14}	2, THG $|\chi^{(3)}|$
LiF	1.4	4.4×10^{-15}	9.0×10^{-17}	1
Si	3.4	2.0×10^{-10}	2.7×10^{-14}	2,THG $|\chi^{(3)}|$
TiO_2	2.48	1.5×10^{-12}	9.4×10^{-15}	1
ZnSe	2.7	4.4×10^{-12}	3.0×10^{-14}	1,1.06 μm
Glasses				
Fused silica	1.47	1.8×10^{-14}	3.2×10^{-16}	1
As_2S_3 glass	2.4	2.9×10^{-11}	2.0×10^{-13}	3
BK-7	1.52	2.0×10^{-14}	3.4×10^{-16}	1
BSC	1.51	3.6×10^{-14}	6.4×10^{-16}	1
Pb Bi gallate	2.3	1.6×10^{-12}	1.3×10^{-14}	4
SF-55	1.73	1.5×10^{-13}	2.0×10^{-15}	1
SF-59	1.953	3.1×10^{-13}	3.3×10^{-15}	1
Nanoparticles				
CdSSe in glass	1.5	1.0×10^{-12}	1.8×10^{-14}	3, nonres.
CS 3-68 glass	1.5	1.3×10^{-8}	2.3×10^{-10}	3, res.
Gold in glass	1.5	1.5×10^{-8}	2.6×10^{-10}	3, res.
Polymers				
Polydiacetylenes				
PTS		6×10^{-10}	$3. \times 10^{-12}$	5, nonres.
PTS		-4×10^{-8}	-2×10^{-10}	6, res.
9BCMU			1.9×10^{-10}	7, $|n_2|$, res.
4BCMU	1.56	-9.2×10^{-12}	-1.5×10^{-13}	8, nonres, $\beta =$ 0.01 cm/MW
Liquids				
Acetone	1.36	1.1×10^{-13}	2.4×10^{-15}	9
Benzene	1.5	6.8×10^{-14}	1.2×10^{-15}	9
Carbon disulfide	1.63	2.2×10^{-12}	3.2×10^{-14}	9, $\tau = 2$ psec
CCl_4	1.45	8.0×10^{-14}	1.5×10^{-15}	9
Diiodomethane	1.69	1.1×10^{-12}	1.5×10^{-14}	9
Ethanol	1.36	3.6×10^{-14}	7.7×10^{-16}	9
Methanol	1.33	3.1×10^{-14}	6.9×10^{-16}	9
Nitrobenzene	1.56	4.1×10^{-12}	6.7×10^{-14}	9
Water	1.33	1.8×10^{-14}	4.1×10^{-16}	9
Other materials				
Air	1.0003	1.2×10^{-17}	5.0×10^{-19}	10
Ag		2.0×10^{-11}		2, THG $|\chi^{(3)}|$
Au		5.4×10^{-11}		2, THG $|\chi^{(3)}|$

TABLE 4.1.2   *(Continued)*

Vacuum	1	2.4×10^{-33}	1.0×10^{-34}	11
Cold atoms	1.0	5.1	0.2	12, (EIT BEC)
Fluorescein dye in glass	1.5	$2 + 2i$	$0.035(1 + i)$	13, $\tau = 0.1s$

[a] This table assumes the definition of the third-order susceptibility $\chi^{(3)}$ used in this book, as given for instance by Eq. (1.1.2) or by Eq. (1.3.21). This definition is consistent with that introduced by N. Bloembergen (*Nonlinear Optics*, Benjamin, New York, 1964). Some workers use an alternative definition which renders their values four times smaller. In compiling this table we have converted the literature values when necessary to the present definition.

The quantity n_2 is the coefficient of the intensity-dependent refractive index which is defined such that $n = n_0 + n_2 I$, where n_0 is the linear refractive index and I is the laser intensity. The relation between n_2 and $\chi^{(3)}$ is consequently $n_2 = 12\pi^2 \chi^{(3)}/n_0^2$. When the intensity is measured in W/cm^2 and $\chi^{(3)}$ is measured in electrostatic units (esu), that is, in cm^2 statvolt^{-2}, the relation between n_2 and $\chi^{(3)}$ becomes $n_2(\text{cm}^2/\text{W}) = 0.0395 \, \chi^{(3)}(\text{esu})/n_0^2$. The quantity β is the coefficient describing two-photon absorption.

[b] References for Table 4.1.2: (1) L. L. Chase and E. W. Van Stryland, Section 8.1 of *CRC Handbook of Laser Science and Technology*, CRC Press, Boca Raton, FL, 1995; (2) N. Bloembergen *et al.*, *Opt. Commun.* **1**, 195 (1969); (3) E. M. Vogel *et al.*, *Phys. Chem. Glasses* **32**, 231 (1991); (4) D. W. Hall *et al.*, *Appl. Phys. Lett.* **54**, 1293 (1989); (5) B. L. Lawrence *et al.*, *Electron. Lett.* **30**, 447 (1994); (6) G. M. Carter *et al.*, *Appl. Phys. Lett*, **47**, 457 (1985); (7) S. Molyneux, A. K. Kar, B. S. Wherrett, T. L. Axon and D. Bloor, *Opt. Lett.* **18**, 2093 (1993); (8) J. E. Erlich *et al.*, *J. Mod. Opt.* **40**, 2151 (1993); (9) R. L. Sutherland, *Handbook of Nonlinear Optics*, Chapter 8, Marcel Dekker, Inc., New York, 1996; (10) D. M. Pennington *et al.*, *Phys. Rev. A* **39**, 3003 (1989); (11) H. Euler and B. Kockel, *Naturwiss enschaften* **23**, 246 (1935); (12) L. V. Hau *et al.*, *Nature* **397**, 594 (1999); (13) M. A. Kramer, W. R. Tompkin, and R. W. Boyd, *Phys. Rev. A* **34**, 2026 (1986).

by considering the general case in which the applied frequencies are arbitrary, and we represent the susceptibility as $\chi_{ijkl} \equiv \chi_{ijkl}^{(3)}(\omega_4 = \omega_1 + \omega_2 + \omega_3)$. Since each of the coordinate axes must be equivalent in an isotropic material, it is clear that the susceptibility possesses the following symmetry properties:

$$\chi_{1111} = \chi_{2222} = \chi_{3333}, \tag{4.2.1a}$$

$$\chi_{1122} = \chi_{1133} = \chi_{2211} = \chi_{2233} = \chi_{3311} = \chi_{3322}, \tag{4.2.1b}$$

$$\chi_{1212} = \chi_{1313} = \chi_{2323} = \chi_{2121} = \chi_{3131} = \chi_{3232}, \tag{4.2.1c}$$

$$\chi_{1221} = \chi_{1331} = \chi_{2112} = \chi_{2332} = \chi_{3113} = \chi_{3223}. \tag{4.2.1d}$$

One can also see that the 21 elements listed are the only nonzero elements of $\chi^{(3)}$, because these are the only elements that possess the property that any cartesian index (1, 2, or 3) that appears at least once appears an even number of times. An index cannot appear an odd number of times, because, for example, χ_{1222} would give the response in the $\hat{x}_1$ direction due to a field applied in the $\hat{x}_2$ direction. This response must vanish in an isotropic material, because there is no reason why the response should be in the $+\hat{x}_1$ direction rather than in the $-\hat{x}_1$ direction.

The four types of nonzero elements appearing in Eqs. (4.2.1) are not independent of one another, and in fact are related by the equation

$$\chi_{1111} = \chi_{1122} + \chi_{1212} + \chi_{1221}. \qquad (4.2.2)$$

One can deduce this result by requiring that the predicted value of the nonlinear polarization be the same when calculated in two different coordinate systems that are rotated with respect to each other by an arbitrary amount. A rotation of 45 degrees about the $\hat{x}_3$ axis is a convenient choice for deriving this relation. The results given by Eqs. (4.2.1) and (4.2.2) can be used to express the nonlinear susceptibility in the compact form

$$\chi_{ijkl} = \chi_{1122}\delta_{ij}\delta_{kl} + \chi_{1212}\delta_{ik}\delta_{jl} + \chi_{1221}\delta_{il}\delta_{jk}. \qquad (4.2.3)$$

This form shows that the third-order susceptibility has three independent elements for the general case in which the field frequencies are arbitrary.

Let us first specialize this result to the case of third-harmonic generation, where the frequency dependence of the susceptibility is taken as $\chi_{ijkl}(3\omega = \omega + \omega + \omega)$. As a consequence of the intrinsic permutation symmetry of the nonlinear susceptibility, the elements of the susceptibility tensor are related by $\chi_{1122} = \chi_{1212} = \chi_{1221}$ and hence Eq. (4.2.3) becomes

$$\chi_{ijkl}(3\omega = \omega + \omega + \omega) = \chi_{1122}(3\omega = \omega + \omega + \omega)$$
$$\times (\delta_{ij}\delta_{kl} + \delta_{ik}\delta_{jl} + \delta_{il}\delta_{jk}). \qquad (4.2.4)$$

Hence there is only one independent element of the susceptibility tensor describing third-harmonic generation.

We next apply the result given in Eq. (4.2.3) to the nonlinear refractive index, that is, we consider the choice of frequencies given by $\chi_{ijkl}(\omega = \omega + \omega - \omega)$. For this choice of frequencies, the condition of intrinsic permutation symmetry requires that χ_{1122} be equal to χ_{1212}, and hence χ_{ijkl} can be represented by

$$\chi_{ijkl}(\omega = \omega + \omega - \omega) = \chi_{1122}(\omega = \omega + \omega - \omega)(\delta_{ij}\delta_{kl} + \delta_{ik}\delta_{jl})$$
$$+ \chi_{1221}(\omega = \omega + \omega - \omega)(\delta_{il}\delta_{jk}). \qquad (4.2.5)$$

The nonlinear polarization leading to the nonlinear refractive index is given in terms of the nonlinear susceptibility by (see also Eq. (1.3.21))

$$P_i(\omega) = 3\sum_{jkl}\chi_{ijkl}(\omega = \omega + \omega - \omega)E_j(\omega)E_k(\omega)E_l(-\omega). \qquad (4.2.6)$$

If we introduce Eq. (4.2.6) into this equation, we find that

$$P_i = 6\chi_{1122}E_i(\mathbf{E}\cdot\mathbf{E}^*) + 3\chi_{1221}E_i^*(\mathbf{E}\cdot\mathbf{E}). \qquad (4.2.7)$$

This equation can be written entirely in vector form as

$$\mathbf{P} = 6\chi_{1122}(\mathbf{E} \cdot \mathbf{E}^*)\mathbf{E} + 3\chi_{1221}(\mathbf{E} \cdot \mathbf{E})\mathbf{E}^*. \qquad (4.2.8)$$

Following the notation of Maker and Terhune (1965), we introduce the coefficients

$$A = 6\chi_{1122} \qquad (\text{or } A = 3\chi_{1122} + 3\chi_{1212}) \qquad (4.2.9a)$$

and

$$B = 6\chi_{1221}, \qquad (4.2.9b)$$

in terms of which the nonlinear polarization of Eq. (4.2.8) can be written as

$$\mathbf{P} = A(\mathbf{E} \cdot \mathbf{E}^*)\mathbf{E} + \tfrac{1}{2}B(\mathbf{E} \cdot \mathbf{E})\mathbf{E}^*. \qquad (4.2.10)$$

We see that the nonlinear polarization consists of two contributions. These contributions have very different physical characters, since the first contribution has the vector nature of $\mathbf{E}$ whereas the second contribution has the vector nature of $\mathbf{E}^*$. The first contribution thus produces a nonlinear polarization with the same handedness as $\mathbf{E}$, whereas the second contribution produces a nonlinear polarization with the opposite handedness. The consequences of this behavior on the propagation of a beam of light through a nonlinear optical medium are described below.

The origin of the different physical characters of the two contributions to $\mathbf{P}$ can be understood in terms of the energy level diagrams shown in Fig. 4.2.1. Here part (a) illustrates one-photon-resonant contributions to the nonlinear coupling. We will show in Eq. (4.3.14) that processes of this sort contribute only to the coefficient A. Part (b) of the figure illustrates two-photon-resonant processes, which in general contribute to both the coefficients A and B (see Eqs. (4.3.13) and (4.3.14)). However, under certain circumstances, such as those described later in connection with Fig. 7.2.9, two-photon-resonant processes contribute only to the coefficient B.

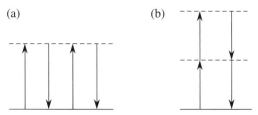

FIGURE 4.2.1   Diagrams (a) and (b) represent the resonant contributions to the nonlinear coefficients A and B, respectively.

For some purposes, it is useful to describe the nonlinear polarization not by Eq. (4.2.10) but rather in terms of an effective linear susceptibility defined by means of the relationship

$$P_i = \sum_j \chi_{ij}^{(eff)} E_j. \tag{4.2.11}$$

Then, as can be verified by direct substitution, Eqs. (4.2.10) and (4.2.11) lead to identical predictions for the nonlinear polarization if the effective linear susceptibility is given by

$$\chi_{ij}^{(eff)} = A'(\mathbf{E} \cdot \mathbf{E}^*)\delta_{ij} + \tfrac{1}{2}B'(E_i E_j^* + E_i^* E_j), \tag{4.2.12a}$$

where

$$A' = A - \tfrac{1}{2}B = 6\chi_{1122} - 3\chi_{1221} \tag{4.2.12b}$$

and

$$B' = B = 6\chi_{1221}. \tag{4.2.12c}$$

The results given in Eq. (4.2.10) or in Eqs. (4.2.12) show that the nonlinear susceptibility tensor describing the nonlinear refractive index of an isotropic material possesses only two independent elements. The relative magnitude of these two coefficients depends upon the nature of the physical process that produces the optical nonlinearity. For some of the physical mechanisms leading to a nonlinear refractive index, these ratios are given by

$$B/A = 6, \quad B'/A' = -3 \quad \text{for molecular orientation,} \tag{4.2.13a}$$

$$B/A = 1, \quad B'/A' = 2 \quad \text{for nonresonant electronic response,} \tag{4.2.13b}$$

$$B/A = 0, \quad B'/A' = 0 \quad \text{for electrostriction.} \tag{4.2.13c}$$

These conclusions will be justified in the discussion that follows; see especially Eq. (4.4.37) for the case of molecular orientation, Eq. (4.3.14) for nonresonant electronic response of bound electrons, and Eq. (9.2.15) for electrostriction. Note also that A is equal to B by definition whenever the Kleinman symmetry condition is valid.

The trace of the effective susceptibility is given by

$$\mathrm{Tr}\,\chi_{ij} \equiv \sum_i \chi_{ii} = (3A' + B')\mathbf{E} \cdot \mathbf{E}^*. \tag{4.2.14}$$

Hence, $\mathrm{Tr}\,\chi_{ij}$ vanishes for the molecular orientation mechanism; the reason for this behavior is discussed in connection with Eq. (4.4.56). For the resonant response of an atomic transition, the ratio of B to A depends upon the angular momentum quantum numbers of the two atomic levels. Formulas for A and B for such a case have been presented by Saikan and Kiguchi (1982).

Propagation through Isotropic Nonlinear Media

Let us next consider the propagation of a beam of light through a material whose nonlinear optical properties are described by Eq. (4.2.10). As we show below, only linearly or circularly polarized light is transmitted through such a medium with its state of polarization unchanged. When elliptically polarized light propagates through such a medium, the orientation of the polarization ellipse rotates as a function of propagation distance due to the nonlinear interaction.

Let us consider a beam of arbitrary polarization propagating in the positive z direction. The electric field vector of such a beam can always be decomposed into a linear combination of left- and right-hand circular components as

$$\mathbf{E} = E_+\hat{\sigma}_+ + E_-\hat{\sigma}_-, \tag{4.2.15}$$

where the circular-polarization unit vectors are illustrated in Fig. 4.2.2 and are defined by

$$\hat{\sigma}_\pm = \frac{\hat{\mathbf{x}} \pm i\hat{\mathbf{y}}}{\sqrt{2}}. \tag{4.2.16}$$

By convention, $\hat{\sigma}_+$ corresponds to left-hand circular and $\hat{\sigma}_-$ to right-hand circular polarization (for a beam propagating in the positive z direction).

We now introduce the decomposition (4.2.15) into Eq. (4.2.10). We find, using the identities

$$\hat{\sigma}_\pm^* = \hat{\sigma}_\mp, \qquad \hat{\sigma}_\pm \cdot \hat{\sigma}_\pm = 0, \qquad \hat{\sigma}_\pm \cdot \hat{\sigma}_\mp = 1,$$

that the products $\mathbf{E}^* \cdot \mathbf{E}$ and $\mathbf{E} \cdot \mathbf{E}$ become

$$\mathbf{E}^* \cdot \mathbf{E} = (E_+^*\hat{\sigma}_+^* + E_-^*\hat{\sigma}_-^*) \cdot (E_+\hat{\sigma}_+ + E_-\hat{\sigma}_-) = E_+^*E_+ + E_-^*E_-$$
$$= |E_+|^2 + |E_-|^2$$

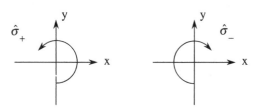

FIGURE 4.2.2   The $\hat{\sigma}_+$ and $\hat{\sigma}_-$ circular polarizations.

and

$$\mathbf{E} \cdot \mathbf{E} = (E_+\hat{\sigma}_+ + E_-\hat{\sigma}_-) \cdot (E_+\hat{\sigma}_+ + E_-\hat{\sigma}_-) = E_+E_- + E_-E_+ = 2E_+E_-,$$

so that Eq. (4.2.10) can be written as

$$\mathbf{P}^{\mathrm{NL}} = A(|E_+|^2 + |E_-|^2)\mathbf{E} + B(E_+E_-)\mathbf{E}^*. \tag{4.2.17}$$

If we now represent $\mathbf{P}_{\mathrm{NL}}$ in terms of its circular components as

$$\mathbf{P}^{\mathrm{NL}} = P_+\hat{\sigma}_+ + P_-\hat{\sigma}_-, \tag{4.2.18}$$

we find that the coefficient P_+ is given by

$$\begin{aligned}
P_+ &= A(|E_+|^2 + |E_-|^2)E_+ + B(E_+E_-)E_-^* \\
&= A(|E_+|^2 + |E_-|^2)E_+ + B|E_-|^2E_+ \\
&= A|E_+|^2E_+ + (A+B)|E_-|^2E_+
\end{aligned} \tag{4.2.19a}$$

and similarly that

$$P_- = A|E_-|^2E_- + (A+B)|E_+|^2E_-. \tag{4.2.19b}$$

These results can be summarized as

$$P_\pm \equiv \chi_\pm^{\mathrm{NL}} E_\pm, \tag{4.2.20a}$$

where we have introduced the effective nonlinear susceptibilities

$$\chi_\pm^{\mathrm{NL}} = A|E_\pm|^2 + (A+B)|E_\mp|^2. \tag{4.2.20b}$$

The expressions (4.2.15) and (4.2.18) for the field and nonlinear polarization are now introduced into the wave equation,

$$\nabla^2 \mathbf{E}(z,t) = \frac{\epsilon^{(1)}}{c^2}\frac{\partial^2 \mathbf{E}(z,t)}{\partial t^2} + \frac{4\pi}{c^2}\frac{\partial^2}{\partial t^2}\mathbf{P}^{\mathrm{NL}}, \tag{4.2.21}$$

where $\tilde{\mathbf{E}}(z,t) = \mathbf{E}\exp(-i\omega t) + $ c.c. and $\tilde{\mathbf{P}}(z,t) = \mathbf{P}\exp(-i\omega t) + $ c.c. We next decompose Eq. (4.2.21) into its $\hat{\sigma}_+$ and $\hat{\sigma}_-$ components. Since, according to Eq. (4.2.20a), $P_\pm$ is proportional to $E_\pm$, the two terms on the right-hand side of the resulting equation can be combined into a single term, so that the wave equation for each circular component becomes

$$\nabla^2 \tilde{E}_\pm(z,t) = \frac{\epsilon_\pm^{(\mathrm{eff})}}{c^2}\frac{\partial^2 \tilde{E}_\pm(z,t)}{\partial t^2}, \tag{4.2.22a}$$

where

$$\epsilon_\pm^{(\mathrm{eff})} = \epsilon^{(1)} + 4\pi\chi_\pm^{\mathrm{NL}}. \tag{4.2.22b}$$

This equation possesses solutions of the form of plane waves propagating with the phase velocity $c/n^\pm$, where $n_\pm = [\epsilon_\pm^{(\mathrm{eff})}]^{1/2}$. Letting $n_0^2 = \epsilon^{(1)}$, we find that

$$n_\pm^2 = n_0^2 + 4\pi \chi_\pm^{\mathrm{NL}} = n_0^2 + 4\pi [A|E_\pm|^2 + (A+B)|E_\mp|^2]$$

$$= n_0^2 \left(1 + \frac{4\pi}{n_0^2}[A|E_\pm|^2 + (A+B)|E_\mp|^2] \right),$$

and hence that

$$n_\pm \simeq n_0 + \frac{2\pi}{n_0}[A|E_\pm|^2 + (A+B)|E_\mp|^2]. \tag{4.2.23}$$

We see that the left- and right-circular components of the beam propagate with different phase velocities. The difference in their refractive indices is given by

$$\Delta n \equiv n_+ - n_- = \frac{2\pi B}{n_0}(|E_-|^2 - |E_+|^2). \tag{4.2.24}$$

Note that this difference depends upon the value of the coefficient B but not that of the coefficient A. Since the left- and right-hand circular components propagate with different phase velocities, the polarization ellipse of the light will rotate as the beam propagates through the nonlinear medium; a similar effect occurs in the linear optics of optically active materials.

In order to determine the angle of rotation, we express the field amplitude as

$$E(z) = E_+\hat{\sigma}_+ + E_-\hat{\sigma}_- = A_+ e^{in_+\omega z/c}\hat{\sigma}_+ + A_- e^{in_-\omega z/c}\hat{\sigma}_-$$
$$= \left(A_+ e^{i(1/2)\Delta n\omega z/c}\hat{\sigma}_+ + A_- e^{-i(1/2)\Delta n\omega z/c}\hat{\sigma}_- \right) e^{i(1/2)(n_+ + n_-)\omega z/c}. \tag{4.2.25}$$

We now introduce the mean propagation constant $k_m = \frac{1}{2}(n_+ + n_-)\omega/c$ and the angle

$$\theta = \frac{1}{2}\Delta n \frac{\omega}{c} z, \tag{4.2.26a}$$

in terms of which Eq. (4.2.26) becomes

$$\mathbf{E}(z) = (A_+\hat{\sigma}_+ e^{i\theta} + A_-\hat{\sigma}_- e^{-i\theta})e^{ik_m z}. \tag{4.2.26b}$$

As illustrated in Fig. 4.2.3, this equation describes a wave whose polarization ellipse is the same as that of the incident wave, but rotated through the angle θ (measured clockwise in the xy plane, in conformity with the sign convention for rotation angles in optical activity). This conclusion can be demonstrated

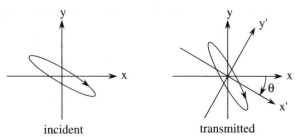

incident   transmitted

FIGURE 4.2.3   Polarization ellipses of the incident and transmitted waves.

by noting that

$$\hat{\sigma}_{\pm} e^{\pm i\theta} = \frac{\hat{x}' \pm i\hat{y}'}{\sqrt{2}},$$ (4.2.27)

where $\hat{x}'$ and $\hat{y}'$ are polarization unit vectors in a new coordinate system, that is,

$$x' = x\cos\theta - y\sin\theta,$$ (4.2.28a)

$$y' = x\sin\theta + y\cos\theta.$$ (4.2.28b)

Measurement of the rotation angle θ provides a sensitive method for determining the nonlinear coefficient B (see also Eqs. (4.2.24) and (4.2.26a)). For the case of propagation through an atomic vapor, saturation effects can lead to interesting new phenomena including polarization rotation through an angle that, for a broad range of intensities, is nearly independent of the laser intensity; see Davis *et al.* (1992) for details.

As mentioned above, there are two cases in which the polarization ellipse does not rotate. One case is that of circularly polarized light. In this case only one of the $\hat{\sigma}_{\pm}$ components is present, and we see from Eq. (4.2.23) that the change in refractive index is given by

$$\delta n_{\text{circular}} = \frac{2\pi}{n_0} A|E|^2,$$ (4.2.29)

which clearly depends on the coefficient A but not upon the coefficient B. The other case in which there is no rotation is that of linearly polarized light. Since linearly polarized light is a combination of equal amounts of left- and right-hand circular components (i.e., $|E_-|^2 = |E_+|^2$), we see directly from Eq. (4.2.24) that the index difference Δn vanishes. If we let E denote the total field amplitude of the linearly polarized radiation, so that $|E|^2 = 2|E_+|^2 = 2|E_-|^2$, we find from Eq. (4.2.23) that for linearly polarized light the change in refractive index is given by

$$\delta n_{\text{linear}} = \frac{2\pi}{n_0}\left(A + \tfrac{1}{2}B\right)|E|^2.$$ (4.2.30)

Note that this change depends on the coefficients $A = 6\chi_{1122}$ and $B = 6\chi_{1221}$ as $A + \frac{1}{2}B$, which according to Eqs. (4.2.2) and (4.2.9a,b) is equal to $3\chi_{1111}$. We see from Eqs. (4.2.29) and (4.2.30) that, for the usual case in which A and B have the same sign, linearly polarized light experiences a larger nonlinear change in refractive index than does circularly polarized light. In general the relative change in refractive index, $\delta n_{\text{linear}}/\delta n_{\text{circular}}$, is equal to $1 + B/2A$, which for the mechanisms described after Eq. (4.2.10) becomes

$$\frac{\delta n_{\text{linear}}}{\delta n_{\text{circular}}} = \begin{cases} 4 & \text{for molecular orientation,} \\ \frac{3}{2} & \text{for nonresonant electronic nonlinearities,} \\ 1 & \text{for electrostriction.} \end{cases} \quad (4.2.31)$$

For the case of two laser beams counterpropagating through a nonlinear material, the theoretical analysis is far more complex than that just presented for the single-beam situation, and a variety of additional phenomena can occur, including polarization bistability and polarization instabilities including chaos. These effects have been described theoretically by Gaeta et al.(1987) and have been observed experimentally by Gauthier et al. (1988, 1990).

4.3. Nonresonant Electronic Nonlinearities

Nonresonant electronic nonlinearities occur as the result of the nonlinear response of bound electrons to an applied optical field. This nonlinearity usually is not particularly large ($\chi^{(3)} \sim 10^{-14}$ esu is typical), but is of considerable importance because it is present in all dielectric materials. Furthermore, recent work has shown that certain organic nonlinear optical materials (such as polydiacetylene) can have nonresonant third-order susceptibilities as large as 10^{-9} esu due to the response of delocalized π electrons.

Nonresonant electronic nonlinearities are extremely fast, since they involve only virtual processes. The characteristic response time of this process is the time required for the electron cloud to become distorted in response to an applied optical field. This response time can be estimated as the orbital period of the electron in its motion about the nucleus, which according to the Bohr model of the atom is given by

$$\tau = 2\pi a_0/v,$$

where $a_0 = 0.5 \times 10^{-8}$ cm is the Bohr radius of the atom and $v \simeq c/137$ is a typical electronic velocity. We hence find that $\tau \simeq 10^{-16}$ s.

Classical, Anharmonic Oscillator Model of Electronic Nonlinearities

A simple model of electronic nonlinearities is the classical, anharmonic oscillator model described in Section 1.4. According to this model, one assumes that the potential well binding the electron to the atomic nucleus deviates from the parabolic potential of the usual Lorentz model. We approximate the actual potential well as

$$U(\mathbf{r}) = \tfrac{1}{2}m\omega_0^2|\mathbf{r}|^2 - \tfrac{1}{4}mb|\mathbf{r}|^4, \tag{4.3.1}$$

where b is a phenomenological nonlinear constant whose value is of the order of ω_0^2/d^2, where d is a typical atomic dimension. By solving the equation of motion for an electron in such a potential well, we obtain expression (1.4.52) for the third-order susceptibility. When applied to the case of the nonlinear refractive index, this expression becomes

$$\chi_{ijkl}^{(3)}(\omega = \omega + \omega - \omega) = \frac{Nbe^4[\delta_{ij}\delta_{kl} + \delta_{ik}\delta_{jl} + \delta_{il}\delta_{jk}]}{3m^3 D(\omega)^3 D(-\omega)}, \tag{4.3.2}$$

where $D(\omega) = \omega_0^2 - \omega^2 - 2i\omega\gamma$. In the notation of Maker and Terhune (Eq. (4.2.10)), this result implies that

$$A = B = \frac{2Nbe^4}{m^3 D(\omega)^3 D(-\omega)}. \tag{4.3.3}$$

Hence, according to the classical, anharmonic oscillator model of electronic nonlinearities, A is equal to B for any value of the optical field frequency (whether resonant or nonresonant). For the case of far-off-resonant excitation (i.e., $\omega \ll \omega_0$), we can replace $D(\omega)$ by ω_0^2 in Eq. (4.3.2). If in addition we set b equal to ω_0^2/d^2, we find that

$$\chi^{(3)} \simeq \frac{Ne^4}{m^3\omega_0^6 d^2}. \tag{4.3.4}$$

For the typical values $N = 4\times 10^{22}$ cm^{-3}, $d = 3\times 10^{-8}$ cm, and $\omega_0 = 7\times 10^{15}$ rad/s, we find that $\chi^{(3)} \simeq 2\times 10^{-14}$ esu.

Quantum-Mechanical Model of Nonresonant Electronic Nonlinearities

Let us now calculate the third-order susceptibility describing the nonlinear refractive index using the laws of quantum mechanics. Since we are interested primarily in the case of nonresonant excitation, we make use of the expression for the nonlinear susceptibility in the form given by Eq. (3.2.33), that is,

$$
\chi_{kjih}^{(3)}(\omega_\sigma, \omega_r, \omega_q, \omega_p)
$$

$$
= \frac{N}{\hbar^3} \mathscr{P}_F \sum_{lmn} \left[\frac{\mu_{gn}^k \mu_{nm}^j \mu_{ml}^i \mu_{lg}^h}{(\omega_{ng} - \omega_\sigma)(\omega_{mg} - \omega_q - \omega_p)(\omega_{lg} - \omega_p)} \right],
\tag{4.3.5}
$$

where $\omega_\sigma = \omega_r + \omega_q + \omega_p$. We want to apply this expression to the case of the nonlinear refractive index, with the frequencies arranged as $\chi_{kjih}^{(3)}$ $(\omega, \omega, \omega, -\omega) = \chi_{kjih}^{(3)}(\omega = \omega + \omega - \omega)$. One sees that Eq. (4.3.6) appears to have divergent contributions for this choice of frequencies, because the factor $\omega_{mg} - \omega_q - \omega_p$ in the denominator vanishes when the dummy index m is equal to g and when $\omega_p = -\omega_q = \pm\omega$. However, in fact this divergence exists in appearance only (Orr and Ward, 1971; Hanna, Yuratich, and Cotter, 1979); one can readily rearrange Eq. (4.3.6) into a form where no divergence appears. We first rewrite Eq. (4.3.6) as

$$
\chi_{kjih}^{(3)}(\omega_\sigma, \omega_r, \omega_q, \omega_p) = \frac{N}{\hbar^3} \mathscr{P}_F \left[\sum_{lmn}' \frac{\mu_{gn}^k \mu_{nm}^j \mu_{ml}^i \mu_{lg}^h}{(\omega_{ng} - \omega_\sigma)(\omega_{mg} - \omega_q - \omega_p)(\omega_{lg} - \omega_p)} \right.
$$

$$
\left. - \sum_{ln} \frac{\mu_{gn}^k \mu_{ng}^j \mu_{gl}^i \mu_{lg}^h}{(\omega_{ng} - \omega_\sigma)(\omega_q + \omega_p)(\omega_{lg} - \omega_p)} \right].
\tag{4.3.6}
$$

Here the prime on the first summation indicates that the terms corresponding to $m = g$ are to be omitted from the summation over m; these terms are displayed explicitly in the second summation. The second summation, which appears to be divergent for $\omega_q = -\omega_p$, is now rearranged. We make use of the identity

$$
\frac{1}{XY} = \frac{1}{(X+Y)Y} + \frac{1}{(X+Y)X},
\tag{4.3.7}
$$

with $X = \omega_q + \omega_p$ and $Y = \omega_{lg} - \omega_p$, to express Eq. (4.3.6) as

$$
\chi_{kjih}^{(3)}(\omega_\sigma, \omega_r, \omega_q, \omega_p) = \frac{N}{\hbar^3} \mathscr{P}_F
$$

$$
\times \left[\sum_{lmn}' \frac{\mu_{gn}^k \mu_{nm}^j \mu_{ml}^i \mu_{lg}^h}{(\omega_{ng} - \omega_\sigma)(\omega_{mg} - \omega_q - \omega_p)(\omega_{lg} - \omega_p)} \right.
$$

$$
\left. - \sum_{ln} \frac{\mu_{gn}^k \mu_{ng}^j \mu_{gl}^i \mu_{lg}^h}{(\omega_{ng} - \omega_\sigma)(\omega_{lg} + \omega_q)(\omega_{lg} - \omega_p)} \right]
\tag{4.3.8}
$$

in addition to the contribution

$$\mathscr{P}_F \sum_{ln} \frac{\mu_{gn}^k \mu_{ng}^j \mu_{gl}^i \mu_{lg}^h}{(\omega_{ng} - \omega_\sigma)(\omega_{lg} + \omega_q)(\omega_q + \omega_p)}. \tag{4.3.9}$$

However, this additional contribution vanishes, because for every term of the form

$$\frac{\mu_{gn}^k \mu_{ng}^j \mu_{gl}^i \mu_{lg}^h}{(\omega_{ng} - \omega_\sigma)(\omega_{lg} + \omega_q)(\omega_q + \omega_p)} \tag{4.3.10a}$$

that appears in Eq. (4.3.9), there is another term with the dummy summation indices n and l interchanged, with the pair $(-\omega_\sigma, k)$ interchanged with (ω_q, i), and with the pair (ω_p, h) interchanged with (ω_r, j); this term is of the form

$$\frac{\mu_{gl}^i \mu_{lg}^h \mu_{gn}^k \mu_{ng}^j}{(\omega_{lg} + \omega_q)(\omega_{ng} - \omega_\sigma)(\omega_r - \omega_\sigma)}. \tag{4.3.10b}$$

Since $\omega_\sigma = \omega_p + \omega_q + \omega_r$, it follows that $(\omega_q + \omega_p) = -(\omega_r - \omega_\sigma)$, and hence the expression (4.3.10a) and (4.3.10b) are equal in magnitude but opposite in sign. The expression (4.3.8) for the nonlinear susceptibility is thus equivalent to Eq. (4.3.6), but is more useful for our present purpose because no apparent divergences are present.

We now specialize Eq. (4.3.8) to the case of the nonlinear refractive index with the choice of frequencies given by $\chi_{kjih}^{(3)}(\omega, \omega, \omega, -\omega)$. When we expand the permutation operator $\mathscr{P}_F$, we find that each displayed term in Eq. (4.3.8) actually represents 24 terms. The resonance nature of each such term can be analyzed by means of diagrams of the sort shown in Fig. 3.2.3.* Rather than considering all 48 terms of the expanded version of Eq. (4.3.8), let us consider only the nearly resonant terms, which would be expected to make the largest contributions to $\chi^{(3)}$. One finds, after detailed analysis of Eq. (4.3.8), that the resonant contribution to the nonlinear susceptibility is given by

$$\chi_{kjih}^{(3)}(\omega, \omega, \omega, -\omega) = \chi_{kjih}^{(3)}(\omega = \omega + \omega - \omega) = \frac{N}{6\hbar^3}$$

$$\times \left(\sum_{lmn}' \frac{\mu_{gn}^k \mu_{nm}^h \mu_{ml}^i \mu_{lg}^j + \mu_{gn}^k \mu_{nm}^h \mu_{ml}^j \mu_{lg}^i + \mu_{gn}^h \mu_{nm}^k \mu_{ml}^i \mu_{lg}^j + \mu_{gn}^h \mu_{nm}^k \mu_{ml}^j \mu_{lg}^i}{(\omega_{ng} - \omega)(\omega_{mg} - 2\omega)(\omega_{lg} - \omega)} \right.$$

$$\left. - \sum_{ln} \frac{\mu_{gn}^k \mu_{ng}^j \mu_{gl}^h \mu_{lg}^i + \mu_{gn}^k \mu_{ng}^i \mu_{gl}^h \mu_{lg}^j + \mu_{gn}^h \mu_{ng}^i \mu_{gl}^k \mu_{lg}^j + \mu_{gn}^h \mu_{ng}^j \mu_{gl}^k \mu_{lg}^i}{(\omega_{ng} - \omega)(\omega_{lg} - \omega)(\omega_{lg} - \omega)} \right). \tag{4.3.11}$$

---

* Note, however, that Fig. 3.2.3 as drawn presupposes that the three input frequencies are all positive, whereas for the case of the nonlinear refractive index two of the input frequencies are positive and one is negative.

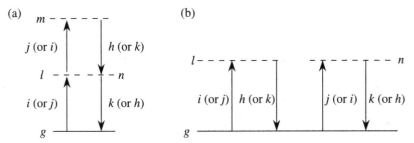

FIGURE 4.3.1   Resonance nature of the first (a) and second (b) summations of Eq. (4.3.11).

Here the first summation represents two-photon-resonant processes and the second summation represents one-photon-resonant processes, in the sense illustrated in Fig. 4.3.1.

We can use Eq. (4.3.11) to obtain explicit expressions for the resonant contributions to the nonvanishing elements of the nonlinear susceptibility tensor for an isotropic medium. We find, for example, that $\chi_{1111}(\omega = \omega + \omega - \omega)$ is given by

$$
\begin{aligned}
\chi_{1111} = {} & \frac{2N}{3\hbar^3} \sum_{lmn}{}' \frac{\mu_{gn}^x \mu_{nm}^x \mu_{ml}^x \mu_{lg}^x}{(\omega_{ng} - \omega)(\omega_{mg} - 2\omega)(\omega_{lg} - \omega)} \\
& - \frac{2N}{3\hbar^3} \sum_{ln} \frac{\mu_{gn}^x \mu_{ng}^x \mu_{gl}^x \mu_{lg}^x}{(\omega_{ng} - \omega)(\omega_{lg} - \omega)(\omega_{lg} - \omega)}.
\end{aligned}
\tag{4.3.12}
$$

Note that both one- and two-photon-resonant terms contribute to this expression. When ω is smaller than any resonant frequency of the material system, the two-photon contribution (the first term) tends to be positive. This contribution is positive because, in the presence of an applied optical field, there is some nonzero probability that the atom will reside in an excited state (state l or n as Fig. 4.3.1a is drawn). Since the (linear) polarizability of an atom in an excited state tends to be larger than that of an atom in the ground state, the effective polarizability of an atom is increased by the presence of an intense optical field; consequently this contribution to $\chi^{(3)}$ is positive. On the other hand, the one-photon contribution to χ_{1111} (the second term of Eq. (4.3.13)) is always negative when ω is smaller than any resonance frequency of the material system, because the product of matrix elements that appears in the numerator of this term is positive definite. We can understand this result from the point of view that the origin of one-photon-resonant contributions to the nonlinear susceptibility is saturation of the atomic response, which in the present case corresponds to a decrease of the positive linear susceptibility. We can also understand this result as a consequence of the ac Stark effect, which (as

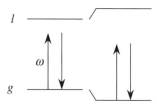

FIGURE 4.3.2   For $\omega < \omega_{lg}$ the ac Stark effect leads to an increase in the energy separation of the ground and excited states.

we shall see in Section 6.5) leads to an intensity-dependent increase in the separation of the lower and upper levels and consequently to a diminished optical response, as illustrated in Fig. 4.3.2.

In a similar fashion, we find that the resonant contribution to χ_{1221} (or to $\frac{1}{6}B$ in the notation of Maker and Terhune) is given by

$$\chi_{1221} = \frac{2}{3}\frac{N}{\hbar^3}\sum_{lmn}{}' \frac{\mu^x_{gn}\mu^x_{nm}\mu^y_{ml}\mu^y_{lg}}{(\omega_{ng} - \omega)(\omega_{mg} - 2\omega)(\omega_{lg} - \omega)}. \qquad (4.3.13)$$

The one-photon-resonant terms do not contribute to χ_{1221}, since these terms involve the summation of the product of two matrix elements of the sort $\mu^x_{gl}\mu^y_{lg}$, and this contribution always vanishes.*

We also find that the resonant contribution to χ_{1122} (or to $\frac{1}{6}A$) is given by

$$\chi_{1122} = \frac{N}{3\hbar^3}\sum_{lmn}{}' \frac{\left(\mu^x_{gn}\mu^y_{nm}\mu^y_{ml}\mu^x_{lg} + \mu^x_{gn}\mu^y_{nm}\mu^x_{ml}\mu^y_{lg}\right)}{(\omega_{ng} - \omega)(\omega_{mg} - 2\omega)(\omega_{lg} - \omega)}$$
$$-\frac{N}{3\hbar^3}\sum_{ln} \frac{\mu^x_{gn}\mu^x_{ng}\mu^y_{gl}\mu^y_{lg}}{(\omega_{ng} - \omega)(\omega_{lg} - \omega)(\omega_{lg} - \omega)}. \qquad (4.3.14)$$

$\chi^{(3)}$ in the Low-Frequency Limit

In practice, one is often interested in determining the value of the third-order susceptibility under highly nonresonant conditions, that is, for the case in which the optical frequency is very much smaller than any resonance frequency of the atomic system. An example would be the nonlinear response of an insulating solid to visible radiation. In such cases, each of the terms in the expansion of the permutation operator in Eq. (4.3.8) makes a comparable contribution to the nonlinear susceptibility, and no simplification such as those leading to Eqs. (4.3.11) through (4.3.14) is possible. It is an experimental

---

* To see that this contribution vanishes, choose x to be the quantization axis. Then if μ^x_{gl} is nonzero, μ^y_{gl} must vanish, and vice versa.

fact that in the low-frequency limit both χ_{1122} and χ_{1221} (and consequently $\chi_{1111} = 2\chi_{1122} + \chi_{1221}$) are positive in sign for the vast majority of optical materials. Also, the Kleinman symmetry condition becomes relevant under conditions of low-frequency excitation, which implies that χ_{1122} is equal to χ_{1221}, or that B is equal to A in the notation of Maker and Terhune.

We can use the results of the quantum-mechanical model to make an order-of-magnitude prediction of the value of the nonresonant third-order susceptibility. If we assume that the optical frequency ω is much smaller than all atomic resonance frequencies, we find from Eq. (4.3.6) that the nonresonant value of the nonlinear optical susceptibility is given by

$$\chi^{(3)} \simeq \frac{8N\mu^4}{\hbar^3 \omega_0^3}, \tag{4.3.15}$$

where μ is a typical value of the dipole matrix element and ω_0 is a typical value of the atomic resonance frequency. It should be noted that while the predictions of the classical model (4.3.4) and the quantum-mechanical model (4.3.15) show different functional dependences on the displayed variables, the two expressions are in fact equal if we identify d with the Bohr radius $a_0 = \hbar^2/me^2$, μ with the atomic unit of electric dipole moment $-ea_0$, and ω_0 with the Rydberg constant in angular frequency units, $\omega_0 = me^4/2\hbar^3$. Hence, the quantum-mechanical model also predicts that the third-order susceptibility is of the order of magnitude of 2×10^{-14} esu. The measured values of $\chi^{(3)}$ and n_2 for several materials that display nonresonant electronic nonlinearities are given in Table 4.3.1.

TABLE 4.3.1   Nonlinear optical coefficient for materials showing electronic nonlinearities[a]

Material	χ_{1111} (esu)	$\bar{n}_2$ (esu)
Diamond	15×10^{-14}	9×10^{-13}
Yttrium aluminum garnet	6×10^{-14}	3×10^{-13}
Sapphire	3×10^{-14}	2×10^{-13}
Borosilicate crown glass	2.5×10^{-14}	1.5×10^{-13}
Fused silica	2×10^{-14}	1.2×10^{-13}
CaF_2	1.6×10^{-14}	1×10^{-13}
LiF	1×10^{-10}	0.6×10^{-13}

[a] Values are obtained from optical frequency mixing experiments and hence do not include electrostrictive contributions, since electrstriction is a slow process that cannot respond at optical frequencies. The value of $\bar{n}_2$ is calculated as $\bar{n}_2 = 3\pi\chi_{1111}/n_0$. (Adapted from R. W. Hellwarth (1977), Tables 7.1 and 9.1.)

4.4. Nonlinearities Due to Molecular Orientation

Liquids that are composed of anisotropic molecules (i.e., molecules having an anisotropic polarizability tensor) typically possess a large value of n_2. The origin of this nonlinearity is the tendency of molecules to become aligned in the electric field of an applied optical wave. The optical wave then experiences a modified value of the refractive index because the average polarizability per molecule has been changed by the molecular alignment.

Consider, for example, the case of carbon disulfide (CS_2), which is illustrated in part (a) of Fig. 4.4.1. Carbon disulfide is a cigar-shaped molecule (i.e., a prolate spheroid), and consequently the polarizability α_3 experienced by an optical field that is parallel to the symmetry axis is larger than the polarizability α_1 experienced by a field that is perpendicular to its symmetry axis, that is,

$$\alpha_3 > \alpha_1. \tag{4.4.1}$$

Consider now what happens when such a molecule is subjected to a dc electric field, as shown in part (b) of the figure. Since α_3 is larger than α_1, the component of the induced dipole moment along the molecular axis will be disproportionately long. The induced dipole moment $\mathbf{p}$ thus will not be parallel to $\mathbf{E}$, but will be offset from it in the direction of the symmetry axis. A torque

$$\tau = \mathbf{p} \times \mathbf{E} \tag{4.4.2}$$

will thus be exerted on the molecule. This torque is directed in such a manner as to twist the molecule into alignment with the applied electric field.

The tendency of the molecule to become aligned in the applied electric field is counteracted by thermal agitation, which tends to randomize the molecular orientation. We calculate the mean degree of molecular orientation through use of the Boltzmann factor. In order to do so, we first calculate the potential energy

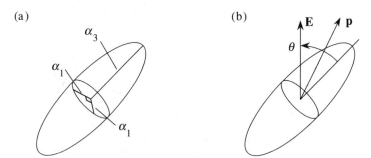

FIGURE 4.4.1   (a) A prolate spheroidal molecule, such as carbon disulfide. (b) The dipole moment $\mathbf{p}$ induced by an electric field $\mathbf{E}$.

of the molecule in the applied electric field. If the applied field is changed by
an amount $d\mathbf{E}$, the orientational potential energy is changed by the amount

$$dU = -\mathbf{p} \cdot d\mathbf{E} = -p_3\,dE_3 - p_1\,dE_1, \tag{4.4.3}$$

where we have decomposed $\mathbf{E}$ into its components along the molecular axis
(E_3) and perpendicular to the molecular axis (E_1). Since

$$p_3 = \alpha_3 E_3 \tag{4.4.4}$$

and

$$p_1 = \alpha_1 E_1, \tag{4.4.5}$$

we find that

$$dU = -\alpha_3 E_3\,dE_3 - \alpha_1 E_1\,dE_1, \tag{4.4.6}$$

which can be integrated to give

$$U = -\tfrac{1}{2}\bigl(\alpha_3 E_3^2 + \alpha_1 E_1^2\bigr). \tag{4.4.7}$$

If we now introduce the angle θ between $\mathbf{E}$ and the molecular axis (see
Fig. 4.4.1b), we find that the orientational potential energy is given by

$$U = -\tfrac{1}{2}[\alpha_3 E^2 \cos^2\theta + \alpha_1 E^2 \sin^2\theta]$$
$$= -\tfrac{1}{2}\alpha_1 E^2 - \tfrac{1}{2}(\alpha_3 - \alpha_1)E^2 \cos^2\theta. \tag{4.4.8}$$

Since $\alpha_3 - \alpha_1$ has been assumed to be positive, this result shows that the
potential energy is lower when the molecule axis is parallel to $\mathbf{E}$ than when it
is perpendicular to $\mathbf{E}$, as illustrated in Fig. 4.4.2.

Our discussion thus far has assumed that the applied field is static. We now
allow the field to vary in time at an optical frequency. For simplicity we assume
that the light is linearly polarized; the general case of elliptical polarization is
treated at the end of the present section. We thus replace $\mathbf{E}$ in Eq. (4.4.9) by

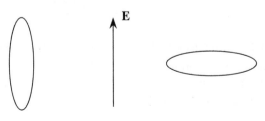

lower potential   higher potential
energy   energy

FIGURE 4.4.2   Alignment energy of a molecule.

the time-varying scalar quantity $\tilde{E}(t)$. The square of $\tilde{E}$ will contain frequency components near zero frequency and components at approximately twice the optical frequency ω. Since orientational relaxation times for molecules are typically of the order of a few picoseconds, the molecular orientation can respond to the frequency components near zero frequency but not to those near 2ω. We can thus formally replace E^2 in Eq. (4.4.9) by $\overline{\tilde{E}^2}$, where the bar denotes a time average over many cycles of the optical field.

We now calculate the intensity-dependent refractive index for such a medium. For simplicity, we first ignore local-field corrections, in which case the refractive index is given by

$$n^2 = 1 + 4\pi\chi = 1 + 4\pi N\langle\alpha\rangle, \qquad (4.4.9)$$

where N is the number density of molecules and where $\langle\alpha\rangle$ denotes the expectation value of the molecular polarizability experienced by the incident radiation. To obtain an expression for $\langle\alpha\rangle$, we note that the mean orientational potential energy is given by $\langle U\rangle = -\frac{1}{2}|E|^2\langle\alpha\rangle$, which by comparison with the average of Eq. (4.4.9) shows that

$$\langle\alpha\rangle = \alpha_3\langle\cos^2\theta\rangle + \alpha_1\langle\sin^2\theta\rangle = \alpha_1 + (\alpha_3 - \alpha_1)\langle\cos^2\theta\rangle. \qquad (4.4.10)$$

Here $\langle\cos^2\theta\rangle$ denotes the expectation value of $\cos^2\theta$ in thermal equilibrium and is given in terms of the Boltzmann distribution as

$$\langle\cos^2\theta\rangle = \frac{\int d\Omega\,\cos^2\theta\,\exp[-U(\theta)/kT]}{\int d\Omega\,\exp[-U(\theta)/kT]}, \qquad (4.4.11)$$

where $\int d\Omega$ denotes an integration over all solid angles. For convenience, we introduce the intensity parameter

$$J = \frac{1}{2}(\alpha_3 - \alpha_1)\overline{\tilde{E}^2}/kT, \qquad (4.4.12)$$

and let $d\Omega = 2\pi\sin\theta d\theta$. We then find that $\langle\cos^2\theta\rangle$ is given by

$$\langle\cos^2\theta\rangle = \frac{\int_0^\pi \cos^2\theta\,\exp(J\cos^2\theta)\sin\theta d\theta}{\int_0^\pi \exp(J\cos^2\theta)\sin\theta d\theta}. \qquad (4.4.13)$$

Equations (4.4.9) through (4.4.13) can be used to determine the refractive index experienced by fields of arbitrary intensity $\overline{\tilde{E}^2}$.

Let us first calculate the refractive index experienced by a weak optical field, by taking the limit $J \to 0$. For this case we find that the average of $\cos^2\theta$ is given by

$$\langle\cos^2\theta\rangle_0 = \frac{\int_0^\pi \cos^2\theta\,\sin\theta d\theta}{\int_0^\pi \sin\theta d\theta} = \frac{1}{3}, \qquad (4.4.14)$$

and that according to Eq. (4.4.10) the mean polarizability is given by

$$\langle \alpha \rangle_0 = \tfrac{1}{3}\alpha_3 + \tfrac{2}{3}\alpha_1. \tag{4.4.15}$$

Using Eq. (4.4.9), we find that the refractive index is given by

$$n_0^2 = 1 + 4\pi N\left(\tfrac{1}{3}\alpha_3 + \tfrac{2}{3}\alpha_1\right). \tag{4.4.16}$$

Note that this result makes good physical sense: in the absence of interactions that tend to align the molecules, the mean polarizability is equal to one-third of that associated with the direction of the symmetry axis of the molecule plus two-thirds of that associated with directions perpendicular to this axis.

For the general case in which an intense optical field is applied, we find from Eqs. (4.4.9) and (4.4.10) that the refractive index is given by

$$n^2 = 1 + 4\pi N[\alpha_1 + (\alpha_3 - \alpha_1)\langle \cos^2 \theta \rangle], \tag{4.4.17}$$

and hence by comparison with Eq. (4.4.16) that the square of the refractive index changes by the amount

$$\begin{aligned}
n^2 - n_0^2 &= 4\pi N\left[\tfrac{1}{3}\alpha_1 + (\alpha_3 - \alpha_1)\langle \cos^2 \theta \rangle - \tfrac{1}{3}\alpha_3\right] \\
&= 4\pi N(\alpha_3 - \alpha_1)\left(\langle \cos^2 \theta \rangle - \tfrac{1}{3}\right).
\end{aligned} \tag{4.4.18}$$

Since $n^2 - n_0^2$ is usually very much smaller than n_0^2, we can express the left-hand side of this equation as

$$n^2 - n_0^2 = (n - n_0)(n + n_0) \simeq 2n_0(n - n_0)$$

and hence find that the refractive index can be expressed as

$$n = n_0 + \delta n, \tag{4.4.19}$$

where the nonlinear change in refractive index is given by

$$\delta n \equiv n - n_0 = \frac{2\pi N}{n_0}(\alpha_3 - \alpha_1)\left(\langle \cos^2 \theta \rangle - \tfrac{1}{3}\right). \tag{4.4.20}$$

The quantity $\langle \cos^2 \theta \rangle$, given by Eq. (4.4.13), can be calculated in terms of a tabulated function (the Dawson integral). Figure 4.4.3 shows a plot of $\langle \cos^2 \theta \rangle - \tfrac{1}{3}$ as a function of the intensity parameter $J = \tfrac{1}{2}(\alpha_3 - \alpha_1)\tilde{E}^2/kT$.

In order to obtain an explicit formula for the change in refractive index, we expand the exponentials appearing in Eq. (4.4.13) and integrate the resulting expression term by term. We find that

$$\langle \cos^2 \theta \rangle = \frac{1}{3} + \frac{4J}{45} + \frac{8J^2}{945} + \cdots. \tag{4.4.21}$$

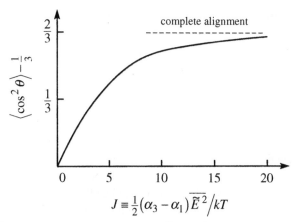

$$J \equiv \tfrac{1}{2}(\alpha_3 - \alpha_1)\overline{\tilde{E}^2}\big/kT$$

FIGURE 4.4.3   Variation of the quantity $(\langle \cos^2 \theta \rangle - \tfrac{1}{3})$, which is proportional to the nonlinear change in refractive index δn, with the intensity parameter J. Note that for $J \lesssim 5$, δn increases nearly linearly with J.

Dropping all terms but the first two, we find from (4.4.20) that the change in the refractive index due to the nonlinear interaction is given by

$$\delta n = \frac{4\pi N}{2n_0}(\alpha_3 - \alpha_1)\frac{4J}{45} = \frac{4\pi}{45}\frac{N}{n_0}(\alpha_3 - \alpha_1)^2\frac{\overline{\tilde{E}^2}}{kT}. \qquad (4.4.22)$$

We can express this result as

$$\delta n = \bar{n}_2 \overline{\tilde{E}^2}, \qquad (4.4.23)$$

where the second-order nonlinear refractive index is given by

$$\bar{n}_2 = \frac{4\pi N}{45n_0}\frac{(\alpha_3 - \alpha_1)^2}{kT}. \qquad (4.4.24)$$

Note that $\bar{n}_2$ is positive both for the case $\alpha_3 > \alpha_1$ (the case that we have been considering explicitly) and for the opposite case where $\alpha_3 < \alpha_1$. The reason for this behavior is that the torque experienced by the molecule is always directed in a manner that tends to align the molecule so that the light sees a *larger* value of the polarizability.

A more accurate prediction of the nonlinear refractive index is obtained by including the effects of local-field corrections. We begin with the Lorentz–Lorentz law (see also Eq. (3.8.8a)),

$$\frac{n^2 - 1}{n^2 + 2} = \frac{4}{3}\pi N \langle \alpha \rangle, \qquad (4.4.25)$$

instead of the approximate relationship (4.4.9). By repeating the derivation leading to Eq. (4.4.24) with Eq. (4.4.9) replaced by Eq. (4.4.25) and with $\tilde{E}^2$ replaced by the square of the Lorentz local field (see the discussion of Section 3.8), we find that the second-order nonlinear refractive index is given by

$$\bar{n}_2 = \frac{4\pi N}{45n_0}\left(\frac{n_0^2 + 2}{3}\right)^4 \frac{(\alpha_3 - \alpha_1)^2}{kT}. \tag{4.4.26}$$

Note that this result is consistent with the general prescription given in Section 3.8, which states that local-field effects can be included by multiplying the results obtained in the absence of local field corrections (that is, Eq. (4.4.24)) by the local-field correction factor $\mathscr{L}^{(3)} = [(n_0^2 + 2)/3]^4$ of Eq. (3.8.25).

Finally, we quote some numerical values relevant to the material carbon disulfide. The maximum possible value of δn is 0.58 and would correspond to a complete alignment of the molecules. The value $J = 1$ corresponds to a field strength of $E \simeq 3 \times 10^7 V/\text{cm}$. The value of $\bar{n}_2$ is hence equal to 1.3×10^{-11} esu. Through use of Eq. (4.1.18) and the value $n_0 = 1.63$, we can convert this result to the value $n_2 = 3 \times 10^{-14}$ cm^2/W.

Tensor Properties of $\chi^{(3)}$ for the Molecular Orientation Effect

Let us now consider the nonlinear response of a collection of anisotropic molecules to light of arbitrary polarization. Close *et al.* (1966) have shown that the mean polarizability in thermal equilibrium for a molecule whose three principal polarizabilities a, b, and c are distinct can be represented as

$$\langle \alpha_{ij} \rangle = \alpha \delta_{ij} + \gamma_{ij}, \tag{4.4.27}$$

where the linear contribution to the mean polarizability is given by

$$\alpha = \tfrac{1}{3}(a + b + c), \tag{4.4.28}$$

and where the lowest-order nonlinear correction term is given by

$$\gamma_{ij} = C \sum_{kl}(3\delta_{ik}\delta_{jl} - \delta_{ij}\delta_{kl})\overline{\tilde{E}_k^{\text{loc}}(t)\tilde{E}_l^{\text{loc}}(t)}. \tag{4.4.29}$$

Here the constant C is given by

$$C = \frac{(a - b)^2 + (b - c)^2 + (a - c)^2}{90kT}, \tag{4.4.30}$$

and $\tilde{E}^{\text{loc}}$ denotes the Lorentz local field. In the appendix to this section, we derive the result given by Eqs. (4.4.27) through (4.4.30) for the special case of an axially symmetric molecule; the derivation for the general case is left as an exercise to the reader. Next, we use these results to determine the form of the third-order susceptibility tensor. We first ignore local-field corrections and replace $\tilde{E}_k^{\mathrm{loc}}(t)$ by the microscopic electric field $\tilde{E}_k(t)$, which we represent as

$$\tilde{E}_k(t) = E_k e^{-i\omega t} + \text{c.c.} \tag{4.4.31}$$

The electric-field-dependent factor appearing in Eq. (4.4.29) thus becomes

$$\overline{\tilde{E}_k^{\mathrm{loc}}(t)\tilde{E}_l^{\mathrm{loc}}(t)} = E_k E_l^* + E_k^* E_l. \tag{4.4.32}$$

Since we are ignoring local-field corrections, we can assume that the polarization is given by

$$P_i = \sum_j N\langle\alpha_{ij}\rangle E_j \tag{4.4.33}$$

and hence that the third-order contribution to the polarization is given by

$$P_i^{(3)} = N\sum_j \gamma_{ij} E_j. \tag{4.4.34}$$

By introducing the form for γ_{ij} given by Eqs. (4.4.29) and (4.4.32) into this expression, we find that

$$P_i^{(3)} = NC\sum_{jkl}(3\delta_{ik}\delta_{jl} - \delta_{ij}\delta_{kl})(E_k E_l^* + E_k^* E_l)E_j,$$

which can be written entirely in vector form as

$$\mathbf{P}^{(3)} = NC[3(\mathbf{E}\cdot\mathbf{E}^*)\mathbf{E} + 3(\mathbf{E}\cdot\mathbf{E})\mathbf{E}^* - (\mathbf{E}\cdot\mathbf{E}^*)\mathbf{E} - (\mathbf{E}\cdot\mathbf{E}^*)\mathbf{E}]$$
$$= NC[(\mathbf{E}\cdot\mathbf{E}^*)\mathbf{E} + 3(\mathbf{E}\cdot\mathbf{E})\mathbf{E}^*]. \tag{4.4.35}$$

This result can be rewritten using the notation of Maker and Terhune (see also Eq. (4.2.10)) as

$$\mathbf{P}^{(3)} = A(\mathbf{E}\cdot\mathbf{E}^*)\mathbf{E} + \tfrac{1}{2}B(\mathbf{E}\cdot\mathbf{E})\mathbf{E}^*, \tag{4.4.36}$$

where the coefficients A and B are given by $B = 6A = 6NC$, which through use of the expression (4.4.30) for C becomes

$$B = 6A = N\left[\frac{(a-b)^2 + (b-c)^2 + (a-c)^2}{15kT}\right]. \tag{4.4.37}$$

This result shows that for the molecular orientation effect the ratio B/A is equal to 6, a result quoted earlier without proof (in (4.2.13a)). As in Eq. (4.4.26), local-field corrections can be included in the present formalism by replacing Eq. (4.4.37) by

$$B = 6A = \left(\frac{n_0^2 + 2}{3}\right)^4 N\left[\frac{(a-b)^2 + (b-c)^2 + (a-c)^2}{15kT}\right]. \tag{4.4.38}$$

Appendix to Section 4.4

The derivation that we presented above of the vector form of the nonlinear polarization due to the molecular orientation effect presupposed the validity of the starting equations (4.4.27) through (4.4.30). Here we derive these starting equations for the special case of an axially symmetric molecule. The derivation follows closely that of Owyoung (1971).

Consider an axially symmetric molecule whose polarizability tensor in the principal-axis coordinate system is described by

$$\alpha^P = \alpha_{ij}^P = \begin{bmatrix} \alpha_1 & 0 & 0 \\ 0 & \alpha_1 & 0 \\ 0 & 0 & \alpha_3 \end{bmatrix}. \tag{4.4.39}$$

We need to express α_{ij}^P in a space-fixed (laboratory) coordinate system. The orientation of the molecule in this system can be described by the three Euler angles θ, ϕ, and ψ illustrated in Fig. 4.4.4. Here θ is the polar angle and ϕ is the azimuthal angle. The angle ψ specifies the rotation angle about the molecular α_3 axis; for the present case of a symmetric molecule ($\alpha_1 - \alpha_2$) this angle need not be specified. In the space-fixed coordinate system, the polarizability tensor is given by

$$\alpha(\theta, \phi) = \mathbf{A}^T \alpha^P \mathbf{A}, \tag{4.4.40}$$

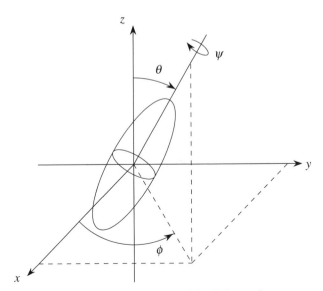

FIGURE 4.4.4 Definition of the Euler angles.

where $\mathbf{A}$ is the transformation matrix

$$\mathbf{A} = \begin{bmatrix} -\cos\psi\sin - \cos\theta\cos\phi\sin\phi & \cos\psi\cos\phi - \cos\theta\sin\phi\sin\psi & \sin\psi\sin\theta \\ \sin\psi\sin\phi - \cos\theta\cos\phi\cos\psi & -\sin\psi\cos\theta - \cos\theta\sin\phi\cos\psi & \cos\psi\sin\theta \\ \sin\theta\cos\phi & \sin\theta\sin\phi & \cos\theta \end{bmatrix},$$

$$(4.4.41)$$

and $\mathbf{A}^{\mathrm{T}}$ is its transpose. Through use of Eqs. (4.4.39) through (4.4.41) we find that

$$\alpha(\theta,\phi)$$
$$= \begin{bmatrix} \alpha_1 + (\alpha_3 - \alpha_1)\sin^2\theta\cos^2\phi & (\alpha_3 - \alpha_1)\sin^2\theta\sin\phi & (\alpha_3 - \alpha_1)\cos\theta\sin\theta\cos\phi \\ (\alpha_3 - \alpha_1)\sin^2\theta\sin\phi & \alpha_1 + (\alpha_3 - \alpha_1)\sin^2\theta\sin^2\phi & (\alpha_3 - \alpha_1)\cos\theta\sin\theta\cos\phi \\ (\alpha_3 - \alpha_1)\sin\theta\cos\theta\cos\phi & (\alpha_3 - \alpha_1)\sin\theta\cos\theta\sin\phi & \alpha_1 + (\alpha_3 - \alpha_1)\cos^2\theta \end{bmatrix}.$$

$$(4.4.42)$$

Note that this expression for the polarizability tensor in the laboratory coordinate system is independent of ψ, since α is symmetric with respect to its principal axes 1 and 2.

We now calculate the mean polarizability in thermal equilibrium for an ensemble of such molecules in the presence of an applied electric field. The probability density that a given molecule will have its major axis oriented at angles (θ,ϕ) is given by

$$P(\theta,\phi) = \frac{\exp[-U(\theta,\phi)/kT]}{\int d\Omega \exp[-U(\theta,\phi)/kT]}, \qquad (4.4.43)$$

where the orientational energy is given by

$$U(\theta,\phi) = -\tfrac{1}{2}\sum_{kl}\alpha_{kl}(\theta,\phi)\overline{\tilde{E}_k^{\mathrm{loc}}(t)\tilde{E}_l^{\mathrm{loc}}(t)}. \qquad (4.4.44)$$

The ensemble-averaged polarizability is then given by

$$\langle\alpha_{ij}\rangle = \int d\Omega\,\alpha_{ij}(\theta,\phi)P(\theta,\phi). \qquad (4.4.45)$$

We assume that the ratio $U(\theta,\phi)/kT$ is much smaller than unity, so that the exponentials can be approximated as

$$\exp\left[-\frac{U(\theta,\phi)}{kT}\right] = 1 - \frac{U(\theta,\phi)}{kT}. \qquad (4.4.46)$$

Equations (4.4.42)–(4.4.45) can then be combined to give

$$
\langle \alpha_{ij} \rangle = \frac{\int \alpha_{ij}(\theta,\phi)d\Omega + \sum_{kl} \frac{\overline{\tilde{E}_k^{loc}(t)\tilde{E}_l^{loc}(t)}}{2kT} \int \alpha_{ij}(\theta,\phi)\alpha_{kl}(\theta,\phi)d\Omega}{\int d\Omega + \sum_{kl} \frac{\overline{\tilde{E}_k^{loc}(t)\tilde{E}_l^{loc}(t)}}{2kT} \int \alpha_{kl}(\Omega,\phi)d\Omega}.
\tag{4.4.47}
$$

We note that $\int d\Omega = 4\pi$ and that the second term in the denominator is much smaller than the first. We thus expand the reciprocal of the denominator as a power series in the ratio of the second to the first terms and find that to lowest order

$$
\langle \alpha_{ij} \rangle = \int \frac{d\Omega}{4\pi}\alpha_{ij}(\theta,\phi) + \sum_{kl} \frac{\overline{\tilde{E}_k^{loc}(t)\tilde{E}_l^{loc}(t)}}{2kT} \int \frac{d\Omega}{4\pi}\alpha_{ij}(\theta,\phi)\alpha_{kl}(\theta,\phi)
$$

$$
- \sum_{kl} \frac{\overline{\tilde{E}_k^{loc}(t)\tilde{E}_l^{loc}(t)}}{2kT} \int \frac{d\Omega}{4\pi}\alpha_{ij}(\theta,\phi) \int \frac{d\Omega}{4\pi}\alpha_{kl}(\theta,\phi).
\tag{4.4.48}
$$

The integrations can be performed explicitly. We let

$$
\int d\Omega \to \int_0^{2\pi} d\phi \int_{-1}^{1} d(\cos\theta)
\tag{4.4.49}
$$

and find that

$$
\int \frac{d\Omega}{4\pi}\alpha_{ij}(\theta,\phi) \equiv \langle \alpha_{ij}\rangle_0 = \alpha\delta_{ij},
\tag{4.4.50}
$$

where

$$
\alpha = \tfrac{2}{3}\alpha_1 + \tfrac{1}{3}\alpha_3.
\tag{4.4.51}
$$

We also find that

$$
\int \frac{d\Omega}{4\pi}\alpha_{ij}(\theta,\phi)\,\alpha_{kl}(\theta,\phi) \equiv \langle \alpha_{ij}\alpha_{kl}\rangle_0
$$

$$
= \begin{cases} \frac{4}{45}(\alpha_3-\alpha_1)^2 + \alpha^2 & \text{for } i=j=k=l, \\ \alpha^2 - \frac{2}{45}(\alpha_3-\alpha_1)^2 & \text{for } i=j\neq k=l, \\ \frac{1}{15}(\alpha_3-\alpha_1)^2 & \text{for } i=k\neq j=l \\ & \text{or } i=l\neq j=k. \end{cases}
\tag{4.4.52}
$$

These results can be combined to represent the polarizability as

$$
\langle \alpha_{ij}\rangle = \langle \alpha_{ij}\rangle_0 + \sum_{kl} \frac{\overline{\tilde{E}_k^{loc}(t)\tilde{E}_l^{loc}(t)}}{2kT}(\langle\alpha_{ij}\alpha_{kl}\rangle_0 - \langle\alpha_{ij}\rangle_0\langle\alpha_{kl}\rangle_0),
\tag{4.4.53}
$$

which can be written as

$$\langle \alpha_{ij} \rangle = \alpha \delta_{ij} + \gamma_{ij}, \tag{4.4.54}$$

where

$$\gamma_{ij} = \sum_{kl} \frac{\overline{\tilde{E}_k^{\mathrm{loc}}(t)\tilde{E}_l^{\mathrm{loc}}(t)}}{2kT} \frac{2}{45}(\alpha_3 - \alpha_1)^2 \left[\tfrac{3}{2}(\delta_{ik}\delta_{jl} + \delta_{il}\delta_{jk}) - \delta_{ij}\delta_{kl}\right]. \tag{4.4.55}$$

Note that since $\tilde{E}_k^{\mathrm{loc}}(t)$ and $\tilde{E}_l^{\mathrm{loc}}(t)$ appear in this last expression only as a symmetric product, we can replace the terms within the parentheses by $2\delta_{ik}\delta_{jl}$. We thus obtain the desired form

$$\gamma_{ij} = \sum_{kl} \frac{\overline{\tilde{E}_k^{\mathrm{loc}}(t)\tilde{E}_l^{\mathrm{loc}}(t)}}{45kT}(\alpha_3 - \alpha_1)^2(3\delta_{ik}\delta_{jl} - \delta_{ij}\delta_{kl}). \tag{4.4.56}$$

Note that γ_{ij} is traceless, that is, that $\sum_i \gamma_{ii} = 0$. Intuitively, we expect γ_{ij} to be traceless, since in applying an optical field to the medium we have not "added" any new polarizability to it; we have simply "rearranged" the polarizability that was initially present among the various tensor components.

4.5. Thermal Nonlinear Optical Effects

Thermal processes can lead to large (and often unwanted) nonlinear optical effects. The origin of thermal nonlinear optical effects is that some fraction of the incident laser power is absorbed in passing through an optical material. The temperature of the illuminated portion of the material consequently increases, which leads to a change in the refractive index of the material. For gases, the refractive index invariably decreases with increasing temperature (at constant pressure), but for condensed matter the refractive index can either increase or decrease with changes in temperature, depending upon details of the internal structure of the material. The time scale for changes in the temperature of the material can be quite long (of the order of seconds), and consequently thermal effects often lead to strongly time-dependent nonlinear optical phenomena.

   Thermal effects can be described mathematically by assuming that the refractive index $\tilde{n}$ varies with temperature according to *

$$\tilde{n} = n_0 + \left(\frac{dn}{dT}\right)\tilde{T}_1 \tag{4.5.1}$$

---

* As elsewhere in this text, a tilde is used to designate an explicitly time-dependent quantity.

TABLE 4.5.1   Thermal properties of various optical materials

Material	$(\rho_0 C)$ (J/cm^3)^a	κ (W/m K)	dn/dT (K^{-1})^b
Diamond	1.76	660	
Ethanol	1.91	0.168	
Fused silica	1.67	1.4	1.2×10^{-5}
Sodium chloride	1.95	6.4	-3.6×10^{-5}
Water (liquid)	4.2	0.56	
Air^c	1.2×10^{-3}	26×10^{-3}	-1.0×10^{-6}

^{a} $(\rho_0 C)$ is the heat capacity per unit volume and κ is the thermal conductivity. More extensive listings of these quantities can be found in the *CRC Handbook of Chemistry and Physics*, Section D, and in the *American Institute of Physics Handbook*, Section 4.

^{b} dn/dT is the temperature coefficient of the refractive index. It can be either positive or negative, and for condensed matter typically lies in the range $\pm 3 \times 10^{-5}$ K^{-1}. See for instance the *American Institute of Physics Handbook*, Section 6b.

^{c} C is measured at constant pressure. Values are quoted at STP. Under other conditions, the values of these quantities can be found by noting that to good approximation $(\rho_0 C)$ is proportional to the density, κ is independent of the density, and that for any ideal gas $dn/dT = -(n-1)/T$.

where the quantity (dn/dT) describes the temperature dependence of the refractive index of a given material and where $\tilde{T}_1$ designates the laser-induced change in temperature. We assume that $\tilde{T}_1$ obeys the heat-transport equation

$$(\rho_0 C)\frac{\partial \tilde{T}_1}{\partial t} - \kappa \nabla^2 \tilde{T}_1 = \alpha \tilde{I}(r). \tag{4.5.2}$$

Here $(\rho_0 C)$ denotes the heat capacity per unit volume, κ denotes the thermal conductivity, and α denotes the linear absorption coefficient of the material. We express the heat capacity in the form $(\rho_0 C)$ because most handbooks tabulate the material density ρ_0 and the heat capacity per unit mass C rather than their product $(\rho_0 C)$, which is the quantity of direct relevance in the present context. Representative values of dn/dT, $(\rho_0 C)$, and κ are shown in Table 4.5.1.

Equation (4.5.2) can be solved as a boundary value problem for any specific physical circumstance, and hence the refractive index at any point in space can be found from Eq. (4.5.1). Note that thermal nonlinear optical effects are nonlocal, because the change in refractive index at some given point will in general depend on the laser intensity at other nearby points. For our present purposes, let us make some simple numerical estimates of the magnitude of the thermal contribution to the change in refractive index for the situation shown in Fig. 4.5.1. We assume that a circular laser beam of intensity I_0 and radius R (and consequently power $P = \pi R^2 I_0$) falls onto a slab of optical material of thickness L and absorption coefficient α.

Let us first estimate the response time τ associated with the change in temperature for this situation. We take τ to be some measure of the time taken

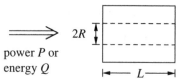

$$2R$$

power P or
energy Q

$$\longmapsto L \longmapsto$$

FIGURE 4.5.1   Geometry for the description of thermal nonlinear optical effects.

for the temperature distribution to reach its new steady state after the laser field is suddenly switched on or switched off. For definiteness we assume the latter situation. We then estimate τ by approximating $\partial \tilde{T}_1 / \partial t$ in Eq. (4.5.2) by T_1/τ and by approximating $\nabla^2 \tilde{T}_1$ as T_1/R^2. Equation (4.5.2) then becomes $(\rho_0 C) T_1/\tau \approx \kappa T_1/R^2$, from which it follows that

$$\tau \approx \frac{(\rho_0 C) R^2}{\kappa}. \tag{4.5.3}$$

We can estimate numerically the response time τ for condensed matter by adopting the typical values $(\rho_0 C) = 10^6$ J/m^3 K, $\kappa = 1$ W/m K, and $R = 1$ mm, and thus find that $\tau \approx 1$ s. Even for a tightly collimated beam with $R = 10$ μm, one finds that $\tau \approx 100$ μs. These response times are much longer than the pulse duration T produced by most pulsed lasers. One thus reaches the conclusion that, in the consideration of thermal effects, the power (or alternatively the intensity) is the relevant quantity for continuous-wave laser beams, but that the pulse energy $Q = PT$ (or alternatively the fluence, the energy per unit cross-sectional area) is the relevant quantity in the consideration of pulsed lasers.

Thermal Nonlinearities with Continuous-Wave Laser Beams

We have just seen that the analysis of thermal effects in nonlinear optics is different for continuous wave than for pulsed radiation. Let us consider first the case of continuous-wave radiation. Under steady-state conditions the equation of heat transport then reduces to

$$-\kappa \nabla^2 \tilde{T}_1 = \alpha \tilde{I}(r). \tag{4.5.4}$$

This equation can be solved explicitly for any assumed laser profile $\tilde{I}(r)$. For our present purposes it suffices to make an order-of-magnitude estimate of the maximum temperature rise $T_1^{(\mathrm{max})}$ at the center of the laser beam. To do so, we replace $\nabla^2 \tilde{T}_1$ by $-T_1^{(\mathrm{max})}/R^2$, and thereby find that

$$T_1^{(\mathrm{max})} = \frac{\alpha I^{(\mathrm{max})} R^2}{\kappa}, \tag{4.5.5}$$

where $I^{(max)}$ is the laser intensity at the center of the laser beam. Then from Eq. (4.5.1) we estimate the maximum change in refractive index as

$$\Delta n = \left(\frac{dn}{dT}\right) \frac{\alpha I^{(max)} R^2}{\kappa}. \tag{4.5.6}$$

We can express this change in terms of an effective nonlinear refractive index coefficient $n_2^{(th)}$ defined through $\Delta n = n_2^{(th)} I^{(max)}$ to obtain

$$n_2^{(th)} = \left(\frac{dn}{dT}\right) \frac{\alpha R^2}{\kappa}. \tag{4.5.7}$$

Note that this quantity is geometry-dependent (through the R^2 factor) and hence is not an intrinsic property of an optical material. Nonetheless, it provides a useful way of quantifying the magnitude of thermal nonlinear optical effects. If we estimate its size through use of the values $(dn/dT) = 10^{-5}$ K^{-1}, $\alpha = 1$ cm^{-1}, $R = 1$ mm, and $\kappa = 1$ W/m K, we find that $n_2^{(th)} = 10^{-5}$ cm^2/W. By way of comparison, recall that for fused silica $n_2 = 3 \times 10^{-16}$ cm^2/W. Even for a much smaller beam size ($R = 10$ μm) and a much smaller absorption coefficient ($\alpha = 0.01$ cm^{-1}), we still obtain a relatively large thermal nonlinear coefficient of $n_2^{(th)} = 10^{-11}$ cm^2/W. The conclusion to be drawn from these numbers is clear: thermal effects are usually the dominant nonlinear optical mechanism for continuous-wave laser beams. Recent experimental investigations of thermal nonlinear optical effects in gases have been reported by Bentley et al. (2000).

Thermal Nonlinearities with Pulsed Laser Beams

As mentioned earlier, for most pulsed lasers the induced change in refractive index is proportional to the pulse energy $Q = \int \tilde{P}(t)dt$ rather than to the instantaneous power $\tilde{P}(t)$ (or alternatively it is proportional to the pulse fluence $F = \int \tilde{I}(t)dt$ rather than to the pulse intensity $\tilde{I}(t)$). For this reason, it is not possible to describe the change in refractive index in terms of a quantity such as $n_2^{(th)}$. Rather, $\Delta\tilde{n}$ increases (or decreases) monotonically during the time extent of the laser pulse. Nonetheless one can develop simple criteria for determining the conditions under which thermal nonlinear optical effects are important. In particular, let us consider the conditions under which the thermal change in refractive index

$$\Delta n^{(th)} = \left(\frac{dn}{dT}\right) T_1^{(max)} \tag{4.5.8}$$

will be greater than or equal to the change resulting from the electronic response

$$\Delta n^{(el)} = n_2^{(el)} I. \tag{4.5.9}$$

We estimate the maximum change in temperature $T_1^{(max)}$ induced by the laser beam as follows: For a short laser pulse (pulse duration t_p much shorter than the thermal response time τ of Eq. (4.5.3)), the heat transport equation (4.5.2) reduces to

$$(\rho_0 C)\frac{\partial \tilde{T}_1}{\partial t} = \alpha \tilde{I}(r); \tag{4.5.10}$$

we have dropped the term $-\kappa \nabla^2 \tilde{T}_1$ because in a time $t_p \ll \tau$ at most a negligible fraction of the absorbed energy can diffuse out of the interaction region. By approximating $\partial \tilde{T}_1/\partial t$ as $T_1^{(max)}/t_p$, we find that

$$T_1^{(max)} = \frac{\alpha I^{(max)} t_p}{(\rho_0 C)}. \tag{4.5.11}$$

By combining Eqs. (4.5.8) through (4.5.11), we find that the thermal contribution to the change in refractive index will exceed the electronic contribution if the laser pulse duration satisfies the inequality

$$t_p \geq \frac{n_2^{(el)}(\rho_0 C)}{(dn/dT)\alpha}. \tag{4.5.12}$$

If we evaluate this expression assuming the typical values $n_2^{(el)} = 3 \times 10^{-16}$ cm^2/W, $(\rho_0 C) = 1 \times 10^6$ J/m^3 K, $(dn/dT) = 1 \times 10^{-5}$ K^{-1}, $\alpha = 1$ cm^{-1}, we find that the condition for the importance of thermal effects becomes

$$t_p \geq 30 \text{ psec.} \tag{4.5.13}$$

We thus see that thermal effects are likely to make a contribution to the nonlinear optical response for all but the shortest ($t_p \ll 30$ psec) laser pulses.

4.6. Semiconductor Nonlinearities

Semiconductor materials play an important role in nonlinear optics both because they produce large nonlinear optical responses and because these materials lend themselves to the construction of integrated devices in which electronic, semiconductor laser, and nonlinear optical components are all constructed on a single semiconductor substrate.

A key feature of semiconductor materials is that their allowed electronic energy states take the form of broad bands separated by forbidden regions. The filled or nearly filled bands are known as valence bands and the empty

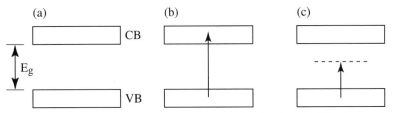

FIGURE 4.6.1 (a) The valence band (VB) are conduction band (CB) of a semiconductor are separated by energy E_g. For $\hbar\omega > E_g$ (b), the nonlinear response results from the transfer of electrons to the conductions band, whereas for $\hbar\omega < E_g$ (c), the nonlinear response involves virtual transitions.

or nearly empty bands are known as conduction bands. The energy separation between the highest valance band and the lowest conduction band is known as the band-gap energy E_g. These concepts are illustrated in Fig. 4.6.1a. A crucial distinction associated with the nonlinear optical properties of a semiconductor material is whether the photon energy $\hbar\omega$ of the laser field is greater than or smaller than the band-gap energy. For $\hbar\omega > E_g$, as illustrated in part (b) of the figure, the nonlinear response results from the transfer of electrons to the conduction band, leading to a modification of the optical properties of the material. For the opposite case $\hbar\omega < E_g$ the nonlinear response is essentially instantaneous and occurs as the result of parametric processes involving virtual levels. We treat these two situations separately.

Nonlinearities Resulting from Band-to-Band Transitions

For $\hbar\omega > E_g$, the nonlinear response occurs as the result of band-to-band transitions. For all but the shortest laser pulses, the nonlinear response can be described in terms of the conduction band population N_c, which can be taken to obey a rate equation of the form

$$\frac{dN_c}{dt} = \frac{\alpha I}{\hbar\omega} - \frac{\left(N_c - N_c^{(0)}\right)}{\tau_R}, \tag{4.6.1}$$

where α is the absorption coefficient of the material at the laser frequency, $N_c^{(0)}$ is the conduction band electron population in thermal equilibrium, and τ_R is the electron–hole recombination time. In steady state this equation possesses the solution

$$N_c = N_c^{(0)} + \frac{\alpha I \tau_R}{\hbar\omega}. \tag{4.6.2}$$

However, for the common situation in which the laser pulse duration is shorter than the material response time τ_R, the conduction-band electron density increases monotonically during the laser pulse.

The change in electron concentration described by Eq. (4.6.1) leads to a change in the optical properties by means of several different mechanisms, which we now describe.

Free-Electron Response.   To first approximation, electrons in the conduction band can be considered to respond freely to an applied optical field. The free electron contribution to the dielectric constant is well known (see Eq. (13.7.3)) and has the form

$$\epsilon(\omega) = \epsilon_0 - \frac{\omega_p^2}{\omega(\omega + i\tau)}, \tag{4.6.3}$$

where ϵ_0 is the contribution to the dielectric constant from all other mechanisms, ω_p^2 is the square of the plasma frequency and is given by $\omega_p^2 = 4\pi N_c e^2 / m$, and τ is an optical response time which in general is not equal to τ_R and is typically much shorter than it. Since N_c increases with laser intensity, $\epsilon(\omega)$ is seen to decrease with laser intensity. In the steady-state limit, we can derive an expression for the change in the real part of the refractive index given by

$$\Delta n = n_2 I \quad \text{where} \quad n_2 = -\frac{\pi e^2 \alpha I \tau_R}{n_0 m \hbar \omega^3}. \tag{4.6.4}$$

Note that n_2 is proportional to ω^{-3}. One thus expects this mechanism to become dominant at long wavelengths. If we evaluate this expression using the characteristic values $m = 0.1\, m_e$ (note that m in Eq. (4.6.4) is the effective mass of the conduction-band electron), $n_0 = 3.5$, $\alpha = 10^4$ cm^{-1}, $\hbar\omega = 0.75$ eV, $\tau_r = 10$ nsec, we find that $n_2 = 3 \times 10^{-6}$ cm^2/W, a reasonably large value.

Modification of Optical Properties by Plasma Screening Effects.   A direct consequence of the presence of electrons in the semiconductor conduction band is that the material becomes weakly conducting. As a result, charges can flow to shield any unbalanced free charges, and the Coulomb interaction between charged particles becomes effectively weakened. In the classical limit in which the electrons obey a Maxwell–Boltzmann distribution, the screened potential energy between two point particles of charge e becomes

$$V = \frac{e^2}{\epsilon r} e^{-\kappa r}, \tag{4.6.5}$$

where ϵ is the (real) dielectric constant of the semiconductor material and where

$$\kappa = \sqrt{\frac{4\pi N_c e^2}{kT}} \qquad (4.6.6)$$

is the Debye–Hückel screening wavenumber.

One consequence of the reduction of the strength of the Coulomb interaction is that excitonic features can disappear at high conduction-band electron densities. Let us recall briefly the nature of excitonic features in semiconductors. An electron in the conduction band will feel a force of attraction to a hole in the valence band as the result of their Coulomb interaction. This attraction can be sufficiently strong that the pair forms a bound state known as an exciton. Excitonic energy levels typically lie slightly below the edge of the conduction band, at an energy given by

$$E_n = E_c - R^*/n^2, \qquad (4.6.7)$$

where n is the principal quantum number, E_c is the energy of the bottom of the conduction band, and $R^* = \hbar^2 (2m_r a_0^{*2})^{-1}$ is the effective Rydberg constant. Here m_r is the reduced mass of the electron–hole pair and $a_0^* = 4\pi\hbar^2 (m_r e^2)^{-1}$ is the effective first Bohr radius. Often only the lowest exciton states contribute significantly to the semiconductor absorption spectrum; the situation in which only the $n = 1$ state is visible is shown in Fig. 4.6.2a. In the presence of a laser beam sufficiently intense to place an appreciable population of electrons into the conduction band, plasma screening effects can lead to the disappearance of these excitonic resonances, leading to an absorption spectrum of the sort shown in part (b) of the figure. Let $\Delta\alpha$ denote the amount by which the absorption coefficient has changed because of the presence of the optical field. The change in absorption coefficient is accompanied by a change in refractive index. This change can be calculated by means of the Kramers–Kronig relations (see Section 1.7), which in the present context we write in the form

$$\Delta n(\omega) = \frac{c}{\pi} \int_0^\infty \frac{\Delta\alpha(\omega')d\omega'}{\omega'^2 - \omega^2}, \qquad (4.6.8)$$

where the principal part of the integral is to be taken. The change in refractive index is shown symbolically in part (c) of Fig. 4.6.2. Note that Δn is positive on the high-frequency side of the exciton resonance and is negative on the low-frequency side. However, the change in refractive index is appreciable only over a narrow range of frequency on either side of the exact resonance.

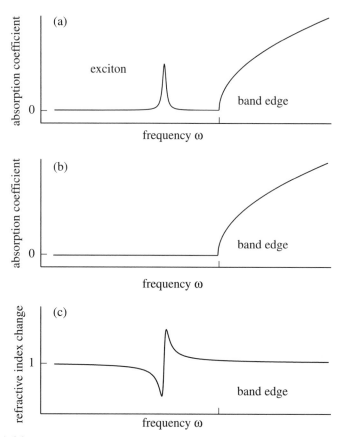

FIGURE 4.6.2   Typical low-temperature absorption spectrum of a semiconductor in the absense (a) and in the presence (b) of an appreciable number of optically excited conduction band electrons. (c) The modification of the refractive index associated with the optically induced change in absorption coefficient.

Change of Optical Properties Due to Band-Filling Effects.   As electrons are transferred from the valence band to the conduction band, the absorption coefficient of a semiconductor must decrease. This effect is in many ways analogous to saturation effects in atomic systems, as described in Chapter 6, but in the present case with the added complexity that the electrons must obey the Pauli principle and thus must occupy a range of energies within the conduction band. This process leads to a lowering of the refractive index for frequencies below the band edge and a raising of the refractive index for frequencies above the band edge. The sense of the change in refractive index is thus the same as that for a two-level atom. The change in refractive index resulting from band filling can be calculated more precisely by means of a Kramers–Kronig analysis of the sort described in the previous paragraph; details are described, for instance, by Peyghambarian et al. (1993), Section 13-4.

Change in Optical Properties Due to Band-Gap Renormalization. For reasons that are rather subtle (exchange and Coulomb correlations), the band-gap energy of most semiconductors decreases at high concentrations of conduction band electrons, with a resulting change in the optical properties.

Nonlinearities Involving Virtual Transitions

Let us next consider the nonlinear response of a semiconductor or insulator under the condition $\hbar\omega < E_g$, as illustrated in part (a) of Fig. 4.6.3. In this situation, the photon energy is too small to allow single-photon absorption to populate the conduction band, and the nonlinear response involves virtual processes such as those shown in parts (b) and (c) of the figure. The "two-photon" process of part (b) usually is much larger than the "one-photon" process of part (c) except for photon energies $\hbar\omega$ approaching the band-gap energy E_g. In the approximation in which only the two-photon process of part (b) is considered, a simple model can be developed to describe the nonlinear response of the material. We will not present the details here, which involve some considerations of the band theory of solids that lie outside the scope of the present work. Sheik-Bahae et al. (1990, 1991) show that the nonlinear refractive index coefficient defined such that $\Delta n = n_2 I$ can be expressed as

$$n_2 = K \frac{\hbar c \sqrt{E_p}}{2n_0^2 E_g^4} G_2(\hbar\omega/E_g) \qquad (4.6.9)$$

where $E_p = 21$ eV, K can be considered to be a single free parameter whose value is found empirically to be 3.1×10^3 in units such that E_p and E_g are

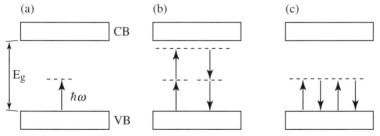

FIGURE 4.6.3 (a) For $\hbar\omega < E_g$, the nonlinear response involves virtual transitions. Under many circumstances, virtual two-photon processes (b) make a larger contribution to the nonlinear response than do one-photon processes (c).

measured in eV and n_2 is measured in cm^2/W, and where G_2 is the universal function

$$G_2(x) = \frac{-2 + 6x - 3x^2 - x^3 - \frac{3}{4}x^4 - \frac{3}{4}x^5 + 2(1 - 2x)^{3/2}\Theta(1 - 2x)}{64x^6},$$

(4.6.10)

where $\Theta(y)$ is the Heaviside step function defined such that $\Theta(y) = 0$ for $y < 0$ and $\Theta(y) = 1$ for $y \geq 0$. In the same approximation, the two-photon absorption coefficient defined such that $\alpha = \alpha_0 + \beta I$ is given by

$$\beta = \frac{K\sqrt{E_p}}{n_0^2 E_g^3} F_2(2\hbar\omega/E_g)$$

(4.6.11)

where F_2 is the universal function

$$F_2(2x) = \frac{(2x - 1)^{3/2}}{(2x)^5} \quad \text{for} \quad 2x > 1$$

(4.6.12)

and $F_2(2x) = 0$ otherwise. These functional forms are illustrated in Fig. 4.6.4. Note that the process of two-photon absoption vanishes for $\hbar\omega < \frac{1}{2}E_g$ for reasons of energetics. Note also that the nonlinear refractive index peaks at $\hbar\omega/E_g \approx 0.54$, vanishes at $\hbar\omega/E_g \approx 0.69$, and is negative for $\hbar\omega/E_g \gtrsim 0.69$. Note also from Eq. (4.6.9) that n_2 scales as E_g^{-4}. Thus narrow band-gap

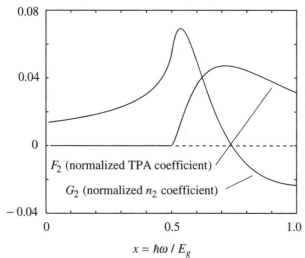

FIGURE 4.6.4   Variation of the nonlinear refraction coefficient n_2 and the two-photon-absorption coefficient with photon energy $\hbar\omega$ according to the model of Sheik-Bahae et al. (1990).

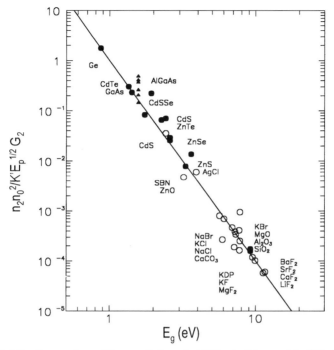

FIGURE 4.6.5   Comparison of the predictions (solid line) of the model of Sheik-Bahae *et al.* (1990) with measured values (data points) of the nonlinear refraction parameter n_2 for a variety of materials.

semiconductors are expected to produce a much larger nonlinear response than large band-gap semiconductors. These predictions are in very good agreement with experimental results; see for instance Fig. 4.6.5.

In general, both the slow, band-to-band nonlinearities considered earlier and the instantaneous nonlinearities considered here occur simultaneously. Said *et al.* (1992) have studied several semiconductors under conditions such that both processes occur simultaneously, and they find that the change in refractive index is well described by the equation

$$\Delta n = n_2 I + \sigma_r N_c, \qquad (4.6.13)$$

where as usual n_2 gives the instantaneous nonlinear response and where σ_r is the change in refractive index per unit conduction band electron density. Their measured values of these quantitites as well as the two-photon-absorption coefficient are given in Table 4.6.1.

TABLE 4.6.1   Nonlinear optical coefficients of several semiconductors

Semiconductor	β(cm/GW)	n_2 (cm^2/W)	σ_r (cm^{-3})
ZnSe at 532 nm	5.8	-6.8×10^{-14}	-0.8×10^{-21}
GaAs at 1064 nm	26	-4.1×10^{-13}	-6.5×10^{-21}
CdTe at 1064 nm	26	-3×10^{-13}	-5×10^{-21}
ZnTe at 1064 nm	4.2	1.2×10^{-13}	-0.75×10^{-21}

After Said *et al.* (1992).

Problems

1. *n_2 for a lossy medium.* Generalize the derivation of Eq. (4.1.19) to allow the linear refractive index to be a complex quantity $\bar{n}_0$.

[Ans: Replace n_0^2 in the denominator of Eq. (4.1.19) by $\bar{n}_0 \, \mathrm{Re} \, \bar{n}_0$.]

2. *Tensor properties of $\chi^{(3)}$ for an isotropic medium.* Derive Eq. (4.2.2).

3. *Ellipse rotation.* A 1-cm-long sample of carbon disulfide is illuminated by elliptically polarized light of intensity $I = 1$ MW/cm^2. Determine how the angle through which the polarization ellipse is rotated depends upon the ellipticity of the light, and calculate numerically the maximum value of the rotation angle. Quantify the ellipticity in terms of the parameter $\delta(-1 \leq \delta \leq 1)$ which defines the polarization unit vector through the relation

$$\hat{\epsilon} = \frac{\hat{\mathbf{x}} + i\delta\hat{\mathbf{y}}}{(1 + \delta^2)^{1/2}}.$$

[Hint: The third-order nonlinear optical response of carbon disulfide is due mainly to molecular orientation.]

4. *Sign of $\chi^{(3)}$.* Verify the statement made in the text that the first term in expression (4.3.13) is positive whenever ω is smaller than any resonance frequency of the atomic system.

5. *Tensor properties of the molecular orientation effect.* Derive the result given by Eqs. (4.4.27) through (4.4.30) for the general case in which a, b, and c are all distinct.

[This problem is extremely challenging.]

6. *Thermal nonlinearities.* In Section 4.5, we basically used dimensional analysis to make an order-of-magnitude estimate of the size of thermal nonlinearities. In this problem, we consider a situation in which the equation of heat transport can be solved exactly.

Consider a laser beam of diameter D_1 and power P propagating through a long glass rod of diameter D_2. The outer surface of the glass rod is held at the fixed temperature T_0. Assume steady-state conditions, and make the simplifying assumption that the transverse intensity profile of the laser beam is uniform. Determine the local temperature T at each point within the glass rod and determine the maximum change in refractive index. Evaluate numerically for realistic conditions.

7. *Nonlinearity due to the magnetic force.* Consider a plane electromagnetic wave incident upon an electron. If the field is strong enough, the electron will acquire sufficient velocity that the magnetic force $\mathbf{F}_M = (-e/c)\mathbf{v} \times \mathbf{B}$ has a noticeable effect on its motion. This is one source of the nonlinear electronic response.
(a) Show that for an optical plane wave with electric field $\mathbf{E} = \hat{\mathbf{x}}$ $\left(E_0 e^{i(kz-\omega t)} + \text{c.c.}\right)$ the electromagnetic force on an electron is

$$\mathbf{F}_{EM} = -e(E_0 e^{-i\omega t} + \text{c.c.})\left[\hat{\mathbf{x}}\left(1 - \frac{\dot{z}}{c}\right) + \hat{\mathbf{z}}\left(\frac{\dot{x}}{c}\right)\right].$$

How large (order of magnitude) can $\dot{x}/c$ become for a free electron in a beam with a peak intensity of 10^{17} W/cm^2?
(b) Derive expressions for $\chi^{(2)}(2\omega)$ and $\chi^{(2)}(0)$ for a collection of free electrons in terms of the electron number density N. (You may assume there are no "frictional" forces.) In what direction(s) will light at 2ω be emitted?
(c) Derive expressions for $\chi^{(3)}(\omega)$ and $\chi^{(3)}(3\omega)$.
(d) Good conductors can often be modeled using the free electron model. Assuming the magnetic force is the only source of optical nonlinearity, make a numerical estimate (order of magnitude) of $\chi^{(3)}(\omega)$ for gold.

8. *Nonlinear phase shift of a focused gaussian beam.* Derive an expression for the nonlinear phase shift experienced by a focused gaussian laser beam of beam-waist radius w_0 carrying power P in passing through a nonlinear optical material characterized by a nonlinear refractive index n_2. Perform this calculation by integrating the on-axis intensity from $z = -\infty$ to $z = +\infty$. Comment on the accuracy of this method of calculation, and speculate regarding computational methods that could provide a more accurate prediction of the nonlinear phase shift.

9. *Nonlinear phase shift of a focused gaussian beam.* Assuming the validity of the procedure used in the previous question, determine numerically the nonlinear phase shift that can be obtained by a focused gaussian laser beam in propagating through optical glass, when the power of the beam is adjusted to be just below the laser damage threshold. Assume initially that the glass is a plate 1 cm thick, but also describe how the phase shift scales with the thickness of the glass plate. For definiteness, assume that bulk (not surface) damage is the limiting process. Take $I(\text{damage}) = 10 \text{ GW/cm}^2$.

10. *Nonlinear phase shift of a focused gaussian beam.* Same as the previous problem, but assume that surface damage is the limiting process.

References

General References

R. G. Brewer, J. R. Lifshitz, E. Garmire, R. Y. Chiao, and C. H. Townes, *Phys. Rev.* **166**, 326 (1968).

R. L. Carman, R. Y. Chiao, and P. L. Kelley, *Phys. Rev. Lett.* **17**, 1281 (1966).

R. Y. Chiao and J. Godine, *Phys. Rev.* **185**, 430 (1969).

R. Y. Chiao, P. L. Kelley, and E. Garmire, *Phys. Rev. Lett.* **17**, 1158 (1966).

D. H. Close, C. R. Giuliano, R. W. Hellwarth, L. D. Hess, and F. J. McClung, *IEEE J. Quantum Electron*, **2**, 553 (1966).

D. C. Hanna, M. A. Yuratich, and D. Cotter, *Nonlinear Optics of Free Atoms and Molecules*, Springer-Verlag, Berlin, 1979.

R. W. Hellwarth, *Prog. Quantum Electron.* **5**, 1-68 (1977).

R. Landauer, *Phys. Lett.* **25A**, 416 (1967).

B. J. Orr and J. F. Ward, *Mol. Phys.* **20**, 513 (1971).

A. Owyoung, *The Origins of the Nonlinear Refractive Indices of Liquids and Glasses*, Ph.D. dissertation, California Institute of Technology, 1971.

O. Svelto, in *Progress in Optics VII*, (E. Wolf, ed.), North Holland, Amsterdam, 1974.

Tensor Nature of the Third-Order Susceptibility

P. N. Butcher, *Nonlinear Optical Phenomena*, Ohio State University, 1965.

W. V. Davis, A. L. Gaeta, and R. W. Boyd, *Opt. Lett.* **17**, 1304 (1992).

A. L. Gaeta, R. W. Boyd, J. R. Ackerhalt, and P. W. Milonni, *Phys. Rev. Lett.* **58**, 2432 (1987).

D. J. Gauthier, M. S. Malcuit, and R. W. Boyd, *Phys. Rev. Lett.* **61**, 1827 (1988).

D. J. Gauthier, M. S. Malcuit, A. L. Gaeta, and R. W. Boyd, *Phys. Rev. Lett.* **64**, 1721 (1990).

P. D. Maker and R. W. Terhune, *Phys. Rev.* **137**, A801 (1965).

P. D. Maker, R. W. Terhune, and C. M. Savage, *Phys. Rev. Lett.* **12**, 507 (1964).

S. Saikan and M. Kiguchi, *Opt. Lett.* **7**, 555 (1982).

Thermal Nonlinear Optical Effects

S. J. Bentley, R. W. Boyd, W. E. Butler, and A. C. Melissinos, *Opt. Lett.* **25**, 1192 (2000).

V. I. Bespalov, A. A. Betin, E. A. Zhukov, O. V. Mitropol'sky, and, N. Y. Rusov *IEEE J. Quantum Electron.* **25**, 360 (1989).

H. J. Hoffman, *J. Opt. Soc. Am. B* **3**, 253 (1986).

G. Martin and R. W. Hellwarth, *Appl. Phys. Lett.* **34**, 371 (1979).

J. O. Tochio, W. Sibbett, and D. J. Bradley, *Opt. Commun.* **37**, 67 (1981).

Semiconductor Nonlinearities

P. N. Butcher and D. Cotter, *The Elements of Nonlinear Optics*, Cambridge University Press, Cambridge, UK, 1990, Chapter 8.

CRC Handbook of Laser Science and Technology, Supplement 2, *Optical Materials* (M. J. Weber, ed.), CRC Press, Boca Raton, FL, Chapter 8. 1995.

t F. Hache, D. Ricard, C. Flytzanis, and U. Kreibig, *Appl. Phys. A* **47**, 347 (1988).

N. Peyghambarian and S. W. Koch, in *Nonlinear Photonics* (H. M. Gibbs, G. Khitrova, and N. Peyghambarian, eds.), Springer Series in Electronics and Photonics, Vol. 30, 1990.

N. Peyghambarian, S. W. Koch, and A. Mysyrowicz, *Introduction to Semiconductor Optics*, Prentice Hall, Englewood Cliffs, NJ, 1993.

A. A. Said, M. Sheik-Bahae, D. J. Hagan, T. H. Wei, J. Wang, J. Young, and E. W. Van Stryland, *J. Opt. Soc. Am. B* **9**, 405 (1992).

M. Sheik-Bahae, D. J. Hagan, and E. W. Van Stryland, *Phys. Rev. Lett.* **65**, 96 (1990).

M. Sheik-Bahae, D. C. Hutchings, D. J. Hagan, and E. W. Van Stryland, *IEEE J. Quantum Electron.* **27**, 1296 (1991).

Chapter 5

Molecular Origin of the Nonlinear Optical Response

Earlier, in Chapter 3, we presented a general quantum-mechanical theory of the nonlinear optical susceptibility. This calculation was based on time-dependent perturbation theory and led to explicit predictions for the complete frequency dependence of the linear and nonlinear optical susceptibilities. Unfortunately, however, these quantum-mechanical expressions are typically far too complicated to be of use for practical calculations.

In the present chapter we review some of the simpler approaches that have been implemented to develop an understanding of the nonlinear optical characteristics of various materials. Many of these approaches are based on understanding the optical properties at the molecular level. In the present chapter we also present brief descriptions of the nonlinear optical charactristics of conjugated polymers, of chiral molecules, and of liquid crystal materials.

5.1. Nonlinear Susceptibilities Calculated Using Time-Independent Perturbation Theory

One approach to the practical calculation of nonlinear optical susceptibilities is based on the use of time-independent perturbation theory (Jha and Bloembergen, 1968). The motivation for using this approach is that time-independent perturbation theory is usually much easier to implement that time-dependent perturbation theory. The justification of the use of this approach is that one is often interested in the study of nonlinear optical interactions in the highly nonresonant limit $\omega \ll \omega_0$ (where ω is the optical frequency and ω_0 is the resonance frequency of the material system), in order to avoid absorption losses. For $\omega \ll \omega_0$, the optical field can to good approximation take to be a quasi-static quantity.

To see how this method proceeds, let us represent the polarization of a material system in the usual form[*]

$$\tilde{P} = \chi^{(1)} \tilde{E} + \chi^{(2)} \tilde{E}^2 + \chi^{(3)} \tilde{E}^3 + \cdots. \tag{5.1.1}$$

We can then calculate the energy stored in polarizing the medium as

$$W = -\int_0^{\tilde{E}} \tilde{P}(\tilde{E}') \, d\tilde{E}' = -\frac{1}{2} \chi^{(1)} \tilde{E}^2 - \frac{1}{3} \chi^{(2)} \tilde{E}^3 - \frac{1}{4} \chi^{(3)} \tilde{E}^3 \cdots$$

$$\equiv W^{(2)} + W^{(3)} + W^{(4)} + \cdots. \tag{5.1.2}$$

The significance of this result is that it shows that if we know W as a function of $\tilde{E}$ (either by calculation or, for instance, from Stark effect measurements), we can use this knowledge to deduce the various orders of susceptibility $\chi^{(n)}$. For instance, if we know W as a power series in $\tilde{E}$ we can determine the susceptibilities as[†]

$$\chi^{(n-1)} = -n \frac{W^{(n)}}{\tilde{E}^n}. \tag{5.1.3}$$

More generally, even if the power series expansion is not known, the nonlinear susceptibilities can be obtained through differentiation as

$$\chi^{(n-1)} = \frac{-1}{(n-1)!} \frac{\partial^n W}{\partial \tilde{E}^n}\bigg|_{E=0}. \tag{5.1.4}$$

Before turning our attention to the general quantum-mechanical calculation of $W^{(n)}$, let us see how to apply the result given by Eq. (5.1.3) to the special case of the hydrogen atom.

Hydrogen Atom

From considerations of the Stark effect, it is well known how to calculate the ground state energy w of the hydrogen atom as a function of the strength E of an applied electric field (Sewell, 1949; Schiff, 1968). We shall not present the

---

[*] As a notational convention, in the present discussion we retain the tilde over P and E both for slowly varying (quasi-static) and for fully static fields.

[†] For time-varying fields, Eq. (5.1.3) still holds, but with $W^{(n)}$ and $\tilde{E}^n$ replaced by their time averages, that is, by $\langle W^{(n)} \rangle$ and $\langle \tilde{E}^n \rangle$. For $\tilde{E} = E e^{-i\omega t} + \text{c.c.}$, one finds that $\tilde{E} = 2E \cos(\omega t + \phi)$, and $\tilde{E}^n = 2^n E^n \cos^n(\omega t + \phi)$, so that $\langle \tilde{E}^n \rangle = 2^n E^n \langle \cos^n(\omega t + \phi) \rangle$. Note that $\langle \cos^2(\omega t + \phi) \rangle = 1/2$ and $\langle \cos^4(\omega t + \phi) \rangle = 3/8$.

details of the calculation here, both because they are readily available in the scientific literature and because the simplest method for obtaining this result makes use of the special symmetry properties of the hydrogen atom and does not readily generalize to other situations. One finds that

$$\frac{w}{2R} = -\frac{1}{2} - \frac{9}{4}\left(\frac{E}{E_{\text{at}}}\right)^2 - \frac{3555}{64}\left(\frac{E}{E_{\text{at}}}\right)^4 + \cdots \qquad (5.1.5)$$

where $R = e^2\hbar^2/mc^2 = 13.6$ eV is the Rydberg constant and where $E_{\text{at}} = e/a_0^2 = m^2 e^5/\hbar^4 = 5.14 \times 10^9$ V/cm is the atomic unit of electric field strength. We now let $W = Nw$ where N is the number density of atoms and introduce Eq. (5.1.5) into Eq. (5.1.3). We thus find that

$$\chi^{(1)} = N\alpha \quad \text{where} \quad \alpha = \frac{9}{2}a_0^3, \qquad (5.1.6a)$$

$$\chi^{(3)} = N\gamma \quad \text{where} \quad \gamma = \frac{3555}{16}\frac{a_0^7}{e^6}, \qquad (5.1.6b)$$

where $a_0 = \hbar^2/me^2$ is the Bohr radius. Note that these results conform with standard scaling laws for nonresonant polarizabilities

$$\alpha \simeq \text{atomic volume } V, \qquad (5.1.7a)$$

$$\gamma \simeq V^{7/3}. \qquad (5.1.7b)$$

General Expression for the Nonlinear Susceptibility in the Quasi-Static Limit

A standard problem in quantum mechanics involves determining how the energy of some state $|\psi_n\rangle$ of an atomic system is modified in response to a perturbation of the atom. To treat this problem mathematically, we assume that the Hamiltonian of the system can be represented as

$$\hat{H} = \hat{H}_0 + \hat{V}, \qquad (5.1.8)$$

where $\hat{H}_0$ represents the total energy of the free atom and $\hat{V}$ represents the quasi-static perturbation due to some external field. For the problem at hand we assume that

$$\hat{V} = -\hat{\mu}\tilde{E}, \qquad (5.1.9)$$

where $\hat{\mu} = -e\hat{x}$ is the electric dipole moment operator and $\tilde{E}$ is an applied quasi static field. We require that the atomic wavefunction obey the time-independent Schrödinger equation

$$\hat{H}|\psi_n\rangle = w_n|\psi_n\rangle. \qquad (5.1.10)$$

For most situations of interest Eqs. (5.1.8)–(5.1.10) cannot be solved in closed form, and must be solved using perturbation theory. One represents the energy w_n and state vector $|\psi_n\rangle$ as power series in the perturbation as

$$w_n = w_n^{(0)} + w_n^{(1)} + w_n^{(2)} + \cdots, \tag{5.1.11a}$$

$$|\psi_n\rangle = \left|\psi_n^{(0)}\right\rangle + \left|\psi_n^{(1)}\right\rangle + \psi_n^{(2)}\rangle + \cdots. \tag{5.1.11b}$$

The details of the procedure are well documented in the scientific literature; see for instance Dalgarno (1961). One finds that the energies are given by

$$w_n^{(1)} = e\tilde{E}\langle n|x|n\rangle, \tag{5.1.12a}$$

$$w_n^{(2)} = e^2\tilde{E}^2\sum_s{}' \frac{\langle n|x|s\rangle\langle s|x|n\rangle}{w_s^{(0)} - w_n^{(0)}}, \tag{5.1.12b}$$

$$w_n^{(3)} = e^3\tilde{E}^3\sum_{st}{}' \frac{\langle n|x|s\rangle\langle s|x|t\rangle\langle t|x|n\rangle}{\left(w_s^{(0)} - w_n^{(0)}\right)\left(w_t^{(0)} - w_n^{(0)}\right)}, \tag{5.1.12c}$$

$$w_n^{(4)} = e^4\tilde{E}^3\sum_{stu}{}' \frac{\langle n|x|s\rangle\langle s|x|t\rangle\langle t|x|u\rangle\langle u|x|n\rangle}{\left(w_s^{(0)} - w_n^{(0)}\right)\left(w_t^{(0)} - w_n^{(0)}\right)\left(w_u^{(0)} - w_n^{(0)}\right)}$$

$$- e^2\tilde{E}^2 w_n^{(2)}\sum_u{}' \frac{\langle n|x|u\rangle\langle u|x|n\rangle}{\left(w_u^{(0)} - w_n^{(0)}\right)^2}. \tag{5.1.12d}$$

The prime following each summation symbol indicates that the state n is to be omitted from the indicated summation. Through use of these expressions one can deduce explicit forms for the linear and nonlinear susceptibilities. We let $W = Nw$, assume that the state of interest is the ground state g, and make use of Eqs. (5.1.3) to find that

$$\chi^{(1)} = N\alpha, \quad \alpha = \alpha_{xx} = \frac{2e^2}{\hbar}\sum_{s\neq g} \frac{x_{gs}x_{sg}}{\omega_{sg}}, \tag{5.1.13a}$$

$$\chi^{(2)} = N\beta, \quad \beta = \beta_{xxx} = \frac{3e^3}{\hbar^2}\sum_{s,t\neq g} \frac{x_{gt}x_{ts}x_{sg}}{\omega_{tg}\omega_{sg}}, \tag{5.1.13b}$$

$$\chi^{(3)} = N\gamma, \quad \gamma = \gamma_{xxxx} = \frac{4e^4}{\hbar^3}\left(\sum_{s,t,u\neq g} \frac{x_{gu}x_{ut}x_{ts}x_{sg}}{\omega_{ug}\omega_{tg}\omega_{sg}} - \sum_{s,t\neq g} \frac{x_{gt}x_{tg}x_{gs}x_{sg}}{\omega_{tg}\omega_{sg}^2}\right),$$

$$\tag{5.1.13c}$$

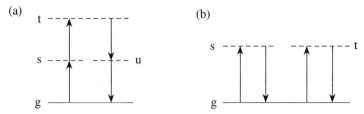

FIGURE 5.1.1 Schematic representation of the two terms appearing in Eq. (5.1.13c).

where $\hbar\omega_{sg} = w_s^{(0)} - w_g^{(0)}$, etc. We see that $\chi^{(3)}$ naturally decomposes into the sum of two terms, which can be represented schematically in terms of the two diagrams shown in Fig. 5.1.1. Note that this result is entirely consistent with the predictions of the model of the nonlinear susceptibility based on time-dependent perturbation theory (see Eq. (4.3.12)), but is more simply predicted by the present formalism.

Equations (5.1.13) constitute the quantum-mechanical predictions for the static values of the linear and nonlinear susceptibilities. Evaluation of these expressions can be quite demanding, as it requires knowledge of all of the resonance frequencies and dipole transition moments connecting to the atomic ground state. Several approximations can be made to simplify these expressions. One example is the Unsold approximation, which entails replacing each resonance frequency (e.g., ω_{sg}) by some average transition frequency ω_0. The expression (5.1.13a) for the linear polarizability then becomes

$$\alpha = \frac{2e^2}{\hbar\omega_0} \sum_s{}' \langle g|x|s\rangle\langle s|x|g\rangle. \tag{5.1.14}$$

We formally rewrite this expression as

$$\alpha = \frac{2e^2}{\hbar\omega_0} \langle g|x\hat{O}x|g\rangle \quad \text{where} \quad \hat{O} = \sum_s{}' |s\rangle\langle s|. \tag{5.1.15}$$

We now replace $\hat{O}$ by the unrestricted sum

$$\hat{O} = \sum_s |s\rangle\langle s|, \tag{5.1.16}$$

which we justify by noting that for states of fixed parity $\langle g|x|g\rangle$ vanishes, and hence it is immaterial whether or not the state g is included in the sum over all s. We next note that

$$\sum_s |s\rangle\langle s| = 1 \tag{5.1.17}$$

by the closure assumption of quantum mechanics. We thus find that

$$\alpha = \frac{2e^2}{\hbar\omega_0}\langle x^2 \rangle. \tag{5.1.18a}$$

This results shows that the linear susceptibility is proportional to the electric quadrupole moment of the ground-state electron distribution. We can apply similar reasoning to the simplification of the expressions for the second- and third-order nonlinear coefficients to find that

$$\beta = -\frac{3e^3}{\hbar^2\omega_0^2}\langle x^3 \rangle, \tag{5.1.18b}$$

$$\gamma = \frac{4e^4}{\hbar^3\omega_0^3}[\langle x^4 \rangle - 2\langle x^2 \rangle^2]. \tag{5.1.18c}$$

These results show that the hyperpolarizabilities can be interpreted as measures of various higher-order moments of the ground state electron distribution. Note that the linear polarizability and hyperpolarizabilities increase rapidly with the physical dimensions of the electron cloud associated with the atomic ground state. Note further that Eqs. (5.1.18a) and (5.1.18c) can be combined to express γ in the intriguing form

$$\gamma = \alpha^2 \frac{g}{\hbar\omega_0} \quad \text{where} \quad g = \left[\frac{\langle x^4 \rangle}{\langle x^2 \rangle^2} - 2 \right]. \tag{5.1.19}$$

Here g is a dimensionless quantity (known in statistics as the kurtosis) that provides a measure of the normalized fourth moment of the ground-state electron distribution.

These expressions can be simplified still further by noting that within the context of the present model the average transition frequency ω_0 can itself be represented in terms of the moments of x. We start with the Thomas–Reiche–Kuhn sum rule (see, for instance, Eq. 61.1 of Bethe and Salpeter, 1977), which states that

$$\frac{2m}{\hbar} \sum_k \omega_{kg}|x_{kg}|^2 = Z, \tag{5.1.20}$$

where Z is the number of optically active electrons. If we now replace ω_{kg} by the average transition frequency ω_0 and perform the summation over k in the same manner as in the derivation of Eq. (5.1.18a), we obtain

$$\omega_0 = \frac{Z\hbar}{2m\langle x^2 \rangle}. \tag{5.1.21}$$

This expression for ω_0 can now be introduced into Eqs. (5.1.18) to obtain

$$\alpha = \frac{4e^2 m}{Z\hbar^2} \langle x^2 \rangle^2, \tag{5.1.22a}$$

$$\beta = -\frac{12e^3 m^2}{Z^2 \hbar^4} \langle x^2 \rangle^2 \langle x^3 \rangle, \tag{5.1.22b}$$

$$\gamma = \frac{32e^4 m^3}{Z^3 \hbar^6} \langle x^2 \rangle^3 (\langle x^4 \rangle - 2\langle x^2 \rangle^2). \tag{5.1.22c}$$

Note that these formulas can be used to infer scaling laws relating the optical constants to the characteristic size L of a molecule. In particular, one finds that $\alpha \sim L^4$, $\beta \sim L^7$, and $\gamma \sim L^{10}$. Note the important result that nonlinear coefficients increase rapidly with the size of a molecule. Note also that α is a measure of the electric quadrupole moment of the ground-state electron distribution, β is a measure of the octopole moment of the ground-state electron distribution, and γ depends on both the hexadecimal pole and the quadrupole moment of the electron ground-state electron distribution.*

5.2. Semiempirical Models of the Nonlinear Optical Susceptibility

We noted earlier in Section 1.4 that Miller's rule can be successfully used to predict the second-order nonlinear optical properties of a broad range of materials. Miller's rule can be generalized to third-order nonlinear optical interactions, where it takes the form

$$\chi^{(3)}(\omega_4, \omega_3, \omega_2, \omega_1) = A\chi^{(1)}(\omega_4)\,\chi^{(1)}(\omega_3)\,\chi^{(1)}(\omega_2)\,\chi^{(1)}(\omega_1), \tag{5.2.1}$$

where $\omega_4 = \omega_1 + \omega_2 + \omega_3$ and A is a quantity which is assumed to be frequency independent and nearly the same for all materials. Wynne (1969) has shown that this generalization of Miller's rule is valid for certain optical materials, such as ionic crystals. However, this generalization is not universally valid.

Wang (1970) has proposed a different relation that seems to be more generally valid. Wang's relation is formulated for the nonlinear optical response in the quasi-static limit and states that

$$\chi^{(3)} = Q'\left(\chi^{(1)}\right)^2 \quad \text{where} \quad Q' = g'/N_{\text{eff}}\,\hbar\omega_0, \tag{5.2.2}$$

---

* There is an additional contribution to the hyperpolarizability β resulting from the difference in permanent dipole moment between the ground and excited states. This contribution is not accounted for by the present model.

where N_{eff} is the product of the molecular number density with the oscillator strength, ω_0 is an average transition frequency, and g' is a dimensionless parameter of the order of unity which is assumed to be nearly the same for all materials. Wang has shown empirically that the predictions of Eq. (5.2.2) are accurate both for low-pressure gases (where Miller's rule does not make accurate predictions) and for ionic crystals (where Miller's rule does make accurate predictions). By comparison of this relation with Eq. (5.1.19), we see that g' is intimately related to the kurtosis of the ground-state electron distribution. There does not seem to be any simple physical argument for why the quantity g' should be the same for all materials.

Model of Boling, Glass, and Owyoung

The formula (Eq. 5.2.2) of Wang serves as a starting point for the model of Boling, Glass, and Owyoung (1978), which allows one to predict the nonlinear refractive index constant n_2 on the basis of linear optical properties. One assumes that the linear refractive index is described by the Lorentz–Lorenz law (see Eq. (3.8.8a)) and Lorentz oscillator model (see Eq. (1.4.17) or Eq. (3.5.25)) as

$$\frac{n^2 - 1}{n^2 + 2} = \frac{4\pi}{3} N\alpha, \tag{5.2.3a}$$

$$\alpha = \frac{f e^2/m}{\omega_0^2 - \omega^2}, \tag{5.2.3b}$$

where f is the oscillator strength of the transition making the dominant contribution to the optical properties. Note that by measuring the refractive index as a function of frequency it is possible through use of these equations to determine both the resonance frequency ω_0 and the effective number density Nf. The nonlinear refractive index is determined from the standard set of equations

$$n_2 = \frac{12\pi^2}{n^2 c} \chi^{(3)}, \qquad \chi^{(3)} = L^4 N\gamma, \qquad L = \frac{n^2 + 2}{3}, \tag{5.2.4a}$$

$$\gamma = \frac{g\alpha^2}{\hbar\omega_0}. \tag{5.2.4b}$$

Equation (5.2.4b) is the microscopic form of Wang's formula (5.2.3b), where g is considered to be a free parameter. If Eq. (5.2.3a) is solved for α, which is then introduced into Eq. (5.2.4b), and use is made of Eqs. (5.2.4a), we find that the expression for n_2 is given by

$$n_2 = \frac{\pi(n^2 + 2)^2(n^2 - 1)^2(gf)}{3n^2 c\,\hbar\omega_0(Nf)}.$$
(5.2.5)

This equation gives a prediction for n_2 in terms of the linear refractive index n, the quantities ω_0 and (Nf) which (as described above) can be deduced from the dispersion in the refractive index, and the combination (gf), which is considered to be a constant quantity for a broad range of optical materials. The value $(gf) = 3$ is found empirically to give good agreement with measured values. A comparison of the predictions of this model with measured values of n_2 has been performed by Adair et al. (1989), and some of their results are shown in Fig. 5.2.1. The two theoretical curves shown in this figure correspond to two different choices of the parameter (gf) of Eq. (5.2.5). Lenz et al. (2000) have described a model related to that of Boling, Glass, and Owyoung that

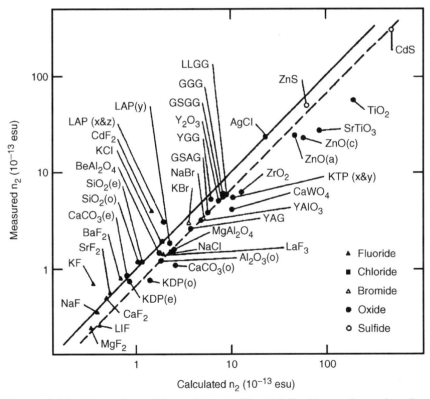

FIGURE 5.2.1 Comparison of the predictions of Eq. (5.2.5) with experimental results. After Adair et al. (1989).

has good predictive value for describing the nonlinear optical properties of chalcogenide glasses.

5.3. Nonlinear Optical Properties of Conjugated Polymers

Certain polymers known as conjugated polymers can possess an extremely large nonlinear optical response. For example, a certain form of polydiacetylene known as PTS possesses a third-order susceptibility of 2.5×10^{-10} esu, as compared to the value of 1.9×10^{-12} esu for carbon disulfide. In this section some of the properties of conjugated polymers are described.

A polymer is said to be conjugated if it contains alternating single and double (or single and triple) bonds. Alternatively, a polymer is said to saturated if it contains only single bonds. A special class of conjugated polymers is the polyenes, which are molecules that contain many double bonds.

Part (a) of Fig. 5.3.1 shows the structure of polyacetylene, a typical chainlike conjugated polymer. According to convention, the single lines in this diagram represent single bonds and double lines represent double bonds. A single bond always has the structure of a σ bond, which is shown schematically in part (b) of the figure. In contrast, a double bond consists of a σ bond and a π bond, as shown in part (c) of the figure. A π bond is made up of the overlap of two p orbitals, one from each atom that is connected by the bond.

The optical response of σ bonds is very different from that of π bonds for the following reason: σ electrons (that is, electrons contained in a σ bond) tend to be localized in space. In contrast, π electrons tend to be delocalized. Because π electrons are delocalized, they tend to be less tightly bound and to respond more freely to an applied optical field, and thus tend to produce larger linear and nonlinear optical responses.

π electrons tend to be delocalized in the sense that a given electron can be found anywhere along the polymer chain. They are delocalized because (unlike the σ electrons) they tend to be located at some distance from the symmetry axis. In addition, even though one conventionally draws a polymer chain in the form shown in part (a) of the figure, for a long chain it would be equally valid to exchange the locations of the single and double bonds, as shown in part (d) of the figure. The actual form of the polymer chain is thus a superposition of the two configurations shown in the figure. This perspective is reinforced by noting that p orbitals extend both to the left and to the right of each carbon atom, and thus there is considerable arbitrariness as to which bonds we should call single bonds and which we should call double bonds. Thus the actual electron distribution might look more like that shown in part (e) of the figure.

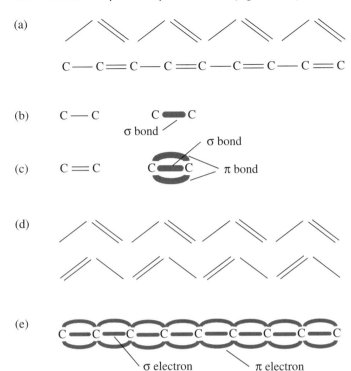

FIGURE 5.3.1   (a) Two common representations of a conjugated chain-like polymer. (b) Standard representation of a single bond (left) and a schematic representation of the electron charge distribution of the single bond (right). (c) Standard representation of a double bond (left) and a schematic represntation of the electron charge distribution of the double bond (right). (d) Two representations of the same polymer chain with the locations of the single and double bonds interchanged, suggesting the arbitrariness of which bond is called the single bond and which a double bond in an actual polymer chain. (e) Representation of the charge distribution of a conjugated chainlike polymer.

As an abstraction, one can model the π electrons of a conjugated chain-like polymer as being entirely free to move in a one-dimensional square well potential whose length L is that of the polymer chain. Rustagi and Ducuing performed such a calculation in 1974 and found that the linear and third-order polarizabilities are given by

$$\alpha = \frac{8L^3}{3a_0\pi^2\mathcal{N}} \quad \text{and} \quad \gamma = \frac{256L^5}{45\,a_0^3\,e^2\pi^6\mathcal{N}^5}, \tag{5.3.1}$$

where $\mathcal{N}$ is the number of electrons per unit length and where a_0 is the Bohr radius. (See also Problem 3 at the end of this chapter.) It should be noted that the linear optical response increases rapidly with the length L of the polymer

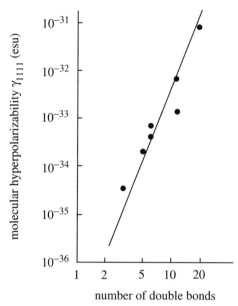

number of double bonds

FIGURE 5.3.2   Measured dependence of the value of the hyperpolarizability γ_{1111} on the number of double bonds in the molecule. The data are from Hermann and Ducuing (1974) and the straight line has a slope of 5 in accordance with Eq. (5.3.1).

chain, and that the nonlinear optical response increases even more rapidly. Of course, for condensed matter the number of polymer chains per unit volume N will decrease with increasing chain length L, so the susceptibilities $\chi^{(1)}$ and $\chi^{(3)}$ will increase less rapidly with L than the dependences given by Eq. (5.3.1). Nonetheless, the present model predicts that conjugated polymers in the form of long chains should possess extremely large values of the nonlinear optical susceptibility. Some experimental results that confirm the L^5 dependence of the hyperpolarizability are shown in Fig. 5.3.2.

5.4. Bond-Charge Model of Nonlinear Optical Properties

In a collection of free atoms, the natural basis set for describing the optical properties of the atomic system is the energy eigenstates of the individual atoms. However, when atoms are arranged into a crystal lattice, it becomes more natural to think of the outer electrons as being localized within the bonds that confine the atoms into the solid. (The inner-core electrons are so tightly bound that they make negligible contribution to the optical response in any case.) Extensive evidence shows that one can ascribe a linear polarizability,

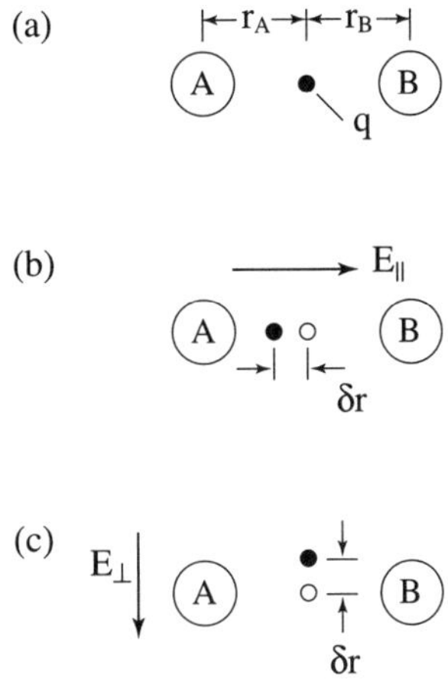

FIGURE 5.4.1   The bond-charge model applied to a chemical bond between constituents A and B. Parts (b) and (c) show how the charge moves in response to applied electric fields.

and higher-order polarizabilities, to each bond in a molecule or crystalline solid. This evidence also shows that the polarizability of one bond is reasonably unaffected by the nature of nearby bonds. Thus the susceptibility of a complex system can be predicted by summing (taking proper account of their orientation) the response of the various bonds present in the material. Bond hyperpolarizabilities can be determined either experimentally or by one of several different theoretical approaches.

The bond-charge model is illustrated in Fig. 5.4.1. Part (a) of this figure shows a bond connecting atoms A and B. As an idealization, the bond is considered to be a point charge of charge q located between the two ions. Here r_A and r_B are the covalent radii of atoms A and B and $d = r_A + r_B$ is known as the bond length. According to Levine (1973), the bond charge is given by

$$q = en_v \left(1/\epsilon + \tfrac{1}{3} f_c \right), \qquad (5.4.1)$$

where n_v is the number of electrons per bond, ϵ is the static dielectric constant of the material, and f_c is the fractional degree of covalency of the bond.

Part (b) of Fig. 5.4.1 shows how the bond charge q moves in the presence of an electric field $\mathbf{E}$ that is applied parallel to the bond axis. The charge is seen to move by an amount $\delta r = \alpha_{\parallel} E/q$, where $\alpha_{\parallel}$ in the polarizability measured along the bond axis, and consequently the ion-to-bond-charge distances r_A and r_B change by amounts

$$-\Delta r_A = \Delta r_B = \delta r = \alpha_{\parallel} E/q. \qquad (5.4.2)$$

Part (c) of the figure shows how the bond charge moves when $\mathbf{E}$ is applied perpendicular to the bond axis. In this case $\delta r = \alpha_\perp E/q$, and to lowest order the distances r_A and r_B change by amounts

$$\Delta r_A = \frac{\delta r^2}{2r_A} = \frac{\alpha_\perp^2 E^2}{2r_A q^2}, \tag{5.4.3a}$$

$$\Delta r_B = \frac{\delta r^2}{2r_B} = \frac{\alpha_\perp^2 E^2}{2r_B q^2}. \tag{5.4.3b}$$

We see that a field parallel to the bond axis can induce a linear change in the distances r_A and r_B, but that a field perpendicular to the axis can induce only a second-order change in these quantities.

Let us now see how to make quantitative predictions using the bond-charge model (Chemla *et al.*, 1974). According to Phillips (1967) and Van Vechten (1969), the (linear) bond polarizability can be represented as

$$\alpha \equiv \tfrac{1}{3}(\alpha_\| + 2\alpha_\perp) = (2a_0)^3 D \frac{E_0^2}{E_g^2}, \tag{5.4.4}$$

where $a_0 = \hbar^2/me^2$, $E_0 = me^4/2\hbar^2$, D is a numerical factor of the order of unity, and E_g is the mean energy gap associated with the bond. This quantity can be represented as

$$E_g^2 = E_h^2 + C^2, \tag{5.4.5}$$

where E_h is the homopolar contribution given by

$$E_h = 40d^{-2.5}, \tag{5.4.6a}$$

and where C is the heteropolar contribution given by

$$C = 1.5e^{-kR}\left(\frac{z_A}{r_A} - \frac{z_B}{r_B}\right)e^2, \tag{5.4.6b}$$

where z_A and z_B are the number of valence electrons on atoms A and B, respectively, and where $\exp(-kR)$ is the Thomas–Fermi screening factor, with $R = \tfrac{1}{2}(r_A + r_B) = \tfrac{1}{2}d$. The numerical factor in Eq. (5.4.6a) presupposes that d is measured in angstroms and E_h in electron volts.

The bond-charge model ascribes the nonlinear optical response of a material system to the variation of the bond polarizability α_{ij} induced by an applied field E_j. Explicitly one expresses the bond dipole moment as

$$p_i = p_i^{(1)} + p_i^{(2)} + p_i^{(3)} + \cdots$$

$$= \left[(\alpha_{il})_0 + \left(\frac{\partial \alpha_{il}}{\partial E_j} \right) E_j + \frac{1}{2} \left(\frac{\partial^2 \alpha_{il}}{\partial E_j \partial E_k} \right) E_j E_k \right] E_l + \cdots$$

$$\equiv (\alpha_{il})_0 E_l + \beta_{ijk} E_j E_l + \gamma_{ijkl} E_j E_k E_l + \cdots. \qquad (5.4.7)$$

Let us now see how to calculate the hyperpolarizabilities β_{ijk} and γ_{ijkl}. Since our model assumes that the bonds are axially symmetric, the only nonvanishing components of the hyperpolarizabilities are

$$\beta_\parallel = \beta_{zzz}, \qquad \beta_\perp = \beta_{xzx}, \qquad (5.4.8a)$$

$$\gamma_\parallel = \gamma_{zzzz}, \qquad \gamma_\perp = \gamma_{xxxx}, \qquad \gamma_{\parallel\perp} = \gamma_{zzxx}, \qquad (5.4.8b)$$

where we have assumed that z lies along the bond axis. We next note that, as a consequence of Eq. (5.4.3), a transverse field $E_\perp$ cannot produce a first-order (or in fact any odd order) change in α_{ij}, that is

$$\left(\frac{\partial}{\partial E_\perp} \right)^q \alpha_{ij} = 0 \quad \text{for } q \text{ odd.} \qquad (5.4.9)$$

We also note that the present model obeys Kleinman symmetry, since it does not consider the frequency dependence of any of the optical properties. Because of Kleinman symmetry, we can express $\beta_\perp \equiv \partial \alpha_{xx} / \partial E_z$ as

$$\beta_\perp = \frac{\partial \alpha_{xz}}{\partial E_x}, \qquad (5.4.10)$$

which vanishes by Eq. (5.4.9). We likewise find that

$$\gamma_{\parallel\perp} = \frac{1}{2} \frac{\partial^2 \alpha_{xz}}{\partial E_x \partial E_z} = 0. \qquad (5.4.11)$$

We thus deduce that the only nonvanishing components are $\beta_\parallel$, $\gamma_\parallel$, and $\gamma_\perp$, which can be expressed as

$$\beta_\parallel = \frac{\partial \alpha_\parallel}{\partial E_\parallel} = 3 \frac{\partial \alpha}{\partial E_\parallel}, \qquad (5.4.12a)$$

$$\gamma_\parallel = \frac{\partial^2 \alpha_\parallel}{\partial E_\parallel^2} = \frac{3}{2} \frac{\partial^2 \alpha}{\partial E_\parallel^2}, \qquad (5.4.12b)$$

$$\gamma_\perp = \frac{\partial^2 \alpha_\perp}{\partial E_\perp^2} = \frac{3}{4} \frac{\partial^2 \alpha}{\partial E_\perp^2}. \qquad (5.4.12c)$$

TABLE 5.4.1   Representative bond hyperpolarizabilities γ in units of 10^{-36} esu[a]

Bond	$\lambda = 1.064 \ \mu m$	$\lambda = 1.907 \ \mu m$
C–Cl	0.90 ± 0.04	0.7725
C–H	0.05 ± 0.04	-0.0275
O–H	0.42 ± 0.02	0.5531
C–C	0.32 ± 0.42	0.6211
C=C	1.03 ± 1.52	0.61
C–O	0.24 ± 0.19	0.30
C=O	0.82 ± 1.1	0.99

[a] After Kajzar and Messier (1985).

The equations just presented provide the basis of the bond-charge model. The application of this model requires extensive numerical computation which will not be reproduced here. In brief summary, the quantities E_h and C of Eqs. (5.4.6) are developed in power series in the applied fields $E_\parallel$ and $E_\perp$ through use of Eqs. (5.4.2) and (5.4.3). Expression (5.4.4) for α can then be expressed in a power series in the applied field, and the hyperpolarizabilities can be extracted from this power series expression through use of Eqs. (5.4.12). Finally, susceptibilities $\chi_{ijk}^{(2)}$ and $\chi_{ijkl}^{(3)}$ are determined by summing over all bonds in a unit volume, taking account of the orientation of each particular bond. This model has been shown to provide good predictive value. For instance, Chemla et al. (1974) have found that this model provides $\sim$30% accuracy in calculating the third-order nonlinear optical response for Ge, Si, and GaAs. Table 5.4.1 gives values of some measured bond hyperpolarizabiliaties. In addition Levine (1973) provides extensive tables comparing the predictions of this model with experimental results.

5.5. Nonlinear Optics of Chiral Media

Special considerations apply to the analysis of the nonlinear optical properties of a medium composed of a collection of chiral molecules. A chiral molecule is a molecule with a "handedness," that is, the mirror image of such a molecule looks different from the molecule itself. By way of example, simple molecules such as CS_2, H_2O, CH_4 are achiral (that is, are not chiral); however, many organic molecules including simple sugars such as dextrose are chiral.

In the field of linear optics, it is well known that chiral media lead to the property of optical activity, that is, the rotation of the direction of linear

(a)   (b)

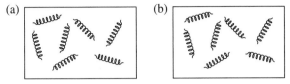

FIGURE 5.5.1   (a) A collection of right-handed spirals and (b) a collection of left-handed spirals. Each medium is isotropic (looks the same in all directions), but neither possesses a center of inversion symmetry.

polarization of a light beam as it propagates through such a material. (See, for instance, Jenkins and White, 1976.) A material is said to be dextrorotatory if the direction of the polarization rotates in a clockwise sense as the beam propagates; if the polarization rotates counterclockwise, the medium is said to be levorotatory. Two molecules that are mirror images of each other are said to be enantiomers. An equal mixture of two enantiomers is said to be a racemic mixture. Optical activity obviously vanishes for a racemic mixture.

Let us now turn to a discussion of the nonlinear optical properties of chiral materials. A liquid composed of chiral molecules is isotropic but nonetheless noncentrosymmetric (see Fig. 5.5.1), and thus it can possess a second-order nonlinear optical response. As we shall see, such a medium can produce sum- or difference-frequency generation, but not second-harmonic generation, and moreover can produce sum- or difference-frequency generation only if the two input beams are non-collinear. The theory of second-order processes in chiral media was developed by Giordmaine (1965) and was confirmed experimentally by Rentzipis *et al.* (1966). More recent research on the nonlinear optics of chiral media includes that of Verbiest *et al.* (1998).

Let us now turn to a theoretical description of second-order processes in chiral materials. We represent the second-order polarization induced in such a material as

$$P_i\left(\omega_\sigma\right) = \sum_{jk} 2\chi_{ijk}^{(2)}\left(\omega_\sigma = \omega_1 + \omega_2\right) E_j F_k, \qquad (5.5.1)$$

where E_j represents a field at frequency ω_1 and F_k represents a field at frequency ω_2 (which can be a negative frequency). We now formally rewrite Eq. (5.5.1) as

$$P_i = \sum_{jk} S_{ijk}(E_j F_k + E_k F_j) + \sum_{jk} A_{ijk}(E_j F_k - E_k F_j), \qquad (5.5.2)$$

where S_{ijk} and A_{ijk} denote the symmetric and antisymmetric parts of $\chi_{ijk}^{(2)}$ and are given by

$$S_{ijk} = \frac{1}{2}\left(\chi_{ijk}^{(2)} + \chi_{ikj}^{(2)}\right),$$ (5.5.3a)

$$A_{ijk} = \frac{1}{2}\left(\chi_{ijk}^{(2)} - \chi_{ikj}^{(2)}\right).$$ (5.5.3b)

Note that A_{ijk} vanishes for second-harmonic generation or more generally whenever the Kleinman symmetry condition is valid.

The tensor properties of the tensors S_{ijk} and A_{ijk} can be deduced using methods analogous to those described in Section 1.5. For the case of an isotropic but noncentrosymmetric medium (which corresponds to point group $\infty\,\infty$) one finds that S_{ijk} vanishes identically and that the only nonvanishing elements of A_{ijk} are

$$A_{123} = A_{231} = A_{312}.$$ (5.5.4)

Consequently the nonlinear polarization can be expressed as

$$\mathbf{P} = A_{123}\,\mathbf{E} \times \mathbf{F}.$$ (5.5.5)

The experimental setup used by Rentzipis *et al.* to study these effects is shown in Fig. 5.5.2. The two input beams are at different frequencies, as required for A_{123} to be nonzero. In addition, they are orthogonally polarized to ensure that $\mathbf{E} \times \mathbf{F}$ is nonzero and are noncollinear to ensure that $\mathbf{P}$ has a transverse component. Generation of a sum-frequency signal at 2314 Å was observed for both dextrorotatory and levorotatory forms of arabinose, but no signal was observed when the cell contained a racemic mixture of the two forms. The measured value of A_{123} was 0.9×10^{-10} esu; for comparison note that $d_\parallel$ (quartz) $= 1.15 \times 10^{-9}$ esu. A detailed reexamination of the second-order nonlinear optical properties of this system has been presented by Belkin *et al.* (2001).

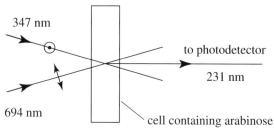

FIGURE 5.5.2   Experimental setup to observe sum-frequency generation in an isotropic, chiral medium.

5.6. Nonlinear Optics of Liquid Crystals

Liquid crystal materials often display large nonlinear optical effects. The time scale for the development of such effects is often quite long (perhaps as long as milliseconds), but even response times this long are adequate for many applications.

Liquid crystals are composed of large, anisotropic molecules. Above a certain transition temperature, which varies significantly among various liquid crystal materials but which might typically be 100°C, these materials exist in an isotropic phase in which they behave like ordinary liquids. Below this transition temperature, liquid crystals exist in a mesotropic phase in which the orientation of adjacent molecules becomes highly correlated, giving rise to the name *liquid crystal*. At still lower temperatures liquid crystal materials undergo another phase transition and behave as ordinary solids.

Several different types of order can occur in the mesotrophic phase. Two of the most common are the nematic phase and the chiral nematic phase (which is also known as the cholesteric phase), which are illustrated in Fig. 5.6.1.

Liquid crystalline materials possess strong nonlinear optical effects in both the isotropic and mesotropic phases.

In the isotropic phase, liquid crystal materials display a molecular- orientation nonlinear response of the sort described in Section 4.4, but typically with a much larger magnitude which is strongly temperature dependent. In one particular case, Hanson *et al.* (1977) find that the nonlinear coefficient $\bar{n}_2$

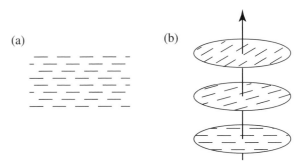

FIGURE 5.6.1 Two examples of ordered-phases (mesophases or mesotropic phases) of liquid crystals. (a) In the nematic phase, the molecules are randomly distributed in space but are aligned such that the long axis of each molecule points in the same direction, known as the director. (b) In the chiral nematic phase, the molecules in each plane are aligned as in the nematic phase, but the director orientation rotates between successive planes.

and the response time τ are given by

$$\bar{n}_2 = \frac{6.35 \times 10^{-9}}{T - T^*} \text{ esu K}, \qquad\qquad T > T^*, \qquad (5.6.1)$$

$$\tau = \frac{e^{2800/T(^\circ\text{K})}}{T - T^*} 7 \times 10^{-11} \text{ ns K}, \qquad T > T^*, \qquad (5.6.2)$$

where $T^* = 77^\circ\text{C}$ is the liquid-crystal transition temperature. In the range of temperatures 130 to 80°C, $\bar{n}_2$ ranges from 12 to 237 $\times 10^{-11}$ esu and τ varies from 1 to 72 nsec. These $\bar{n}_2$ values are 10 to 200 times larger than those of carbon disulfide.

Liquid crystal materials possess even stronger nonlinear optical properties in the mesophase. Once again, the mechanism is one of molecular orientation, but in this case the interaction involves the collective orientation of many interacting molecules. The effective nonlinear response can be as much as 10^9 times larger than that of carbon disulfide.

Experimental studies of nonlinear optical processes in nematic liquid crystals are often performed with the molecules anchored at the walls of the cell that contains the liquid crystal material, as shown in Fig. 5.6.2.

The analysis of such a situation proceeds by considering the angle $\theta + \beta$ between the director and the propagation vector **k** of the laser beam. Here β is this angle in the absence of the laser field and θ is the reorientation angle induced by the laser beam. It can be shown (Khoo, 1995) that this quantity obeys the relation (for definiteness we assume the geometry of Fig. 5.6.2b)

$$K_1 \frac{d^2\theta}{dz^2} + \left(n_e^2 - n_o^2\right) \frac{|A|^2}{4\pi} \sin 2(\theta + \theta_0) = 0. \qquad (5.6.3)$$

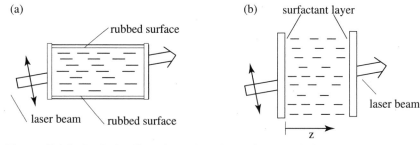

FIGURE 5.6.2   Typical cell configurations for studying optical processes in nematic liquid crystals. (a) Planar alignment: The molecules are induced to anchor at the upper and lower glass walls by rubbing these surfaces to induce small scratches into which the molecules attach. (b) Homeotropic alignment: A surfactant is applied to the cell windows to induce the molecules to align perpendicular to the plane of the window.

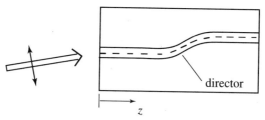

FIGURE 5.6.3   Nature of director reorientation and typical molecular alignment of a homeotropic-alignment, nematic-liquid-crystal cell in the presence of an intense laser beam.

Here K_1 is an elastic constant of the liquid crystal and n_o and n_e are the ordinary and extraordinary values of the refractive index of the nematic liquid crystal in the absence of the influence of the incident laser beam. This equation is to be solved subject to the boundary conditions at the input ($z = 0$) and output ($z = d$) planes of the cell. Khoo (1995) shows that if this procedure is carried through one finds that the director orientation might have the form shown in Fig. 5.6.3 and that the resulting change in refractive index, averaged over the length of the cell, can be expressed as $\Delta n = n_2 I$ where

$$n_2 = \frac{\left(n_e^2 - n_0^2\right)^2 \sin^2(2\beta)\, d^2}{24 K_1 c}. \tag{5.6.4}$$

This expression can be evaluated for the conditions $d = 100\,\mu\mathrm{m}$, $n_e^2 - n_0^2 = 0.6$, $K_1 = 10^{-6}$(cgs), $\beta = 45°$, giving

$$n_2 = 5 \times 10^{-3} \text{ cm}^2/\text{W}. \tag{5.6.5}$$

Problems

1. *Stark shift in hydrogen.* Verify Eq. (5.1.5).

2. *Nonlinear response of the square-well potential.* Making use of the formalism of Section 5.1, calculate the linear and third-order susceptibilities of a collection of electrons confined in a one-dimensional, infinitely deep, square-well potential. Note that this calculation constitutes a simple model of the optical response of a conjugated polymer. (Hint: See Rustagi and Ducuing, 1974.)

3. *Classical calculation of the second-order response of chiral materials.* Consider an anharmonic oscillator for which the potential is of the form

$$V = \tfrac{1}{2}(k_a x^2 + k_b y^2 + k_c z^2) + A xyz.$$

Calculate the response of such an oscillator to an applied field of the form

$$E(t) = E_1 e^{-i\omega_1 t} + E_2 e^{-i\omega_2 t} + \text{c.c.}$$

Then by assuming that there is a randomly oriented distribution of such oscillators, derive an expression for $\chi^{(2)}$ of such a material. Does it possess both symmetric and antisymmetric contributions? Show that the antisymmetric contribution can be expressed as

$$P = \chi_{NL} E_1 \times E_2.$$

References

Books on Molecular Nonlinear Optics

D. S. Chemla and J. Zyss, *Nonlinear Optical Properties of Organic Molecules and Crystals*, Vols. 1 and 2, Academic Press, New York, 1987.

M. G. Kuzyk and C. W. Dirk, eds., *Characterization Techniques and Tabulation for Organic Nonlinear Optical Materials*, Marcel Dekker, Inc., 1998.

P. N. Prasad and D. J. Williams, *Introduction of Nonlinear Opitcal Effects in Molecules and Polymers*, John Wiley and Sons, New York, 1991.

Section 5.1.   Nonlinear Susceptibility . . . Time-Independent Perturbation Theory

H. A. Bethe and E. A. Salpeter, *Quantum Mechanics of One- and Two-Electron Atoms*, Plenum, New York, 1977.

A. Dalgarno, in *Quantum Theory* (D. R. Bates, ed.), Academic Press, New York, 1961.

J. Ducuing, in *Proceedings of the International School of Physics "Enrico Fermi," Course LXIV* (N. Bloembergen, ed.), North Holland, Amsterdam, 1977.

S. S. Jha and N. Bloembergen, *Phys. Rev.* **171**, 891 (1968).

G. L. Sewell, *Proc. Cam. Phil. Soc.* **45**, 678 (1949).

L. I. Schiff, *Quantum Mechanics*, 3rd ed., McGraw Hill, New York, 1968. See especially Eq. (33.9).

Section 5.2.   Semiempirical Models

R. Adair, L. L. Chase, and S. A. Payne, *Phys. Rev. B* **39**, 3337 (1989).

N. L. Boling, A. J. Glass, and A. Owyoung, *IEEE J. Quantum Electron* **14**, 601 (1978).

G. Lenz, J. Zimmermann, T. Katsufuji, M. E. Lines, H. Y. Hwang, S. Spälter, R. E. Slusher, and S.-W. Cheong, *Opt. Lett.* **25**, 254 (2000).

C. C. Wang, *Phys. Rev. B* **2**, 2045 (1970).

J. J. Wynne, *Phys. Rev.* **178**, 1295 (1969).

Section 5.3. Nonlinear Optics of Conjugated Polymers

W. J. Blau, H. J. Byrne, D. J. Cardin, T. J. Dennis, J. P. Hare, H. W. Kroto, R. Taylor, and D. R. M. Walton, *Phys. Rev. Lett.* **67**, 1423 (1991); see also R. J. Knize and J. P. Partanen, *Phys. Rev. Lett.* **68**, 2704 (1992) and Z. H. Kafafi, F. J. Bartoli, J. R. Lindle, and R. G. S. Pong, *Phys. Rev. Lett*, **68**, 2705 (1992).

J. Ducuing, in *Procedings of the International School of Physics, "Enrico Fermi," Course LXIV* (N. Bloembergen, ed.), North Holland, Amsterdam, 1977.

J. P. Hermann and J. Ducuing, *J. Appl. Phys.* **45**, 5100 (1974).

K. C. Rustagi and J. Ducuing, *Opt. Commun.* **10**, 258 (1974).

Section 5.4. Bond Charge Model

D. S. Chemla, *Phys. Rev. Lett.* **26**, 1441 (1971).

D. S. Chemla, R. F. Begley, and R. L. Byer, *IEEE Jr. Quantum Electron.* **10**, 71 (1974).

F. Kajzar and J. Messier, *Phys. Rev. A* **32**, 2352 (1985).

B. F. Levine, *Phys. Rev. Lett.* **22**, 787 (1969).

B. F. Levine, *Phys. Rev. B* **7**, 2600 (1973).

J. C. Phillips, *Phys. Rev. Lett.* **19**, 415 (1967).

J. A. Van Vechten, *Phys. Rev.* **182**, 891 (1969).

Section 5.5. Nonlinear Optics of Chiral Media

M. A. Belkin, S. H. Han, X. Wei, and Y. R. Shen, *Phys. Rev. Lett.* **87**, 113001 (2001).

J. A. Giordmaine, *Phys. Rev.* **138**, A1599 (1965).

F. A. Jenkins and H. E. White, *Fundamentals of Optics*, McGraw-Hill, New York, 1976.

P. M. Rentzipis, J. A. Giordmaine, and K. W. Wecht, *Phys. Rev. Lett.* **16**, 792 (1966).

T. Verbiest, S. Van Elshocht, M. Kauranen, L. Hellemans, J. Snauwaert, C. Nuckolls, T. J. Katz, and A. Persoons, *Science* **282**, 913 (1998).

Section 5.6. Liquid Crystal Nonlinear Optics

E. G. Hanson, Y. R. Shen, and G. K. L. Wang, *Appl. Phys.* **14**, 65 (1977).

I. C. Khoo, *Liquid Crystals*, John Wiley, New York, 1995.

I. C. Khoo and Y. R. Shen, *Opt. Eng.* **24**, 579 (1985).

M. Peccianti, A. De Rossi, G. Assanto, A. De Luca, C. Umeton, and I. C. Khoo, *Appl. Phys. Lett.* **77**, 7 (2000).

G. K. L. Wang and Y. R. Shen, *Phys. Rev. Lett.* **30**, 895 (1973) and *Phys. Rev. A* **10**, 1277 (1974).

Chapter 6

Nonlinear Optics in the Two-Level Approximation

6.1. Introduction

Our treatment of nonlinear optics in the previous chapters has for the most part made use of power series expansions to relate the response of a material system to the strength of the applied optical field. In simple cases, this relation can be taken to be of the form

$$\tilde{P}(t) = \chi^{(1)} \tilde{E}(t) + \chi^{(2)} \tilde{E}(t)^2 + \chi^{(3)} \tilde{E}(t)^3 + \cdots. \qquad (6.1.1)$$

However, there are circumstances under which such a power series expansion does not converge, and under such circumstances different methods must be employed to describe nonlinear optical effects. One example is that of a saturable absorber, where the absorption coefficient α is related to the intensity $I = nc|E|^2/2\pi$ of the applied optical field by the relation

$$\alpha = \frac{\alpha_0}{1 + I/I_s}, \qquad (6.1.2)$$

where α_0 is the weak-field absorption coefficient and I_s is an optical constant called the saturation intensity. We can expand this equation in a power series to obtain

$$\alpha = \alpha_0[1 - (I/I_s) + (I/I_s)^2 - (I/I_s)^3 + \cdots]. \qquad (6.1.3)$$

However, this series converges only for $I < I_s$, and hence only in this limit can saturable absorption be described by means of a power series of the sort given by Eq. (6.1.1).

It is primarily under conditions such that a transition of the material system is resonantly excited that perturbation techniques fail to provide an adequate description of the response of the system to an applied optical field. However, under such conditions it is usually adequate to deal only with the two atomic levels that are resonantly connected by the optical field. The increased complexity entailed in describing the atomic system in a nonperturbative manner is thus compensated in part by the ability to make the two-level approximation. When only two levels are included in the theoretical analysis, there is no need to perform the sums over *all* atomic states that appear in the general quantum-mechanical expressions for $\chi^{(3)}$ given in Chapter 3.

In the present chapter we shall for the most part study how a monochromatic beam of frequency ω interacts with a collection of two-level atoms. The treatment is thus an extension of that of chapter 4, which treated the interaction of a monochromatic beam with a nonlinear medium in terms of the third-order susceptibility $\chi^{(3)}(\omega = \omega + \omega - \omega)$. In addition, in the last two sections of the present chapter we generalize the treatment by studying nondegenerate four-wave mixing involving a collection of two-level atoms.

Even though the two-level model ignores many of the features present in real atomic systems, there is still an enormous richness in the physical processes that are described within the two-level approximation. Some of the processes that can occur and that are described in the present chapter include saturation effects, power broadening, Rabi oscillations, and optical Stark shifts. Parallel treatments of optical nonlinearities in two-level atoms can be found in the books of Allen and Eberly (1975) and Cohen-Tannoudji, Dupont-Roc, and Grynberg (1989) and in the reviews of Sargent (1978) and Boyd and Sargent (1988).

6.2. Density Matrix Equations of Motion for a Two-Level Atom

We first consider the density matrix equations of motion for a two-level system in the absence of damping effects. Since damping mechanisms can be very different under different physical conditions, there is no *unique* way to include damping in the model. The present treatment thus serves as a starting point for the inclusion of damping by any mechanism.

The interaction we are treating is illustrated in Fig. 6.2.1. The lower atomic level is denoted a and the upper level b. We represent the Hamiltonian for this system as

$$\hat{H} = \hat{H}_0 + \hat{V}(t), \tag{6.2.1}$$

where $\hat{H}_0$ denotes the atomic Hamiltonian and $\hat{V}(t)$ denotes the energy of interaction of the atom with the electromagnetic field. We denote the energies

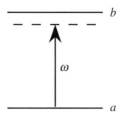

FIGURE 6.2.1   Near-resonant excitation of a two-level atom.

of the states a and b as

$$E_a = \hbar\omega_a \quad \text{and} \quad E_b = \hbar\omega_b. \tag{6.2.2}$$

The Hamiltonian $\hat{H}_0$ can thus be represented by the diagonal matrix whose elements are given by

$$H_{0,nm} = E_n \delta_{nm}. \tag{6.2.3}$$

We assume that the interaction energy can be adequately described in the electric dipole approximation, in which case the interaction Hamiltonian has the form

$$\hat{V}(t) = -\hat{\mu}\tilde{E}(t). \tag{6.2.4}$$

We also assume that the atomic wave functions corresponding to states a and b have definite parity so that the diagonal matrix elements of $\hat{\mu}$ vanish, that is, we assume that $\mu_{aa} = \mu_{bb} = 0$ and hence that

$$V_{aa} = V_{bb} = 0. \tag{6.2.5}$$

The only nonvanishing elements of $\tilde{V}$ are hence V_{ba} and V_{ab}, which are given explicitly by

$$V_{ba} = V_{ab}^* = -\mu_{ba}\tilde{E}(t). \tag{6.2.6}$$

We describe the state of this system by means of the density matrix, which is given explicitly by

$$\hat{\rho} = \begin{bmatrix} \rho_{aa} & \rho_{ab} \\ \rho_{ba} & \rho_{bb} \end{bmatrix} \tag{6.2.7}$$

where $\rho_{ba} = \rho_{ab}^*$. The time evolution of the density matrix is given, still in the absense of damping effects, by Eq. (3.3.21) as

$$\dot{\rho}_{nm} = \frac{-i}{\hbar}[\hat{H}, \hat{\rho}]_{nm} = \frac{-i}{\hbar}[(\hat{H}\hat{\rho})_{nm} - (\hat{\rho}\hat{H})_{nm}]$$
$$= \frac{-i}{\hbar}\sum_{v}(H_{nv}\rho_{vm} - \rho_{nv}H_{vm}). \tag{6.2.8}$$

We now introduce the decomposition of the Hamiltonian into atomic and interaction parts (Eqs. (6.2.1)) into this expression to obtain

$$\dot{\rho}_{nm} = -i\omega_{nm}\rho_{nm} - \frac{i}{\hbar}\sum_\nu (V_{n\nu}\rho_{\nu m} - \rho_{n\nu}V_{\nu m}), \qquad (6.2.9)$$

where we have introduced the transition frequency $\omega_{nm} = (E_n - E_m)\hbar$. For the case of the two-level atom, the indices n, m, and ν can take on the values a or b only, and the equations of motion for the density matrix elements are given explicitly as

$$\dot{\rho}_{ba} = -i\omega_{ba}\rho_{ba} + \frac{i}{\hbar}V_{ba}(\rho_{bb} - \rho_{aa}), \qquad (6.2.10a)$$

$$\dot{\rho}_{bb} = \frac{i}{\hbar}(V_{ba}\rho_{ab} - \rho_{ba}V_{ab}), \qquad (6.2.10b)$$

$$\dot{\rho}_{aa} = \frac{i}{\hbar}(V_{ab}\rho_{ba} - \rho_{ab}V_{ba}). \qquad (6.2.10c)$$

It can be seen by inspection that

$$\dot{\rho}_{bb} + \dot{\rho}_{aa} = 0, \qquad (6.2.11)$$

which shows that the total population $\rho_{bb} + \rho_{aa}$ is a conserved quantity. From the definition of the density matrix, we know that the diagonal elements of $\hat{\rho}$ represent probabilities of occupation, and hence that

$$\rho_{aa} + \rho_{bb} = 1. \qquad (6.2.12)$$

No separate equation of motion is required for ρ_{ab}, because of the relation $\rho_{ab} = \rho_{ba}^*$.

Equations (6.2.10) constitute the density matrix equations of motion for a two-level atom in the absence of relaxation processes. These equations provide an adequate description of resonant nonlinear optical processes under conditions where relaxation processes can be neglected, such as excitation with short pulses whose duration is much less than the material relaxation times. We next see how these equations are modified in the presence of relaxation processes.

Closed Two-Level Atom

Let us first consider relaxation processes of the sort illustrated schematically in Fig. 6.2.2. We assume that the upper level b decays to the lower level at a rate Γ_{ba} and therefore that the lifetime of the upper level is given by $T_1 = 1/\Gamma_{ba}$. Typically, the decay of the upper level would be due to spontaneous emission. This system is called closed, because any population that leaves the upper

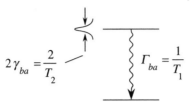

FIGURE 6.2.2   Relaxation processes of the closed two-level atom.

level enters the lower level. We also assume that the atomic dipole moment is dephased in the characteristic time T_2, leading to a transition linewidth (for weak applied fields) of characteristic width $\gamma_{ba} = 1/T_2$.*

We can describe these relaxation processes mathematically by adding decay terms phenomenologically to Eqs. (6.2.10); the modified equations are given by

$$\dot{\rho}_{ba} = -\left(i\omega_{ba}\rho_{ba} + \frac{1}{T_2}\right)\rho_{ba} + \frac{i}{\hbar}V_{ba}(\rho_{bb} - \rho_{aa}), \qquad (6.2.13a)$$

$$\dot{\rho}_{bb} = \frac{-\rho_{bb}}{T_1} - \frac{i}{\hbar}(V_{ba}\rho_{ab} - \rho_{ba}V_{ab}), \qquad (6.2.13b)$$

$$\dot{\rho}_{aa} = \frac{\rho_{bb}}{T_1} + \frac{i}{\hbar}(V_{ba}\rho_{ab} - \rho_{ba}V_{ab}). \qquad (6.2.13c)$$

We can see by inspection that the condition

$$\dot{\rho}_{bb} + \dot{\rho}_{aa} = 0 \qquad (6.2.14)$$

is still satisfied.

Since Eq. (6.2.13a) depends on the populations ρ_{bb} and ρ_{aa} only in terms of the population difference, $\rho_{bb} - \rho_{aa}$, it is useful to consider the equation of motion satisfied by this difference. We subtract Eq. (6.2.13c) from Eq. (6.2.13b) to find that

$$\frac{d}{dt}(\rho_{bb} - \rho_{aa}) = \frac{-2\rho_{bb}}{T_1} - \frac{2i}{\hbar}(V_{ba}\rho_{ab} - \rho_{ba}V_{ab}). \qquad (6.2.15)$$

The first term on the right-hand side can be rewritten using the relation $2\rho_{bb} = (\rho_{bb} - \rho_{aa}) + 1$ (which follows from Eq. (6.2.12)) to obtain

$$\frac{d}{dt}(\rho_{bb} - \rho_{aa}) = -\frac{(\rho_{bb} - \rho_{aa}) + 1}{T_1} - \frac{2i}{\hbar}(V_{ba}\rho_{ab} - \rho_{ba}V_{ab}). \qquad (6.2.16)$$

This relation is often generalized by allowing the possibility that the population difference $(\rho_{bb} - \rho_{aa})^{\text{eq}}$ in thermal equilibrium can have some value other

---

* In fact, one can see from Eq. (6.3.25) that the full width at half maximum in angular frequency units of the absorption line in the limit of weak fields is equal to $2\gamma_{ba}$.

than -1, the value taken above by assuming that only downward spontaneous transitions could occur. This generalized version of Eq. (6.2.16) is given by

$$\frac{d}{dt}(\rho_{bb} - \rho_{aa}) = -\frac{(\rho_{bb} - \rho_{aa}) - (\rho_{bb} - \rho_{aa})^{\text{eq}}}{T_1} - \frac{2i}{\hbar}(V_{ba}\rho_{ab} - \rho_{ba}V_{ab}).$$

(6.2.17)

We therefore see that for a closed two-level system the density matrix equations of motion reduce to just two coupled equations, Eqs. (6.2.13a) and (6.2.17).

In order to justify the choice of relaxation terms used in Eqs. (6.2.13a) and (6.2.17), let us examine the nature of the solutions to these equations in the absence of an applied field, i.e., for $V_{ba} = 0$. The solution to Eq. (6.2.17) is

$$\rho_{bb}(t) - \rho_{aa}(t) = (\rho_{bb} - \rho_{aa})^{\text{eq}} + \{[\rho_{bb}(0) - \rho_{aa}(0)] - (\rho_{bb} - \rho_{aa})^{\text{eq}}\}e^{-t/T_1}.$$

(6.2.18)

This equation shows that the population inversion $\rho_{bb}(t) - \rho_{aa}(t)$ relaxes from its initial value $\rho_{bb}(0) - \rho_{aa}(0)$ to its equilibrium value $(\rho_{bb} - \rho_{aa})^{\text{eq}}$ in a time of the order of T_1. For this reason, T_1 is called the population relaxation time.

Similarly, the solution to Eq. (6.2.13a) for the case $V_{ba} = 0$ is of the form

$$\rho_{ba}(t) = \rho_{ba}(0)e^{-(i\omega_{ba}+1/T_2)t}.$$

(6.2.19)

We can interpret this result more directly by considering the expectation value of the induced dipole moment, which is given by

$$\langle \tilde{\mu}(t) \rangle = \mu_{ab}\rho_{ba}(t) + \mu_{ba}\rho_{ab}(t) = \mu_{ab}\rho_{ba}(0)e^{-(i\omega_{ba}+1/T_2)t} + \text{c.c.}$$

$$= [\mu_{ab}\rho_{ba}(0)e^{-i\omega_{ba}t} + \text{c.c.}]e^{-t/T_2}.$$

(6.2.20)

This result shows that, for an undriven atom, the dipole moment oscillates at frequency ω_{ba} and decays to zero in the characteristic time T_2, which is hence known as the dipole dephasing time.

For reasons that were discussed in relation to Eq. (3.3.25), T_1 and T_2 are related to the collisional dephasing rate γ_c by

$$\frac{1}{T_2} = \frac{1}{2T_1} + \gamma_c.$$

(6.2.21a)

For an atomic vapor, γ_c is usually described accurately by the formula

$$\gamma_c = C_s N + C_f N_f,$$

(6.2.21b)

where N is the number density of atoms having resonance frequency ω_{ba}, and N_f is the number density of any "foreign" atoms of a different atomic species having a different resonance frequency. The parameters C_s and C_f are coefficients describing self-broadening and foreign-gas broadening, respectively. As an example, for the resonance line (i.e., the $3s \rightarrow 3p$ transition) of atomic sodium, T_1 is equal to 16 nsec, $C_s = 1.50 \times 10^{-7} \text{cm}^3/\text{sec}$, and, for the case of foreign-gas broadening by collisions with argon atoms, $C_f = 2.53 \times 10^{-9} \text{cm}^3/\text{sec}$. The values of T_1, C_s, and C_f for other transitions are tabulated, for example, by Miles and Harris (1973).

Open Two-Level Atom

The open two-level atom is shown schematically in Fig. 6.2.3. Here the upper and lower levels are allowed to exchange population with associated reservoir levels. These levels might, for example, be magnetic sublevels or hyperfine levels associated with states a and b. The system is called open because the population that leaves the upper level does not necessarily enter the lower level. This model is often encountered in connection with laser theory, in which case the upper level or both levels are assumed to acquire population at some controllable pump rates, which we take to be λ_b and λ_a for levels b and a, respectively. In order to account for relaxation and pumping processes of the sort just described, the density matrix equations (6.2.10) are modified to become

$$\dot{\rho}_{ba} = -\left(i\omega_{ba} + \frac{1}{T_2}\right)\rho_{ba} + \frac{i}{\hbar}V_{ba}(\rho_{bb} - \rho_{aa}), \qquad (6.2.22a)$$

$$\dot{\rho}_{bb} = \lambda_b - \Gamma_b\left(\rho_{bb} - \rho_{bb}^{eq}\right) - \frac{i}{\hbar}(V_{ba}\rho_{ab} - \rho_{ba}V_{ab}), \qquad (6.2.22b)$$

$$\dot{\rho}_{aa} = \lambda_a - \Gamma_a\left(\rho_{aa} - \rho_{aa}^{eq}\right) + \frac{i}{\hbar}(V_{ba}\rho_{ab} - \rho_{ba}V_{ab}) \qquad (6.2.22c)$$

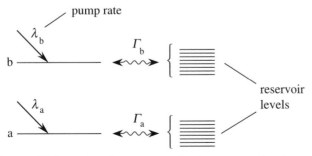

FIGURE 6.2.3   Relaxation processes for the open two-level atom.

Note that in this case the total population contained in the two levels a and b is not conserved, and that in general all three equations must be considered. The relaxation rates are related to the collisional dephasing rate γ_c and population rates Γ_b and Γ_a by

$$\frac{1}{T_2} = \tfrac{1}{2}(\Gamma_b + \Gamma_a) + \gamma_c. \tag{6.2.23}$$

Two-Level Atom with a Non-Radiatively Coupled Third Level

The energy level scheme shown in Fig. 6.2.4 is often used to model a saturable absorber. Population spontaneously leaves the optically excited level b at a rate $\Gamma_{ba} + \Gamma_{bc}$, where Γ_{ba} is the rate of decay to the ground state a, and Γ_{bc} is the rate of decay to level c. Level c acts as a trap level; population decays from level c back to the ground state at a rate Γ_{ca}. In addition, any dipole moment associated with the transition between levels a and b is damped at a rate γ_{ba}. These relaxation processes are modeled by modifying Eqs. (6.2.10) to

$$\dot{\rho}_{ba} = -(i\omega_{ba} + \gamma_{ba})\rho_{ba} + \frac{i}{\hbar}V_{ba}(\rho_{bb} - \rho_{aa}), \tag{6.2.24a}$$

$$\dot{\rho}_{bb} = -(\Gamma_{ba} + \Gamma_{bc})\rho_{bb} - \frac{i}{\hbar}(V_{ba}\rho_{ab} - \rho_{ba}V_{ab}), \tag{6.2.24b}$$

$$\dot{\rho}_{cc} = \Gamma_{bc}\rho_{bb} - \Gamma_{ca}\rho_{cc}, \tag{6.2.24c}$$

$$\dot{\rho}_{aa} = \Gamma_{ba}\rho_{bb} + \Gamma_{ca}\rho_{cc} + \frac{i}{\hbar}(V_{ba}\rho_{ab} - \rho_{ba}V_{ab}). \tag{6.2.24d}$$

It can be seen by inspection that the population in the three levels is conserved, that is, that

$$\dot{\rho}_{aa} + \dot{\rho}_{bb} + \dot{\rho}_{cc} = 0.$$

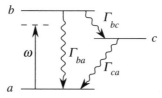

FIGURE 6.2.4   Relaxation processes for a two-level atom with a non-radiatively coupled third level.

6.3. Steady-State Response of a Two-Level Atom to a Monochromatic Field

We next examine the nature of the solution to the density matrix equations of motion for a two-level atom in the presence of a monochromatic, steady-state field. For definiteness, we treat the case of a closed two-level atom, although our results would be qualitatively similar for any of the models described above (see Problem 1 at the end of this chapter). For the closed two-level atomic system, the density matrix equations were shown above (Eqs. (6.2.13a) and (6.2.17)) to be of the form

$$\frac{d}{dt}\rho_{ba} = -\left(i\omega_{ba} + \frac{1}{T_2}\right)\rho_{ba} + \frac{i}{\hbar}V_{ba}(\rho_{bb} - \rho_{aa}), \qquad (6.3.1)$$

$$\frac{d}{dt}(\rho_{bb} - \rho_{aa}) = -\frac{(\rho_{bb} - \rho_{aa}) - (\rho_{bb} - \rho_{aa})^{\text{eq}}}{T_1} - \frac{2i}{\hbar}(V_{ba}\rho_{ab} - \rho_{ba}V_{ab}). \qquad (6.3.2)$$

In the electric dipole approximation, the interaction Hamiltonian for an applied field in the form of a monochromatic wave of frequency ω is given by

$$\hat{V} = -\hat{\mu}\tilde{E}(t) = -\hat{\mu}(Ee^{-i\omega t} + E^*e^{i\omega t}), \qquad (6.3.3)$$

and the matrix elements of the interaction Hamiltonian are then given by

$$V_{ba} = -\mu_{ba}(Ee^{-i\omega t} + E^*e^{i\omega t}). \qquad (6.3.4)$$

Equations (6.3.1) and (6.3.2) cannot be solved exactly for V_{ba} given by Eq. (6.3.4). However, they can be solved in an approximation known as the rotating wave approximation. We recall from the discussion of Eq. (6.2.20) that, in the absence of a driving field, ρ_{ba} tends to evolve in time as $\exp(-i\omega_{ba}t)$. For this reason, when ω is approximately equal to ω_{ba}, the part of V_{ba} that oscillates as $e^{-i\omega t}$ acts as a far more effective driving term for ρ_{ba} than does the part that oscillates as $e^{i\omega t}$. It is thus a good approximation to take V_{ba} not as Eq. (6.3.4) but instead as

$$V_{ba} = -\mu_{ba}Ee^{-i\omega t}. \qquad (6.3.5)$$

This approximation is called the rotating wave approximation. Within this approximation, the density matrix equations of motion (6.3.1) and (6.3.2) become

$$\frac{d}{dt}\rho_{ba} = -\left(i\omega_{ba} + \frac{1}{T_2}\right)\rho_{ba} - \frac{i}{\hbar}\mu_{ba}Ee^{-i\omega t}(\rho_{bb} - \rho_{aa}), \quad (6.3.6)$$

$$\frac{d}{dt}(\rho_{bb} - \rho_{aa}) = -\frac{(\rho_{bb} - \rho_{aa}) - (\rho_{bb} - \rho_{aa})^{\text{eq}}}{T_1}$$

$$+ \frac{2i}{\hbar}(\mu_{ba}Ee^{-i\omega t}\rho_{ab} - \mu_{ab}E^*e^{i\omega t}\rho_{ba}). \quad (6.3.7)$$

Note that (in the rotating wave approximation) ρ_{ba} is driven only at nearly its resonance frequency ω_{ba}, and $\rho_{bb} - \rho_{aa}$ is driven only at nearly zero frequency, which is its natural frequency.

We next find the steady-state solution to Eqs. (6.3.6) and (6.3.7), that is, the solution that is valid long after the transients associated with the turn-on of the driving field have died out. We do so by introducing the slowly varying quantity σ_{ba}, defined by

$$\rho_{ba}(t) = \sigma_{ba}(t)e^{-i\omega t}. \quad (6.3.8)$$

Equations (6.3.6) and (6.3.7) then become

$$\frac{d}{dt}\sigma_{ba} = \left[i(\omega - \omega_{ba}) - \frac{1}{T_2}\right]\sigma_{ba} - \frac{i}{\hbar}\mu_{ba}E(\rho_{bb} - \rho_{aa}), \quad (6.3.9)$$

$$\frac{d}{dt}(\rho_{bb} - \rho_{aa}) = -\frac{(\rho_{bb} - \rho_{aa}) - (\rho_{bb} - \rho_{aa})^{\text{eq}}}{T_1} + \frac{2i}{\hbar}(\mu_{ba}E\sigma_{ab} - \mu_{ab}E^*\sigma_{ba}).$$

$$(6.3.10)$$

The steady-state solution can now be obtained by setting the left-hand sides of Eqs. (6.3.9) and (6.3.10) equal to zero. We thereby obtain two coupled equations, which we solve algebraically to obtain

$$\rho_{bb} - \rho_{aa} = \frac{(\rho_{bb} - \rho_{aa})^{\text{eq}}\left[1 + (\omega - \omega_{ba})^2 T_2^2\right]}{1 + (\omega - \omega_{bs})^2 T_2^2 + (4/\hbar^2)|\mu_{ba}|^2|E|^2 T_1 T_2}, \quad (6.3.11)$$

$$\rho_{ba} = \sigma_{ba}e^{-i\omega t} = \frac{\mu_{ba}Ee^{-i\omega t}(\rho_{bb} - \rho_{aa})}{\hbar(\omega - \omega_{ba} + i/T_2)}. \quad (6.3.12)$$

We now use this result to calculate the polarization (i.e., the dipole moment per unit volume), which is given in terms of the off-diagonal elements of the density matrix by

$$\tilde{P}(t) = N\langle\tilde{\mu}\rangle = N\text{Tr}(\hat{\tilde{\rho}}\hat{\tilde{\mu}}) = N(\mu_{ab}\rho_{ba} + \mu_{ba}\rho_{ab}), \quad (6.3.13)$$

where N is the number density of atoms. We introduce the complex amplitude

P of the polarization through the relation

$$\tilde{P}(t) = Pe^{-i\omega t} + \text{c.c.}, \tag{6.3.14}$$

and we define the susceptibility χ as the constant of proportionality relating P and E according to

$$P = \chi E. \tag{6.3.15}$$

We hence find from Eqs. (6.3.12) through (6.3.15) that the susceptibility is given by

$$\chi = \frac{N|\mu_{ba}|^2(\rho_{bb} - \rho_{aa})}{\hbar(\omega - \omega_{ba} + i/T_2)}, \tag{6.3.16}$$

where $\rho_{bb} - \rho_{aa}$ is given by Eq. (6.3.11). We introduce this expression for $\rho_{bb} - \rho_{aa}$ into Eq. (6.3.16) and rationalize the denominator to obtain the result

$$\chi = \frac{N(\rho_{bb} - \rho_{aa})^{\text{eq}}|\mu_{ba}|^2(\omega - \omega_{ba} - i/T_2)T_2^2/\hbar}{1 + (\omega - \omega_{ba})^2 T_2^2 + (4/\hbar^2)|\mu_{ba}|^2|E|^2 T_1 T_2}. \tag{6.3.17}$$

Note that this expression gives the total susceptibility, including both its linear and nonlinear contributions.

We next introduce new notation to simplify this expression for the susceptibility. We introduce the quantity

$$\Omega = 2|\mu_{ba}||E|/\hbar, \tag{6.3.18}$$

which is known as the on-resonance Rabi frequency, and the quantity

$$\Delta = \omega - \omega_{ba}, \tag{6.3.19}$$

which is known as the detuning factor, so that the susceptibility can be expressed as

$$\chi = \left[N(\rho_{bb} - \rho_{aa})^{\text{eq}}|\mu_{ba}|^2 \frac{T_2}{\hbar} \right] \frac{\Delta T_2 - i}{1 + \Delta^2 T_2^2 + \Omega^2 T_1 T_2}. \tag{6.3.20}$$

Next, we express the combination of factors set off in square brackets in terms of the normal (i.e., linear) absorption coefficient of the material system, which is a directly measurable quantity. The absorption coefficient is given in general by

$$\alpha = \frac{2\omega}{c} \text{Im} \, n = \frac{2\omega}{c} \text{Im}[(1 + 4\pi\chi)^{1/2}], \tag{6.3.21a}$$

and whenever the condition $|\chi| \ll 1$ is valid, the absorption coefficient can be expressed by

$$\alpha = \frac{4\pi\omega}{c} \text{Im} \, \chi. \tag{6.3.21b}$$

If we let $\alpha_0(\Delta)$ denote the absorption coefficient experienced by a *weak* optical wave detuned from the atomic resonance by an amount Δ, we find by ignoring the contribution $\Omega^2 T_1 T_2$ to the denominator of Eq. (6.3.20) that $\alpha_0(\Delta)$ can be expressed as

$$\alpha_0(\Delta) = \frac{\alpha_0(0)}{1 + \Delta^2 T_2^2}, \qquad (6.3.22a)$$

where the unsaturated, line-center absorption coefficient is given by

$$\alpha_0(0) = -\frac{4\pi \omega_{ba}}{c}\left[N(\rho_{bb} - \rho_{aa})^{eq}|\mu_{ba}|^2 \frac{T_2}{\hbar} \right]. \qquad (6.3.22b)$$

By introducing this last expression into Eq. (6.3.20), we find that the susceptibility can be expressed as

$$\chi = -\frac{\alpha_0(0)}{4\pi \omega_{ba}/c} \frac{\Delta T_2 - i}{1 + \Delta^2 T_2^2 + \Omega^2 T_1 T_2}. \qquad (6.3.23)$$

In order to interpret this result, it is useful to express the susceptibility as $\chi = \chi' + i\chi''$ with its real and imaginary parts given by

$$\chi' = -\frac{\alpha_0(0)}{4\pi \omega_{ba}/c} \frac{1}{\sqrt{1 + \Omega^2 T_1 T_2}} \frac{\Delta T_2/\sqrt{1 + \Omega^2 T_1 T_2}}{1 + \Delta^2 T_2^2/(1 + \Omega^2 T_1 T_2)}, \qquad (6.3.24a)$$

$$\chi'' = \frac{\alpha_0(0)}{4\pi \omega_{ba}/c} \left(\frac{1}{1 + \Omega^2 T_1 T_2} \right) \frac{1}{1 + \Delta^2 T_2^2/(1 + \Omega^2 T_1 T_2)}. \qquad (6.3.24b)$$

We see from these expressions that, even in the presence of an intense laser field, χ' has a standard dispersive lineshape and χ'' has a Lorentzian lineshape. However, each of these lines has been broadened with respect to its weak-field width by the factor $(1 + \Omega^2 T_1 T_2)^{1/2}$. In particular, the width of the absorption line (full width at half maximum) is given by

$$\Delta\omega_{FWHM} = \frac{2}{T_2}(1 + \Omega^2 T_1 T_2)^{1/2}. \qquad (6.3.25)$$

The tendency of spectral lines to become broadened when measured using intense optical fields is known as power broadening. We also see (e.g., from Eq. (6.3.24b)) that the line center value of χ'' (and consequently of the absorption coefficient α) is decreased with respect to its weak-field value by the factor $(1 + \Omega^2 T_1 T_2)^{1/2}$. The tendency of the absorption to decrease when measured using intense optical fields is known as saturation. This behavior is illustrated in Fig. 6.3.1.

It is convenient to define, by means of the relation

$$\Omega^2 T_1 T_2 = \frac{|E|^2}{|E_s^o|^2}, \qquad (6.3.26)$$

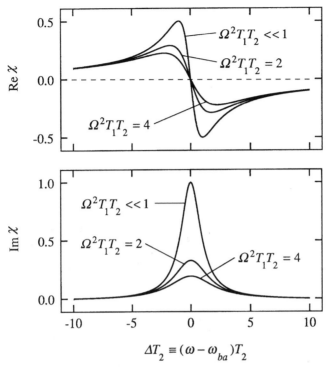

$$\Delta T_2 \equiv (\omega - \omega_{ba})T_2$$

FIGURE 6.3.1 Real and imaginary parts of the susceptibility χ (in units of $\alpha_0 c/4\pi\omega_{ba}$) plotted as functions of the optical frequency ω for several values of the saturation parameter $\Omega^2 T_1 T_2$.

the quantity E_s^o, which is known as the line-center saturation field strength. Through use of Eq. (6.3.18) we find that E_s^o is given explicitly by

$$\left|E_s^o\right|^2 = \frac{\hbar^2}{4|\mu_{ba}|^2 T_1 T_2}. \tag{6.3.27}$$

The expression (6.3.23) for the susceptibility can be rewritten in terms of the saturation field strength as

$$\chi = \frac{-\alpha_0(0)}{4\pi\omega_{ba}/c} \frac{\Delta T_2 - i}{1 + \Delta^2 T_2^2 + |E|^2/\left|E_s^o\right|^2}. \tag{6.3.28}$$

We see from this expression that the significance of E_s^o is that the absorption experienced by an optical wave tuned to line center (which is proportional to $\mathrm{Im}\chi$ evaluated at $\Delta = 0$) drops to one-half its weak-field value when the optical field has a strength of E_s^o. We can analogously define a saturation field strength for a wave of arbitrary detuning, which we denote E_s^Δ, by means of the relation

$$\left|E_s^\Delta\right|^2 = \left|E_s^o\right|^2\left(1 + \Delta^2 T_2^2\right). \tag{6.3.29}$$

We then see from Eq. (6.3.28) that Im χ drops to one-half its weak-field value when a field of detuning Δ has a field strength of E_s^Δ.

It is also useful to define the saturation *intensity* for a wave at line center (assuming that $|n - 1| \ll 1$) as

$$I_s^o = \frac{c}{2\pi}\left|E_s^o\right|^2, \tag{6.3.30}$$

and the saturation intensity for a wave of arbitrary detuning as

$$I_s^\Delta = \frac{c}{2\pi}\left|E_s^\Delta\right|^2 = I_s^o\left(1 + \Delta^2 T_2^2\right). \tag{6.3.31}$$

In order to relate our present treatment of the nonlinear optical susceptibility to the perturbative treatment that we have used in the previous chapters, we next calculate the first- and third-order contributions to the susceptibility of a collection of two-level atoms. By performing a power series expansion of Eq. (6.3.28) in the quantity $|E|^2/|E_s^o|^2$ and retaining only the first and second terms, we find that the susceptibility can be approximated as

$$\chi \simeq \frac{-\alpha_0(0)}{4\pi\omega_{ba}/c}\left(\frac{\Delta T_2 - i}{1 + \Delta^2 T_2^2}\right)\left(1 - \frac{1}{1 + \Delta^2 T_2^2}\frac{|E|^2}{|E_s^o|^2}\right). \tag{6.3.32}$$

We now equate this expression with the usual power series expansion $\chi = \chi^{(1)} + 3\chi^{(3)}|E^2|$ (where $\chi^{(3)} \equiv \chi^{(3)}(\omega = \omega + \omega - \omega)$) to find that the first- and third-order susceptibilities are given by

$$\chi^{(1)} = \frac{-\alpha_0(0)}{4\pi\omega_{ba}/c}\frac{\Delta T_2 - i}{1 + \Delta^2 T_2^2}, \tag{6.3.33a}$$

$$\chi^{(3)} = \frac{\alpha_0(0)}{12\pi\omega_{ba}/c}\left[\frac{\Delta T_2 - i}{\left(1 + \Delta^2 T_2^2\right)^2}\right]\frac{1}{\left|E_s^o\right|^2}. \tag{6.3.33b}$$

For some purposes, it is useful to express the nonlinear susceptibility in terms of the line-center saturation intensity as

$$\chi^{(3)} = \frac{\alpha_0(0)}{24\pi^2\omega_{ba}/c^2}\left[\frac{\Delta T_2 - i}{\left(1 + \Delta^2 T_2^2\right)^2}\right]\frac{1}{I_s^o} \tag{6.3.34a}$$

or, through use of Eqs. (6.3.22a) and (6.3.31), in terms of the saturation intensity and absorption coefficient at the laser frequency as

$$\chi^{(3)} = \frac{-\alpha_0(\Delta)}{24\pi^2\omega_{ba}/c^2}\frac{\Delta T_2 - i}{I_s^\Delta}. \tag{6.3.34b}$$

Note also that the third-order susceptibility can be related to the linear susceptibility by

$$\chi^{(3)} = \frac{-\chi^{(1)}}{3\left(1 + \Delta^2 T_2^2\right)\left|E_s^0\right|^2} = \frac{-\chi^{(1)}}{3\left|E_s^\Delta\right|^2}. \tag{6.3.35}$$

Furthermore, through use of Eqs. (6.3.22b) and (6.3.27), the first- and third-order susceptibilities can be expressed in terms of microscopic quantities as

$$\chi^{(1)} = \left[N(\rho_{bb} - \rho_{aa})^{\mathrm{eq}}|\mu_{ba}|^2 \frac{T_2}{\hbar}\right] \frac{\Delta T_2 - i}{1 + \Delta^2 T_2^2}, \tag{6.3.36a}$$

$$\chi^{(3)} = -\tfrac{4}{3} N(\rho_{bb} - \rho_{aa})^{\mathrm{eq}}|\mu_{ba}|^4 \frac{T_1 T_2^2}{\hbar^3} \frac{\Delta T_2 - i}{\left(1 + \Delta^2 T_2^2\right)^2}. \tag{6.3.36b}$$

In the limit $\Delta T_2 \gg 1$, the expression for $\chi^{(3)}$ reduces to

$$\chi^{(3)} = -\tfrac{4}{3} N(\rho_{bb} - \rho_{aa})^{\mathrm{eq}}|\mu_{ba}|^4 \frac{1}{\hbar^3 \Delta^3} \frac{T_1}{T_2}. \tag{6.3.37}$$

Let us consider the magnitudes of some of the physical quantities we have introduced in this section. Since (for $n = 1$) the intensity of an optical wave with field strength E is given by $I = c|E|^2/2\pi$, the Rabi frequency of Eq. (6.3.18) can be expressed as

$$\Omega = \frac{2|\mu_{ba}|}{\hbar}\left(\frac{2\pi I}{c}\right)^{1/2}. \tag{6.3.38}$$

Assuming that $|\mu_{ba}| = 2.5\, ea_0 = 5.5 \times 10^{-18}$ esu (as it is for the $3s \to 3p$ transition of atomic sodium) and that I is measured in W/cm^2, this relationship gives the numerical result

$$\Omega[\mathrm{rad/sec}] = 2\pi(1 \times 10^9)\left(\frac{I[\mathrm{W/cm}^2]}{127}\right)^{1/2}. \tag{6.3.39}$$

Hence, whenever the intensity I exceeds 127 W/cm^2, $\Omega/2\pi$ is greater than 1 GHz, which is a typical value of the Doppler-broadened linewidth of an atomic transition; such intensities are available from the focused output of even low-power, cw lasers.

The saturation intensity of an atomic transition can be quite small. Again using $|\mu_{ba}| = 5.5 \times 10^{-18}$ esu, and assuming that $T_1 = 16$ nsec (the value for the $3p \to 3s$ transition of atomic sodium) and that $T_2/T_1 = 2$ (the ratio for a radiatively broadened transition; see Eq. (6.2.21a)), we find from Eq. (6.3.30) that

$$I_s^0 = 5.27 \times 10^4 \, \frac{\mathrm{erg}}{\mathrm{cm}^2\mathrm{sec}} = 5.27 \, \frac{\mathrm{mW}}{\mathrm{cm}^2}. \tag{6.3.40}$$

Lastly, let us consider the magnitude of $\chi^{(3)}$ under conditions of the near-resonant excitation of an atomic transition. We take the typical values $N = 10^{14}$ cm^{-3}, $(\rho_{bb} - \rho_{aa})^{eq} = -1$, $\mu_{ba} = 5.5 \times 10^{-18}$ esu, $\Delta = \omega - \omega_{ba} = 2\pi c$ $(1\ cm^{-1}) = 6\pi \times 10^{10}$ rad/sec, and $T_2/T_1 = 2$, in which case we find from Eq. (6.3.37) that $\chi^{(3)} = 1.5 \times 10^{-8}$ esu. Note that this value is very much larger than the values of the nonresonant susceptibilities discussed in Chapter 4.

6.4. Optical Bloch Equations

In the previous two sections, we have treated the response of a two-level atom to an applied optical field by working directly with the density matrix equations of motion. We chose to work with the density matrix equations in order to establish a connection with the calculation of the second- and third-order susceptibilities presented in Chapter 3. However, in theoretical quantum optics the response of a two-level atom is often treated through use of the optical Bloch equations or through related theoretical formalisms. Although these various formalisms are equivalent in their predictions, the equations of motion look very different within different formalisms, and consequently different intuition regarding the nature of resonant optical nonlinearities is obtained. In this section, we review several of these formalisms.

We have seen (Eqs. (6.3.9) and (6.3.10)) that the density matrix equations describing the interaction of a closed two-level atomic system with the optical field,

$$\tilde{E}(t) = E(t)e^{-i\omega t} + \text{c.c.}, \qquad (6.4.1)$$

can be written in the rotating wave approximation as

$$\frac{d}{dt}\sigma_{ba} = \left[i(\omega - \omega_{ba}) - \frac{1}{T_2}\right]\sigma_{ba} - \frac{i}{\hbar}\mu_{ba}E(\rho_{bb} - \rho_{aa}), \qquad (6.4.2a)$$

$$\frac{d}{dt}(\rho_{bb} - \rho_{aa}) = -\frac{(\rho_{bb} - \rho_{aa}) - (\rho_{bb} - \rho_{aa})^{eq}}{T_1} + \frac{2i}{\hbar}(\mu_{ba}E\sigma_{ab} - \mu_{ab}E^*\sigma_{ba}),$$

$$(6.4.2b)$$

where the slowly varying, off-diagonal density matrix component $\sigma_{ba}(t)$ is defined by

$$\rho_{ba}(t) = \sigma_{ba}(t)e^{-i\omega t}. \qquad (6.4.3)$$

The form of Eqs. (6.4.2) can be greatly simplified by introducing the following quantities:

1. The population inversions

$$w = \rho_{bb} - \rho_{aa} \quad \text{and} \quad w^{\text{eq}} = (\rho_{bb} - \rho_{aa})^{\text{eq}} \qquad (6.4.4a)$$

2. The detuning of the optical field from resonance,*

$$\Delta = \omega - \omega_{ba} \qquad (6.4.4b)$$

3. The atom-field coupling constant

$$\kappa = 2\mu_{ba}/\hbar \qquad (6.4.4c)$$

We also drop the subscripts on σ_{ba} for compactness. The density matrix equations of motion (6.4.2) then take the simpler form

$$\frac{d}{dt}\sigma = \left(i\Delta - \frac{1}{T_2}\right)\sigma - \tfrac{1}{2}i\kappa E w, \qquad (6.4.5a)$$

$$\frac{d}{dt}w = -\frac{w - w^{\text{eq}}}{T_1} + i(\kappa E \sigma^* - \kappa^* e^* \sigma). \qquad (6.4.5b)$$

It is instructive to consider the equation of motion satisfied by the complex amplitude of the induced dipole moment. We first note that the expectation value of the induced dipole moment is given by

$$\langle \hat{\mu} \rangle = \rho_{ba}\mu_{ab} + \rho_{ab}\mu_{ba} = \sigma_{ba}\mu_{ab}e^{-i\omega t} + \sigma_{ab}\mu_{ba}e^{i\omega t}. \qquad (6.4.6)$$

If we define the complex amplitude p of the dipole moment $\langle \hat{\mu} \rangle$ through the relation

$$\langle \hat{\mu} \rangle = pe^{-i\omega t} + \text{c.c.}, \qquad (6.4.7)$$

we find by comparison with Eq. (6.4.6) that

$$p = \sigma_{ba}\mu_{ab}. \qquad (6.4.8)$$

Equations (6.4.5) can hence be rewritten in terms of the dipole amplitude p as

$$\frac{dp}{dt} = \left(i\Delta - \frac{1}{T_2}\right)p - \frac{\hbar}{4}i|\kappa|^2 E w, \qquad (6.4.9a)$$

$$\frac{dw}{dt} = -\frac{w - w^{\text{eq}}}{T_2} - \frac{4}{\hbar}\text{Im}(Ep^*). \qquad (6.4.9b)$$

These equations illustrate the nature of the coupling between the atom and the optical field. Note that they are linear in the atomic variables p and w and in the applied field amplitude E. However, the coupling is parametric: the dipole moment p is driven by a term that depends on the product of E with

---

* Note that some authors use the opposite sign convention for Δ.

the inversion w, and likewise the inversion is driven by a term that depends on the product of E with p.

For those cases in which the field amplitude E can be taken to be a real quantity, the density matrix equations (6.4.5) can be simplified in a different way. We assume that the phase convention for describing the atomic energy eigenstates has been chosen such that μ_{ba} and hence κ are real quantities. It is then useful to express the density matrix element σ in terms of two real quantities u and v as

$$\sigma = \tfrac{1}{2}(u - iv). \tag{6.4.10}$$

The factor of one-half and the minus sign are used here to conform with convention (Allen and Eberly, 1975). This definition is introduced into Eq. (6.4.5a), which becomes

$$\frac{d}{dt}(u - iv) = \left(i\Delta - \frac{1}{T_2}\right)(u - iv) - i\kappa E w.$$

This equation can be separated into its real and imaginary parts as

$$\frac{d}{dt}u = \Delta v - \frac{u}{T_2}, \tag{6.4.11a}$$

$$\frac{d}{dt}v = -\Delta u - \frac{v}{T_2} + \kappa E w. \tag{6.4.11b}$$

Similarly, Eq. (6.4.5b) becomes

$$\frac{d}{dt}w = -\frac{w - w^{\text{eq}}}{T_1} - \kappa E v. \tag{6.4.11c}$$

The set (6.4.11) is known as the optical Bloch equations.

We next show that in the absence of relaxation processes (i.e., in the limit $T_1, T_2 \to \infty$) the variables u, v, and w obey the conservation law

$$u^2 + v^2 + w^2 = 1. \tag{6.4.12}$$

First, we note that the time derivative of $u^2 + v^2 + w^2$ vanishes:

$$\frac{d}{dt}(u^2 + v^2 + w^2) = 2u\frac{du}{dt} + 2v\frac{dv}{dt} + 2w\frac{dw}{dt}$$
$$= 2u\Delta v - 2v\Delta u + 2v\kappa E w - 2w\kappa E v \tag{6.4.13}$$
$$= 0,$$

where we have used Eqs. (6.4.11) in obtaining expressions for the time derivatives. We hence see that $u^2 + v^2 + w^2$ is a constant. Next, we note that before the optical field is applied the atom must be in its ground state and hence that $w = -1$ and $u = v = 0$ (as there can be no probability amplitude to be in the upper level). In this case we see that $u^2 + v^2 + w^2$ is equal to 1, but since the quantity $u^2 + v^2 + w^2$ is conserved, it must have this value at all times. We also note that since all of the damping terms in Eqs. (6.4.11) have negative signs associated with them, it must generally be true that

$$u^2 + v^2 + w^2 \leq 1. \tag{6.4.14}$$

Harmonic Oscillator Form of the Density Matrix Equations

Still different intuition regarding the nature of resonant optical nonlinearities can be obtained by considering the equation of motion satisfied by the expectation value of the dipole moment induced by the applied field (rather than considering the equation satisfied by its complex amplitude). This quantity is given by

$$\tilde{M} \equiv \langle \hat{\tilde{\mu}} \rangle = \rho_{ba}\mu_{ab} + \text{c.c.} \tag{6.4.15}$$

For simplicity of notation, we have introduced the new symbol $\tilde{M}$ rather than continuing to use $\langle \hat{\tilde{\mu}} \rangle$. Note that $\tilde{M}$ is a real quantity that oscillates at an optical frequency.

We take the density matrix equations of motion in the form

$$\dot{\rho}_{ba} = -\left(i\omega_{ba} + \frac{1}{T_2}\right)\rho_{ba} - \frac{i}{\hbar}\mu_{ba}\tilde{E}w, \tag{6.4.16a}$$

$$\dot{w} = -\frac{w - w^{\text{eq}}}{T_1} - \frac{4\tilde{E}}{\hbar}\,\text{Im}(\mu_{ab}\rho_{ba}), \tag{6.4.16b}$$

where the dot denotes a time derivative. These equations follow from Eqs. (6.2.6), (6.2.13a), and (6.2.17) and the definition $w = \rho_{bb} - \rho_{aa}$. Here $\tilde{E}$ is the real, time-varying optical field; note that we have not made the rotating-wave approximation. We find by direct time differentiation of Eq. (6.4.15) and subsequent use of Eq. (6.4.16a) that the time derivative of $\tilde{M}$ is given by

$$\dot{\tilde{M}} = \dot{\rho}_{ba}\mu_{ab} + \text{c.c.}$$

$$= -\left(i\omega_{ba} + \frac{1}{T_2}\right)\rho_{ba}\mu_{ab} - \frac{i}{\hbar}|\mu_{ba}|^2\tilde{E}w + \text{c.c.} \tag{6.4.17}$$

$$= -\left(i\omega_{ba} + \frac{1}{T_2}\right)\rho_{ba}\mu_{ab} + \text{c.c.}$$

We have dropped the second term in the second-to-last form because it is imaginary and disappears when added to its complex conjugate. Next, we calculate the second time derivative of $\tilde{M}$ by taking the time derivative of

Eq. (6.4.17) and introducing expression (6.4.16a) for $\dot{\rho}_{ba}$:

$$\ddot{\tilde{M}} = -\left(i\omega_{ba} + \frac{1}{T_2}\right)\dot{\rho}_{ba}\mu_{ab} + \text{c.c.}$$

$$= \left(i\omega_{ba} + \frac{1}{T_2}\right)^2 \rho_{ba}\mu_{ab} + \frac{i}{\hbar}\left(i\omega_{ba} + \frac{1}{T_2}\right)|\mu_{ba}|^2 \tilde{E} w + \text{c.c.}$$

or

$$\ddot{\tilde{M}} = -\left(-\omega_{ba}^2 + \frac{2i\omega_{ba}}{T_2} + \frac{1}{T_2^2}\right)\rho_{ba}\mu_{ab} - \frac{\omega_{ba}}{\hbar}|\mu_{ba}|^2 \tilde{E} w + \text{c.c.} \quad (6.4.18)$$

If we now introduce Eqs. (6.4.15) and (6.4.17) into this expression, we find that $\tilde{M}$ obeys the equation

$$\ddot{\tilde{M}} + \frac{2}{T_2}\dot{\tilde{M}} + \omega_{ba}^2\tilde{M} = \frac{-\tilde{M}}{T_2^2} - \frac{2\omega_{ba}}{\hbar}|\mu_{ba}|^2 \tilde{E} w. \quad (6.4.19)$$

Since ω_{ba}^2 is much larger than $1/T_2^2$ in all physically realistic circumstances, we can drop the first term on the right-hand side of this expression to obtain the result

$$\ddot{\tilde{M}} + \frac{2}{T_2}\dot{\tilde{M}} + \omega_{ba}^2\tilde{M} = -\frac{2\omega_{ba}}{\hbar}|\mu_{ba}|^2 \tilde{E} w. \quad (6.4.20)$$

This is the equation of a damped, driven harmonic oscillator. Note that the driving term is proportional to the product of the applied field strength $\tilde{E}(t)$ with the inversion w.

We next consider the equation of motion satisfied by the inversion w. In order to simplify Eq. (6.4.16b), we need an explicit expression for $\text{Im}(\rho_{ba}\mu_{ab})$. To find such an expression, we rewrite Eq. (6.4.17) as

$$\dot{\tilde{M}} = -\left(i\omega_{ba} + \frac{1}{T_2}\right)\rho_{ba}\mu_{ab} + \text{c.c.}$$

$$= -i\omega_{ba}(\rho_{ba}\mu_{ab} - \text{c.c.}) - \frac{1}{T_2}(\rho_{ba}\mu_{ab} + \text{c.c.}) \quad (6.4.21)$$

$$= 2\omega_{ba}\,\text{Im}(\rho_{ba}\mu_{ab}) - \frac{\tilde{M}}{T_2},$$

which shows that

$$\text{Im}(\rho_{ba}\mu_{ab}) = \frac{1}{2\omega_{ba}}\left(\dot{\tilde{M}} + \frac{\tilde{M}}{T_2}\right). \quad (6.4.22)$$

This result is now introduced into Eq. (6.4.16b), which becomes

$$\dot{w} = -\frac{w - w^{\text{eq}}}{T_1} - \frac{2\tilde{E}}{\hbar\omega_{ba}}\left(\dot{\tilde{M}} + \frac{\tilde{M}}{T_2}\right). \tag{6.4.23}$$

Since $\dot{\tilde{M}}$ oscillates at an optical frequency (which is much larger than $1/T_2$), the term $\tilde{M}/T_2$ can be omitted, yielding the result

$$\dot{w} = -\frac{w - w^{\text{eq}}}{T_1} - \frac{2}{\hbar\omega_{ba}}\tilde{E}\dot{\tilde{M}}. \tag{6.4.24}$$

We see that the inversion w is driven by the product of $\tilde{E}$ with $\dot{\tilde{M}}$, which is proportional to the part of $\tilde{M}$ that is 90 degrees out of phase with $\tilde{E}$. We also see that w relaxes to its equilibrium value w^{eq} (which is typically equal to -1) in a time of the order of T_1.

Equations (6.4.20) and (6.4.24) provide a description of the two-level atomic system. Note that each equation is linear in the atomic variables $\tilde{M}$ and w. The origin of the nonlinear response of atomic systems lies in the fact that the coupling to the optical field depends parametrically on the atomic variables. A linear harmonic oscillator, for example, would be described by Eq. (6.4.20) with the inversion w held fixed at the value -1. The fact that the coupling depends on the inversion w, whose value depends on the applied field strength as described by Eq. (6.4.24), leads to nonlinearities.

Adiabatic Following Limit

The treatment of Section 6.3 considered the steady-state response of a two-level atom to a cw laser field. The adiabatic following limit (Grischkowsky, 1970) is another limit in which it is relatively easy to obtain solutions to the density matrix equations of motion. The nature of the adiabatic following approximation is as follows: We assume that the optical field is in the form of a pulse whose length τ_p obeys the condition

$$\tau_p \ll T_1, T_2; \tag{6.4.25}$$

we thus assume that essentially no relaxation occurs during the extent of the optical pulse. In addition, we assume that the laser is detuned sufficiently far from resonance that

$$|\omega - \omega_{ba}| \gg T_2^{-1}, \tau_p^{-1}, \mu_{ba}E/\hbar, \tag{6.4.26}$$

that is, we assume that the detuning is greater than the transition linewidth, that no Fourier component of the pulse extends to the transition frequency, and that the transition is not power-broadened into resonance with the pulse.

These conditions ensure that no appreciable population is excited to the upper level by the laser pulse.

To simplify the following analysis, we introduce the (complex) Rabi frequency

$$\Omega(t) = 2\mu_{ba}E(t)/\hbar \qquad (6.4.27)$$

where $E(t)$ gives the time evolution of the pulse envelope. The density matrix equations of motion (6.4.5) then become, in the limit $T_1 \to \infty$, $T_2 \to \infty$,

$$\frac{d\sigma}{dt} = i\Delta\sigma - \tfrac{1}{2}i\Omega w, \qquad (6.4.28a)$$

$$\frac{dw}{dt} = -i(\Omega^*\sigma - \Omega\sigma^*). \qquad (6.4.28b)$$

We note that the quantity $w^2 + 4\sigma\sigma^*$ is a constant of the motion whose value is given by

$$w^2(t) + 4|\sigma(t)|^2 = 1. \qquad (6.4.29)$$

This conclusion is verified by means of a derivation analogous to that leading to Eq. (6.4.12).

We now make the adiabatic following approximation, that is, we assume that for all times the atomic response is nearly in steady state with the applied field. We thus set $d\sigma/dt$ and dw/dt equal to zero in Eqs. (6.4.28). The simultaneous solution of these equations (which in fact is just the solution to (6.4.28a)) is given by

$$\sigma(t) = \frac{w(t)\Omega(t)}{2\Delta}. \qquad (6.4.30)$$

Since $w(t)$ is a real quantity, this result shows that $\sigma(t)$ is always in phase with the driving field $\Omega(t)$. We now combine Eqs. (6.4.29) and (6.4.30) to obtain the equation

$$w(t)^2 + \frac{w(t)^2|\Omega|^2}{\Delta^2} = 1, \qquad (6.4.31)$$

which can be solved for $w(t)$ to obtain

$$w(t) = \frac{-|\Delta|}{\sqrt{\Delta^2 + |\Omega(t)|^2}}. \qquad (6.4.32)$$

This expression can now be substituted back into Eq. (6.4.30) to obtain the result

$$\sigma(t) = -\frac{\Delta}{|\Delta|} \frac{\frac{1}{2}\Omega(t)}{\sqrt{\Delta^2 + |\Omega(t)|^2}}. \tag{6.4.33}$$

We now use these results to deduce the value of the nonlinear susceptibility. As in Eqs. (6.3.11) through (6.3.17), the polarization P is related to $\sigma(t)$ (recall that $\sigma = \sigma_{ba}$) through

$$P = N\mu_{ab}\sigma, \tag{6.4.34}$$

which through use of Eq. (6.4.33) becomes

$$P = -\frac{\Delta}{|\Delta|} \frac{\frac{1}{2}N\mu_{ab}\Omega(t)}{\sqrt{\Delta^2 + |\Omega(t)|^2}}. \tag{6.4.35}$$

Our derivation has assumed that the condition $|\Delta| \gg |\Omega|$ is valid. We can thus expand Eq. (6.4.35) in a power series in the small quantity $|\Omega|/\Delta$ to obtain

$$P = \frac{\Delta\Omega}{\Delta^2} \frac{-\frac{1}{2}N\mu_{ab}}{(1 + |\Omega|^2/\Delta^2)^{1/2}} = -\frac{\Delta\Omega}{\Delta^2} \frac{1}{2}N\mu_{ab}\left(1 - \frac{1}{2}\frac{|\Omega|^2}{\Delta^2} + \cdots\right). \tag{6.4.36}$$

The contribution to P that is third-order in the applied field is thus given by

$$\frac{|\Omega|^2\Omega\Delta n\mu_{ab}}{4\Delta^4} = \frac{2N|\mu_{ab}|^4}{\hbar^3\Delta^3}|E|^2E, \tag{6.4.37}$$

where, in obtaining the second form, we have used the fact that $\Omega = 2\mu_{ba}E/\hbar$. By convention, the coefficient of $|E|^2E$ is $3\chi^{(3)}$, and hence we find that

$$\chi^{(3)} = \frac{2N|\mu_{ba}|^4}{3\hbar^3\Delta^3}. \tag{6.4.38}$$

Note that this prediction is identical to that of the steady-state theory (Eq. (6.3.37)) in the limit $\Delta T_2 \gg 1$ for the case of a radiatively broadened transition (i.e., $T_2/T_1 = 2$) for which $(\rho_{bb} - \rho_{aa})^{eq} = -1$.

6.5. Rabi Oscillations and Dressed Atomic States

In this section we consider the response of a two-level atom to an optical field sufficiently intense to remove a significant fraction of the population from the atomic ground state. One might think that only consequence of a field this intense would be to lower the overall response of the atom. Such is not the case, however. Stark shifts induced by the laser field profoundly modify the energy-level structure of the atom, leading to new resonances in the optical susceptibility. In the present section, we explore some of the processes that occur in the presence of a strong driving field.

Rabi Solution of the Schrödinger Equation

Let us consider the solution to the Schrödinger equation for a two-level atom in the presence of an intense optical field.[*] We describe the state of the system in terms of the atomic wave function ψ, which obeys the Schrödinger equation

$$i\hbar \frac{\partial \psi}{\partial t} = \hat{H}\psi \tag{6.5.1}$$

with the Hamiltonian operator $\hat{H}$ given by

$$\hat{H} = \hat{H}_0 + \hat{V}(t). \tag{6.5.2}$$

Here $\hat{H}_0$ represents the Hamiltonian of a free atom, and $\hat{V}(t)$ represents the energy of interaction with the applied field. In the electric dipole approximation, $\hat{V}(t)$ is given by

$$\hat{V}(t) = \hat{\mu}\tilde{E}(t), \tag{6.5.3}$$

where the dipole moment operator is given by $\hat{\mu} = -e\hat{r}$.

We assume that the applied field is given by $\tilde{E}(t) = Ee^{-i\omega t} +$ c.c. with E constant, and that the field is nearly resonant with an allowed transition between the atomic ground state a and some other level b, as shown in Fig. 6.2.1. Since the effect of the interaction is to mix states a and b, the atomic wavefunction in the presence of the applied field can be represented as

$$\psi(\mathbf{r}, t) = C_a(t)u_a(\mathbf{r})e^{-i\omega_a t} + C_b(t)u_b(\mathbf{r})e^{-i\omega_b t}. \tag{6.5.4}$$

Here $u_a(\mathbf{r})e^{-i\omega_a t}$ represents the wavefunction of the atomic ground state a, and $u_b(\mathbf{r})e^{-i\omega_b t}$ represents the wavefunction of the excited state b. We assume that these wavefunctions are orthonormal in the sense that

$$\int d^3r\, u_i^*(\mathbf{r})\, u_j(\mathbf{r}) = \delta_{ij}. \tag{6.5.5}$$

The quantities $C_a(t)$ and $C_b(t)$ that appear in Eq. (6.5.4) can be interpreted as the probability amplitudes that at time t the atom is in state a or state b, respectively.

We next derive the equations of motion for $C_a(t)$ and $C_b(t)$, using methods analogous to those used in Section 3.2. By introducing Eq. (6.5.4) into the Schrödinger equation (6.5.1), multiplying the resulting equation by u_a^*, and

---

[*] See also Sargent *et al.* (1974), p. 26, or Dicke and Wittke (1960), p. 203.

integrating this equation over all space, we find that

$$\dot{C}_a = \frac{1}{i\hbar} C_b V_{ab} e^{-i\omega_{ba}t}, \tag{6.5.6}$$

where we have introduced the resonance frequency $\omega_{ba} = \omega_b - \omega_a$ and the interaction matrix element

$$V_{ab} = V_{ba}^* = \int d^3r\, u_a^* \hat{V} u_b. \tag{6.5.7}$$

Similarly, by multiplying instead by u_b^* and again integrating over all space, we find that

$$\dot{C}_b = \frac{1}{i\hbar} C_a V_{ba} e^{i\omega_{ba}t}. \tag{6.5.8}$$

We now explicitly introduce the form of the interaction Hamiltonian and represent the interaction matrix elements as

$$V_{ab}^* = V_{ba} = -\mu_{ba}\tilde{E}(t) = -\mu_{ba}(Ee^{-i\omega t} + E^*e^{i\omega t}). \tag{6.5.9}$$

Equations (6.5.6) and (6.5.8) then become

$$\dot{C}_a = \frac{-\mu_{ab}}{i\hbar} C_b \left(E^* e^{-i(\omega_{ba}-\omega)t} + E^* e^{-i(\omega_{ba}+\omega)t} \right) \tag{6.5.10a}$$

and

$$\dot{C}_b = \frac{-\mu_{ba}}{i\hbar} C_a \left(E e^{i(\omega_{ba}-\omega)t} + E^* e^{i(\omega_{ba}+\omega)t} \right). \tag{6.5.10b}$$

We next make the rotating wave approximation, that is, we drop the rapidly oscillating second terms in these equations and retain only the first terms.* We also introduce the detuning factor

$$\Delta = \omega - \omega_{ba}. \tag{6.5.11}$$

The coupled equations (6.5.10) then reduce to the set

$$\dot{C}_a = i\frac{\mu_{ab}E^*}{\hbar} C_b e^{i\Delta t}, \tag{6.5.12a}$$

$$\dot{C}_b = i\frac{\mu_{ab}E}{\hbar} C_a e^{-i\Delta t}. \tag{6.5.12b}$$

This set of equations can be readily solved by adopting a trial solution of the form

$$C_a = Ke^{-\lambda t}. \tag{6.5.13}$$

---

* See also the discussion preceding Eq. (6.3.5).

This expression is introduced into Eq. (6.5.12a), which shows that C_b must be of the form

$$C_b = \frac{-\hbar \lambda K}{\mu_{ab} E^*} e^{-i(\lambda + \Delta)t}. \tag{6.5.14}$$

This form for C_b and the trial solution (6.5.13) for C_a are now introduced into Eq. (6.5.12b), which shows that the characteristic frequency λ must obey the equation

$$\lambda(\lambda + \Delta) = \frac{|\mu_{ba}|^2 |E|^2}{\hbar^2}. \tag{6.5.15}$$

The solutions of this equation are given by

$$\lambda_\pm = -\tfrac{1}{2}\Delta \pm \tfrac{1}{2}\Omega', \tag{6.5.16}$$

where we have introduced the generalized (or detuned) Rabi frequency

$$\Omega' = (|\Omega|^2 + \Delta^2)^{1/2} \tag{6.5.17}$$

and where, as before, $\Omega = 2\mu_{ba} E/\hbar$ denotes the complex Rabi frequency. The general solution to Eqs. (6.5.12) for $C_a(t)$ can thus be expressed as

$$C_a(t) = e^{(1/2)i\Delta t}\left(A_+ e^{-(1/2)i\Omega' t} + A_- e^{(1/2)i\Omega' t}\right), \tag{6.5.18a}$$

where A_+ and A_- are constants of integration whose values depend on the initial conditions. The corresponding expression for $C_b(t)$ is obtained by introducing this result into Eq. (6.5.12a):

$$
\begin{aligned}
C_b(t) &= \frac{-\hbar \dot{C}_a}{\mu_{ab} E^*} e^{-i\Delta t} \\
&= e^{-(1/2)i\Delta t}\left(\frac{\Delta - \Omega'}{\Omega^*} A_+ e^{-(1/2)i\Omega' t} + \frac{\Delta + \Omega'}{\Omega^*} A_- e^{(1/2)i\Omega' t}\right).
\end{aligned}
\tag{6.5.18b}
$$

Equations (6.5.18) give the general solution to Eqs. (6.5.12). Next, we find the specific solution for two different sets of initial conditions.

Solution for an Atom Initially in the Ground State

One realistic set of initial conditions is that of an atom known to be in the ground state at time $t = 0$, so that

$$C_a(0) = 1 \quad \text{and} \quad C_b(0) = 0. \tag{6.5.19}$$

Equation (6.5.18a) evaluated at $t = 0$ then shows that

$$A_+ + A_- = 1, \tag{6.5.20}$$

while Eq. (6.5.18b) evaluated at $t = 0$ shows that

$$(\Delta - \Omega')A_+ + (\Delta + \Omega')A_- = 0. \tag{6.5.21}$$

These equations are solved algebraically to find that

$$A_+ = 1 - A_- = \frac{\Omega' + \Delta}{2\Omega'}. \tag{6.5.22}$$

The probability amplitudes $C_a(t)$ and $C_b(t)$ are now determined by introducing these expressions for A_+ and A_- into Eqs. (6.5.18), to obtain

$$C_a(t) = e^{(1/2)i\Delta t}\left[\left(\frac{\Omega' + \Delta}{2\Omega'}\right)e^{-(1/2)i\Omega't} + \left(\frac{\Omega' + \Delta}{2\Omega'}\right)e^{(1/2)i\Omega't}\right]$$
$$= e^{(1/2)i\Delta t}\left[\cos\left(\tfrac{1}{2}\Omega't\right) - \frac{i\Delta}{\Omega'}\sin\left(\tfrac{1}{2}\Omega't\right)\right] \tag{6.5.23}$$

and

$$C_b(t) = e^{-(1/2)i\Delta t}\left(\frac{-\Omega}{2\Omega'}e^{-(1/2)i\Omega't} + \frac{\Omega}{2\Omega'}E^{(1/2)i\Omega't}\right)$$
$$= ie^{-(1/2)i\Delta t}\left[\frac{\Omega}{\Omega'}\sin\left(\tfrac{1}{2}\Omega't\right)\right]. \tag{6.5.24}$$

The probability that the atom is in level a at time t is hence given by

$$|C_a|^2 = \cos^2\left(\tfrac{1}{2}\Omega't\right) + \frac{\Delta^2}{\Omega'^2}\sin^2\left(\tfrac{1}{2}\Omega't\right), \tag{6.5.25}$$

while the probability of being in level b is given by

$$|C_b|^2 = \frac{|\Omega|^2}{\Omega'^2}\sin^2\left(\tfrac{1}{2}\Omega't\right). \tag{6.5.26}$$

Note that (since $\Omega'^2 = |\Omega|^2 + \Delta^2$)

$$|C_a|^2 + |C_b|^2 = 1, \tag{6.5.27}$$

which shows that probability is conserved.

For the case of exact resonance ($\Delta = 0$), Eqs. (6.5.25) and (6.5.26) reduce to

$$|C_a|^2 = \cos^2\left(\tfrac{1}{2}|\Omega|t\right), \tag{6.5.28a}$$

$$|C_b|^2 = \sin^2\left(\tfrac{1}{2}|\Omega|t\right), \tag{6.5.28b}$$

and the probabilities oscillate between zero and one in the simple manner illustrated in Fig. 6.5.1. Note that, since the probability amplitude C_a oscillates at angular frequency $|\Omega|/2$, the probability $|C_a|^2$ oscillates at angular frequency

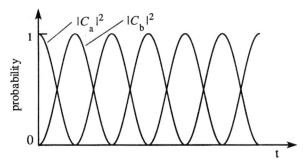

FIGURE 6.5.1   Rabi oscillations of the populations in the ground ($|C_a|^2$) and excited ($|C_b|^2$) states for the case of exact resonance ($\Delta = 0$).

$|\Omega|$, that is, at the Rabi frequency. As the detuning Δ is increased, the angular frequency at which the population oscillates increases, since the generalized Rabi frequency is given by $\Omega' = [|\Omega|^2 + \Delta^2]^{1/2}$, but the amplitude of the oscillation decreases, as shown in Fig. 6.5.2.

Next we calculate the expectation value of the atomic dipole moment for an atom known to be in the atomic ground state at time $t = 0$. This quantity is given by

$$\langle \tilde{\mu} \rangle = \langle \psi | \tilde{\mu} | \psi \rangle, \tag{6.5.29}$$

where $\psi(\mathbf{r}, t)$ is given by Eq. (6.5.4). We assume as before that $\langle a | \tilde{\mu} | a \rangle = \langle b | \tilde{\mu} | b \rangle = 0$, and we denote the nonvanishing matrix elements of $\tilde{\mu}$ by

$$\mu_{ab} = \langle a | \tilde{\mu} | b \rangle = \langle b | \tilde{\mu} | a \rangle^* = \mu_{ba}^*. \tag{6.5.30}$$

We thus find that the induced dipole moment is given by

$$\langle \tilde{\mu} \rangle = C_a^* C_b \mu_{ab} e^{-i\omega_{ba} t} + \text{c.c.} \tag{6.5.31}$$

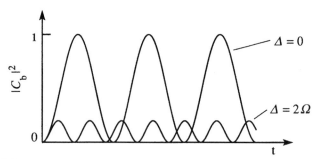

FIGURE 6.5.2   Rabi oscillations of the excited state population for two values of $\Delta \equiv \omega - \omega_{ba}$.

or, introducing Eqs. (6.5.23) and (6.5.24) for C_a and C_b, by

$$\langle \tilde{\mu} \rangle = \mu_{ab} \frac{\Omega}{\Omega'} \left[\frac{-\Delta}{2\omega'} e^{-i\omega t} + \frac{1}{4} \left(\frac{\Delta}{\Omega'} - 1 \right) e^{-i(\omega + \Omega')t} \right.$$
$$\left. + \frac{1}{4} \left(\frac{\Delta}{\Omega'} + 1 \right) e^{-i(\omega - \Omega')t} \right] + \text{c.c.} \tag{6.5.32}$$

This result shows that the atomic dipole oscillates not only at the driving frequency ω but also at the Rabi sideband frequencies $\omega + \Omega'$ and $\omega - \Omega'$. We can understand the origin of this effect by considering the frequencies that are present in the atomic wavefunction. We recall that the wavefunction is given by Eq. (6.5.4), where (according to Eqs. (6.5.23) and (6.5.24)) $C_a(t)$ contains frequencies $-\frac{1}{2}(\Delta \pm \Omega')$ and $C_b(t)$ contains frequencies $\frac{1}{2}(\Delta \pm \Omega')$. Figure 6.5.3 shows graphically the frequencies that are present in the atomic wavefunction. Note that the frequencies at which the atomic dipole oscillates

(a)

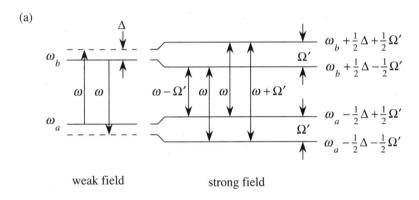

weak field   strong field

(b)

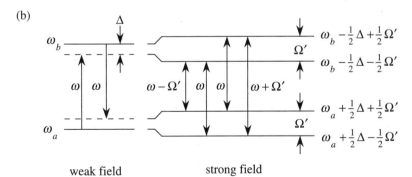

weak field   strong field

FIGURE 6.5.3   Frequency spectrum of the atomic wavefunction given by Eq. (6.5.6) (with $C_a(t)$ and $C_b(t)$ given by Eqs. (6.5.28) and (6.5.29)) for the case of (a) positive detuning ($\Delta > 0$) and (b) negative detuning ($\Delta < 0$).

correspond to differences of the various frequency components of the wavefunction.

Dressed States

Another important solution to the Schrödinger equation for a two-level atom is that corresponding to the dressed atomic states (Autler and Townes, 1955; Cohen-Tannoudji and Reynaud, 1977). The characteristic feature of these states is that the probability to be in atomic level a (or b) is constant in time. As a consequence, the probability amplitudes $C_a(t)$ and $C_b(t)$ can depend on time only in terms of exponential phase factors. Recall, however, that in general $C_a(t)$ and $C_b(t)$ are given by Eqs. (6.5.18).

There are two ways in which the solution of Eqs. (6.5.18) can lead to time-independent probabilities of occupancy for levels a and b. One such solution, which we designate as ψ_+, corresponds to the case in which the integration constants A_+ and A_- have the values

$$A_+ = 1, \quad A_- = 0 \quad (\text{for } \psi_+); \tag{6.5.33a}$$

the other solution, which we designate as ψ_-, corresponds to the case in which

$$A_+ = 0, \quad A_- = 1 \quad (\text{for } \psi_-). \tag{6.5.33b}$$

Explicitly, the atomic wavefunction corresponding to each of these solutions is given, through use of Eqs. (6.5.4), (6.5.18), and (6.5.33), as

$$\psi_\pm = N_\pm \left\{ u_a(\mathbf{r}) \exp\left[-i\left(\omega_a - \tfrac{1}{2}\Delta \pm \tfrac{1}{2}\Omega'\right)t\right] \right.$$
$$\left. + \frac{\Delta \mp \Omega'}{\Omega^*} u_b(\mathbf{r}) \exp\left[-i\left(\omega_b + \tfrac{1}{2}\Delta \pm \tfrac{1}{2}\Omega'\right)t\right] \right\}, \tag{6.5.34}$$

where $N_\pm$ is a normalization constant. The value of this constant is determined by requiring that

$$\int |\psi_\pm|^2 \, d^3r = 1. \tag{6.5.35}$$

By introducing Eq. (6.5.34) into this expression and performing the integrations, we find that

$$|N_\pm|^2 \left[1 + \frac{(\Delta \mp \Omega')^2}{|\Omega|^2} \right] = 1. \tag{6.5.36}$$

For future convenience, we choose the phases of $N_\pm$ such that $N_\pm$ are given by

$$N_\pm = \frac{\Omega^*}{\Omega'} \left[\frac{\Omega'}{2(\Omega' \mp \Delta)} \right]^{1/2}. \tag{6.5.37}$$

The normalized dressed-state wavefunctions are hence given by

$$\psi_\pm = \frac{\Omega^*}{\Omega'}\left[\frac{\Omega'}{2(\Omega' \mp \Delta)}\right]^{1/2} u_a(\mathbf{r})\exp\left[-i\left(\omega_a - \tfrac{1}{2}\Delta \pm \tfrac{1}{2}\Omega'\right)t\right]$$

$$\mp \left[\frac{\Omega' \mp \Delta}{2\Omega'}\right]^{1/2} u_b(\mathbf{r})\exp\left[-i\left(\omega_b - \tfrac{1}{2}\Delta \pm \tfrac{1}{2}\Omega'\right)t\right]. \tag{6.5.38}$$

We next examine some of the properties of the dressed states. The probability amplitude for an atom in the dressed state $\psi_\pm$ to be in the atomic level a is given by

$$\langle a|\psi_\pm\rangle = \frac{\Omega^*}{\Omega'}\left[\frac{\Omega'}{2(\Omega' \mp \Delta)}\right]^{1/2}\exp\left[-i\left(\omega_a - \tfrac{1}{2}\Delta \pm \tfrac{1}{2}\Omega'\right)t\right], \tag{6.5.39}$$

and hence the probability of finding the atom in the state a is given by

$$|\langle a|\psi_\pm\rangle|^2 = \frac{|\Omega|^2}{\Omega'^2}\frac{\Omega'}{2(\Omega' \mp \Delta)} = \frac{|\Omega|^2}{2\Omega'(\Omega' \mp \Delta)}. \tag{6.5.40}$$

Similarly, the probability amplitude of finding the atom in state b is given by

$$\langle b|\psi_\pm\rangle = \mp\left(\frac{\Omega' \mp \Delta}{2\Omega'}\right)^{1/2}\exp\left[-i\left(\omega_b + \tfrac{1}{2}\Delta \pm \tfrac{1}{2}\Omega'\right)t\right], \tag{6.5.41}$$

and hence the probability of finding the atom in the state b is given by

$$|\langle b|\psi_\pm\rangle|^2 = \frac{\Omega' \pm \Delta}{2\Omega'}. \tag{6.5.42}$$

Note that these probabilities of occupancy are indeed constant in time; in this sense the dressed states constitute the stationary states of the coupled atom–field system.

The dressed states $\psi_\pm$ are solutions of Schrödinger's equation in the presence of the total Hamiltonian $\hat{H} = \hat{H}_0 + \hat{V}(t)$. Thus, if the system is known to be in state ψ_+ (or ψ_-) at the time $t = 0$, the system will remain in this state, even though the system is subject to the interaction Hamiltonian $\hat{V}$. They are stationary states in the sense mentioned above, that the probability of finding the atom in either of the atomic states a or b is constant in time. Although the states $\psi_\pm$ are stationary states, they are not energy eigenstates, because the Hamiltonian $\hat{H}$ depends explicitly on time.

It is easy to demonstrate that the dressed states are orthogonal, that is, that

$$\langle\psi_+|\psi_-\rangle = 0. \tag{6.5.43}$$

The expectation value of the induced dipole moment for an atom in a dressed state is given by

$$\langle \psi_{\pm} | \hat{\mu} | \psi_{\pm} \rangle = \mp \frac{\Omega}{2\Omega'} \mu_{ab} e^{-i\omega t} + \text{c.c.} \tag{6.5.44}$$

Thus the induced dipole moment of an atom in a dressed state oscillates only at the driving frequency. However, the dipole transition moment between the dressed states is nonzero:

$$\langle \psi_{\pm} | \hat{\mu} | \psi_{\mp} \rangle = \pm \mu_{ab} \frac{\Omega}{2\Omega'} \left(\frac{\Omega' \pm \Delta}{\Omega' \mp \Delta} \right)^{1/2} e^{-i(\omega \mp \Omega')t}$$

$$\mp \mu_{ba} \frac{\Omega^*}{2\Omega'} \left(\frac{\Omega' \mp \Delta}{\Omega' \pm \Delta} \right)^{1/2} e^{i(\omega \pm \Omega')t}. \tag{6.5.45}$$

The properties of the dressed states are summarized in the frequency level diagram shown for the case of positive Δ in Fig. 6.5.4a and for the case of negative Δ in Fig. 6.5.4b.

Next, we consider the limiting form of the dressed states for the case of a weak applied field, i.e., for $|\Omega| \ll |\Delta|$. In this limit, we can approximate the generalized Rabi frequency Ω' as

$$\Omega' = (|\Omega|^2 + \Delta^2)^{1/2} = |\Delta| \left(1 + \frac{|\Omega|^2}{\Delta^2} \right)^{1/2}$$

$$\simeq |\Delta| \left(1 + \frac{1}{2} \frac{|\Omega|^2}{\Delta^2} \right). \tag{6.5.46}$$

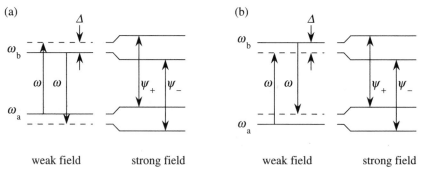

FIGURE 6.5.4   The dressed atomic states ψ_+ and ψ_- for Δ positive (a) and negative (b).

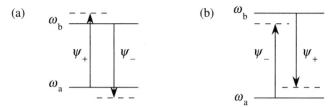

FIGURE 6.5.5   The weak field limit of the dressed states ψ_+ and ψ_- for the case of (a) positive and (b) negative detuning Δ.

Using this result, we can approximate the dressed-state wavefunctions of Eq. (6.5.38) for the case of positive Δ as

$$\psi_+ = \frac{\Omega^*}{|\Omega|}u_a e^{-i\omega_a t} - \frac{|\Omega|}{2\Delta}u_b e^{i(\omega_b+\Delta)t}, \qquad (6.5.47a)$$

$$\psi_- = \frac{\Omega^*}{2\Delta}u_a e^{-i(\omega_a-\Delta)t} + u_b e^{-i\omega_b t}. \qquad (6.5.47b)$$

We note that in this limit ψ_+ is primarily ψ_a and ψ_- is primarily ψ_b. The smaller contribution to ψ_+ can be identified with the virtual level induced by the transition. For the case of negative Δ, we obtain

$$\psi_+ = -\frac{\Omega^*}{2\Delta}u_a e^{-i(\omega_a-\Delta)t} - u_b e^{-i\omega_b t}, \qquad (6.5.48a)$$

$$\psi_- = \frac{\Omega^*}{|\Omega|}u_a e^{-i\omega_a t} - \frac{|\Omega|}{2\Delta}u_b e^{-i(\omega_b+\Delta)t}. \qquad (6.5.48b)$$

Now ψ_+ is primarily ψ_b, and ψ_- is primarily ψ_a. These results are illustrated in Fig. 6.5.5. Note that these results have been anticipated in drawing certain of the levels as dashed lines in the weak-field limit of the diagrams shown in Figs. 6.5.3 and 6.5.4.

Inclusion of Relaxation Phenomena

In the absence of damping phenomena, it is adequate to treat the response of a two-level atom to an applied optical field by solving Schrödinger's equation for the time evolution of the wavefunction. We have seen that under such circumstances the population inversion oscillates at the generalized Rabi frequency $\Omega' = (\Omega^2 + \Delta^2)^{1/2}$. If damping effects are present, we expect that these Rabi oscillations will eventually become damped out and that the population difference will approach some steady-state value. In order to treat this behavior, we need to solve the density matrix equations of motion with the inclusion of damping effects. We take the density matrix equations in the form

$$\dot{p} = \left(i\Delta - \frac{1}{T_2} \right) p - \frac{i}{\hbar} |\mu|^2 E w, \tag{6.5.49a}$$

$$\dot{w} = -\frac{w+1}{T_1} - \frac{2i}{\hbar} (pE^* - p^*E), \tag{6.5.49b}$$

and we assume that at $t = 0$ the atom is in its ground state, that is, that

$$p(0) = 0, \qquad w(0) = -1, \tag{6.5.50}$$

and that the field $\tilde{E}(t)$ is turned on at $t = 0$ and oscillates harmonically there-after (i.e., $E = 0$ for $t < 0$, $E = $ constant for $t \geq 0$).

Equations (6.5.49) can be solved in general under the conditions given above (see Problem 4 at the end of this chapter). For the special case in which $T_1 = T_2$, the form of the solution is considerably simpler than in the general case. The solution to Eqs. (6.5.49) for the population inversion for this special case is given by

$$w(t) = w_0 - (1 + w_0) \cos \Omega' t e^{-t/T_2} \left[\cos \Omega' t + \frac{1}{\Omega' T_2} \sin \Omega' t \right], \tag{6.5.51a}$$

where

$$w_0 = \frac{-\left(1 + \Delta^2 T_2^2 \right)}{1 + \Delta^2 T_2^2 + \Omega^2 T_1 T_2}. \tag{6.5.51b}$$

The nature of this solution is shown in Fig. 6.5.6. Note that the Rabi oscillations are damped out in a time of the order of T_2. Once the Rabi oscillations have damped out, the system enters one of the dressed states of the coupled atom–field system.

In summary, we have just seen that, in the absence of damping effects, the population inversion of a strongly driven two-level atom oscillates at the generalized Rabi frequency $\Omega' = (\Omega^2 + \Delta^2)^{1/2}$ and that consequently the

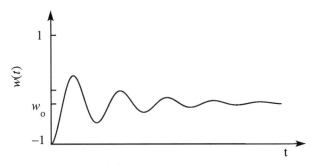

FIGURE 6.5.6   Damped Rabi oscillations.

induced dipole moment oscillates at the applied frequency ω and also at the Rabi sideband frequencies $\omega \pm \Omega'$. In the presence of dephasing processes, the Rabi oscillations die out in a characteristic time given by the dipole dephasing time T_2. Hence, Rabi oscillations are not present in the steady state.

In the following section, we explore the nature of the response of the atom to a strong field at frequency ω and a weak field at frequency $\omega + \delta$. If the frequency difference δ (or its negative $-\delta$) between these two fields is nearly equal to the generalized Rabi frequency Ω', the beat frequency between the two applied fields can act as a source term to drive the Rabi oscillation. We shall find that, in the presence of such a field, the population difference oscillates at the beat frequency δ, and that the induced dipole moment contains the frequency components ω and $\omega \pm \delta$.

6.6.   Optical Wave Mixing in Two-Level Systems

In the present section we consider the response of a collection of two-level atoms to the simultaneous presence of a strong optical field (which we call the pump field) and one or more weak optical fields (which we call probe fields). These latter fields are considered weak in the sense that they alone cannot saturate the response of the atomic system.

An example of such an occurrence is saturation spectroscopy, using a setup of the sort shown in Fig. 6.6.1. In such an experiment, one determines how the response of the medium to the probe wave is modified by the presence of the pump wave. Typically, one might measure the transmission of the probe wave as a function of the frequency ω and intensity of the pump wave and of the frequency detuning δ between the pump and probe waves. The results of such experiments can be used to obtain information regarding the dipole transition moments and the relaxation times T_1 and T_2.

Another example of the interactions considered in this section is the multi-wave mixing experiment shown in part (a) of Fig. 6.6.2. Here the pump wave

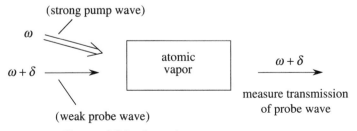

FIGURE 6.6.1   Saturation spectroscopy setup.

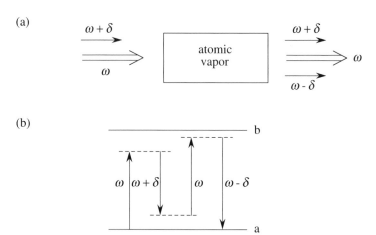

FIGURE 6.6.2   (a) Forward four-wave mixing. (b) Energy-level description of the four-wave mixing process, drawn for clarity for the case in which δ is negative.

at frequency ω and the probe wave at frequency $\omega + \delta$ are copropagating (or nearly copropagating) through the medium. For this geometry, the four-wave mixing process shown in part (b) of the figure becomes phase-matched (or nearly phase-matched), and this process leads to the generation of the symmetric sideband at frequency $\omega - \delta$.

At low intensities of the pump laser, the response of the atomic system at the frequencies $\omega + \delta$ and $\omega - \delta$ can be calculated using perturbation theory of the sort developed in Chapter 3. In this limit, one finds that the absorption (and dispersion) experienced by the probe wave in the geometry of Fig. 6.6.1 is somewhat reduced by the presence of the pump wave. One also finds that, for the geometry of Fig. 6.6.2, the intensity of the generated sideband at frequency $\omega - \delta$ increases quadratically as the pump intensity is increased.

In this section we show that the character of these nonlinear processes is profoundly modified when the intensity of the pump laser is increased to the extent that perturbation theory is not sufficient to describe the interaction. These higher-order processes become important when the Rabi frequency Ω associated with the pump field is greater than both the detuning Δ of the pump wave from the atomic resonance and the transition linewidth $1/T_2$. Under this condition, the atomic energy levels are strongly modified by the pump field, leading to new resonances in the absorptive and mixing responses. In particular, we shall find that these new resonances can be excited when the pump-probe detuning δ is approximately equal to $\pm\Omega'$, where Ω' is the generalized Rabi frequency.

Solution of the Density Matrix Equations for a Two-Level Atom in the Presence of Pump and Probe Fields

We have seen in Section 6.4 that the dynamical behavior of a two-level atom in the presence of the optical field

$$\tilde{E}(t) = Ee^{-i\omega t} + \text{c.c.} \tag{6.6.1}$$

can be described in terms of equations of motion for the population inversion $w = \rho_{bb} - \rho_{aa}$ and the complex dipole amplitude $p = \mu_{ab}\sigma_{ba}$, which is related to the expectation value $\tilde{p}(t)$ of the atomic dipole moment by

$$\tilde{p}(t) = pe^{-i\omega t} + \text{c.c.} \tag{6.6.2}$$

The equations of motion for p and w are given explicitly by

$$\frac{dp}{dt} = \left(i\Delta - \frac{1}{T_2}\right)p - \frac{i}{\hbar}|\mu_{ba}|^2 Ew, \tag{6.6.3}$$

$$\frac{dw}{dt} = -\frac{w - w^{\text{eq}}}{T_1} + \frac{4}{\hbar}\,\text{Im}(pE^*), \tag{6.6.4}$$

where $\Delta = \omega - \omega_{ba}$. For the problem at hand, we represent the amplitude of the applied optical field as

$$E = E_0 + E_1 e^{-i\delta t}, \tag{6.6.5}$$

where we assume that $|E_1| \ll |E_0|$. By introducing Eq. (6.6.5) into Eq. (6.6.1), we find that the electric field can alternatively be expressed as

$$\tilde{E}(t) = E_0 e^{-i\omega t} + E_1 e^{-i(\omega+\delta)t} + \text{c.c.}; \tag{6.6.6}$$

hence E_0 and E_1 represent the complex amplitudes of the pump and probe waves, respectively.

Equations (6.6.3) and (6.6.4) cannot readily be solved exactly for the field given in Eq. (6.6.5). Instead, our strategy will be to find a solution that is exact for an applied strong field E_0 and is correct to lowest order in the amplitude E_1 of the weak field. We hence require that the steady-state solution of Eqs. (6.6.3) and (6.6.4) be of the form

$$p = p_0 + p_1 e^{-i\delta t} + p_{-1} e^{i\delta t} \tag{6.6.7}$$

and

$$w = w_0 + w_1 e^{-i\delta t} + w_{-1} e^{i\delta t}, \tag{6.6.8}$$

where p_0 and w_0 denote the solution for the case in which only the pump field E_0 is present, and where the other terms are assumed to be small in the sense that

$$|p_1|, |p_{-1}| \ll |p_0|, \qquad |w_1|, |w_{-1}| \ll |w_0|. \qquad (6.6.9)$$

Note that, to lowest order in the amplitude E_1 of the probe field, 0 and $\pm\delta$ are the only frequencies that can be present in the solution of Eqs. (6.6.3) and (6.6.4). Note also that, in order for $w(t)$ to be a real quantity, w_{-1} must be equal to w_1^*. Hence $w(t)$ is of the form $w(t) = w_0 + 2|w_1|\cos(\delta t + \phi)$, where ϕ is the phase of w. Thus, in the simultaneous presence of pump and probe fields, the population difference oscillates harmonically at the pump–probe frequency difference, and w_1 represents the complex amplitude of the population oscillation.

We now introduce the trial solution (6.6.7) and (6.6.8) into the density matrix equations (6.6.3) and (6.6.4) and equate terms with the same time dependence. In accordance with our perturbation assumptions, we drop any term that contains the product of more than one small quantity. Then, for example, the zero-frequency part of the equation of motion for dipole amplitude, Eq. (6.6.3), becomes

$$0 = \left(i\Delta - \frac{1}{T_2} \right) p_0 - \frac{i}{\hbar} |\mu_{ba}|^2 E_0 w_0,$$

whose solution is

$$p_0 = \frac{\hbar^{-1} |\mu_{ba}|^2 E_0 w_0}{\Delta + i/T_2}. \qquad (6.6.10)$$

Likewise, the part of Eq. 6.6.3) oscillating as $e^{i\delta t}$ is

$$-i\delta p_1 = \left(i\Delta - \frac{1}{T_2} \right) p_1 - \frac{i}{\hbar} |\mu_{ba}|^2 (E_0 w_1 + E_1 w_0),$$

which can be solved to obtain

$$p_1 = \frac{\hbar^{-1} |\mu_{ba}|^2 (E_0 w_1 + E_1 w_0)}{(\Delta - \delta) + i/T_2}; \qquad (6.6.11)$$

the part of Eq. (6.6.3) oscillating as $e^{i\delta t}$ is

$$i\delta p_{-1} = \left(i\Delta - \frac{1}{T_2} \right) p_{-1} - \frac{i}{\hbar} |\mu_{ba}|^2 (E_0 w_{-1}),$$

which can be solved to obtain

$$p_{-1} = \frac{\hbar^{-1} |\mu_{ba}|^2 E_0 w_{-1}}{(\Delta - \delta) + i/T_2}. \qquad (6.6.12)$$

Next, we consider the solution of the inversion equation (6.6.4). We introduce the trial solution (6.6.7) and (6.6.8) into this equation. The zero-frequency part of the resulting expression is

$$0 = -\frac{w_0 - w^{\text{eq}}}{T_1} + \frac{4}{\hbar}\,\text{Im}(p_0 E_0^*). \tag{6.6.13}$$

We now introduce the expression (6.6.10) for p_0 into this expression to obtain

$$\frac{w_0 - w^{\text{eq}}}{T_1} = \Omega^2 w_0\,\text{Im}\left(\frac{\Delta - i/T_2}{\Delta^2 + 1/T_2^2}\right) = \frac{-\Omega^2 w_0/T_2}{\Delta^2 + 1/T_2^2}, \tag{6.6.14}$$

where we have introduced the on-resonance Rabi frequency $\Omega = 2|\mu E|/\hbar$. We now solve Eq. (6.6.14) algebraically for w_0 to obtain

$$w_0 = \frac{w^{\text{eq}}\left(1 + \Delta^2 T_2^2\right)}{1 + \Delta^2 T_2^2 + \Omega^2 T_1 T_2}. \tag{6.6.15}$$

We next consider the oscillating part of Eq. (6.6.4). The part of $\text{Im}(pE^*)$ oscillating at frequencies $\pm\delta$ is given by

$$\text{Im}(pE^*) = \text{Im}(p_0 E_1^* e^{i\delta t} + p_1 E_0^* e^{-i\delta t} + p_{-1} E_0^* e^{i\delta t})$$

$$= \frac{1}{2i}(p_0 E_1^* e^{i\delta t} + p_1 E_0^* e^{-i\delta t} + p_{-1} E_0^* e^{i\delta t} \tag{6.6.16}$$

$$- p_0^* E_1 e^{-i\delta t} - p_1^* E_0 e^{i\delta t} - p_{-1}^* E_0 e^{-i\delta t}),$$

where in obtaining the second form we have used the identity $\text{Im}\,z = (z - z^*)/2i$. We now introduce this result into Eq. (6.6.4). The part of the resulting expression which varies as $e^{i\delta t}$ is

$$-i\delta w_1 = \frac{-w_1}{T_1} - \frac{2i}{\hbar}(p_1 E_0^* - p_0^* E_1 - p_{-1}^* E_0).$$

This expression is solved for w_1 to obtain

$$w_1 = \frac{2\hbar^{-1}(p_1 E_0^* - p_0^* E_1 - p_{-1}^* E_0)}{\delta + i/T_1}. \tag{6.6.17}$$

We similarly find from the part of Eq. (6.6.4) oscillating as $e^{i\delta t}$ that

$$w_{-1} = \frac{2\hbar^{-1}(p_1^* E_0 - p_0 E_1^* - p_{-1} E_0^*)}{\delta - i/T_1}. \tag{6.6.18}$$

Note that $w_{-1} = w_1^*$, as required from the condition that $w(t)$ as given by Eq. (6.6.8) be real.

At this point we have a set of six coupled equations [(6.6.10), (6.6.11), (6.6.12), (6.6.15), (6.6.17), (6.6.18)] for the six quantities p_0, p_1, p_{-1}, w_0, w_1, w_{-1}. We note that w_0 is given by Eq. (6.6.15) in terms of known quantities. Our strategy will thus be to solve next for w_1, since the other unknown quantities are simply related to w_0 and w_1. We thus introduce the expressions for p_1, p_0, and p_{-1} into Eq. (6.6.17), which becomes

$$\left(\delta + \frac{i}{T_1}\right)w_1 = \frac{2|\mu_{ba}|^2}{\hbar^2}$$

$$\times \left(\frac{|E_0|^2 w_1}{\Delta + \delta + i/T_2} + \frac{E_1 E_0^* w_0}{\Delta + \delta + i/T_2} - \frac{E_1 E_0^* w_0}{\Delta - i/T_2} - \frac{|E_0|^2 w_1}{\Delta - \delta - i/T_2}\right).$$

This equation is now solved algebraically for w_1, yielding

$$w_1 = -\frac{w_0^2|\mu_{ba}|^2 E_1 E_0^*\hbar^{-2}\dfrac{(\delta - \Delta + i/T_2)(\delta + 2i/T_2)}{\Delta - i/T_2}}{(\delta + i/T_1)(\delta - \Delta + i/T_2)(\Delta + \delta + i/T_2) - \Omega^2(\delta + i/T_2)}.$$

$$(6.6.19)$$

The combination of terms that appears in the denominator of this expression appears repeatedly in the subsequent equations. For convenience we define the quantity

$$D(\delta) = \left(\delta + \frac{i}{T_1}\right)\left(\delta - \Delta + \frac{i}{T_2}\right)\left(\delta + \Delta + \frac{i}{T_2}\right) - \Omega^2\left(\delta + \frac{i}{T_2}\right),$$

$$(6.6.20)$$

so that Eq. (6.6.19) can be written as

$$w_1 = -2w_0|\mu_{ba}|^2 E_1 E_0^*\hbar^{-2}\frac{(\delta - \Delta + i/T_2)(\delta + 2i/T_2)}{(\Delta - i/T_2)D(\delta)}. \quad (6.6.21)$$

Note that w_1 (and consequently p_1 and p_{-1}) shows a resonance whenever the pump wave is tuned to line center so that $\Delta = 0$, or whenever a zero occurs in the function $D(\delta)$. We thus examine the resonance nature of the function $D(\delta)$. We first consider the limit $\Omega^2 \to 0$, that is, the $\chi^{(3)}$ perturbation theory limit. In this limit $D(\delta)$ is automatically factored into the product of three terms as

$$D(\delta) = \left(\delta + \frac{i}{T_1}\right)\left(\delta - \Delta + \frac{i}{T_2}\right)\left(\Delta + \delta + \frac{i}{T_2}\right), \quad (6.6.22)$$

and we see by inspection that zeros of $D(\delta)$ occur near

$$\delta = 0, \pm\Delta. \quad (6.6.23)$$

The positions of these frequencies are indicated in part (a) of Fig. 6.6.3. However, inspection of Eq. (6.6.21) shows that no resonance in w_1 occurs at $\delta = \Delta$, because the factor $\delta - \Delta + i/T_2$ in the numerator exactly cancels the same factor in the denominator. However, a resonance occurs *near* $\delta = \Delta$

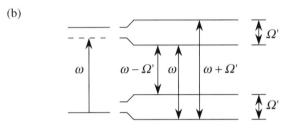

FIGURE 6.6.3   Resonances in the response of a two-level atom to pump and probe fields, as given by the function $D(\delta)$, (a) in the limit $\Omega^2 \to 0$, and (b) in the general case.

when the term containing Ω^2 in Eq. (6.6.20) is not ignored. $\chi^{(5)}$ is the lowest-order contribution to this resonance.

In the general case in which Ω^2 is not small, the full form of Eq. (6.6.20) must be used. In order to determine its resonance structure, we write $D(\delta)$ in terms of its real and imaginary parts as

$$D(\delta) = \delta\left(\delta^2 - \Omega'^2 - \frac{1}{T_2^2} - \frac{2}{T_1 T_2}\right)$$
$$+ i\left(\frac{\delta^2 - \Delta^2}{T_1} + \frac{2\delta^2}{T_2} - \frac{\Omega^2}{T_2} - \frac{1}{T_1 T_2^2}\right),$$
(6.6.24)

where we have introduced the detuned Rabi frequency $\Omega' = (\Omega^2 + \Delta^2)^{1/2}$. We see by inspection that the real part of D vanishes for

$$\delta = 0, \quad \delta = \pm\left(\Omega'^2 + \frac{1}{T_2^2} + \frac{2}{T_1 T_2}\right)^{1/2}. \tag{6.6.25}$$

If we now assume that $\Omega' T_2$ is much greater than unity, these three resonances will be well separated, and we can describe their properties separately. In this limit, the function $D(\delta)$ becomes

$$D(\delta) = \delta(\delta^2 - \Omega'^2) + i\left(\frac{\delta^2 - \Delta^2}{T_1} + \frac{2\delta^2 - \Omega^2}{T_2}\right), \tag{6.6.26}$$

and the three resonances occur at

$$\delta = 0, \pm\Omega'. \tag{6.6.27}$$

Near the resonance at $\delta = 0$, and $D(\delta)$ can be approximated as

$$D(\delta) = -\Omega'^2(\delta + i\Gamma_0), \tag{6.6.28a}$$

where

$$\Gamma_0 = \frac{\Delta^2/T_1 + \Omega^2/T_2}{\Delta^2 + \Omega^2} \tag{6.6.28b}$$

represents the width of this resonance. Likewise, near the resonances at $\delta = \mp\Omega'$, $D(\delta)$ can be approximated as

$$D(\delta) = 2\Omega'^2[(\delta \pm \Omega') + i\Gamma_\pm], \tag{6.6.29a}$$

where

$$\Gamma_\pm = \frac{\Omega^2/T_1 + (2\Delta^2 + \Omega^2)/T_2}{2(\Omega^2 + \Delta^2)} \tag{6.6.29b}$$

represents the width of these resonances. Note that the positions of these resonances can be understood in terms of the energies of the dressed atomic states, as illustrated in Fig. 6.6.3b. Note also that, for the case of weak optical excitation (i.e., for $\Omega^2 \ll \Delta^2$), Γ_0 approaches the population decay rate $1/T_1$, and $\Gamma_\pm$ approach the dipole dephasing rate $1/T_2$. In the limit of strong optical excitation (i.e., for $\Omega^2 \gg \Delta^2$), Γ_0 approaches the limit $1/T_2$ and $\Gamma_\pm$ approach the limit $\frac{1}{2}(1/T_1 + 1/T_2)$.

We next calculate the response of the atomic dipole at the sideband frequencies $\pm\delta$. We introduce the expression (6.6.19) for w_1 into Eq. (6.6.11) for p_1 and obtain

$$p_1 = \frac{\hbar^{-1}|\mu_{ba}|^2 w_0 E_1}{\Delta + \delta + i/T_2}$$

$$\times \left[1 - \frac{\frac{1}{2}\Omega^2(\delta - \Delta + i/T_2)(\delta + 2i/T_2)/(\Delta - i/T_2)}{(\delta + i/T_1)(\delta - \Delta + i/T_2)(\Delta + \delta + i/T_2) - \Omega^2(\delta + i/T_2)} \right].$$

$$\tag{6.6.30}$$

Written in this form, we see that the response at the probe frequency $\omega + \delta$ can be considered to be the sum of two contributions. The first is the result of the zero-frequency part of the population difference w. The second is the result of population oscillations. The first term is resonant only at $\delta = -\Delta$, whereas the second term contains the additional resonances associated with the function $D(\delta)$. Sargent (1978) has pointed out that the second term obeys the relation

$$
\int_{-\infty}^{\infty} \frac{-\frac{1}{2}\Omega^2 \dfrac{(\delta - \Delta + i/T_2)(\delta + 2i/T_2)}{\Delta - i/T_2}}{(\delta + i/T_1)(\delta - \Delta + i/T_2)(\Delta + \delta + i/T_2) - \Omega^2(\delta + i/T_2)}\, d\delta = 0.
$$

$$(6.6.31)$$

Thus, the second term, which results from population oscillations, does not modify the integrated absorption of the atom in the presence of a pump field; it simply leads to a spectral redistribution of probe-wave absorption.

A certain simplification of Eq. (6.6.30) can be obtained by combining the two terms algebraically so that p_1 can be expressed as

$$
p_1 = \frac{\hbar^{-1}|\mu_{ba}|^2 w_0 E_1}{D(\delta)}\left[\left(\delta + \frac{i}{T_1}\right)\left(\delta - \Delta + \frac{i}{T_2}\right) - \frac{1}{2}\Omega^2 \frac{\delta}{\Delta - i/T_2}\right].
$$

$$(6.6.32)$$

Finally, we calculate the response at the sideband opposite to the applied probe wave through use of Eqs. (6.6.12) and (6.6.21) and the fact that $w_{-1} = w_1^*$, as noted in the discussion following Eq. (6.6.18). We obtain the result

$$
p_{-1} = \frac{2w_0|\mu_{ba}|^4 E_0^2 E_1^* \dfrac{(\delta - \Delta - i/T_2)(-\delta + 2i/T_2)}{\Delta + i/T_2}}{\hbar^3(\Delta - \delta + i/T_2)D^*(\delta)}.
$$

$$(6.6.33)$$

Nonlinear Susceptibility and Coupled-Amplitude Equations

Let us now use these results to determine the forms of the nonlinear polarization and the nonlinear susceptibility. Since p_1 is the complex amplitude of the dipole moment at frequency $\omega + \delta$ induced by a probe wave at this frequency, the polarization at this frequency is $P(\omega + \delta) = Np_1$. If we set $P(\omega + \delta)$ equal to $\chi_{\text{eff}}^{(1)}(\omega + \delta)E_1$, we find that $\chi_{\text{eff}}^{(1)}(\omega + \delta) = Np_1/E_1$, or through use of Eq. (6.6.32) that

$$
\chi_{\text{eff}}^{(1)}(\omega + \delta) = \frac{N|\mu_{ba}|^2 w_0}{\hbar D(\delta)}\left[\left(\delta + \frac{i}{T_1}\right)\left(\delta - \Delta + \frac{i}{T_2}\right) - \frac{1}{2}\Omega^2 \frac{\delta}{\Delta - i/T_2}\right].
$$

$$(6.6.34)$$

We have called this quantity an effective linear susceptibility because it depends on the intensity of the pump wave. Similarly, the part of the nonlinear polarization oscillating at frequency $\omega - \delta$ is given by $P(\omega - \delta) = Np_{-1}$. If we set this quantity equal to $3\chi_{\text{eff}}^{(3)}[\omega - \delta = \omega + \omega - (\omega + \delta)]E_0^2 E_1^*$, we find through use of Eq. (6.6.33) that

$$\chi_{\text{eff}}^{(3)}[\omega - \delta = \omega + \omega - (\omega + \delta)]$$

$$= \frac{2N w_0 |\mu_{ba}|^4 \dfrac{(\delta - \Delta - i/T_2)(-\delta + 2i/T_2)}{\Delta + i/T_2}}{3\hbar^3 (\Delta - \delta + i/T_2) D^*(\delta)}. \qquad (6.6.35)$$

We have called this quantity an effective third-order susceptibility, because it too depends on the laser intensity.

The calculation just presented has assumed that E_1 (the field at frequency $\omega + \delta$) is the only weak wave that is present. However, for the geometry of Fig. 6.5.2, a weak wave at frequency $\omega - \delta$ is generated by the interaction, and the response of the medium to this wave must also be taken into consideration. If we let E_{-1} denote the complex amplitude of this new wave, we find that we can represent the total response of the medium through the equations

$$P(\omega + \delta) = \chi_{\text{eff}}^{(1)}(\omega + \delta)E_1 + 3\chi_{\text{eff}}^{(3)}[\omega + \delta = \omega + \omega - (\omega - \delta)]E_0^2 E_{-1}^*,$$

$$(6.6.36a)$$

$$P(\omega - \delta) = \chi_{\text{eff}}^{(1)}(\omega - \delta)E_{-1} + 3\chi_{\text{eff}}^{(3)}[\omega - \delta = \omega + \omega - (\omega + \delta)]E_0^2 E_1^*.$$

$$(6.6.36b)$$

Formulas for the new quantities $\chi_{\text{eff}}^{(3)}(\omega - \delta)$ and $\chi_{\text{eff}}^{(3)}[\omega + \delta = \omega + \omega - (\omega - \delta)]$ can be obtained by formally replacing δ by $-\delta$ in Eqs. (6.6.34) and (6.6.35).

The nonlinear response of the medium as described by Eqs. (6.6.36) will of course influence the propagation of the weak waves at frequencies $\omega \pm \delta$. We can describe the propagation of these waves by means of coupled-amplitude equations that we derive using methods described in Chapter 2. We introduce the slowly varying amplitudes $A_{\pm 1}$ of the weak waves by means of the equation

$$E_{\pm 1} = A_{\pm 1} e^{ik_{\pm 1}z}, \qquad (6.6.37a)$$

where the propagation constant is given by

$$k_{\pm 1} = n_{\pm 1}(\omega \pm \delta)/c. \qquad (6.6.37b)$$

Here $n_{\pm 1}$ is the real part of the refraction index experienced by each of the sidebands and is given by

$$n_{\pm 1}^2 = 1 + 4\pi \operatorname{Re} \chi_{\text{eff}}^{(1)}(\omega \pm \delta). \qquad (6.6.37c)$$

We now introduce the nonlinear polarization of Eqs. (6.6.36) and the field decomposition of Eq. (6.6.37a) into the wave equation in the form of Eq. (2.1.22), and assume the validity of the slowly-varying amplitude approximation.

We find that the slowly varying amplitudes must obey the set of coupled equations

$$\frac{dA_1}{dz} = -\alpha_1 A_1 + \kappa_1 A_{-1}^* e^{i \Delta k z} \tag{6.6.38a}$$

$$\frac{dA_{-1}}{dz} = -\alpha_{-1} A_{-1} + \kappa_{-1} A_1^* e^{i \Delta k z} \tag{6.6.38b}$$

where we have introduced the nonlinear absorption coefficients

$$\alpha_{\pm 1} = -2\pi \frac{\omega \pm \delta}{n_{\pm 1} c} \, \text{Im} \, \chi_{\text{eff}}^{(1)} (\omega \pm \delta), \tag{6.6.39a}$$

the nonlinear coupling coefficients

$$\kappa_{\pm 1} = -6\pi i \frac{\omega \pm \delta}{n_{\pm 1} c} \chi_{\text{eff}}^{(3)} [\omega \pm \delta = \omega + \omega - (\omega \mp \delta)] A_0^2, \tag{6.6.39b}$$

and the wavevector mismatch

$$\Delta k = 2k_0 - k_1 - k_{-1}, \tag{6.6.39c}$$

where k_0 is the magnitude of the wavevector of the pump wave.* The coupled wave equations given by Eqs. (6.6.38) can be solved explicitly for arbitrary boundary conditions. We shall not present the solution here; it is formally equivalent to the solution presented in Chapter 10 to the equations describing Stokes–anti-Stokes coupling in stimulated Raman scattering. The nature of the solution to Eqs. (6.6.38) for the case of a two-level atomic system has been described in detail by Boyd *et al.* (1981). These authors find that significant amplification of the A_1 and A_{-1} waves can occur in the near-forward direction as a consequence of the four-wave mixing processes described by Eqs. (6.6.38). They also find that the gain is particularly large when the detuning δ (or its negative $-\delta$) is approximately equal to the generalized Rabi frequency Ω'. These effects have been studied experimentally by Harter *et al.* (1981).

Let us consider the nature of the solutions of Eqs. (6.6.38) for the special case of the geometry shown in Fig. 6.6.1. For this geometry, because of the large angle θ between the pump and probe beams, the magnitude Δk of the

---

* We have arbitrarily placed the real part of $\chi_{\text{eff}}^{(1)}$ in $n_{\pm 1}$ and the imaginary part in $\alpha_{\pm 1}$. We could equivalently have placed all of $\chi_{\text{eff}}^{(1)}$ in a complex absorption coefficient $\alpha_{\pm 1}$ and set Δk equal to zero, or could have placed all of $\chi_{\text{eff}}^{(1)}$ in a complex refraction index $n_{\pm 1}$ and set $\alpha_{\pm 1}$ equal to zero. We have chosen the present convention because it illustrates most clearly the separate effects of absorption and of wavevector mismatch.

wavevector mismatch is very large, and as a result the coupled-amplitude equations (6.6.38a) and (6.6.38b) decouple into the two equations

$$\frac{dA_1}{dz} = -\alpha_1 A_1, \qquad \frac{dA_{-1}}{dz} = -\alpha_{-1} A_{-1}. \qquad (6.6.40)$$

Recall that $\alpha_{\pm 1}$ denotes the absorption coefficient experienced by the probe wave at frequency $\omega \pm \delta$, and that $\alpha_{\pm 1}$ depends on the probe–pump detuning δ, on the detuning Δ of the pump wave from the atomic resonance, and on the intensity I of the pump wave.

The dependence of α_1 on the probe–pump detuning δ is illustrated for one representative case in part (a) of Fig. 6.6.4. We see that three features appear in the probe absorption spectrum. One of these features is centered on the laser frequency, and the other two occur at the *Rabi sidebands* of the laser frequency, that is, they occur at frequencies detuned from the laser frequency by the generalized Rabi frequency $\Omega' = (\Omega^2 + \Delta^2)^{1/2}$ associated with the atomic response. Note that α_1 can become negative for two of these features;

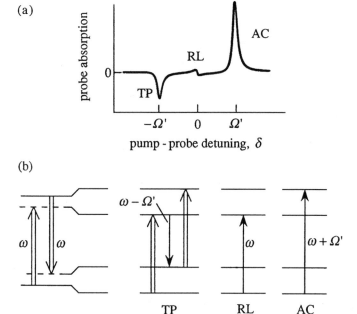

FIGURE 6.6.4   (a) Absorption spectrum of a probe wave in the presence of a strong pump wave for the case $\Delta T_2 = -3$, $\Omega T_2 = 8$, and $T_2/T_1 = 2$. (b) Each of the features in the spectrum shown in part (a) is identified by the corresponding transition between dressed states of the atom. TP denotes the three-photon resonance, RL denotes the Rayleigh resonance, and AC denotes the ac-Stark-shifted atomic resonance.

the gain associated with these features was predicted by Mollow (1972) and has been observed experimentally by Wu *et al.* (1977) and by Gruneisen *et al.* (1988, 1989). The gain feature that occurs near $\delta = 0$ can be considered to be a form of stimulated Rayleigh scattering (see also Chapter 9). The gain associated with these features has been utilized to construct optical parametric oscillators (Grandclement *et al.*, 1987).

Part (b) of Fig. 6.6.4 shows the origin of each of the features shown in part (a). The leftmost portion of this figure shows how the dressed states of the atom are related to the unperturbed atomic energy states. The next diagram, labeled TP, shows the origin of the three-photon resonance. Here the atom makes a transition from the lowest dressed level to the highest dressed level by the simultaneous absorption of two pump photons and the emission of a photon at the Rabi sideband frequency $\omega - \Omega'$. This process can amplify a wave at the Rabi sideband frequency, as indicated by the region of negative absorption labeled TP in part (a). The third diagram of part (b), labeled RL, shows the origin of the stimulated Rayleigh resonance. The Rayleigh resonance corresponds to a transition from the lower level of the lower doublet to the upper doublet. Each of these transitions is centered on the frequency of the pump laser. The final diagram of part (b) of the figure, labeled AC, corresponds to the usual absorptive resonance of the atom as modified by the ac Stark effect. For the sign of the detuning used in the diagram, the atomic absorption is shifted to higher frequencies. Note that this last feature can lead only to absorption, whereas the first two features can lead to amplification. The theory of optical wave mixing has been generalized by Agarwal and Boyd (1988) to treat the quantum nature of the optical field; this theory shows how quantum fluctuations can initiate the the four-wave mixing process described in this section.

Problems

1. *Alternative relaxation models.* Determine how the saturated absorption of an atomic transition depends on the intensity of the incident (monochromatic) laser field for the case of an open two-level atom and for a two-level atom with a non-radiatively coupled intermediate level, and compare these results with those derived in Section 6.3 for a closed two-level atom.

2. $\chi^{(3)}$ *for an impurity-doped solid.* One is often interested in determining the third-order susceptibility of a collection of two-level atoms contained in a medium of constant (i.e., wavelength-independent and non-intensity-dependent) refractive index n_0. Show that the third-order susceptibility of such a system is given by Eq. (6.3.36b) in the form shown, or by Eq. (6.3.33b) with a factor of n_0 introduced in the numerator, or by Eq. (6.3.34a) or (6.3.34b) with a factor of n_0^2 introduced in the numerator. In cases in which I_s^o, I_s^Δ, or $\alpha_0(\Delta)$ appears in the expression, it is to be understood that the expressions (6.3.30) and (6.3.31) for I_s^o and I_s^Δ should each be multiplied by a factor of n_0 and the expression (6.3.22b) for $\alpha_0(0)$ should be divided by a factor of n_0.

3. *Orthogonality of dressed states.* Verify Eq. (6.5.43).

4. *Damping of Rabi oscillations.* The intent of this problem is to determine the influence of T_1- and T_2-type relaxation processes on Rabi oscillations of the sort predicted by the solution to the Schrödinger equation for an atom in the presence of an intense, near-resonant driving field. In particular, you are to solve the Bloch equation in the form of Eqs. (6.5.49) for the time evolution of an atom known to be in the ground state at time $t = 0$ and subject to a field $Ee^{-i\omega t} +$ c.c. that is turned on at time $t = 0$. In addition, sketch the behavior of w and of p as functions of time.

[Hint: At a certain point in the calculation, the mathematical complexity will be markedly reduced by assuming that $T_1 = T_2$. Make this simplification only when it becomes necessary.]

5. *Response times.* Consider the question of estimating the response time of non-resonant electronic nonlinearities of the sort described in Section 4.3. Student A argues that it is well known that the response time under such conditions is of the order of the reciprocal of the detuning of the laser field from the nearest atomic resonance. Student B argues that only relaxation processes can allow a system to enter the steady state, and that consequently the response time is of the order of the longer of T_1 and T_2, that is, is of the order of T_1. Who is right, and in what sense is each of them correct?

[Hint: Consider how the graph shown in Fig. 6.5.6 and the analogous graph of $p(t)$ would look in the limit of $\Delta \gg \Omega$, $\Delta T_2 \gg 1$.] [Partial answer: The nonlinearity turns on in a time Δ^{-1} but does not reach its steady-state value until a time of the order of T_1.]

6. *Identity pertaining to population oscillations.* Verify Eq. (6.6.31).

References

G. S. Agarwal and R. W. Boyd, *Phys. Rev. A* **38**, 4019 (1988).

L. D. Allen and J. H. Eberly, *Optical Resonance and Two-Level Atoms*, Wiley, New York, 1975.

S. H. Autler and C. H. Townes, *Phys. Rev.* **100**, 703 (1955).

R. W. Boyd, M. G. Raymer, P. Narum, and D. J. Harter, *Phys. Rev.* **A24**, 411 (1981).

R. W. Boyd and M. Sargent III, *J. Opt. Soc. Am. B*, **5**, 1 (1988).

C. Cohen-Tannoudji and S. Reynaud, *J. Phys.* **B10**, (1977); **10**, 365 (1977); **10**, 2311 (1977).

C. Cohen-Tannoudji, J. Dupont-Roc, and G. Grynberg, *Photons and Atoms*, Wiley, New York, 1989; *Atom–Photon Interactions*, Wiley, New York, 1991.

R. H. Dicke and J. P. Wittke, *Introduction to Quantum Mechanics*, Addison-Wesley, Reading, MA, 1960.

D. Grandclement, G. Grynberg, and M. Pinard, *Phys. Rev. Lett.*, **59**, 44 (1987); see also D. Grandclement, D. Pinard, and G. Grynberg, *IEEE J. Quantum Electron*, **25**, 580 (1989).

D. Grischkowsky, *Phys. Rev. Lett.* **24**, 866 (1970); see also D. Grischkowsky and J. A. Armstrong, *Phys. Rev.* **A6**, 1566 (1972); D. Grischkowsky, *Phys. Rev.* **A7**, 2096 (1973); D. Grischkowsky, E. Courtens, and J. A. Armstrong, *Phys. Rev. Lett.* **31**, 422 (1973).

M. T. Gruneisen, K. R. MacDonald, and R. W. Boyd, *J. Opt. Soc. Am.* **B5**, 123 (1988); M. T. Gruneisen, K. R. MacDonald, A. L. Gaeta, R. W. Boyd, and D. J. Harter, *Phys. Rev.* **A40**, 3464 (1989).

D. J. Harter, P. Narum, M. G. Raymer, and R. W. Boyd, *Phys. Rev. Lett.* **46**, 1192 (1981).

R. B. Miles and S. E. Harris, *IEEE J. Quantum Electron.* **9**, 470 (1973).

B. R. Mollow, *Phys. Rev.* **A5**, 2217 (1972).

M. Sargent III, *Phys. Rep.* **43**, 223 (1978).

M. Sargent III, M. O. Scully, and W. E. Lamb, Jr., *Laser Physics*, Addison-Wesley, Reading, MA, 1974.

F. Y. Wu, S. Ezekiel, M. Ducloy, and B. R. Mollow, *Phys. Rev. Lett.* **38**, 1077 (1977).

Chapter 7

Processes Resulting from the Intensity-Dependent Refractive Index

In this chapter, we explore several processes that occur as a result of the nonlinear refractive index.

7.1. Self-Focusing of Light and Other Self-Action Effects

Self-focusing of light is the process in which an intense beam of light modifies the optical properties of a material medium in such a manner that the beam is caused to come to a focus within the material. This circumstance is shown schematically in Fig. 7.1.1a. Here we have assumed that n_2 is positive. As a result, the laser beam induces a refractive index variation within the material with a larger refractive index at the center of the beam than at its periphery. Thus the material acts as if it were a positive lens, causing the beam to come to a focus within the material. More generally, one refers to self-action effects as effects in which a beam of light modifies its own propagation by means of the nonlinear response of a material medium.

Another self-action effects is the self-trapping of light, which is illustrated in Fig. 7.1.1b. In this process a beam of light propagates with a constant diameter as a consequence of an exact balance between self-focusing and diffraction effects. An analysis of this circumstance, which is presented below, shows that self-trapping can occur only if the power carried by the beam is exactly equal to the so-called critical power for self-trapping

$$P_{\mathrm{cr}} = \frac{\pi (0.61)^2 \lambda_0^2}{8 n_0 n_2},$$

(7.1.1)

311

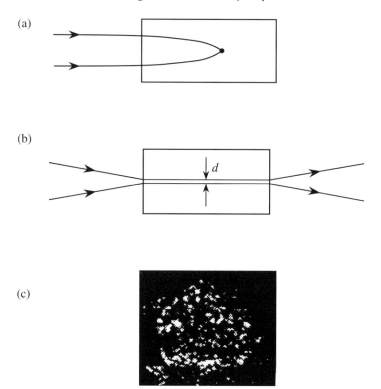

FIGURE 7.1.1   Schematic illustration of three self-action effects: (a) self-focusing of light, (b) self-trapping of light, and (c) laser beam filamentation, showing the transverse distribution of intensity of a beam that has undergone filamentation.

where λ_0 is the vacuum wavelength of the laser radiation. This line of reasoning leads to the conclusion that self-focusing can occur only if the beam power P is greater than P_{cr}.

The final self-action effect shown in Fig. 7.1.1 as part (c) is laser beam filamentation. This process occurs only for $P \gg P_{cr}$ and leads to the breakup of the beam into many components each carrying approximately power P_{cr}. This process occurs as a consequence of the growth of imperfections of the laser wavefront by means of forward four-wave mixing amplification.

Let us begin our analysis of self-action effects by developing a simple model of the self-focusing process. For the present, we ignore the effects of diffraction; these effects are introduced below. The neglect of diffraction is justified if the beam diameter and/or beam intensity is sufficiently large. Figure 7.1.2 shows a collimated beam of light of characteristic radius w_0 and an on-axis intensity I_0 falling onto a nonlinear optical material for which n_2 is positive. We describe

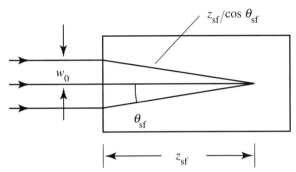

FIGURE 7.1.2 Prediction of the self-focusing distance z_{sf} by means of Fermat's principle. The curved ray trajectories within the nonlinear material are approximated as straight lines.

the location of the self-focus through use of Fermat's principle, which states that the optical path length $\int n(\mathbf{r})\,dl$ of all rays traveling from a wavefront at the input face to the self-focus must be equal. As a first estimate, we take the refractive index along the marginal ray to be the linear refractive index n_0 of the medium and the refractive index along the central ray to be $n_0 + n_2 I_0$. Fermat's principle then tells us that

$$(n_0 + n_2 I)\, z_{sf} = n_0 z_{sf} / \cos\theta_{sf}. \tag{7.1.2}$$

If we approximate $\cos\theta_{sf}$ as $1 - \frac{1}{2}\theta_{sf}^2$ and solve the resulting expression for θ_{sf}, we find that

$$\theta_{sf} = \sqrt{2 n_2 I / n_0}. \tag{7.1.3}$$

This quantity is known as the self-focusing angle and in general can be interpreted as the characteristic angle through which a beam of light is deviated as a consequence of self-action effects. The ratio $n_2 I / n_0$ of nonlinear to linear refractive index is invariably a very small quantity. As the self-focusing angle varies as the square root of this quantity, it can be large enough to be appreciable. In terms of the self-focusing angle, we can calculate the characteristic self-focusing distance as $z_{sf} = w_0 / \theta_{sf}$ or as

$$z_{sf} = w_0 \sqrt{\frac{n_0}{2 n_2 I}} = \frac{2 n_0 w_0^2}{\lambda_0} \frac{1}{\sqrt{P/P_{cr}}} \quad (P \gg P_{cr}), \tag{7.1.4}$$

where in writing the result in the second form we have made use of expression (7.1.1).

The derivation leading to the result given by Eq. (7.1.4) ignores the effects of diffraction, and thus might be expected to be valid when self-action effects overwhelm those of diffraction, that is, for $P \gg P_{cr}$. For smaller laser powers, the self-focusing distance can be estimated by arguing that the beam convergence angle is reduced by diffraction effects and is given by $\theta = (\theta_{sf}^2 - \theta_{dif}^2)^{1/2}$, where

$$\theta_{dif} = 0.61\lambda_0/n_0 d \qquad (7.1.5)$$

is the diffraction angle of a beam of diameter d and vacuum wavelength λ_0. Then, once again arguing that $z_{sf} = w_0/\theta$, we find that

$$z_{sf} = \frac{2nw_0^2}{\lambda_0} \frac{1}{\sqrt{P/P_{cr} - 1}} \qquad (7.1.6)$$

Yariv (1975) has shown that for the still more general case in which the beam has arbitrary power and arbitrary beam-waist position, the distance from the entrance face to the position of the self-focus is given by the formula

$$z_{sf} = \frac{\frac{1}{2}kw^2}{(P/P_{cr} - 1)^{1/2} + 2z_{min}/kw_0^2}, \qquad (7.1.7)$$

where $k = n_0\omega/c$. The beam radius parameters w and w_0 (which have their conventional meanings) and z_{min} are defined in Fig. 7.1.3.

Self-Trapping of Light

Let us next consider the conditions under which self-trapping of light can occur. One expects self-trapping to occur when the tendency of a beam to spread as a consequence of diffraction is precisely balanced by the tendency of the beam to contract as a consequence of self-focusing effects. The condition for self-trapping can thus be expressed mathematically as a statement that the diffraction angle of Eq. (7.1.5) be equal to the self-focusing angle of Eq. (7.1.3), that is, that

$$\theta_{dif} = \theta_{sf}. \qquad (7.1.8)$$

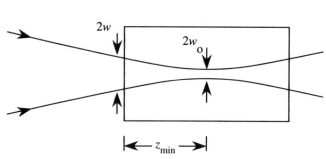

FIGURE 7.1.3   Definition of the parameters w, w_0, and z_{min}. The rays are shown as unmodified by the nonlinear interaction.

By introducing Eqs. (7.1.3) and (7.1.5) into this equality, we find that self-trapping will occur only if the intensity of the light within the filament is given by

$$I = \frac{(0.61)^2 \lambda_0^2}{2 n_2 n_0 d^2}. \qquad (7.1.9)$$

Since the power contained in such a filament is given by $P = (\pi/4) d^2 I$, we also see that self-trapping occurs only if the power contained in the beam has the critical value

$$P_{\text{cr}} = \frac{\pi (0.61)^2 \lambda_0^2}{8 n_0 n_2} \approx \frac{\lambda_0^2}{8 n_0 n_2}. \qquad (7.1.10)$$

This result was stated above (Eq. (7.1.1)) without proof. Note that according to the present model, a self-trapped filament can have any beam diameter d, and that for any value of d the power contained in the filament has the same value, given by Eq. (7.1.10). The value of the numerical coefficient appearing in this formula depends on the detailed assumptions of the mathematical model of self-focusing; this point has been discussed in detail by Fibich and Gaeta (2000).

The process of laser beam self trapping can be described perhaps more physically in terms of an argument presented by Chiao *et al.* (1964). One makes the simplifying assumption that the laser beam has a flat-top intensity distribution, as shown in Fig. 7.1.4a. The refractive index distribution within the nonlinear medium then has the form shown in part (b) of the figure, which shows a cut

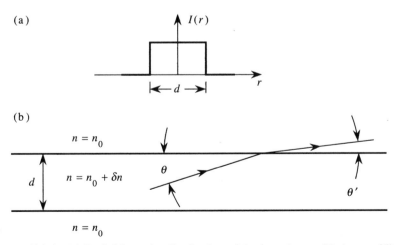

FIGURE 7.1.4   (a) Radial intensity distribution of the laser beam. (b) A ray of light incident on the boundary formed by the edge of the laser beam.

316 7 ◊ Processes Resulting from the Intensity-Dependent Refractive Index

through the medium that includes the symmetry axis of the laser beam. Here the refractive index of the bulk of the material is denoted by n_0 and the refractive index of that part of the medium exposed to the laser beam is denoted by $n_0 + \delta n$, where δn is the nonlinear contribution to the refractive index. Also, shown in part (b) of the figure is a ray of light incident on the boundary between the two regions. It is one ray of the bundle that makes up the laser beam. This ray will remain trapped within the laser beam if it undergoes total internal reflection at the boundary between the two regions. Total internal reflection occurs if θ is less than the critical angle θ_0 for total internal reflection, which is given by the equation

$$\cos\theta_0 = \frac{n_0}{n_0 + \delta n}. \tag{7.1.11}$$

Since δn is very much smaller than n_0 for most nonlinear optical materials, and consequently θ_0 is much smaller than unity, Eq. (7.1.11) can be approximated by

$$1 - \tfrac{1}{2}\theta_0^2 = 1 - \frac{\delta n}{n_0},$$

which shows that the critical angle is related to the nonlinear change in refractive index by

$$\theta_0 = (2\delta n/n_0)^{1/2}. \tag{7.1.12}$$

A laser beam of diameter d will contain rays within a cone whose maximum angular extent is of the order of magnitude of characteristic diffraction angle $\theta_{\text{dif}} = 0.61\lambda_0/n_0 d$, where λ_0 is the wavelength of the light in vacuum. We expect that self-trapping will occur if total internal reflection occurs for all of the rays contained within the beam, that is, if $\theta_{\text{dif}} = \theta_0$. By comparing Eqs. (7.1.12) and (7.1.5), we see that self-trapping will occur if

$$\delta n = \tfrac{1}{2}n_0(0.61\lambda_0/dn_0)^2, \tag{7.1.13a}$$

or equivalently, if

$$d = 0.61\lambda_0(2n_0\delta n)^{-1/2}. \tag{7.1.13b}$$

If we now replace δn by $n_2 I$, we see that the diameter of a self-trapped filament is related to the intensity of the light within the filament by

$$d = 0.61\lambda_0(2n_0 n_2 I)^{-1/2}. \tag{7.1.14}$$

The power contained in a filament whose diameter is given by Eq. (7.1.14) is as before given by

$$P_{cr} = \frac{\pi}{4}d^2 I = \frac{\pi (0.61)^2 \lambda_0^2}{8 n_0 n_2}. \tag{7.1.15}$$

Note that the *power*, not the *intensity*, of the laser beam is crucial in determining whether self-focusing will occur.

When the power P greatly exceeds the critical power P_{cr} and self-focusing does occur, the beam will usually break up into several filaments, each of which contains power P_{cr}. The theory of filament formation has been described by Bespalov and Talanov (1966) and is described more fully in a following subsection.

It is instructive to determine the numerical values of the various physical quantities introdcued in this section. For carbon disulfide (CS_2), n_2 for linearly polarized light is equal to 3.2×10^{-14} cm^2/W, n_0 is equal to 1.7, and P_{cr} at a wavelength of 1 μm is equal to 27 kW. For typical crystals and glasses, n_2 is in the range 5×10^{-16} to 5×10^{-15} cm^2/W and P_{cr} is in the range 0.2 to 2 MW. We can also estimate the self-focusing distance of Eq. (7.1.4). A fairly modest Q-switched Nd:YAG laser operating at a wavelength of 1.06 μm might produce an output pulse containing 10 mJ of energy with a pulse duration of 10 nsec, and hence with a peak power of the order of 1 MW. If we take w_0 equal to 100 μm, Eq. (7.1.4) predicts that $z_{sf} = 1$ cm for carbon disulfide.

Mathematical Description of Self-Action Effects

The treatment of self-action effects just presented has been of a somewhat qualitative nature. Self-action effects can be described more rigorously by means of the nonlinear optical wave equation.

For the present we consider steady-state conditions only, as would apply for excitation with a continuous-wave laser beam. The paraxial wave equation under these conditions is given by

$$2 i k_0 \frac{\partial A}{\partial z} + \nabla_T^2 A = -4\pi \frac{\omega^2}{c^2} P_{NL}, \tag{7.1.16}$$

where for a purely third-order nonlinear optical response the amplitude of the nonlinear polarization is given by

$$P_{NL} = 3\chi^{(3)} |A|^2 A. \tag{7.1.17}$$

Steady-state self trapping can be described by these equations.

We consider first the solution of Eqs. (7.1.16) and (7.1.17) for a beam that is allowed to vary in only one transverse dimension. Such a situation could be realized experimentally by studying self-action effects of light constrained to propagate in a planar waveguide. In this case these equations reduce to

$$2ik_0\frac{\partial A}{\partial z} + \frac{\partial^2 A}{\partial x^2} = -12\pi\chi^{(3)}\frac{\omega^2}{c^2}|A|^2A \qquad (7.1.18)$$

where A is now a function of x and z only. This equation possesses a solution of the form

$$A(x,z) = A_0\,\mathrm{sech}(x/x_0)e^{i\gamma z}, \qquad (7.1.19)$$

where the width of the field distribution is given by

$$x_0 = \frac{1}{k_0}\sqrt{n_0/2\bar{n}_2|A_0|^2} \qquad (7.1.20)$$

and the rate of nonlinear phase acquisition is given by

$$\gamma = k_0\bar{n}_2|A_0|^2/n_0, \qquad (7.1.21)$$

where, as in Section 4.1, $\bar{n}_2 = 3\pi\chi^{(3)}/n_0$. The solution given by Eq. (7.1.19) is sometimes referred to as a spatial soliton, because it describes a field that can propagate for long distances with an invariant transverse profile. Behavior of this sort has been observed experimentally by Barthelemy *et al.* (1985) and by Aitchison *et al.* (1991).

For a beam that varies in both transverse directions, Eqs. (7.1.16) and (7.1.17) cannot be solved analytically and only numerical results are known. The lowest-order solution for a beam with cylindrical symmetry was reported by Chiao *et al.* (1964) and is of the form of a bell-shaped curve of approximately gaussian shape. Detailed analysis shows that in two transverse dimensions spatial solitons are unstable in a pure Kerr medium (that is, one described by an $\bar{n}_2$ nonlinearity), but that they can propagate stably in a saturable nonlinear medium. Such behavior has been observed experimentally by Bjorkholm and Ashkin (1974). Higher-order solutions have been reported by Haus (1966).

Laser Beam Filamentation

We mentioned above that filamentation occurs as a consequence of the growth by forward four-wave-mixing amplification of irregularities initially present on the laser wavefront. This occurrence is illustrated schematically in Fig. 7.1.5. Filamentation typically leads to the generation of a beam with a random intensity distribution, of the sort shown in part (c) of Fig. 7.1.1. However, under certain circumstances, the filamentation process can produce beams with a transverse structure in the form of highly regular geometrical patters; see for instance Bennink *et al.* (2002).

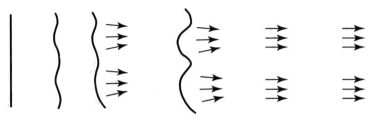

FIGURE 7.1.5   Illustration of laser beam filamentation by the growth of wavefront perturbations.

Let us now present a mathematical description of the process of laser beam filamentation. Our derivation follows closely that of the original description of Bespalov and Talanov (1966). We begin by expressing the field within the nonlinear medium as (see also Fig. 7.1.6)

$$\tilde{E}(\mathbf{r}, t) = E(\mathbf{r})e^{-i\omega t} + \text{c.c.}, \tag{7.1.22}$$

where it is convenient to express the electric field amplitude as the sum of three plane-wave components as

$$E(\mathbf{r}) = E_0(\mathbf{r}) + E_1(\mathbf{r}) + E_{-1}(\mathbf{r}) = [A_0(z) + A_1(\mathbf{r}) + A_{-1}(\mathbf{r})]\, e^{ikz} \tag{7.1.23}$$
$$= [A_0(z) + a_1(z)e^{i\mathbf{q}\cdot\mathbf{r}} + a_{-1}(z)e^{-i\mathbf{q}\cdot\mathbf{r}}]\, e^{ikz},$$

where $k = n_0\omega/c$. Also E_0 represents the strong central component of the laser field and E_1 and E_{-1} represent weak, symmetrically displaced spatial sidemodes; at various points in the calculation it will prove useful to introduce the related quantities A_0, $A_{\pm 1}$ and $a_{\pm 1}$. The latter quantities are defined in relation to the transverse component $\mathbf{q}$ of the optical wavevector of the off-axis modes. We next calculate the nonlinear polarization in the usual manner:

$$P = 3\chi^{(3)}|E|^2 E \equiv P_0 + P_1 + P_{-1}, \tag{7.1.24}$$

(a)   (b)

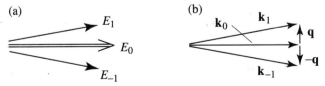

FIGURE 7.1.6   (a) Filamentation occurs by the growth of the spatial sidemodes E_1 and E_{-1} at the expense of the strong central component E_0. (b) Wavevectors of the interacting waves.

where the part of the polarization that is phase matched to the strong central component is given by

$$P_0 = 3\chi^{(3)}|E_0|^2 E_0 = 3\chi^{(3)}|A_0|^2 A_0 \, e^{ikz} \equiv p_0 e^{ikz}, \qquad (7.1.25)$$

and where the part of the polarization that is phase matched to the sidemodes is given by

$$P_{\pm 1} = 3\chi^{(3)} \left(2|E_0|^2 E_{\pm 1} + E_0^2 E_{\mp 1}^* \right) \equiv p_{\pm 1} e^{ikz}. \qquad (7.1.26)$$

Let us first solve the wave equation for the spatial evolution of A_0, which is given by

$$2ik \frac{\partial A_0}{\partial z} + \nabla_\perp^2 A_0 = -\frac{4\pi\omega^2}{c^2} p_0. \qquad (7.1.27)$$

The solution of this equation is

$$A_0(z) = A_{00} e^{i\gamma z}, \qquad (7.1.28)$$

where

$$\gamma = \frac{6\pi\omega\chi^{(3)}}{n_0 c} |A_{00}|^2 = n_2 k_{\text{vac}} I \qquad (7.1.29)$$

denotes the spatial rate of nonlinear phase acquisition and where for simplicity, and without loss of generality, we assume that A_{00} is a real quantity. This solution expresses the expected result that the strong central component simply acquires a nonlinear phase shift as it propagates. We now use this result with Eq. (7.1.26) to find that the part of the nonlinear polarization which couples to the sidemodes is given by

$$p_{\pm 1} = 3\chi^{(3)} \left[2|A_{00}|^2 A_{\pm 1} + A_{00}^2 e^{2i\gamma z} A_{\mp 1}^* \right]. \qquad (7.1.30)$$

We next consider the wave equation for the off-axis modes. Starting with

$$2ik \frac{\partial A_{\pm 1}}{\partial z} + \nabla_\perp^2 A_{\pm 1} = -\frac{4\pi\omega^2}{c^2} p_{\pm 1}, \qquad (7.1.31)$$

we introduce $A_{\pm 1} = a_{\pm 1} \exp(\pm i\mathbf{q} \cdot \mathbf{r})$ and expression (7.1.30) for $P_{\pm 1}$ to obtain

$$2ik \frac{\partial a_{\pm 1}}{\partial z} - q^2 a_{\pm 1} = -\frac{4\pi\omega^2}{c^2} 3\chi^{(3)} |A_{00}|^2 [2a_{\pm 1} + a_{\mp 1}^* e^{2i\gamma z}]. \quad (7.1.32)$$

This equation is now rearranged, and expression (7.1.29) for γ is introduced to obtain

$$\frac{da_{\pm 1}}{dz} + \frac{iq^2}{2k} a_{\pm 1} = i\gamma (2a_{\pm 1} + a_{\mp 1}^* e^{2i\gamma z}) \qquad (7.1.33)$$

We next perform a change of variables to remove the unwanted exponential phase factor appearing in the last term in this equation. In particular, we define

$$a_{\pm 1} = a'_{\pm 1} e^{i\gamma z}. \tag{7.1.34}$$

In terms of the new "primed" variables, Eq. (7.1.33) becomes

$$\frac{d}{dz} a'_{\pm 1} = i(\gamma - q^2/2k) a'_{\pm 1} + i\gamma a'^{*}_{\mp 1}. \tag{7.1.35}$$

This set of equations now possesses constant coefficients and can be solved directly. Perhaps the simplest way to solve these equations is to express them in matrix form as

$$\frac{d}{dz} \begin{bmatrix} a'_1 \\ a'^{*}_{-1} \end{bmatrix} = \begin{bmatrix} i(\gamma - \beta) & i\gamma \\ -i\gamma & -i(\gamma - \beta) \end{bmatrix} \begin{bmatrix} a'_1 \\ a'^{*}_{-1} \end{bmatrix}, \tag{7.1.36}$$

where $\beta \equiv q^2/2k$. We seek the eigensolutions of this equation, that is, solutions of the form

$$\begin{bmatrix} a'_1(z) \\ a'^{*}_{-1}(z) \end{bmatrix} = \begin{bmatrix} a'_1(0) \\ a'^{*}_{-1}(0) \end{bmatrix} e^{\Lambda z}. \tag{7.1.37}$$

This assumed solution is substituted into Eq. (7.1.36) which then becomes

$$\begin{bmatrix} i(\gamma - \beta) - \Lambda & i\gamma \\ -i\gamma & -i(\gamma - \beta) - \Lambda \end{bmatrix} \begin{bmatrix} a'_1(0) \\ a'^{*}_{-1}(0) \end{bmatrix} = 0. \tag{7.1.38}$$

This equation possesses nonvanishing solutions only if the determinant of the two-by-two matrix appearing in this equation vanishes. This condition leads to the result that

$$\Lambda = \pm\sqrt{\beta(2\gamma - \beta)}. \tag{7.1.39}$$

Note that this system of equations can produce gain (Re $\Lambda > 0$) only for $\gamma > \frac{1}{2}\beta$, which shows immediately that n_2 must be positive in order for filamentation to occur. More explicitly, Fig. 7.1.7 shows a plot of the forward-four-wave-mixing gain coefficient Λ as a function of the transverse wavevector magnitude q. We see that the maximum gain is numerically equal to the non-linear phase shift γ experienced by the pump wave. We also see that the gain vanishes for all values of q greater than $q_{max} = 2\sqrt{k\gamma}$ and reaches its maximum value for wavevector $q_{opt} = q_{max}/\sqrt{2}$. There is consequently a characteristic angle at which the filamentation process occurs, which is given by

$$\theta_{opt} = q_{opt}/k. \tag{7.1.40}$$

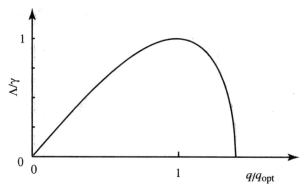

FIGURE 7.1.7   Variation of the gain coefficient Λ of the forward four-wave mixing process that leads to laser-beam filamentation with transverse wavevector magnitude q.

This angle has a direct physical interpretation, as described originally by Chiao *et al.* (1966). In particular θ_{opt} is the direction in which the near-forward four-wave-mixing process becomes phase matched, when account is taken of the nonlinear contributions to the wavevectors of the on- and off-axis waves.

It is extremely instructive to calculate the characteristic power carried by each of the filaments created by the filamentation process. This power P_{fil} is of the order of the initial intensity I of the laser beam times the characteristic cross sectional area of one of the filaments. If we identify this area with the square of the characteristic transverse distance scale associated with the filamentation process, that is, with $w_{\text{eff}}^2 = (\pi/q)^2$, we find that

$$P_{\text{fil}} = \frac{\lambda^2}{8n_0 n_2}, \qquad (7.1.41)$$

which is of the same order of magnitude as the critical power for self focusing P_{cr} introduced in Eq. (7.1.1). We thus see that the filamentation process is one in which the laser beam breaks up into a large number of individual components, each of which carries power of the order of P_{cr}.

Conditions for the Occurrence of Filamentation.   Let us next determine the conditions under which laser filamentation is expected to occur. This is actually a quite subtle question, for at least two reasons. First, filamentation grows from perturbation initially present on the laser wavefront, and thus a very clear laser beam will have a much higher filamentation threshold than a "dirty" beam. Second, whereas the gain of the filamentation process depends directly on laser intensity, the properties of whole-beam self-focusing depend on the intensity and beam spot size in a more complicated manner.

To address this question, let us define a filamentation distance z_{fil} by $n_2 k_{\text{vac}} I z_{\text{fil}} = G$, where G is a numerical factor (of the order of 3 to 10) that quantifies the level of gain that must be present in order for filamentation to occur. Clearly

$$z_{\text{fil}} = \frac{G}{n_2 k_{\text{vac}} I}. \tag{7.1.42}$$

This distance is to be compared to the self-focusing distance

$$z_{\text{sf}} = \frac{2 n_0 w_0^2}{\lambda_0} \frac{1}{\sqrt{P/P_{\text{cr}} - 1}} \tag{7.1.43}$$

(see Eq. (7.1.4) derived earlier), with $P_{\text{cr}} = \pi (0.61)^2 \lambda_0^2 / 8 n_0 n_2$. The condition for the occurrence of filamentation then can be stated as $z_{\text{fil}} < L$, where L is the interaction path length, and $z_{\text{fil}} < z_{\text{sf}}$. These conditions state that the filament process must occur within the length of the interaction region, and that the competing process of whole-beam self-focusing does *not* occur. Note that z_{fil} decreases more rapidly with increasing laser power (or intensity) than does z_{sf}, and thus filamentation can always be induced through use of a sufficiently large laser power. Let us calculate the value of the laser power under conditions such that z_{fil} is exactly equal to z_{sf}. We find, using Eq. (7.1.43) in the limit $P \gg P_{\text{cr}}$, that

$$P/P_{\text{cr}} = 4G^2. \tag{7.1.44}$$

For the representative value $G = 5$, we find that filamentation is expected only for

$$P > 100 P_{cr}. \tag{7.1.45}$$

Self-Action Effects with Pulsed Laser Beams

For simplicity, the preceding discussion has dealt with continuous-wave laser beams. Self-action effects are quite different in character when excited using pulsed radiation. Here some general comments are presented. A more detailed account of self-action effects when excited by ultrashort optical pulses is presented in a later chapter.

Moving Focus Model.   The moving focus model was developed by Loy and Shen (1973) to describe the properties of self-focusing when excited with nanosecond laser pulses. To understand this model, one notes that for pulsed radiation, the self-focusing distance z_{sf} of Eq. (7.1.4) (that is, the distance from the input face of the nonlinear medium to the self-focus point) will vary according to the value of the instantaneous intensity $I(t)$ at the input face. Thus the focal point will sweep through the material as it follows the temporal evolution of the pulse intensity. Under many circumstances, damage will occur at the point of peak intensity, and thus the damage tracks observed by early works (Hercher, 1964) can be interpreted as the locus of focal points for all values of the input intensity $I(t)$. Some aspects of the moving focus model are quite subtle. For instance, because of transit time effects, there are typically two self-focal points within the material at any given time. One of these occurs closer to the entrance face of the material and is a consequence of intense light near the peak of the pulse, whereas another focus occurs at greater distances into the material and occurs as a consequence of earlier, weaker parts of the pulse.

Transient Self-Focusing.   Transient self-focusing occurs when the laser pulse duration τ_p is comparable to or shorter than the turn-on time of the material response. In this situation, the nonlinear response develops during the time extent of the laser pulse, and consequently the nonlinear response is stronger for the trailing edge of the pulse than for the leading edge. Thus the trailing edge is more strongly self-focused than is the leading edge, leading to significant distortion of the pulse intensity distribtion in both space and time. This process has been described in detail by Shen (1975). Transient self-focusing can be observed through use of picosecond laser pulses propagating through liquids in which the dominant nonlinearity is the molecular orientation effect.

7.2.   Optical Phase Conjugation

Optical phase conjugation is a process that can be used to remove effects of aberrations from certain types of optical systems. The nature of the phase conjugation process is illustrated in Fig. 7.2.1. Part (a) of the figure shows an optical wave falling at normal incidence onto an ordinary metallic mirror. We see that the most advanced portion of the incident wavefront remains the most advanced after reflection has occurred. Part (b) of the figure shows the same wavefront falling onto a phase-conjugate mirror. In this case the most advanced portion turns into the most retarded portion in the reflection process. For this reason, optical phase conjugation is sometimes referred to as wavefront reversal. Note, however, that the wavefront is reversed only with respect to normal geometrical reflection; in fact the generated wavefront exactly replicates the incident wavefront but propagates in the opposite direction. For this reason, optical phase conjugation is sometimes referred to as the generation of a time-reversed wavefront.

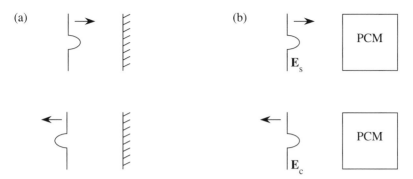

FIGURE 7.2.1 Reflection from (a) an ordinary mirror and (b) a phase-conjugate mirror.

The reason why the process illustrated in part (b) of Fig. 7.2.1 is called phase conjugation can be understood by introducing a mathematical description of the process. We represent the wave incident on the phase-conjugate mirror (called the signal wave) as

$$\tilde{\mathbf{E}}_s(\mathbf{r}, t) = \mathbf{E}_s(\mathbf{r})e^{-i\omega t} + \text{c.c.} \qquad (7.2.1)$$

When illuminated by such a wave, a phase-conjugate mirror produces a reflected wave, called the phase-conjugate wave, described by

$$\tilde{\mathbf{E}}_c(\mathbf{r}, t) = r\mathbf{E}_s^*(\mathbf{r})e^{-i\omega t} + \text{c.c.}, \qquad (7.2.2)$$

where r represents the amplitude reflection coefficient of the mirror. In order to determine the significance of replacing $\mathbf{E}_s(\mathbf{r})$ by $\mathbf{E}_s^*(\mathbf{r})$ in the reflection process, it is useful to represent $\mathbf{E}_s(\mathbf{r})$ as the product

$$\mathbf{E}_s(\mathbf{r}) = \hat{\epsilon}_s A_s(\mathbf{r})e^{i\mathbf{k_s}\cdot\mathbf{r}}, \qquad (7.2.3)$$

where $\hat{\epsilon}_s$ represents the polarization unit vector, $A_s(\mathbf{r})$ the slowly varying field amplitude, and $\mathbf{k_s}$ the mean wavevector of the incident light. The complex conjugate of Eq. (7.2.3) is given explicitly by

$$\mathbf{E}_s^*(\mathbf{r}) = \hat{\epsilon}_s^* A_s^*(\mathbf{r})e^{-i\mathbf{k_s}\cdot\mathbf{r}}. \qquad (7.2.4)$$

We thus see that the action of an ideal phase-conjugate mirror is threefold:

1. The complex polarization unit vector of the incident radiation is replaced by its complex conjugate. For example, right-hand circular light remains right-hand circular in reflection from a phase-conjugate mirror rather than being converted into left-hand circular light, as is the case in reflection at normal incidence from a metallic mirror.

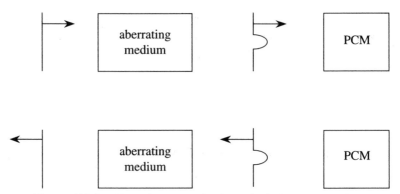

FIGURE 7.2.2   Aberration correction by optical phase conjugation.

2. $A_s(\mathbf{r})$ is replaced by $A_s^*(\mathbf{r})$, implying that the wavefront is reversed in the sense illustrated in Fig. 7.2.1b.*

3. $\mathbf{k_s}$ is replaced by $-\mathbf{k_s}$, showing that the incident wave is reflected back into its direction of incidence. From the point of view of ray optics, this result shows that each ray of the incident beam is precisely reflected back onto itself.

Note further that Eqs. (7.2.1) through (7.2.4) imply that

$$\tilde{\mathbf{E}}_c(\mathbf{r}, t) = r\tilde{\mathbf{E}}_s(\mathbf{r}, -t). \tag{7.2.5}$$

This result shows that the phase conjugation process can be thought of as the generation of a time-reversed wavefront.

It is important to note that the description given here refers to an *ideal* phase-conjugate mirror. Many physical devices that are ordinarily known as phase-conjugate mirrors are imperfect either in the sense that they do not possess all three properties listed above or in the sense that they possess these properties only approximately. For example, many phase-conjugate mirrors are highly imperfect in their polarization properties, even though they are nearly perfect in their ability to perform wavefront reversal.

Aberration Correction by Phase Conjugation

The process of phase conjugation is able to remove the effects of aberrations under conditions such that a beam of light passes twice in opposite directions through an aberrating medium. The reason why optical phase conjugation leads to aberration correction is illustrated in Fig. 7.2.2. Here an initially plane wavefront propagates through an aberrating medium. The aberration may be

* Because of this property, the phase conjugation process displays special quantum noise characteristics. These characteristics have been described by Gaeta and Boyd (1988).

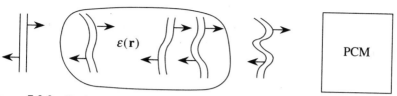

FIGURE 7.2.3 Conjugate waves propagating through an inhomogeneous optical medium.

due to turbulence in the earth's atmosphere, inhomogeneities in the refractive index of a piece of glass, or a poorly designed optical system. The wavefront of the light leaving the medium therefore becomes distorted in the manner shown schematically in the figure. If this aberrated wavefront is now allowed to fall onto a phase-conjugate mirror, a conjugate wavefront will be generated, and the sense of the wavefront distortion will be inverted in this reflected wave. As a result, when this wavefront passes through the aberrating medium again, an undistorted output wave will emerge.

Let us now see how to demonstrate mathematically that optical phase conjugation leads to aberration correction. (Our treatment here is similar to that of Yariv and Fisher in Fisher, 1983.) We consider a wave $\tilde{E}(\mathbf{r}, t)$ propagating through a lossless material of nonuniform refractive index $n(\mathbf{r}) = [\epsilon(\mathbf{r})]^{1/2}$, as shown in Fig. 7.2.3.

We assume that the spatial variation of $\epsilon(\mathbf{r})$ occurs on a scale that is much larger than an optical wavelength. The optical field in this region must obey the wave equation, which we write in the form

$$\nabla^2 \tilde{E} - \frac{\epsilon(\mathbf{r})}{c^2} \frac{\partial^2 \tilde{E}}{\partial t^2} = 0. \tag{7.2.6}$$

We represent the field propagating to the right through this region as

$$\tilde{E}(\mathbf{r}, t) = A(\mathbf{r}) e^{i(kz - \omega t)} + \text{c.c.}, \tag{7.2.7}$$

where the field amplitude $A(\mathbf{r})$ is assumed to be a slowly varying function of $\mathbf{r}$. Since we have singled out the z direction as the mean direction of propagation, it is convenient to express the laplacian operator which appears in Eq. (7.2.6) as

$$\nabla^2 = \frac{\partial^2}{\partial z^2} + \nabla_T^2, \tag{7.2.8}$$

where $\nabla_T^2 = \partial^2/\partial x^2 + \partial^2/\partial y^2$ is called the transverse laplacian. Equations (7.2.7) and (7.2.8) are now introduced into Eq. (7.2.6), which becomes

$$\nabla_T^2 A + \left[\frac{\omega^2 \epsilon(\mathbf{r})}{c^2} - k^2 \right] A + 2ik \frac{\partial A}{\partial z} = 0. \tag{7.2.9}$$

In writing this equation in the form shown, we have omitted the term $\partial^2 A/\partial z^2$ because $A(\mathbf{r})$ has been assumed to be slowly varying. Since this equation is generally valid, so is its complex conjugate, which is given explicitly by

$$\nabla_T^2 A^* + \left[\frac{\omega^2 \epsilon(\mathbf{r})}{c^2} - k^2\right] A^* - 2ik\frac{\partial A^*}{\partial z} = 0. \qquad (7.2.10)$$

However, this equation describes the wave

$$\tilde{E}_c(\mathbf{r}, t) = A^*(\mathbf{r})e^{i(-kz-\omega t)} + \text{c.c.}, \qquad (7.2.11)$$

which is a wave propagating in the negative z direction whose complex amplitude is *everywhere* the complex conjugate of the forward-going wave. This proof shows that if the phase-conjugate mirror can generate a backward-going wave whose amplitude is the complex conjugate of that of the forward-going wave at any one plane (say the input face of the mirror), then the field amplitude of the backward-going wave will be the complex conjugate of that of the forward-going wave at *all* points in front of the mirror. In particular, if the forward-going wave is a plane wave before entering the aberrating medium, then the backward-going (i.e., conjugate) wave emerging from the aberrating medium will also be a plane wave.

The phase conjugation process is directly suited for removing the effects of aberrations in double pass, but under special circumstances can be used to perform single-pass aberration correction; see for instance MacDonald *et al.* (1988).

Phase Conjugation by Degenerate Four-Wave Mixing

Let us now consider a physical process that can produce a phase conjugate wavefront. It has been shown by Hellwarth (1977) and by Yariv and Pepper (1977) that the phase conjugate of an incident wave can be created by the process of degenerate four-wave mixing (DFWM) using the geometry shown in Fig. 7.2.4. This four-wave mixing process is degenerate in the sense that

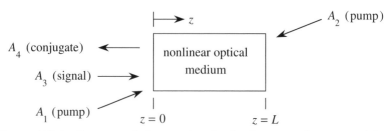

FIGURE 7.2.4   Geometry of phase conjugation by degenerate four-wave mixing.

all four interacting waves have the same frequency. In this process, a lossless nonlinear medium characterized by a third-order nonlinear susceptibility $\chi^{(3)}$ is illuminated by two strong counterpropagating pump waves E_1 and E_2 and by a signal wave E_3. The pump waves are usually taken to be plane waves, although in principle they can possess any wavefront structure as long as their amplitudes are complex conjugates of one another. The signal wave is allowed to have an arbitrary wavefront. In this section we show that, as a result of the nonlinear coupling between these waves, a new wave E_4 is created that is the phase conjugate of E_3. We also derive an expression (Eq. (7.2.37b)) that describes the efficiency with which the conjugate wave is generated.

Since the mathematical development that follows is somewhat involved, it is useful to consider first in simple terms why the interaction illustrated in Fig. 7.2.4 leads to the generation of a conjugate wavefront. We represent the four interacting waves by

$$\tilde{E}_i(\mathbf{r}, t) = E_i(\mathbf{r})e^{-i\omega t} + \text{c.c.}$$
$$= A_i(\mathbf{r})E^{i(\mathbf{k_i}\cdot\mathbf{r}-\omega t)} + \text{c.c.}$$

(7.2.12)

for $i = 1, 2, 3, 4$, where the $A_i(\mathbf{r})$ are slowly varying quantities. The nonlinear polarization produced within the medium by the three input waves will have, in addition to a large number of other terms, a term of the form

$$P^{\text{NL}} = 6\chi^{(3)}E_1E_2E_3^* = 6\chi^{(3)}A_1A_2A_3^*e^{i(\mathbf{k_1}+\mathbf{k_2}-\mathbf{k_3})\cdot\mathbf{r}}. \qquad (7.2.13)$$

Since we have assumed that the pump waves E_1 and E_2 are counterpropagating, their wavevectors are related by

$$\mathbf{k_1} + \mathbf{k_2} = 0, \qquad (7.2.14)$$

and hence Eq. (7.2.13) becomes

$$P^{\text{NL}} = 6\chi^{(3)}A_1A_2A_3^*e^{-i\mathbf{k_3}\cdot\mathbf{r}}. \qquad (7.2.15)$$

We see that this contribution to the nonlinear polarization has a spatial dependence that allows it to act as a phase-matched source term for a conjugate wave (E_4) having wavevector $-\mathbf{k_3}$, and hence we see that the wavevectors of the signal and conjugate waves are related by

$$\mathbf{k_3} + \mathbf{k_4} = 0. \qquad (7.2.16)$$

The field amplitude of the wave generated by the nonlinear polarization of Eq. (7.2.15) will be proportional to $A_1A_2A_3^*$. This wave will be the phase conjugate of A_3 whenever the phase of the product A_1A_2 is spatially invariant, either because A_1 and A_2 both represent plane waves and hence are each constant or because A_1 and A_2 are phase conjugates of one another (because if A_2 is proportional to A_1^*, then $A_1 A_2$ will be proportional to the real quantity $|A_1|^2$).

We can also understand the interaction shown in Fig. 7.2.4 from the following point of view. The incoming signal wave of amplitude A_3 interferes with one of the pump waves (e.g., the forward-going pump wave of amplitude A_1) to form a spatially varying intensity distribution. As a consequence of the nonlinear response of the medium, a refractive index variation accompanies this interference pattern. This variation acts as a volume diffraction grating, which scatters the other pump wave to form the outgoing conjugate wave of amplitude A_4.

Let us now treat the degenerate four-wave mixing process more rigorously. The total field amplitude within the nonlinear medium is given by

$$E = E_1 + E_2 + E_3 + E_4. \qquad (7.2.17)$$

This field produces a nonlinear polarization within the medium, which is given by

$$P = 3\chi^{(3)} E^2 E^* \qquad (7.2.18)$$

where $\chi^{(3)} = \chi^{(3)}(\omega = \omega + \omega - \omega)$. The product $E^2 E^*$ that appears on the right-hand side of this equation contains a large number of terms with different spatial dependences. Those terms with spatial dependence of the form

$$e^{i\mathbf{k_i} \cdot \mathbf{r}} \quad \text{for} \quad i = 1, 2, 3, 4 \qquad (7.2.19)$$

are particularly important because they can act as phase-matched source terms for one of the four interacting waves. The polarization amplitudes associated with these phase-matched contributions are as follows:

$$P_1 = 3\chi^{(3)}\left[E_1^2 E_1^* + 2E_1 E_2 E_2^* + 2E_1 E_3 E_3^* + 2E_1 E_4 E_4^* + 2E_3 E_4 E_2^*\right],$$

$$P_2 = 3\chi^{(3)}\left[E_2^2 E_2^* + 2E_2 E_1 E_1^* + 2E_2 E_3 E_3^* + 2E_2 E_4 E_4^* + 2E_3 E_4 E_1^*\right],$$

$$P_3 = 3\chi^{(3)}\left[E_3^2 E_3^* + 2E_3 E_1 E_1^* + 2E_3 E_2 E_2^* + 2E_3 E_4 E_4^* + 2E_1 E_2 E_4^*\right],$$

$$P_4 = 3\chi^{(3)}\left[E_4^2 E_4^* + 2E_4 E_1 E_1^* + 2E_4 E_2 E_2^* + 2E_4 E_3 E_3^* + 2E_1 E_2 E_3^*\right].$$

$$(7.2.20)$$

We next assume that the fields E_3 and E_4 are much weaker than the pump fields E_1 and E_2. In the above expressions we therefore drop those terms that contain more than one weak-field amplitude. We hence obtain

$$P_1 = 3\chi^{(3)}\left[E_1^2 E_1^* + 2E_1 E_2 E_2^*\right],$$

$$P_2 = 3\chi^{(3)}\left[E_2^2 E_2^* + 2E_2 E_1 E_1^*\right],$$

$$P_3 = 3\chi^{(3)}[2E_3 E_1 E_1^* + 2E_3 E_2 E_2^* + 2E_1 E_2 E_4^*],$$

$$P_4 = 3\chi^{(3)}[2E_4 E_1 E_1^* + 2E_4 E_2 E_2^* + 2E_1 E_2 E_3^*].$$

(7.2.21)

Note that, at the present level of approximation, the E_3 and E_4 fields are each driven by a polarization that depends on the amplitudes of all of the fields, but that the polarizations driving the E_1 and E_2 fields depend only upon E_1 and E_2 themselves. We thus consider first the problem of calculating the spatial evolution of the pump field amplitudes E_1 and E_2. We can then later use these known amplitudes when we calculate the spatial evolution of the signal and conjugate waves.

We assume that each of the interacting waves obeys the wave equation in the form

$$\nabla^2 \tilde{E}_i - \frac{\epsilon}{c^2}\frac{\partial^2 \tilde{E}_i}{\partial t^2} = \frac{4\pi}{c^2}\frac{\partial^2}{\partial t^2}\tilde{P}_i.$$

(7.2.22)

We now introduce Eqs. (7.2.12) and (7.2.21) into this equation and make the slowly-varying amplitude approximation. Also, we let z' be the spatial coordinate measured in the direction of propagation of the E_1 field, and we assume for simplicity that the pump waves have plane wavefronts. We then find that the pump field A_1 must obey the equation

$$\left[\left(-k_1^2 + 2ik_1\frac{d}{dz'} + \frac{\epsilon\omega^2}{c^2}\right)A_1\right]e^{i(k_1 z' - \omega t)}$$

$$= -\frac{4\pi\omega^2}{c^2}\,3\chi^{(3)}[|A_1|^2 + 2|A_2|^2]A_1 e^{i(k_1 z' - \omega t)},$$

which, after simplification, becomes

$$\frac{dA_1}{dz'} = \frac{6\pi i\omega}{nc}\chi^{(3)}[|A_1|^2 + 2|A_2|^2]A_1 \equiv i\kappa_1 A_1.$$

(7.2.23a)

We similarly find that the backward-going pump wave is described by the equation

$$\frac{dA_2}{dz'} = \frac{-6\pi i\omega}{nc}\chi^{(3)}[|A_2|^2 + 2|A_1|^2]A_2 \equiv i\kappa_2 A_2.$$

(7.2.23b)

Since κ_1 and κ_2 are real quantities, these equations show that A_1 and A_2 each undergo phase shifts as they propagate through the nonlinear medium. The phase shift experienced by each wave depends both on its own intensity and on that of the other wave. Note that each wave shifts the phase of the other wave by twice as much as it shifts its own phase, in consistency with the general result described in the discussion following Eq. (4.1.14). These phase shifts can induce a phase mismatch into the process that generates the phase-conjugate signal. Note that since only the phases (and not the amplitudes) of the pump waves are affected by the nonlinear coupling, the quantities $|A_1|^2$ and $|A_2|^2$ are spatially invariant, and hence the quantities κ_1 and κ_2 that appear in Eqs. (7.2.23) are in fact constants. These equations can therefore be solved directly to obtain

$$A_1(z') = A_1(0)e^{i\kappa_1 z'}, \tag{7.2.24a}$$

$$A_2(z') = A_2(0)e^{-i\kappa_2 z'}. \tag{7.2.24b}$$

The product $A_1 A_2$ that appears in the expression (7.2.15) for the nonlinear polarization responsible for producing the phase-conjugate wave therefore varies spatially as

$$A_1(z')A_2(z') = A_1(0)A_2(0)e^{i(\kappa_1 - \kappa_2)z'}; \tag{7.2.25}$$

the factor $e^{i(\kappa_1 - \kappa_2)z'}$ shows the effect of wavevector mismatch. If the two pump beams have equal intensities, so that $\kappa_1 = \kappa_2$, the product $A_1 A_2$ becomes spatially invariant, so that

$$A_1(z')A_2(z') = A_1(0)A_2(0), \tag{7.2.26}$$

and in this case the interaction is perfectly phase-matched. We shall henceforth assume that the pump intensities are equal.

We next consider the coupled-amplitude equations describing the signal and conjugate fields, $\tilde{E}_3$ and $\tilde{E}_4$. We assume for simplicity that the incident signal wave has plane wavefronts. This is actually not a restrictive assumption, because an arbitrary signal field can be decomposed into plane-wave components, each of which will couple to a plane-wave component of the conjugate field $\tilde{E}_4$. Under this assumption, the wave equation (7.2.22) applied to the signal and conjugate fields leads to the coupled-amplitude equations

$$\frac{dA_3}{dz} = \frac{12\pi i \omega}{nc}\chi^{(3)}\left[(|A_1|^2 + |A_2|^2)A_3 + A_1 A_2 A_4^*\right], \tag{7.2.27a}$$

$$\frac{dA_4}{dz} = -\frac{12\pi i \omega}{nc}\chi^{(3)}\left[(|A_1|^2 + |A_2|^2)A_4 + A_1 A_2 A_3^*\right]. \tag{7.2.27b}$$

For convenience, we write these equations as

$$\frac{dA_3}{dz} = i\kappa_3 A_3 + i\kappa A_4^*, \tag{7.2.28a}$$

$$\frac{dA_4}{dz} = -i\kappa_3 A_4 - i\kappa A_3^*, \tag{7.2.28b}$$

where we have introduced the coupling coefficients

$$\kappa_3 = \frac{12\pi\omega}{nc}\chi^{(3)}(|A_1|^2 + |A_2|^2), \tag{7.2.29a}$$

$$\kappa = \frac{12\pi\omega}{nc}\chi^{(3)}A_1 A_2. \tag{7.2.29b}$$

The set of equations (7.2.28) can be simplified through a change of variables. We let

$$A_3 = A_3' e^{i\kappa_3 z}, \tag{7.2.30a}$$

$$A_4 = A_4' e^{-i\kappa_3 z}. \tag{7.2.30b}$$

Note that the primed and unprimed variables coincide at the input face of the interaction region, that is, at the plane $z = 0$. We introduce these relations into Eq. (7.2.28a), which becomes

$$i\kappa_3 A_3' e^{i\kappa_3 z} + \frac{dA_3'}{dz} e^{i\kappa_3 z} = i\kappa_3 A_3' e^{i\kappa_3 z} + i\kappa A_4'^* e^{i\kappa_3 z},$$

or

$$\frac{dA_3'}{dz} = i\kappa A_4'^*. \tag{7.2.31a}$$

We similarly find that Eq. (7.2.28b) becomes

$$\frac{dA_4'}{dz} = -i\kappa A_3'^*. \tag{7.2.31b}$$

This set of equations shows why degenerate four-wave mixing leads to phase conjugation: The generated field A_4' is driven only by the complex conjugate of the input field amplitude. We note that this set of equations is *formally* identical to the set that we would have obtained if we had taken the driving polarizations of Eq. (7.2.21) to be simply

$$P_1 = P_2 = 0, \qquad P_3 = 6\chi^{(3)}E_1 E_2 E_4^*, \qquad P_4 = 6\chi^{(3)}E_1 E_2 E_3^*, \tag{7.2.32}$$

that is, if we had ignored the modification of the pump waves due to the non-linear interaction.

Next, we solve the set of equations (7.2.31). We take the derivative of Eq. (7.2.31b) with respect to z and introduce Eq. (7.2.31a) to obtain*

$$\frac{d^2 A_4'}{dz^2} + |\kappa|^2 A_4' = 0. \tag{7.2.33}$$

This result shows that the spatial dependence of A_4' must be of the form

$$A_4'(z) = B \sin|\kappa|z + C \cos|\kappa|z. \tag{7.2.34}$$

In order to determine the constants B and C, we must specify the boundary conditions for each of the two weak waves at their respective input planes. In particular, we assume that $A_3'^*(0)$ and $A_4'(L)$ are specified. In this case, the solution of Eq. (7.2.33) is

$$A_3'^*(z) = -\frac{i|\kappa|}{\kappa} \frac{\sin|\kappa|z}{\cos|\kappa|L} A_4'(L) + \frac{\cos[|\kappa|(z-L)]}{\cos|\kappa|L} A_3'^*(0), \tag{7.2.35a}$$

$$A_4'(z) = \frac{\cos|\kappa|z}{\cos|\kappa|L} A_4'(L) - \frac{i\kappa}{|\kappa|} \frac{\sin[|\kappa|(z-L)]}{\cos|\kappa|L} A_3'^*(0). \tag{7.2.35b}$$

However, for the case of four-wave mixing for optical phase conjugation, we can usually assume that there is no conjugate wave injected into the medium at $z = L$, that is, we can assume that

$$A_4'(L) = 0. \tag{7.2.36}$$

Furthermore, we are usually interested only in the output values of the two interacting fields. These output field amplitudes are then given by

$$A_3'^*(L) = \frac{A_3'^*(0)}{\cos |\kappa|L}, \tag{7.2.37a}$$

$$A_4'(0) = \frac{i\kappa}{|\kappa|} (\tan |\kappa|L) A_3'^*(0). \tag{7.2.37b}$$

Note that the transmitted signal wave $A_3'^*(L)$ is always more intense than the incident wave. Note also that the output conjugate wave $A_4(0)$ can have any intensity ranging from zero to infinity, the actual value depending on the particular value of $|\kappa|L$. The reflectivity of a phase-conjugate mirror based on degenerate four-wave mixing can exceed 100% because the mirror is actively pumped by externally applied waves, which can supply energy.

From the point of view of energetics, we can describe the process of degenerate four-wave mixing as a process in which one photon from each of the

---

* We are assuming throughout this discussion that $\chi^{(3)}$ and hence κ are real; we have written the equation in the form shown for generality and for consistency with other cases where κ is complex.

(a)   (b)   (c)

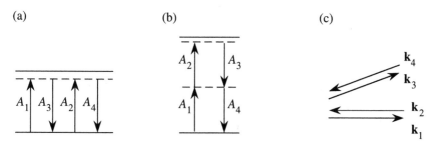

FIGURE 7.2.5   Parts (a) and (b) are energy-level diagrams describing two different interactions that can lead to phase conjugation by degenerate four-wave mixing. In either case, the interaction involves the simultaneous annihilation of two pump photons with the creation of signal and conjugate photons. Diagram (a) describes the dominant interaction if the applied field frequency is nearly resonant with a one-photon transition of the material system, whereas (b) describes the dominant interaction under conditions of two-photon-resonant excitation. Part (c) shows the wavevectors of the four interacting waves. Since $\mathbf{k}_1 + \mathbf{k}_2 - \mathbf{k}_3 - \mathbf{k}_4 = 0$, the process is perfectly phase-matched.

pump waves is annihilated and one photon is added to each of the signal and conjugate waves, as shown in Fig. 7.2.5. Hence, the conjugate wave A_4 is created, and the signal wave A_3 is amplified. The degenerate four-wave mixing process with counterpropagating pump waves is automatically phase-matched (when the two pump waves have equal intensity or whenever we can ignore the nonlinear phase shifts experienced by each wave). We see that this is true because no phase-mismatch terms of the sort $e^{\pm i \Delta \kappa z}$ appear on the right-hand sides of Eqs. (7.2.31). The fact that degenerate four-wave mixing in the phase conjugation geometry is automatically phase-matched has a very simple physical interpretation. Since this process entails the annihilation of two pump photons and the creation of a signal and conjugate photon, the total input energy is $2\hbar\omega$ and the total input momentum is $\hbar(\mathbf{k}_1 + \mathbf{k}_2) = 0$; likewise the total output energy is $2\hbar\omega$ and the total output momentum is $\hbar(\mathbf{k}_3 + \mathbf{k}_4) = 0$. If the two pump beams are not exactly counterpropagating, then $\hbar(\mathbf{k}_1 + \mathbf{k}_2)$ does not vanish and the phase-matching condition is not automatically satisfied.

The first experimental demonstration of phase conjugation by degenerate four-wave mixing was performed by Bloom and Bjorklund (1977). Their experimental setup is shown in Fig. 7.2.6. They observed that the presence of the aberrating glass plate did not lower the resolution of the system when the mirror was aligned to retroreflect the pump laser beam onto itself. However, when this mirror was partially misaligned, the return beam passed through a different portion of the aberrating glass and the resolution of the system was degraded.

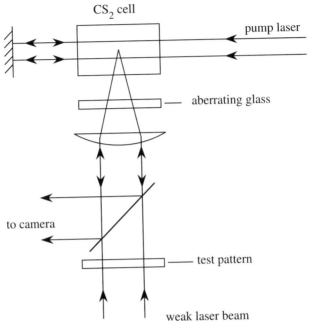

FIGURE 7.2.6   Experimental setup for studying phase conjugation by degenerate four-wave mixing.

Degenerate four-wave mixing is usually performed using the geometry of Fig. 7.2.4, although it can also be performed using the surface nonlinearity of the interface between a linear and nonlinear medium; see for instance Maki *et al.* (1992) for details.

Polarization Properties of Phase Conjugation

Our discussion thus far has treated phase conjugation in the scalar approximation and has shown that phase conjugation can be used to remove the effects of wavefront aberrations. It is often desirable that phase conjugation be able to remove the effects of polarization distortions as well. An example is shown in Fig. 7.2.7. Here a beam of light that initially is linearly polarized passes through a stressed optical component. As a result of stress-induced birefringence, the state of polarization of the beam becomes distorted nonuniformly over the cross section of the beam. This beam then falls onto a phase-conjugate mirror. If this mirror is ideal in the sense that the polarization unit vector $\hat{\epsilon}$ of the incident light is replaced by its complex conjugate in the reflected beam, the effects of the polarization distortion will be removed in the second pass

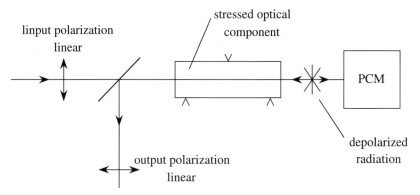

linput polarization
linear

stressed optical
component

PCM

depolarized
radiation

output polarization
linear

FIGURE 7.2.7   Polarization properties of phase conjugation.

through the stressed optical component, and the beam will be returned to its initial state of linear polarization. A phase-conjugate mirror that produces a reflected beam that is both a wavefront conjugate and a polarization conjugate is often called a vector phase-conjugate mirror.

In order to describe the polarization properties of the degenerate four-wave mixing process, we consider the geometry shown in Fig. 7.2.8, where $\mathbf{F}$, $\mathbf{B}$, and $\mathbf{S}$ denote the amplitudes of the forward- and backward-going pump waves and of the signal wave, respectively. The total applied field is thus given by

$$\mathbf{E} = \mathbf{F} + \mathbf{B} + \mathbf{S}. \tag{7.2.38}$$

We assume that the angle θ between the signal and forward-going pump wave is much smaller than unity, so that only the x and y components of the incident fields have appreciable amplitudes. We also assume that the nonlinear optical material is isotropic, so that the third-order nonlinear optical susceptibility $\chi_{ijkl}^{(3)} = \chi_{ijkl}^{(3)}(\omega = \omega + \omega - \omega)$ is given by Eq. (4.2.5) as

$$\chi_{ijkl}^{(3)} = \chi_{1122}(\delta_{ij}\delta_{kl} + \delta_{ik}\delta_{jl}) + \chi_{1221}\delta_{il}\delta_{jk}, \tag{7.2.39}$$

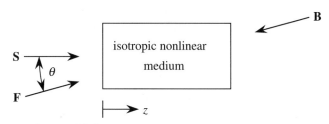

$\mathbf{S}$

θ

$\mathbf{F}$

isotropic nonlinear
medium

$\mathbf{B}$

z

FIGURE 7.2.8   Geometry of vector phase conjugation.

and so that the nonlinear polarization can be expressed as

$$\mathbf{P} = 6\chi_{1122}(\mathbf{E} \cdot \mathbf{E}^*)\mathbf{E} + 3\chi_{1221}(\mathbf{E} \cdot \mathbf{E})\mathbf{E}^*$$

$$= A(\mathbf{E} \cdot \mathbf{E}^*)\mathbf{E} + \tfrac{1}{2}B(\mathbf{E} \cdot \mathbf{E})\mathbf{E}^*. \tag{7.2.40}$$

If we now introduce Eq. (7.2.38) into Eq. (7.2.40), we find that the phase-matched contribution to the nonlinear polarization that acts as a source for the conjugate wave is given by

$$\begin{bmatrix} P_x \\ P_y \end{bmatrix} = 6 \begin{bmatrix} \chi_{1111}B_xF_x + \chi_{1221}B_yF_y & \chi_{1122}(B_xF_y + B_yF_x) \\ \chi_{1122}(B_yF_x + B_xF_y) & \chi_{1111}B_yF_y + \chi_{1221}B_xF_x \end{bmatrix} \begin{bmatrix} S_x^* \\ S_y^* \end{bmatrix}, \tag{7.2.41}$$

where $\chi_{1111} = 2\chi_{1122} + \chi_{1221}$. The polarization properties of the phase-conjugation process will be ideal (i.e., vector phase conjugation will be obtained) whenever the two-by-two transfer matrix of Eq. (7.2.41) is a multiple of the identity matrix. Under these conditions, both cartesian components of the incident field are reflected with equal efficiency and no coupling between orthogonal components occurs.

There are two different ways in which the matrix in Eq. (7.2.41) can be made to reduce to a multiple of the identity matrix. One way is for $A = 6\chi_{1122}$ to vanish identically. In this case Eq. (7.2.41) becomes

$$\begin{bmatrix} P_x \\ P_y \end{bmatrix} = 6\chi_{1221}(B_xF_x + B_yF_y) \begin{bmatrix} 1 & 0 \\ 0 & 1 \end{bmatrix} \begin{bmatrix} S_x^* \\ S_y^* \end{bmatrix}$$

$$= 6\chi_{1221}(B_xF_x + B_yF_y) \begin{bmatrix} S_x^* \\ S_y^* \end{bmatrix}, \tag{7.2.42}$$

and hence the nonlinear polarization is proportional to the complex conjugate of the signal amplitude for *any* choice of the polarization vectors of the pump waves. This result can be understood directly in terms of Eq. (7.2.40), which shows that $\mathbf{P}$ has the vector character of $\mathbf{E}^*$ whenever χ_{1122} vanishes. However, χ_{1122} (or A) vanishes identically only under very unusual circumstances. The only known case for this condition to occur is that of degenerate four-wave mixing in an atomic system utilizing a two-photon resonance between certain atomic states. This situation has been analyzed by Grynberg (1984) and studied experimentally by Malcuit *et al.* (1988). The analysis can be described most simply for the case of a transition between two S states of an atom with zero electron spin. The four-wave mixing process can then be described graphically by the diagram shown in Fig. 7.2.9. Since the lower and upper levels each possess zero angular momentum, the sum of the angular momenta of the signal

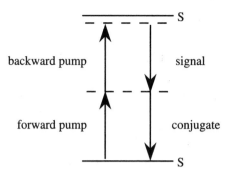

FIGURE 7.2.9   Phase conjugation by degenerate four-wave mixing using a two-photon transition.

and conjugate photons must be zero, and this condition implies that the polarization unit vectors of the two waves must be related by complex conjugation.

For most physical mechanisms giving rise to optical nonlinearities, the coefficient A does not vanish. (Recall that for molecular orientation $B/A = 6$, for electrostriction $B/A = 0$, and for nonresonant electronic response $B/A = 1$.) For the general case in which A is not equal to 0, vector phase conjugation in the geometry in Fig. 7.2.8 can be obtained only when the pump waves are circularly polarized and counterrotating. By counterrotating, we mean that if the forward-going wave is described by

$$\tilde{\mathbf{F}}(z, t) = F \, \frac{\hat{x} - i\hat{y}}{\sqrt{2}} \, e^{i(kz - \omega t)} + \text{c.c.}, \qquad (7.2.43a)$$

then the backward-going wave is described by

$$\tilde{\mathbf{B}}(z, t) = B \, \frac{\hat{x} + i\hat{y}}{\sqrt{2}} \, e^{i(-kz - \omega t)} + \text{c.c.} \qquad (7.2.43b)$$

These waves are counterrotating in the sense that, for any fixed value of z, $\tilde{\mathbf{F}}$ rotates clockwise in time in the xy plane and $\tilde{\mathbf{B}}$ rotates counterclockwise in time. However, both waves are right-hand circularly polarized, since, by convention, the handedness of a wave is the sense of rotation as determined when looking into the beam.

In the notation of Eq. (7.2.41), the amplitudes of the fields described by Eqs. (7.2.43) are given by

$$F_x = \frac{F}{\sqrt{2}} e^{ikz}, \qquad F_y = -i\frac{F}{\sqrt{2}} e^{ikz},$$

$$B_x = \frac{B}{\sqrt{2}} e^{-ikz}, \qquad B_y = i\frac{B}{\sqrt{2}} e^{-ikz}, \qquad (7.2.44)$$

and hence Eq. (7.2.41) becomes

$$\begin{bmatrix} P_x \\ P_y \end{bmatrix} = 3FB(\chi_{1111} + \chi_{1221}) \begin{bmatrix} 1 & 0 \\ 0 & 1 \end{bmatrix} \begin{bmatrix} S_x^* \\ S_y^* \end{bmatrix}. \tag{7.2.45}$$

We see that the transfer matrix is again a multiple of the identity matrix and hence that the nonlinear polarization vector is proportional to the complex conjugate of the signal field vector. The fact that degenerate four-wave mixing excited by counterrotating pump waves leads to vector phase conjugation was predicted theoretically by Zel'dovich and Shkunov (1979) and was verified experimentally by Martin *et al.* (1980).

The reason why degenerate four-wave mixing with counterrotating pump waves leads to vector phase conjugation can be understood in terms of the conservation of linear and angular momentum. As described above, phase conjugation can be visualized as a process in which one photon from each pump wave is annihilated and a signal and conjugate photon are simultaneously created. Since the pump waves are counterpropagating and counterrotating, the total linear and angular momenta of the two input photons must vanish. Then conservation of linear and angular momentum requires that the conjugate wave must be emitted in a direction opposite to the direction of propagation of the signal wave and that its polarization vector must rotate in a sense opposite to that of the signal wave.

7.3. Optical Bistability and Optical Switching

Certain nonlinear optical systems can possess more than one output state for a given input state. The term *optical bistability* refers to the situation in which two different output intensities are possible for a given input intensity, and the more general term *optical multistability* is used to describe the circumstance in which two or more stable output states are possible. Interest in optical bistability stems from its potential usefulness as a switch for use in optical communication and in optical computing.

Optical bistability was first described theoretically by Szöke *et al.* (1969) and was first observed experimentally by Gibbs *et al.* (1976). The bistable optical device described in these works consists of a nonlinear medium placed inside of a Fabry–Perot resonator. Such a device is illustrated schematically in Fig. 7.3.1. Here A_1 denotes the field amplitude of the incident wave, A_1' denotes that of the reflected wave, A_2 and A_2' denote the amplitudes of the forward- and backward-going waves within the interferometer, and A_3 denotes the amplitude of the transmitted wave. The cavity mirrors are assumed to be identical and

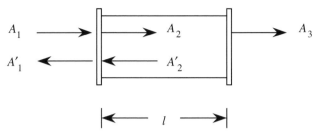

FIGURE 7.3.1   Bistable optical device in the form of a Fabry–Perot interferometer containing a nonlinear medium.

lossless, with amplitude reflectance ρ and transmittance τ which are related to the intensity reflectance R and transmittance T through

$$R = |\rho|^2 \quad \text{and} \quad T = |\tau|^2 \tag{7.3.1a}$$

with

$$R + T = 1. \tag{7.3.1b}$$

The incident and internal fields are related to each other through boundary conditions of the form

$$A'_2 = \rho A_2 e^{2ikl-\alpha l}, \tag{7.3.2a}$$

$$A_2 = \tau A_1 + \rho A'_2. \tag{7.3.2b}$$

In these equations, we assume that the field amplitudes are measured at the inner surface of the left-hand mirror. The propagation constant $k = n\omega/c$ and intensity absorption coefficient α are taken to be real quantities, which include both their linear and nonlinear contributions. In writing Eq. (7.3.2a) in the form shown, we have implicitly made the *mean-field approximation*, that is, we have assumed that the quantities k and α are spatially invariant; if such is not the case, the exponent should be replaced by $\int_0^l dz[2ik(z) - \alpha(z)]$. For simplicity we also assume that the nonlinear material and the medium surround the resonator have the same linear refractive indices.

Equations (7.3.2) can be solved algebraically by eliminating A'_2 to obtain

$$A_2 = \frac{\tau A_1}{1 - \rho^2 e^{2ikl-\alpha l}}, \tag{7.3.3}$$

which is known as Airy's equation and which describes the properties of a Fabry–Perot interferometer. If k or α (or both) is a sufficiently nonlinear function of the intensity of the light within the interferometer, this equation predicts bistability in the intensity of the transmitted wave. In general, both k and α can display nonlinear behavior; however, we an obtain a better understanding of the nature of optical bistability by considering in turn the limiting cases in which either the absorptive or the refractive contribution dominates.

Absorptive Bistability

Let us first examine the case in which only the absorption coefficient α depends nonlinearly on the field intensity. The wavevector magnitude k is hence assumed to be constant. To simplify the following analysis, we assume that the mirror separation l is adjusted so that the cavity is tuned to resonance with the applied field; in such a case the factor $\rho^2 e^{2ikl}$ that appears in the denominator of Eq. (7.3.3) is equal to the real quantity R. We also assume that $\alpha l \ll 1$, so that we can ignore the spatial variation of the intensity of the field inside the cavity, which justifies the use of the mean-field approximation. Under these conditions, Airy's equation (7.3.3) reduces to

$$A_2 = \frac{\tau A_1}{1 - R(1 - \alpha l)}. \tag{7.3.4}$$

The analogous equation relating the incident and circulating intensities $I_i = (n/c2\pi)|A_i|^2$ is given by

$$I_2 = \frac{T I_1}{[1 - R(1 - \alpha l)]^2}. \tag{7.3.5}$$

This equation can be simplified by introducing the dimensionless parameter C (known as the cooperation number),

$$C = \frac{R\alpha l}{1 - R}, \tag{7.3.6}$$

which becomes (since $1 + C = (1 - R + R\alpha l)/(1 - R) = [1 - R(1 - \alpha l)]/T$)

$$I_2 = \frac{1}{T} \frac{I_1}{(1 + C)^2}. \tag{7.3.7}$$

We now assume that the absorption coefficient α and hence the value of the parameter C depend upon the intensity of the light within the interferometer. For simplicity, we assume that the absorption coefficient obeys the relation valid for a two-level saturable absorber,

$$\alpha = \frac{\alpha_0}{1 + I/I_s}, \tag{7.3.8}$$

where α_0 denotes the unsaturated absorption coefficient, I the local value of the intensity, and I_s the saturation intensity. For simplicity we also ignore the standing-wave nature of the field within the interferometer and take I equal to $I_2 + I_2' \approx 2I_2$. It is only approximately valid to ignore standing-wave effects

nonlinear medium

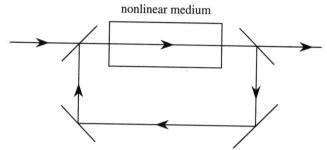

FIGURE 7.3.2 Bistable optical device in the form of a traveling wave interferometer containing a nonlinear medium.

for the interferometer of Fig. 7.3.1, but it is strictly valid for the traveling-wave interferometer shown in Fig. 7.3.2. Under the assumption that the absorption coefficient depends on the intensity of the internal fields according to Eq. (7.3.8) with $I = 2I_2$, the parameter C is given by

$$C = \frac{C_0}{1 + 2I_2/I_s} \tag{7.3.9}$$

with $C_0 = R\alpha_0 l/(1 - R)$. The relation between I_1 and I_2 given by Eq. (7.3.7) can be rewritten using this expression for C as

$$I_1 = TI_2\left(1 + \frac{C_0}{1 + 2I_2/I_s}\right)^2. \tag{7.3.10}$$

Finally, the output intensity I_3 is related to I_2 by

$$I_3 = TI_2. \tag{7.3.11}$$

The input–output relation implied by Eqs. (7.3.10) and (7.3.11) is illustrated graphically in Fig. 7.3.3 for several different values of the weak-field parameter C_0. For C_0 greater than 8, more than one output intensity can occur for certain values of the input intensity, which shows that the system possesses multiple solutions.

The input–output characteristics for a system showing optical bistability are shown schematically in Fig. 7.3.4a. The portion of the curve that has negative slope is shown by a dashed line. This portion corresponds to the branch of the solution to Eq. (7.3.10) for which the output intensity increases as the input intensity decreases. As might be expected on intuitive grounds, and as can be verified by means of a linear stability analysis, this branch of the solution is unstable; if the system is initially in this state, it will rapidly switch to one of the stable solutions through the growth of small perturbations.

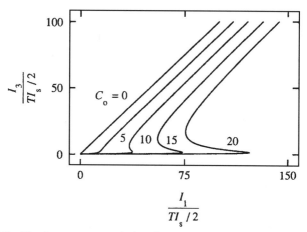

FIGURE 7.3.3   The input–output relation for a bistable optical device described by Eqs. (7.3.10) and (7.3.11).

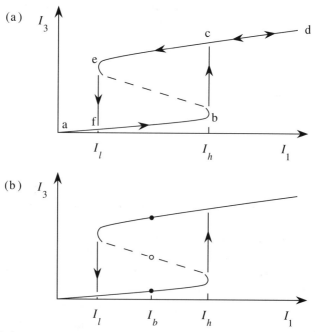

FIGURE 7.3.4   Schematic representation of the input–output characteristics of a system showing optical bistability.

The solution shown in Fig. 7.3.4a displays hysteresis in the following sense. We imagine that the input intensity I_1 is initially zero and is slowly increased. As I_1 is increased from zero to I_h (the high jump point), the output intensity is given by the *lower branch* of the solution, that is, by the segment terminated by points a and b. As the input intensity is increased still further, the output intensity must jump to point c and trace out that portion of the curve labeled $c–d$. If the intensity is now slowly decreased, the system will remain on the upper branch and the output intensity will be given by the curve segment $e–d$. As the input intensity passes through the value I_l (the low jump point), the system makes a transition to point f and traces out the curve of $f–a$ as the input intensity is decreased to zero.

The use of such a device as an optical switch is illustrated in part (b) of Fig. 7.3.4. If the input intensity is held fixed at the value I_b (the bias intensity), the two stable output points indicated by the filled dots are possible. The state of the system can be used to store binary information. The system can be forced to make a transition to the upper state by injecting a pulse of light so that the total input intensity exceeds I_h; the system can be forced to make a transition to the lower state by momentarily blocking the input beam.

Refractive Bistability

Let us now consider the case in which the absorption coefficient vanishes but in which the refractive index n depends nonlinearly on the optical intensity. For $\alpha = 0$, Eq. (7.3.3) becomes

$$A_2 = \frac{\tau A_1}{1 - \rho^2 e^{2ikl}} = \frac{\tau A_1}{1 - R e^{i\delta}}. \tag{7.3.12}$$

In obtaining the second form of this equation, we have written ρ^2 in terms of its amplitude and phase as

$$\rho^2 = R e^{i\phi} \tag{7.3.13}$$

and have introduced the total phase shift δ acquired in a round trip through the cavity. This phase shift is the sum

$$\delta = \delta_0 + \delta_2 \tag{7.3.14}$$

of a linear contribution

$$\delta_0 = \phi + 2n_0 \frac{\omega}{c} l \tag{7.3.15}$$

and a nonlinear contribution

$$\delta_2 = 2n_2 I \frac{\omega}{c} l, \tag{7.3.16}$$

where

$$I = I_2 + I_2' \simeq 2I_2. \tag{7.3.17}$$

Equation (7.3.12) can be used to relate the intensities $I_i = (nc/2\pi)|A_i|^2$ of the incident and internal fields as

$$\begin{aligned}
I_2 &= \frac{T I_1}{(1 - Re^{i\delta})(1 - Re^{-i\delta})} = \frac{T I_1}{1 + R^2 - 2R\cos\delta} \\
&= \frac{T I_1}{(1 - R)^2 + 4R\sin^2\frac{1}{2}\delta} = \frac{T I_1}{T^2 + 4R\sin^2\frac{1}{2}\delta} \\
&= \frac{I_1/T}{1 + (4R/T^2)\sin^2\frac{1}{2}\delta},
\end{aligned} \tag{7.3.18}$$

which shows that

$$\frac{I_2}{I_1} = \frac{1/T}{1 + (4R/T^2)\sin^2\frac{1}{2}\delta}, \tag{7.3.19}$$

where, according to Eqs. (7.3.14) though (7.3.17), the phase shift is given by

$$\delta = \delta_0 + (4n_2\omega l/c)\, I_2. \tag{7.3.20}$$

In order to determine the conditions under which bistability can occur, we solve Eqs. (7.3.19) and (7.3.20) for the internal intensity I_2 as a function of the incident intensity I_1. This procedure is readily performed graphically by plotting each side of Eq. (7.3.19) as a function of I_2. Such a plot is shown in Fig. 7.3.5. We see that the system can possess one, three, five, or more solutions

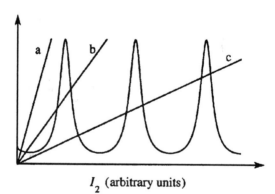

I_2 (arbitrary units)

FIGURE 7.3.5   Graphical solution to Eq. (7.3.19). The oscillatory curve represents the right-hand side of this equation, and the straight lines labeled a though c represent the left-hand side for increasing values of the input intensity I_1.

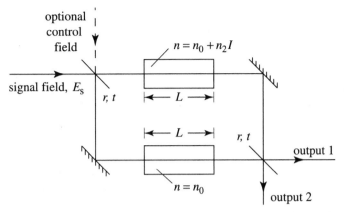

FIGURE 7.3.6   Configuration of an all-optical switch in the form of a Mach–Zehnder interferometer containing a nonlinear element. The input signal field is routed to either output 1 or 2 depending on its intensity and/or on the intensity of the control field.

depending on the value of I_1. For the case in which three solutions exist for the range of input intensities I_1 that are available, a plot of I_3 versus I_1 looks very much like the curves shown in Fig. 7.3.4. Hence, the qualitative discussion of optical bistability given above is applicable in this case as well.

More detailed treatments of optical bistability can be found in Lugiato (1984) and Gibbs (1985).

Optical Switching

Let us now analyze a prototypical all-optical switching device, as illustrated in Fig. 7.3.6. For simplicity, in the present analysis we assume that only a signal field is applied to the device; we shall show that this signal beam is directed to one or the other of the output ports depending on its intensity. Such an application of this device is illustrated in Fig. 7.3.7. A more general situation, in which both signal and control fields are applied to the device, can be treated by a similar but somewhat more complicated calculation, with the conclusion that the control field can be used to route the signal beam to either output port.

We assume that a signal field of amplitude E_s is incident upon the device and that the beam splitters are symmetric (have the same amplitude and reflection coefficients r and t for beams incident on the beam splitter from either side) with coefficients given by[*]

$$r = i\sqrt{R} \qquad t = \sqrt{T} \qquad (7.3.21)$$

[*] This form of the beam-splitter relation ensures that the transfer characteristics obey a unitarity condition or equivalently that they obey the Stokes relations.

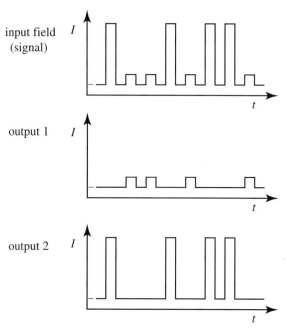

FIGURE 7.3.7   Illustration of the use of the device of Fig. 7.3.6 (without a control field) as a pulse sorter.

with

$$R + T = 1. \tag{7.3.22}$$

The field at output port 1 is then seen to be given by

$$E_1 = E_s \, (rt + rt \, e^{i\phi_{\mathrm{NL}}}) \tag{7.3.23}$$

where

$$\phi_{\mathrm{NL}} = n_2(\omega/c)IL = n_2(\omega/c)|t|^2(n_0c/2\pi)|E_s|^2L. \tag{7.3.24}$$

The intensity at output port 1 is thus proportional to

$$|E_1|^2 = |E_s|^2|r|^2|t|^2(1 + e^{i\phi_{\mathrm{NL}}})(1 + e^{-i\phi_{\mathrm{NL}}})$$
$$= 2|E_s|^2RT(1 + \cos\phi_{\mathrm{NL}}). \tag{7.3.25}$$

We similarly find that the output at port 2 is given by

$$E_2 = E_s \, (r^2 + t^2 \, e^{i\phi_{\mathrm{NL}}}) \tag{7.3.26}$$

with an intensity proportional to

$$|E_2|^2 = |E_s|^2[R^2 + T^2 - 2RT\cos\phi_{\mathrm{NL}}]. \tag{7.3.27}$$

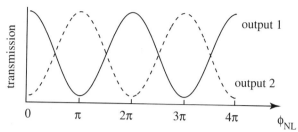

FIGURE 7.3.8   Plot of the transfer relations described by Eqs. (7.3.25) and (7.3.27).

Note that

$$|E_1|^2 + |E_2|^2 = |E_s|^2 \qquad (7.3.28)$$

as required by conservation of energy. These relations are illustrated in Fig. 7.3.8 and lead to the sort of behavior shown qualitatively in Fig. 7.3.7.

Even though the calculation just presented is somewhat simplistic in that it considers the situation in which there is only a single input beam, it illustrates a crucial point: a nonlinear phase shift of π radians is required to produce high-contrast all-optical switching. The requirement that the nonlinear phase shift be as large as π radians is generic to a broad class of all-optical switching devices.

Let us therefore examine more carefully the conditions under which a nonlinear phase shift of π radians can be achieved. Let us first examine the consequences of using a nonlinear optical material that displays linear absorption. Under this circumstance the nonlinear phase shift is given by

$$\phi_{\mathrm{NL}} = n_2(\omega/c) \int_0^L I(z)\,dz \qquad (7.3.29)$$

where

$$I(z) = I_0 e^{-\alpha z}. \qquad (7.3.30)$$

Straightforward integration leads to the result

$$\phi_{\mathrm{NL}} = n_2(\omega/c) I_0 L_{\mathrm{eff}} \qquad (7.3.31\mathrm{a})$$

where

$$L_{\mathrm{eff}} = \frac{1 - e^{-\alpha L}}{\alpha}. \qquad (7.3.31\mathrm{b})$$

Note that

$$L_{\mathrm{eff}} \to L \qquad \text{for } \alpha L \ll 1, \qquad (7.3.32\mathrm{a})$$

$$L_{\mathrm{eff}} \to 1/\alpha \qquad \text{for } \alpha L \gg 1. \qquad (7.3.32\mathrm{b})$$

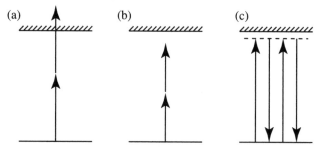

FIGURE 7.3.9   Two-photon absorption (shown in a) can be prevented by utilizing a material (b) such that the laser frequency lies below the half-band-gap energy. This strategy, however, precludes the use of one-photon-resonant nonlinearities (c).

Thus for a strongly absorbing nonlinear optical material the effective interaction length can be much shorter than the physical length of the nonlinear medium. We also note that optical damage (see also Chapter 11) imposes a limit on how large a value of I_0 can be used for a particular material. Thus certain material cannot even in principle be used for all-optical switching.

When the optical material displays two-photon absorption as well as linear absorption, the absorption coefficient appearing in Eq. (7.3.30) should be replaced by

$$\alpha = \alpha_0 + \beta I \qquad (7.3.33)$$

where β is the two-photon absorption coefficient.

Two-photon absorption is often a significant problem in the design of all-optical switching devices because it occurs at the same order of nonlinearly as the intensity-dependent refractive index n_2 (because these processes are proportional to the imaginary and real parts of $\chi^{(3)}$ respectively). Two-photon absorption can be eliminated entirely by choosing a material for which the lowest-lying excited state lies more than $2\hbar\omega$ above the ground state, as illustrated schematically in Fig. 7.3.9.

A good summary of all-optical switching has been presented by Stegeman and Miller (1993).

7.4. Two-Beam Coupling

Let us consider the situation shown in Fig. 7.4.1, in which two beams of light (which in general have different frequencies) interact in a nonlinear material. Under certain conditions, the two beams interact in such a manner that energy is transferred from one beam to the other; this phenomenon is known as two-beam coupling. Two-beam coupling is a process that is automatically phase-matched.

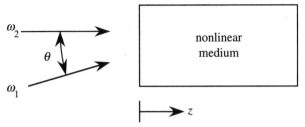

FIGURE 7.4.1 Two-beam coupling.

Consequently the efficiency of the process does not depend critically upon the angle θ between the two beams. The reason why this process is automatically phase-matched will be clarified by the following analysis; for the present it is perhaps helpful to note that the origin of two-beam coupling is that the refractive index experienced by either wave is modified by the *intensity* of the other wave.

Two-beam coupling occurs under several different circumstances in nonlinear optics. We saw in Chapter 6 that the nonlinear response of a two-level atom to pump and probe fields can lead to amplification of the probe wave. Furthermore, we shall see in Chapters 9 and 10 that gain occurs for various scattering processes such as stimulated Brillouin scattering and stimulated Raman scattering. Furthermore, in Chapter 11 we shall see that two-beam coupling occurs in many photorefractive materials. In the present section, we examine two-beam coupling from a general point of view that elucidates the conditions under which such energy transfer can occur. Our analysis is similar to that of Silberberg and Bar-Joseph (1982, 1984).

We describe the total optical field within the nonlinear medium as

$$\tilde{E}(\mathbf{r}, t) = A_1 e^{i(\mathbf{k}_1 \cdot \mathbf{r} - \omega_1 t)} + A_2 e^{i(\mathbf{k}_2 \cdot \mathbf{r} - \omega_2 t)} + \text{c.c.}, \qquad (7.4.1)$$

where $k_i = n_0 \omega_i / c$, with n_0 denoting the linear part of the refractive index experienced by each wave. We now consider the intensity distribution associated with the interference between the two waves. The intensity is given in general by

$$I = \frac{n_0 c}{4\pi} \overline{\tilde{E}^2}, \qquad (7.4.2)$$

where the overbar denotes an average over a time interval of many optical periods. The intensity distribution for $\tilde{E}$ given by Eq. (7.4.1) is hence given by

$$
\begin{aligned}
I &= \frac{n_0 c}{2\pi} \left\{ A_1 A_1^* + A_2 A_2^* + \left[A_1 A_2^* e^{i(\mathbf{k}_1 - \mathbf{k}_2) \cdot \mathbf{r} - i(\omega_1 - \omega_2)t} + \text{c.c.} \right] \right\} \\
&= \frac{n_0 c}{2\pi} \left\{ A_1 A_1^* + A_2 A_2^* + \left[A_1 A_2^* e^{i(\mathbf{q} \cdot \mathbf{r} - \delta t)} + \text{c.c.} \right] \right\}
\end{aligned}
\qquad (7.4.3)
$$

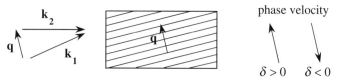

FIGURE 7.4.2   Interference pattern formed by two interacting waves.

where we have introduced the wavevector difference (or "grating" wavevector)

$$\mathbf{q} = \mathbf{k}_1 - \mathbf{k}_2 \qquad (7.4.4a)$$

and the frequency difference

$$\delta = \omega_1 - \omega_2. \qquad (7.4.4b)$$

For the geometry of Fig. 7.4.1, the interference pattern has the form shown in Fig. 7.4.2, where we have assumed that $|\delta| \ll \omega_1$. Note that the pattern moves upward for $\delta > 0$, moves downward for $\delta < 0$, and is stationary for $\delta = 0$.

A particularly simple example is the special case in which $\theta = 180$ degrees. Then, again assuming that $|\delta| \ll \omega_1$, we find that the wavevector difference is given approximately by

$$\mathbf{q} \simeq -2\mathbf{k}_2 \qquad (7.4.5)$$

and hence that the intensity distribution is given by

$$I = \frac{n_0 c}{2\pi} \left\{ A_1 A_1^* + A_2 A_2^* + \left[A_1 A_2^* e^{i(-2kz - \delta t)} + \text{c.c.} \right] \right\}. \qquad (7.4.6)$$

The interference pattern is hence of the form shown in Fig. 7.4.3. If δ is positive, the interference pattern moves to the left, and if δ is negative it moves to the right, in either case with phase velocity $|\delta|/2k$.

Since the material system is nonlinear, a refractive index variation accompanies this intensity variation. Each wave is scattered by this index variation, or grating. We shall show below that no energy transfer accompanies this interaction for the case of a nonlinear material that responds instantaneously to the applied field. In order to allow the possibility of energy transfer, we assume that the nonlinear part of the refractive index (n_{NL}) obeys a Debye

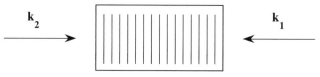

FIGURE 7.4.3   Interference pattern formed by two counterpropagating beams.

relaxation equation of the form

$$\tau \frac{dn_{\mathrm{NL}}}{dt} + n_{\mathrm{NL}} = n_2 I. \tag{7.4.7}$$

Note that this equation predicts that, in steady state, the nonlinear contribution to the refractive index is given simply by $n_{\mathrm{NL}} = n_2 I$, in consistency with Eq. (4.1.15). However, under transient conditions it predicts that the nonlinearity develops in a time interval of the order of τ.

Equation (7.4.7) can be solved (for example, by the method of variation of parameters or by the Green's function method) to give the result

$$n_{\mathrm{NL}} = \frac{n_2}{\tau} \int_{-\infty}^{t} I(t') e^{(t'-t)/\tau} dt'. \tag{7.4.8}$$

The expression (7.4.3) for the intensity $I(t)$ is next introduced into this equation. We find, for example, that the part of $I(t)$ that varies as $e^{-i\delta t}$ leads to an integral of the form

$$\int_{-\infty}^{t} e^{-i\delta t'} e^{(t'-t)/\tau} dt' = e^{-t/\tau} \int_{-\infty}^{t} e^{(-i\delta+1/\tau)t'} dt' = \frac{e^{-i\delta t}}{-i\delta + 1/\tau}. \tag{7.4.9}$$

Equation (7.4.8) hence shows that the nonlinear contribution to the refractive index is given by

$$n_{\mathrm{NL}} = \frac{n_0 n_2 c}{2\pi} \left[(A_1 A_1^* + A_2 A_2^*) + \frac{A_1 A_2^* e^{i(\mathbf{q}\cdot\mathbf{r}-\delta t)}}{1 - i\delta\tau} + \frac{A_1^* A_2 e^{-i(\mathbf{q}\cdot\mathbf{r}-\delta t)}}{1 + i\delta\tau} \right]. \tag{7.4.10}$$

Because of the complex nature of the denominators, the refractive index variation is not in general in phase with the intensity distribution.

In order to determine the degree of coupling between the two fields, we require that the field given by Eq. (7.4.1) satisfy the wave equation

$$\nabla^2 \tilde{E} - \frac{n^2}{c^2} \frac{\partial^2 \tilde{E}}{\partial t^2} = 0, \tag{7.4.11}$$

where we take the refractive index to have the form

$$n = n_0 + n_{\mathrm{NL}}. \tag{7.4.12}$$

We make the physical assumption that $|n_{\mathrm{NL}}| \ll n_0$, in which case it is a good approximation to express n^2 as

$$n^2 = n_0^2 + 2n_0 n_{\mathrm{NL}}. \tag{7.4.13}$$

Let us consider the part of Eq. (7.4.11) that shows a spatial and temporal dependence given by $\exp[i(\mathbf{k}_2 \cdot \mathbf{r} - \omega_2 t)]$. Using Eqs. (7.4.1), (7.4.10), and

(7.4.13), we find that this portion of Eq. (7.4.11) is given by

$$
\frac{d^2 A_2}{dz^2} + 2ik_2 \frac{dA_2}{dz} - k_2^2 A_2 + \frac{n_0^2 \omega_2^2}{c_2} A_2
$$

$$
= -\frac{n_0^2 N_2 \omega_2^2}{\pi c}(|A_1|^2 + |A_2|^2)A_2 - \frac{n_0^2 n_2 \omega_1^2}{\pi c}\frac{|A_1|^2 A_2}{1 + i\delta\tau}.
$$

(7.4.14)

Note that the origin of the last term on the right-hand side is the scattering of the field $A_1 \exp[i(\mathbf{k_1} \cdot \mathbf{r} - \omega_1 t)]$ from the time-varying refractive index distribution (i.e., the moving grating)

$$
\frac{n_0 c}{2\pi} n_2 A_1^* A_2 \frac{e^{-i(\mathbf{q} \cdot \mathbf{r} - \delta t)}}{1 + i\delta\tau},
$$

whereas the origin of the first term on the right-hand side is the scattering of the field $A_2 \exp[i(\mathbf{k_2} \cdot \mathbf{r} - \omega_2 t)]$ from the stationary refractive index variation

$$
\frac{n_0 c}{2\pi} n_2 (A_1 A_1^* + A_2 A_2^*).
$$

We next drop the first term on the left-hand side of Eq. (7.4.14) by making the slowly-varying amplitude approximation, and we note that the third and fourth terms exactly cancel. The equation then reduces to

$$
\frac{dA_2}{dz} = i\frac{n_0 n_2 \omega}{2\pi}(|A_1|^2 + |A_2|^2)A_2 + i\frac{n_0 n_2 \omega}{2\pi}\frac{|A_1|^2 A_2}{1 + i\delta\tau},
$$

(7.4.15)

where, to good approximation, we have replaced ω_1 and ω_2 by ω. We now calculate the rate of change of intensity of the ω_2 field. We introduce the intensities

$$
I_1 = \frac{n_0 c}{2\pi} A_1 A_1^* \quad \text{and} \quad I_2 = \frac{n_0 c}{2\pi} A_2 A_2^*
$$

(7.4.16)

and note that the spatial variation of I_2 is given by

$$
\frac{dI_2}{dz} = \frac{n_0 c}{2\pi}\left(A_2^* \frac{dA_2}{dz} + A_2 \frac{dA_2^*}{dz}\right).
$$

(7.4.17)

We then find from Eqs. (7.4.15) through (7.4.17) that

$$
\frac{dI_2}{dz} = \frac{2n_2 \omega}{c}\frac{\delta\tau}{1 + \delta^2\tau^2} I_1 I_2.
$$

(7.4.18)

Note that only the last term on the right-hand side of Eq. (7.4.15) contributes to energy transfer.

For the case of a positive value of n_2 (for example, for the molecular orientation Kerr effect, for electrostriction, for a two-level atom with the optical

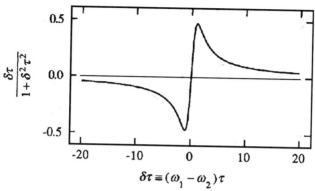

FIGURE 7.4.4 Frequency dependence of the gain for two-beam coupling.

frequencies above the resonance frequency), Eq. (7.4.18) predicts gain for positive δ, i.e., for $\omega_2 < \omega_1$. The frequency dependence of Eq. (7.4.18) is shown in Fig. 7.4.4.

Note that the ω_2 wave experiences maximum gain for $\delta\tau = 1$, in which case Eq. (7.4.18) becomes

$$\frac{dI_2}{dz} = n_2 \frac{\omega}{c} I_1 I_2. \tag{7.4.19}$$

Note also from Eq. (7.4.18) that in the limit of an infinitely fast nonlinearity, that is, in the limit $\tau \to 0$, the coupling of intensity between the two waves vanishes. The reason for this behavior is that only the imaginary part of the (total) refractive index can lead to a change in intensity of the ω_2 wave. We see from Eq. (7.4.10) that (for a real n_2) the only way in which n_{NL} can become complex is if τ is nonzero. When τ is nonzero, the response can lag in phase behind the driving term, leading to a complex value of the nonlinear contribution to the refractive index.

The theory just presented predicts that there will be no energy coupling if the product $\delta\tau$ vanishes, either because the nonlinearity has a fast response or because the input waves are at the same frequency. However, two-beam coupling can occur in certain photorefractive crystals even between beams of the same frequency (Feinberg, 1983). In such cases, energy transfer occurs as a result of a *spatial* phase shift between the nonlinear index grating and the optical intensity distribution. The direction of energy flow depends upon the orientation of the wave vectors of the optical beams with respect to some symmetry axis of the photorefractive crystal. The photorefractive effect is described in greater detail in Chapter 11.

7.5.   Pulse Propagation and Temporal Solitons

In this section we study some of the nonlinear optical effects that can occur when short optical pulses propagate through dispersive nonlinear optical media. We shall see that the spectral content of the pulse can become modified by the nonlinear optical process of self-phase modulation. This process is especially important for pulses of high peak intensity. We shall also see that (even for the case of a medium with a linear response) the shape of the pulse can become modified by means of propagation effects such as dispersion of the group velocity within the medium. This process is especially important for very short optical pulses, which necessarily have a broad spectrum. In general, self-phase modulation and group-velocity dispersion occur simultaneously, and both tend to modify the shape of the optical pulse. However, under certain circumstances, which are described below, an exact cancellation of these two effects can occur, allowing a special type of pulse known as an optical soliton to propagate through large distances with no change in shape.

Self-Phase Modulation

Self-phase modulation is the change in the phase of an optical pulse resulting from the nonlinearity of the refractive index of the material medium. In order to understand the origin of this effect, let us consider the propagation of the optical pulse

$$\tilde{E}(z, t) = \tilde{A}(z, t)e^{i(k_0 z - \omega_0 t)} + \text{c.c.} \tag{7.5.1}$$

through a medium characterized by a nonlinear refractive index of the sort

$$n(t) = n_0 + n_2 I(t), \tag{7.5.2}$$

where $I(t) = (n_0 c / 2\pi)|\tilde{A}(z, t)|^2$. Note that for the present we are assuming that the medium can respond essentially instantaneously to the pulse intensity. We also assume that the nonlinear medium is sufficiently short that no reshaping of the optical pulse can occur within the medium; the only effect of the medium is to change the phase of the transmitted pulse by the amount

$$\phi_{\text{NL}}(t) = -n_2 I(t)\omega_0 L/c. \tag{7.5.3}$$

As a result of the time-varying phase of the wave, the spectrum of the transmitted pulse will be modified and typically will be broader than that of the incident pulse. From a formal point of view, we can determine the spectral content of the transmitted pulse by calculating its energy spectrum

$$S(\omega) = \left| \int_{-\infty}^{\infty} \tilde{A}(t) e^{-i\omega_0 t - i\phi_{\mathrm{NL}}(t)} e^{i\omega t} dt \right|^2. \tag{7.5.4}$$

However, it is more intuitive to describe the spectral content of the transmitted pulse by introducing the concept of the instantaneous frequency $\omega(t)$ of the pulse, which is described by

$$\omega(t) = \omega_0 + \delta\omega(t) \tag{7.5.5a}$$

where

$$\delta\omega(t) = \frac{d}{dt}\phi_{\mathrm{NL}}(t) \tag{7.5.5b}$$

denotes the variation of the instantaneous frequency. The instantaneous frequency is a well-defined concept and is given by Eqs. (7.5.5) whenever the amplitude $\tilde{A}(t)$ varies slowly compared to an optical period.

As an example of the use of these formulas, we consider the case illustrated in part (a) of Fig. 7.5.1, in which the pulse shape is given by the form

$$I(t) = I_0 \operatorname{sech}^2(t/\tau_0). \tag{7.5.6}$$

We then find from Eq. (7.5.3) that the nonlinear phase shift is given by

$$\phi_{\mathrm{NL}}(t) = -n_2 \frac{\omega_0}{c} L I_0 \operatorname{sech}^2(t/\tau_0), \tag{7.5.7}$$

and from Eq. (7.5.5b) that the change in instantaneous frequency is given by

$$\delta\omega(t) = 2n_2 \frac{\omega_0}{c\tau_0} L I_0 \operatorname{sech}^2(t/\tau_0) \tanh(t/\tau_0). \tag{7.5.8}$$

The variation in the instantaneous frequency is illustrated in part (b) of Fig. 7.5.1, under the assumption that n_2 is positive. We see that the leading edge of the pulse is shifted to lower frequencies and that the trailing edge is shifted to higher frequencies. This conclusion is summarized schematically in part (c) of the figure. The maximum value of the frequency shift is of the order of

$$\delta\omega_{\max} \simeq \frac{\Delta\phi_{\mathrm{NL}}^{(\max)}}{\tau_0}, \quad \text{where} \quad \Delta\phi_{\mathrm{NL}}^{(\max)} \simeq n_2 \frac{\omega_0}{c} I_0 L. \tag{7.5.9}$$

We expect that spectral broadening due to self-phase modulation will be important whenever $\delta\omega_{\max}$ exceeds the spectral width of the incident pulse, which for the case of a smooth pulse is of the order of $1/\tau_0$. We thus expect self-phase modulation to be important whenever $\Delta\phi_{\mathrm{NL}}^{(\max)} \geq 2\pi$.

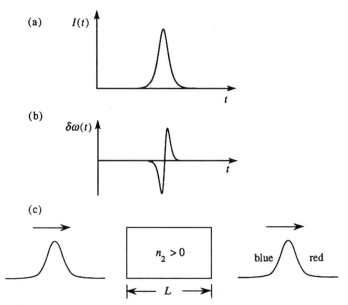

FIGURE 7.5.1   (a) Time dependence of the incident pulse. (b) Change in instantaneous frequency of the transmitted pulse. (c) Experimental arrangement to observe self-phase modulation.

Self-phase modulation of the sort just described was studied initially by Brewer (1967), by Shimizu (1967), and by Cheung *et al.* (1968).

Pulse Propagation Equation

Let us next consider the equations that govern the propagation of the pulse

$$\tilde{E}(z, t) = \tilde{A}(z, t)e^{i(k_0 z - \omega_0 t)} + \text{c.c.}, \qquad (7.5.10)$$

where $k_0 = n_{\text{lin}}(\omega_0)\omega_0/c$, through a dispersive, nonlinear optical medium. In particular, we seek an equation that describes how the pulse envelope function $\tilde{A}(z, t)$ propagates through the medium. We begin with the wave equation in the form (see also Eq. (2.1.9a))

$$\frac{\partial^2 \tilde{E}}{\partial z^2} - \frac{1}{c^2}\frac{\partial^2 \tilde{D}}{\partial t^2} = 0, \qquad (7.5.11)$$

where $\tilde{D}$ represents the total displacement field, including both linear and nonlinear contributions. We now introduce the Fourier transforms of $\tilde{E}(z, t)$ and

$\tilde{D}(z, t)$ by the equations

$$\tilde{E}(z, t) = \int_{-\infty}^{\infty} E(z, \omega) e^{-i\omega t} \frac{d\omega}{2\pi}, \qquad \tilde{D}(z, t) = \int_{-\infty}^{\infty} D(z, \omega) e^{-i\omega t} \frac{d\omega}{2\pi}.$$

$$(7.5.12)$$

The Fourier amplitudes $E(z, \omega)$ and $D(z, \omega)$ are related by

$$D(z, \omega) = \epsilon(\omega) E(z, \omega), \tag{7.5.13}$$

where $\epsilon(\omega)$ is the effective dielectric constant that describes both the linear and nonlinear contributions to the response.

Equations (7.5.12) and (7.5.13) are now introduced into the wave equation (7.5.11), which leads to the result that each Fourier component of the field must obey the equation

$$\frac{\partial^2 E(z, \omega)}{\partial z^2} + \epsilon(\omega) \frac{\omega^2}{c^2} E(z, \omega) = 0. \tag{7.5.14}$$

We now write this equation in terms of the Fourier transform of $\tilde{A}(z, t)$, which is given by

$$A(z, \omega') = \int_{-\infty}^{\infty} \tilde{A}(z, t) e^{i\omega' t} dt, \tag{7.5.15}$$

and which is related to $E(z, \omega)$ by

$$E(z, \omega) = A(z, \omega - \omega_0) e^{ik_0 z} + A^*(z, \omega + \omega_0) e^{-ik_0 z}$$

$$\simeq A(z, \omega - \omega_0) e^{ik_0 z}, \tag{7.5.16}$$

where the second, approximate form is obtained by noting that a quantity such as $\tilde{A}(z, t)$ which varies slowly in time cannot possess high-frequency Fourier components. This expression for $E(z, \omega)$ is now introduced into Eq. (7.5.14), and the slowly-varying amplitude approximation is made, so that the term containing $\partial^2 A/\partial z^2$ can be dropped. One obtains

$$2ik_0 \frac{\partial A}{\partial z} + \left(k^2 - k_0^2\right) A = 0, \tag{7.5.17}$$

where

$$k(\omega) = \sqrt{\epsilon(\omega)} \omega/c. \tag{7.5.18}$$

In practice, k typically differs from k_0 by only a small fraction amount, and thus to good approximation $k^2 - k_0^2$ can be replaced by $2k_0(k - k_0)$, so that

Eq. (7.5.17) becomes

$$\frac{\partial A(z, \omega - \omega_0)}{\partial z} - i(k - k_0)A(z, \omega - \omega_0) = 0. \qquad (7.5.19)$$

Recall that the propagation constant k depends both on the frequency and (through the intensity dependence of ϵ) on the intensity of the optical wave. It is often adequate to describe this dependence in terms of a truncated power series expansion of the form

$$k = k_0 + \Delta k_{\mathrm{NL}} + k_1(\omega - \omega_0) + \tfrac{1}{2}k_2(\omega - \omega_0)^2. \qquad (7.5.20)$$

In this expression, we have introduced the nonlinear contribution to the propagation constant, given by

$$\Delta k_{\mathrm{NL}} = \Delta n_{\mathrm{NL}}\omega_0/c = n_2 I \omega_0/c, \qquad (7.5.21)$$

with $I = [n_{\mathrm{lin}}(\omega_0)c/2\pi]|\tilde{A}(z, t)|^2$, and have introduced the quantities

$$k_1 = \left(\frac{dk}{d\omega}\right)_{\omega=\omega_0} = \frac{1}{c}\left[n_{\mathrm{lin}}(\omega) + \omega\frac{dn_{\mathrm{lin}}(\omega)}{d\omega}\right]_{\omega=\omega_0} \equiv \frac{1}{v_g(\omega_0)} \qquad (7.5.22)$$

and

$$k_2 = \left(\frac{d^2k}{d\omega^2}\right)_{\omega=\omega_0} = \frac{d}{d\omega}\left[\frac{1}{v_g(\omega)}\right]_{\omega=\omega_0} = \left(-\frac{1}{v_g^2}\frac{dv_g}{d\omega}\right)_{\omega=\omega_0}. \qquad (7.5.23)$$

Here k_1 is the reciprocal of the group velocity, and k_2 is a measure of the dispersion of the group velocity. As illustrated in Fig. 7.5.2, the long-wavelength components of an optical pulse propagate faster than the short-wavelength components when the group velocity dispersion parameter k_2 is positive, and vice versa.

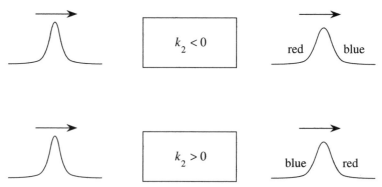

FIGURE 7.5.2   Pulse spreading resulting from group velocity dispersion.

The expression (7.5.20) for k is next introduced into the reduced wave equation (7.5.19), which becomes

$$\frac{\partial A}{\partial z} - i\,\Delta k_{\mathrm{NL}}A - ik_1(\omega - \omega_0)A - \tfrac{1}{2}ik_2(\omega - \omega_0)^2 A = 0. \quad (7.5.24)$$

This equation is now transformed from the frequency domain to the time domain. To do so, we multiply each term by the factor $\exp[-i(\omega - \omega_0)t]$ and integrate the resulting equation over all values of $\omega - \omega_0$. We evaluate the resulting integrals as follows:

$$\int_{-\infty}^{\infty} A(z, \omega - \omega_0)e^{-i(\omega-\omega_0)t}\frac{d(\omega - \omega_0)}{2\pi} = \tilde{A}(z, t), \quad (7.5.25a)$$

$$\int_{-\infty}^{\infty} (\omega - \omega_0)A(z, \omega - \omega_0)e^{-i(\omega-\omega_0)t}\frac{d(\omega - \omega_0)}{2\pi}$$

$$(7.5.25b)$$

$$= \frac{1}{-i}\frac{\partial}{\partial t}\int_{-\infty}^{\infty} A(z, \omega - \omega_0)e^{-i(\omega-\omega_0)t}\frac{d(\omega - \omega_0)}{2\pi} = i\frac{\partial}{\partial t}\tilde{A}(z, t),$$

$$\int_{-\infty}^{\infty} (\omega - \omega_0)^2 A(z, \omega - \omega_0)e^{-i(\omega-\omega_0)t}\frac{d(\omega - \omega_0)}{2\pi} = -\frac{\partial^2}{\partial t^2}\tilde{A}(z, t). \quad (7.5.25c)$$

Equation (7.5.24) then becomes

$$\frac{\partial \tilde{A}}{\partial z} + k_1\frac{\partial \tilde{A}}{\partial t} + \tfrac{1}{2}ik_2\frac{\partial^2 \tilde{A}}{\partial t^2} - i\,\Delta k_{\mathrm{NL}}\tilde{A} = 0. \quad (7.5.26)$$

This equation can be simplified by means of a coordinate transformation. In particular we introduce the retarded time τ by the substitution

$$\tau = t - \frac{z}{v_g} = t - k_1 z, \quad (7.5.27)$$

and we describe the optical pulse by the function $\tilde{A}_s(z, \tau)$, which is related to the function $\tilde{A}(z, t)$ by

$$\tilde{A}_s(z, \tau) = \tilde{A}(z, t). \quad (7.5.28)$$

We next use the chain rule of differentiation to show that

$$\frac{\partial \tilde{A}}{\partial z} = \frac{\partial \tilde{A}_s}{\partial z} + \frac{\partial \tilde{A}_s}{\partial \tau}\frac{\partial \tau}{\partial z} = \frac{\partial \tilde{A}_s}{\partial z} - k_1\frac{\partial \tilde{A}_s}{\partial \tau}, \quad (7.5.29a)$$

$$\frac{\partial \tilde{A}}{\partial t} = \frac{\partial \tilde{A}_s}{\partial z}\frac{\partial z}{\partial t} + \frac{\partial \tilde{A}_s}{\partial \tau}\frac{\partial \tau}{\partial t} = \frac{\partial \tilde{A}_s}{\partial \tau}, \quad (7.5.29b)$$

and analogously that $\partial^2 \tilde{A}/\partial t^2 = \partial^2 \tilde{A}_s/\partial \tau^2$. These expressions are now introduced into Eq. (7.5.26), which becomes

$$\frac{\partial \tilde{A}_s}{\partial z} + \frac{1}{2}ik_2 \frac{\partial^2 \tilde{A}_s}{\partial \tau^2} - i\,\Delta k_{\mathrm{NL}}\tilde{A}_s = 0. \qquad (7.5.30)$$

Finally, we express the nonlinear contribution to the propagation constant as

$$\Delta k_{\mathrm{NL}} = n_2 \frac{\omega_0}{c} I = \frac{n_0 n_2 \omega_0}{2\pi}|\tilde{A}_s|^2 \equiv \gamma |\tilde{A}_s|^2, \qquad (7.5.31)$$

so that Eq. (7.5.30) can be expressed as

$$\frac{\partial \tilde{A}_s}{\partial z} + \frac{1}{2}ik_2 \frac{\partial^2 \tilde{A}_s}{\partial \tau^2} = i\gamma |\tilde{A}_s|^2 \tilde{A}_s. \qquad (7.5.32)$$

This equation describes the propagation of optical pulses through dispersive, nonlinear optical media. Note that the second term on the left-hand side shows how pulses tend to spread because of group velocity dispersion, and that the term on the right-hand side shows how pulses tend to spread because of self-phase modulation. Equation (7.5.32) is sometimes referred to as the nonlinear Schrödinger equation.

Temporal Optical Solitons

Note from the form of the pulse propagation equation (7.5.32) that it is possible for the effects of group velocity dispersion to compensate for the effects of self-phase modulation. In fact, under appropriate circumstances the degree of compensation can be complete, and optical pulses can propagate through a dispersive, nonlinear optical medium with an invariant shape. Such pulses are known as temporal optical solitons.

As an example of a soliton, note that Eq. (7.5.32) is solved identically by a pulse whose amplitude is of the form

$$\tilde{A}_s(z, \tau) = A_s^0 \,\mathrm{sech}(\tau/\tau_0)e^{i\kappa z} \qquad (7.5.33a)$$

where the pulse amplitude A_s^0 and pulse width τ_0 must be related according to

$$\left|A_s^0\right|^2 = \frac{-k_2}{\gamma \tau_0^2} = \frac{-2\pi k_2}{n_0 n_2 \omega_0 \tau_0^2} \qquad (7.5.33b)$$

and where

$$\kappa = -k_2/2\tau_0^2 = \tfrac{1}{2}\gamma\left|A_s^0\right|^2 \qquad (7.5.33c)$$

represents the phase shift experienced by the pulse upon propagation. One can verify by direct substitution that Eqs. (7.5.33) do in fact satisfy the pulse propagation equation (7.5.32) (see Problem 15 at the end of this chapter).

Note that condition (7.5.33b) shows that k_2 and n_2 must have opposite signs in order for Eq. (7.5.33a) to represent a physical pulse in which the intensity $|A_s^0|^2$ and the square of the pulse width τ_0^2 are both positive. We can see from Eq. (7.5.32) that in fact k_2 and γ must have opposite signs in order for group velocity dispersion to compensate for self-phase modulation (because $\tilde{A}_s^{-1}(\partial^2 \tilde{A}_s/\partial \tau^2)$ will be negative near the peak of the pulse, where the factor $|\tilde{A}_s|^2 \tilde{A}_s$ is most important).

Expressions (7.5.33) give what is known as the fundamental soliton solution to the pulse propagation equation (7.5.32). Higher-order soliton solutions also are known. These solutions were first obtained through use of inverse scattering methods by Zakharov and Shabat (1972) and are described in more detail by Agrawal (1989).

One circumstance under which k_2 and γ have opposite signs occurs in fused-silica optical fibers. In this case, the nonlinearity in the refractive index occurs as the result of electronic polarization, and n_2 is consequently positive. The group velocity dispersion parameter k_2 is positive for visible light, but becomes negative for wavelengths longer than approximately 1.3 μm. This effect is illustrated in Fig. 7.5.3, in which the linear refractive index n_{lin} and the group index $n_g \equiv c/v_g$ are plotted as functions of the vacuum wavelength of the incident radiation. Optical solitons of the sort described by Eq. (7.5.33a) have been observed by Mollenauer *et al.* (1980) in the propagation of light pulses at a wavelength of 1.55 μm obtained from a color center laser.

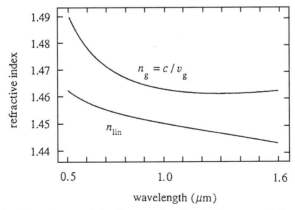

FIGURE 7.5.3 Dependence of the linear refractive index n_{lin} and the group index $n_g = c/v_g$ on the vacuum wavelength for fused silica.

Problems

1. *Spatial solitons.* Verify that Eqs. (7.1.19) through (7.1.21) do indeed satisfy Eq. (7.1.18).

2. *Self-focusing and filamentation.* Read carefully the subsection that includes Eqs. (7.1.42) through (7.1.45) and write a short essay (one or two paragraphs) describing how you would proceed to observe filamentation in the laboratory making use of a laser pulse of 10 nsec duration (assumed for simplicity to have a square-top time evolution) and making use of the nonlinear response of carbon disulfide. Describe issues such as the pulse energy that is required, the length of the interaction region you would use, and the focusing characteristics of your laser beam.

3. *Z-scan.* In this problem we shall develop a mathematical description of the z-scan procedure for measuring the nonlinear refractive index. The basis of this procedure is that a sample is translated longitudinally through the beam-waist region of a focused gaussian laser beam, and the variation of the on-axis intensity in the far field is measured as a function of sample position. The on-axis intensity in the far field is usually determined by measuring the power transmitted through a small aperture placed on the system axis. The variation of measured power with sample position is found to be proportional to the nonlinear phase shift experienced in passing through the sample, and from this measured phase shift the value of n_2 of the sample can be determined from the known sample thickness L and laser intensity at the sample.

In detail, you are to derive an expression for the dependence of the fractional change in on-axis intensity $\Delta I/I$ on the sample position z relative to the position z_0 of the beam waist of the incident laser beam. This expression will also depend on the values of n_2, of L, and on the parameters of the incident laser beam. For simplicity, assume that the sample is thin both in the sense that L is much smaller than the confocal parameter b of the incident laser beam and in that the maximum nonlinear phase shift acquired in passing through the sample is much smaller than unity.

Here are some suggestions on how to proceed. Begin with the expression for a gaussian laser beam. Determine first how the beam will be modified in passing thought the nonlinear material. Note that under the assumed conditions the beam diameter at the sample will be unchanged, but that the wavefront curvature will be modified by nonlinear refraction. Note that the modified beam itself approximates a gaussian beam, but with a different value of the wavefront radius of curvature. By assuming that this beam propagates according to the same laws that govern the propagation of a gaussian beam, determine its on-axis intensity in the far field as a function of $z - z_0$.

Note that a more detailed analysis of the z-scan procedure (Sheik-Bahae *et al.*, *IEEE J. Quantum Electron.* **26**, 760, (1990)) leads to the somewhat different result

$$\frac{\Delta I}{I} = \frac{4\Phi_{max}x}{(x^2 + 1)(x^2 + 9)}, \quad x = 2(z - z_0)/b, \quad (7.5.34)$$

where Φ_{max} is the nonlinear phase shift on-axis when the sample is at the beam waist of the incident laser beam. Plot the functional dependence of both your result and the literature result.

4. *Optical phase conjugation.* Solve the following coupled equations for the boundary conditions that $A_3(0)$ and $A_4(L)$ are arbitrary:

$$\frac{dA_3}{dz} = -\alpha_3 A_3 - i\kappa_4^* A_4^*,$$

$$\frac{dA_4}{dz} = \alpha_4 A_4 + i\kappa_3^* A_3^*.$$

(These equations generalize Eqs. (7.2.31) and describe four-wave mixing in the usual phase conjugation geometry for the case of a lossy medium.)

5. *Optical phase conjugation.* Same as Problem 1, but with the inclusion of phase-mismatch factors so that the coupled equations are given by

$$\frac{dA_3}{dz} = -\alpha_3 A_3 - i\kappa_4^* A_4^* e^{i\Delta kz},$$

$$\frac{dA_4}{dz} = \alpha_4 A_4 + i\kappa_3^* A_3^* e^{-i\Delta kz}$$

where $\Delta k = (\mathbf{k}_1 + \mathbf{k}_2 - \mathbf{k}_3 - \mathbf{k}_4) \cdot \hat{\mathbf{z}}$, and $\mathbf{k}_1$ and $\mathbf{k}_2$ are the wavevectors of the two pump waves.

6. *Optical phase conjugation.* Derive an expression for the phase-conjugate reflectivity obtained by degenerate four-wave mixing utilizing the nonlinear response of a collection of "two-level" atoms. You may make the rotating wave and slowly-varying amplitude approximations and may assume that the amplitudes of the strong pump waves are not modified by the nonlinear interaction.

[Hint: This problem can be solved using the formalism developed in Section 6.3. This problem has been solved in the scientific literature, and the solution is given by R. L. Abrams and R. C. Lind, *Opt. Lett.* **2**, 94 (1978) and **3**, 205 (1978).]

7. *Polarization properties of phase conjugation.* Verify Eq. (7.2.41).

8. *Optical bistability.* The discussion of absorptive optical bistability presented in the text assumed that the incident laser frequency was tuned to a cavity resonance.

Generalize this treatment by allowing the cavity to be mistuned from resonance, so that the factor $\rho^2 e^{2ikl}$ appearing in the denominator of Eq. (7.3.3) can be set equal to $Re^{i\delta_0}$, where δ_0 is the cavity mistuning in radians.

[Ans.: Eq. (7.3.10) must be replaced by

$$TI_1 = I_2 \left[1 - 2R\left(1 - \frac{C_0 T/R}{1 + 2I_2/I_s}\right)\cos\delta_0 + R^2\left(1 - \frac{C_0 T/R}{1 + 2I_2/I_s}\right)^2 \right].$$

Examination of this expression shows that larger values of C_0 and I_2 are required in order to obtain optical bistability for $\delta_0 \neq 0$.]

9. *Optical bistability.* The treatment of absorptive bistability given in the text assumed that the absorption decreased with increasing laser intensity according to

$$\alpha = \frac{\alpha_0}{1 + I/I_s}.$$

In fact, many saturable absorbers are imperfect in that they do not saturate all the way to zero; the absorption can better be represented by

$$\alpha = \frac{\alpha_0}{1 + I/I_s} + \alpha_1,$$

where α_1 is constant. How large can α_1 be (for given α_0) and still allow the occurrence of bistability? Use the same approximations used in the text, namely that $\alpha \ll 1$ and that the cavity is tuned to exact resonance. How are the requirements on the intensity of the incident laser beam modified by a nonzero value of α_1?

10. *Optical bistability.* By means of a graphical analysis of the sort illustrated in Fig. 7.3.5, make a plot of the transmitted intensity I_3 as a function of the incident intensity I_1. Note that more than two stable solutions can occur for a device that displays refractive bistability.

11. *Optical bistability.* Consider refractive bistability in a nonlinear Fabry–Perot interferometer. Assume that the nonlinear material also displays (linear) absorption. How are the intensity requirements for switching modified by the inclusion of loss, and how large can the absorption be and still alllow the existence of bistability?

12. *Two-beam coupling.* According to Section 7.4, the equations governing the growth of two beams subject to two-beam coupling can be written

$$\frac{dI_1}{dz} = \beta I_1 I_2,$$

$$\frac{dI_2}{dz} = -\beta I_1 I_2.$$

Solve this system for $I_2(z)$ in terms of $I_2(0)$ and the total intensity $I = I_1 + I_2$. (This is not difficult, but you may need to refer to a textbook on differential equations.) Make a sketch of $I_2(z)$ for the cases $\beta > 0$ and $\beta < 0$. What is β, both mathematically and physically?

13. *Self-phase modulation.* The analysis of self-phase modulation that led to Fig. 7.5.1 assumed that the medium had instantaneous response and that the temporal evolution of the pulse had a symmetric waveform. In this case the spectrum of the pulse is seen to broaden symmetrically. How is the spectrum modified for a medium with a sluggish response (given, for example, by Eq. (7.4.7) with τ much longer than the pulse duration)? How is the spectrum modified if the pulse waveform is not symmetric (a sawtooth waveform, for example)?

14. *Pulse propagation.* How is the pulse propagation equation (7.5.32) modified if the quantity $k^2 - k_0^2$ is not approximated by $2k_0(k - k_0)$, as was done in going from Eq. (7.5.18) to Eq. (7.5.19)?

[Ans.: k_2 in Eq. (7.5.32) must be replaced by $(k_1^2 + 2k_0k_2)/2k_0$.]

Why is it that this new equation seems to predict that pulses will spread as they propagate, even when both k_2 and γ vanish?

15. *Temporal solitons.* Verify that the solution given by Eqs. (7.5.33) does in fact satisfy the pulse propagation equation (7.5.32).

16. *Temporal solitons.* Calculate the peak power and energy of an optical soliton with $\tau_0 = 10$ psec propagating in a silica-core optical fiber of 8 μm core diameter at a wavelength of 1.55 μm.

[Solution: Using the typical values $k_2 = -20$ psec^2/km and $n_2 = 3 \times 10^{-16}$ cm^2/W, we find that $P = 80$ mW and $Q = 0.8$ pJ. Note also that the full width of the pulse measured at half intensity points is equal to $1.76\tau_0$.]

17. *Self-induced transparency.* Optical solitons can also be formed as a consequence of the resonant nonlinear optical response of a collection of two-level atoms. Show that, in the absence of damping effects and for the case of exact resonance, the equations describing the propagation of an optical pulse through such a medium are of the form

$$\frac{\partial \tilde{A}}{\partial z} + \frac{1}{c}\frac{\partial \tilde{A}}{\partial t} = \frac{2\pi i \omega N}{c}p,$$

$$\frac{dp}{dt} = -\frac{i}{\hbar}|\mu|^2 \tilde{A}w, \qquad \frac{dw}{dt} = \frac{-4i}{\hbar}\tilde{A}p$$

(where the atomic response is described as in Section 6.4). Show that these equations yield soliton-like solutions of the form

$$\tilde{A}(z, t) = \frac{\hbar}{\mu \tau_0} \text{sech}\left(\frac{t - z/v}{\tau_0}\right),$$

$$w(z, t) = -1 + 2 \, \text{sech}^2\left(\frac{t - z/v}{\tau_0}\right),$$

$$p(z, t) = -i\mu \, \text{sech}\left(\frac{t - z/v}{\tau_0}\right) \tanh\left(\frac{t - z/v}{\tau_0}\right)$$

as long as the pulse width and pulse velocity are related by

$$\frac{c}{v} = 1 + \frac{2\pi N \mu^2 \omega \tau_0^2}{\hbar}.$$

What is the value (and the significance) of the quantity

$$\int_{-\infty}^{\infty} \frac{2\mu}{\hbar} \tilde{A}(z, t) dt?$$

(For the case of an inhomogeneously broadened medium, the equations are still satisfied by a sech pulse, but the relation between v and τ_0 is different. See, for example, Allen and Eberly, 1975.)

18. *Modulational instability.* The intent of this problem is to determine the conditions under which the propagation of a monochromatic laser field inside an optical fiber is unstable to the growth of new frequency components. Base your analysis on the nonlinear Schrödinger equation (NLSE) in the form of Eq. (7.5.32). First note that the solution to the NLSE to an input in the form of a cw monochromatic wave is the input field multiplied by an exponential phase factor describing a nonlinear phase shift.

One next wants to determine if this solution is stable to growth of weak perturbations. Assume that the total field within the fiber has the form of the strong component of amplitude A_0 and frequency ω and two weak sidebands symmetrically displaced by frequency δ such that the total field within the fiber can be represented as

$$A(z, \tau) = A_0(z) + A_1(z)e^{-i\delta\tau} + A_2(z)e^{i\delta\tau}.$$

Derive the differential equations satisfied by each of the field amplitudes by linearizing the equations in A_1 and A_2. Note that A_1 and A_2 are coupled by four-wave mixing interactions. Determine the conditions for instability to occur by determining when the simultaneous solution to the equations for the A_1 and A_2 fields will experience exponential growth. For what relative sign of n_2 and k_2 can this instability exist? Sketch the dependence of the gain on the sideband detuning δ for various values of the pump amplitude A_0. Also determine the "eigenvector" associated with the exponentially growing solution, that is, the particular linear combination of A_1 and A_2 that experiences exponential growth. Describe the nature of the modulation present on the transmitted field for this particular eigenvector.

Thought question: Why does your solution depend on the group velocity, not the phase velocity, considering that we have analyzed this situation under continuous-wave conditions?

References

Section 7.1   Self-Focusing of Light and Other Self-Action Effects

J. S. Aitchison *et al.*, *J. Opt. Soc. Am. B* **8**, 1290 (1991).

S. A. Akhmanov, R. V. Khokhlov and A. P. Sukhorukov, in *Laser Handbook* (F. T. Arecchi and E. O. Schulz-Dubois, eds.), North-Holland, 1972.

A. Barthelemy, S. Maneuf, and C. Froehly, *Opt. Commun.* **55**, 201 (1985).

V. I. Bespalov and V. I. Talanov, *JETP Lett.* **3**, 471 (1966).

R. S. Bennink, V. Wong, A. M. Marino, D. L. Aronstein, R. W. Boyd, C. R. Stroud, Jr., S. Lukishova, and D. J. Gauthier, *Phys. Rev. Lett.* **88**, 113901 (2002).

J. E. Bjorkholm and A. Ashkin, *Phys. Rev. Lett.* **32**, 129 (1974).

R. Y. Chiao, E. Garmire, and C. H. Townes, *Phys. Rev. Lett.* **13**, 479 (1964).

R. Y. Chiao, P. L. Kelley, and E. Garmire, *Phys. Rev. Lett.* **17**, 1158 (1966).

G. Fibich and A. L. Gaeta, *Opt. Lett.* **25**, 335 (2000).

H. A. Haus, *Appl. Phys. Lett.* **8**, 128 (1966).

M. Hercher, *J. Opt. Soc. Am.* **54**, 563 (1964).

P. L. Kelly, *Phys. Rev. Lett.* **15**, 1005 (1965).

M. M. T. Loy and Y. R. Shen, *IEEE J. Quantum Electron.* **9**, 409 (1973).

J. H. Marburger, *Prog. Quantum Electron* **4**, 35 (1975).

Y. R. Shen, *Prog. Quantum Electron* **4**, 1 (1975).

O. Svelto, in *Progress in Optics XII* (E. Wolf, ed.), North-Holland, 1974.

A. Yariv, *Quantum Electronics*, Wiley, New York, 1975, p. 498.

Section 7.2   Optical Phase Conjugation

D. M. Bloom and G. C. Bjorklund, *Appl. Phys. Lett.* **31**, 592 (1977).

R. W. Boyd and G. Grynberg, "Optical Phase Conjugation," in *Contemporary Nonlinear Optics* (G. P. Agrawal and R. W. Boyd, eds.), Academic Press, Boston, 1992.

R. A. Fisher, ed., *Optical Phase Conjugation,* Academic Press, New York, 1983.

A. L. Gaeta and R. W. Boyd, *Phys. Rev. Lett.* **60**, 2618 (1988).

R. W. Hellwarth, *J. Opt. Soc. Am.* **67**, 1 (1977).

K. R. MacDonald, W. R. Tompkin, and R. W. Boyd, *Opt. Lett.* **13**, 663 (1988).

J. J. Maki, W. V. Davis, R. W. Boyd, and J. E. Sipe, *Phys. Rev. A* **46**, 7155 (1992).

A. Yariv and D. M. Pepper, *Opt. Lett.* **1**, 16 (1977).

B. Ya. Zel'dovich, N. F. Pilipetsky and V. V. Shkunov, *Principles of Phase Conjugation*, Springer-Verlag, Berlin, 1985.

Polarization Properties of Phase Conjugation.

M. Ducloy and D. Bloch, *Phys. Rev.* **A30**, 3107 (1984).

G. Grynberg, *Opt. Commun.* **48**, 432 (1984).

M. Kauranen, D. J. Gauthier, M. S. Malcuit, and R. W. Boyd, *Phys. Rev. A* **40**, 1908 (1989).

M. S. Malcuit, D. J. Gauthier, and R. W. Boyd, *Opt. Lett.* **13**, 663 (1988).

G. Martin, L. K. Lam, and R. W. Hellwarth, *Opt. Lett.* **5**, 186 (1980).

S. Saikan, *J. Opt. Soc. Am.* **68**, 1185 (1978).

S. Saikan, *J. Opt. Soc. Am.* **72**, 515 (1982).

S. Saikan and M. Kiguchi, *Opt. Lett.* **7**, 555 (1982).

W. R. Tompkin, M. S. Malcuit, R. W. Boyd, and J. E. Sipe, *J. Opt. Soc. Am. B* **6**, 757 (1989).

B. Ya. Zel'dovich and V. V. Shkunov, *Sov. J. Quantum Electron*, **9**, 379 (1979).

Section 7.3   Optical Bistability

H. M. Gibbs, *Optical Bistability*, Academic Press, New York, 1985.

H. M. Gibbs, S. L. McCall and T. N. Venkatesan, *Phys. Rev. Lett.* **36**, 113 (1976).

L. A. Lugiato, "Theory of Optical Bistability," in *Progress in Optics XXI* (E. Wolf, ed.), North-Holland, 1984.

G. I. Stegeman and A. Miller, "Physics of All-Optical Switching Devices," in *Photonic Switching*, Vol. 1, J. E. Midwinter, ed., Academic Press, Boston, 1993.

A. Szöke, V. Daneu, J. Goldhar and N. A. Kurnit, *Appl. Phys. Lett.* **15**, 376 (1969).

Section 7.4   Two-Beam Coupling

J. Feinberg, in *Optical Phase Conjugation* (R. A. Fisher, ed.), Academic Press, New York, 1983.

Y. Silberberg and I. Bar-Joseph, *Phys. Rev. Lett.* **48**, 1541 (1982).

Y. Silberberg and I. Bar-Joseph, *J. Opt. Soc. Am. B* **1**, 662 (1984).

Section 7.5   Pulse Propagation and Temporal Solitons

G. P. Agrawal, *Nonlinear Fiber Optics*, Academic Press, Boston, 1989.

L. D. Allen and J. H. Eberly, *Optical Resonance and Two-Level Atoms*, Wiley, New York, 1975.

R. G. Brewer, *Phys. Rev. Lett.* **19**, 8 (1967).

A. C. Cheung, D. M. Rank, R. Y. Chiao, and C. H. Townes, *Phys. Rev. Lett.* **20**, 786 (1968).

L. F. Mollenauer, R. H. Stolen, and J. P. Gordon, *Phys. Rev. Lett.* **45**, 1095 (1980).

F. Shimizu, *Phys. Rev. Lett.* **19**, 1097 (1967).

V. E. Zakharov and A. B. Shabat, *Sov. Phys. JETP* **34**, 63 (1972).

Chapter 8

Spontaneous Light Scattering and Acoustooptics

8.1. Features of Spontaneous Light Scattering

In this chapter, we describe spontaneous light scattering; Chapters 8 and 9 present descriptions of various stimulated light scattering processes. By spontaneous light scattering, we mean light scattering under conditions such that the optical properties of the material system are unmodified by the presence of the incident light beam. We shall see in the following two chapters that the character of the light-scattering process is profoundly modified whenever the intensity of the incident light is sufficiently large to modify the optical properties of the material system.

Let us first consider the light scattering experiment illustrated in part (a) of Fig. 8.1.1. Under the most general circumstances, the spectrum of the scattered light has the form shown in part (b) of the figure, in which Raman, Brillouin, Rayleigh, and Rayleigh-wing features are present. By definition, those components of the scattered light that are shifted to lower frequencies are known as Stokes components, and those components that are shifted to higher frequencies are known as anti-Stokes components. Table 8.1.1 lists some of the physical processes that can lead to light scattering of the sort shown in the figure and gives some of the physical parameters that describe these processes.

One of the processes is Raman scattering. Raman scattering results from the interaction of light with the vibrational modes of the molecules constituting the scattering medium. Raman scattering can equivalently be described as the scattering of light from optical phonons.

(a)

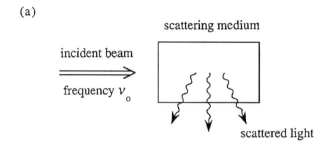

(b)

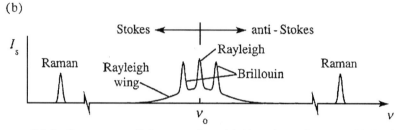

FIGURE 8.1.1   Spontaneous light scattering. (a) Experimental setup. (b) Typical observed spectrum.

Brillouin scattering is the scattering of light from sound waves, that is, from propagating pressure (and hence density) waves. Brillouin scattering can also be considered to be the scattering of light from acoustic phonons.

Rayleigh scattering (or Rayleigh-center scattering) is the scattering of light from nonpropagating density fluctuations. Formally, it can be described as scattering from entropy fluctuations. It is known as quasielastic scattering because it induces no frequency shift.

Rayleigh-wing scattering (that is, scattering in the wing of the Rayleigh line) is scattering from fluctuations in the orientation of anisotropic molecules. Since the molecular reorientation process is very rapid, this component is spectrally

TABLE 8.1.1   Typical values of the parameters describing several light-scattering processes

Process	Shift (cm^{-1})	Linewidth (cm^{-1})	Relaxation time (sec)	Gain[a] (cm/MW)
Raman	1000	5	10^{-12}	5×10^{-3}
Brillouin	0.1	5×10^{-3}	10^{-9}	10^{-2}
Rayleigh	0	5×10^{-4}	10^{-8}	10^{-4}
Rayleigh-wing	0	5	10^{-12}	10^{-3}

[a] Gain of the stimulated version of the process.

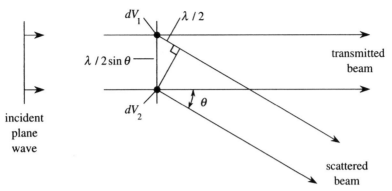

FIGURE 8.1.2   Light scattering cannot occur in a completely homogeneous medium.

very broad. Rayleigh-wing scattering does not occur for molecules with an isotropic polarizability tensor.

Fluctuations as the Origin of Light Scattering

Light scattering occurs as a consequence of fluctuations in the optical properties of a material medium; a completely homogeneous material can scatter light only in the forward direction (see, for example, Fabelinskii, 1968). This conclusion can be demonstrated with the aid of Fig. 8.1.2, which shows a completely homogeneous medium being illuminated by a plane wave. We suppose that the volume element dV_1 scatters light into the θ direction. However, for any direction except the exact forward direction ($\theta = 0$) there must be a nearby volume element (labeled dV_2) whose scattered field interferes destructively with that from dV_1. Since the same argument can be applied to any volume element in the medium, we conclude that there can be no scattering in any direction except $\theta = 0$. Scattering in the direction $\theta = 0$ is known as coherent forward scattering and is the origin of the index of refraction. (See, for example, the discussion in Section 31 of Feynman et al., 1963.)

Note that the argument that scattering cannot occur (except in the forward direction) requires that the medium be *completely* homogeneous. Scattering can occur as the result of fluctuations in any of the optical properties of the medium. For example, if the density of the medium is nonuniform, then the total number of molecules in the volume element dV_1 may not be equal to the number of molecules in dV_2, and consequently the destructive interference between the fields scattered by these two elements will not be exact.

Since light scattering results from fluctuations in the optical properties of a material medium, it is useful to represent the dielectric tensor of the medium (which for simplicity we assume to be isotropic in its average properties) as (Landau and Lifshitz, 1960)

$$\epsilon_{ik} = \epsilon_0 \delta_{ik} + \Delta\epsilon_{ik}, \qquad (8.1.1)$$

where ϵ_0 represents the mean dielectric constant of the medium and where $\Delta\epsilon_{ik}$ represents the (temporally and/or spatially varying) fluctuations in the dielectric tensor that lead to light scattering. It is convenient to decompose the fluctuation $\Delta\epsilon_{ik}$ in the dielectric tensor into the sum of a scalar contribution $\Delta\epsilon\delta_{ik}$ and a (traceless) tensor contribution $\Delta\epsilon_{ik}^{(t)}$ as

$$\Delta\epsilon_{ik} = \Delta\epsilon\delta_{ik} + \Delta\epsilon_{ik}^{(t)}. \qquad (8.1.2)$$

The scalar contribution $\Delta\epsilon$ arises from fluctuations in thermodynamic quantities such as the pressure, entropy, density, or temperature. In a chemical solution it also has a contribution from fluctuations in concentration. Scattering that results from $\Delta\epsilon$ is called scalar light scattering; examples of scalar light scattering include Brillouin and Rayleigh scattering.

Scattering that results from $\Delta\epsilon_{ik}^{(t)}$ is called tensor light scattering. The tensor $\Delta\epsilon_{ik}^{(t)}$ can be taken to be traceless (i.e., $\sum_i \Delta\epsilon_{ii}^{(t)} = 0$), since the scalar contribution $\Delta\epsilon$ has been separated out. It is useful to express $\Delta\epsilon_{ik}^{(t)}$ as

$$\Delta\epsilon_{ik}^{(t)} = \Delta\epsilon_{ik}^{(s)} + \Delta\epsilon_{ik}^{(a)}, \qquad (8.1.3)$$

where $\Delta\epsilon_{ik}^{(s)}$ is the symmetric part of $\Delta\epsilon_{ik}^{(t)}$ (symmetric in the sense that $\Delta\epsilon_{ik}^{(s)} = \Delta\epsilon_{ki}^{(s)}$) and gives rise to Rayleigh-wing scattering, and where $\Delta\epsilon_{ik}^{(a)}$ is the antisymmetric part of $\Delta\epsilon_{ik}^{(t)}$ (that is, $\Delta\epsilon_{ik}^{(a)} = -\Delta\epsilon_{ki}^{(a)}$) and gives rise to Raman scattering.

It can be shown that the fluctuations $\Delta\epsilon$, $\Delta\epsilon_{ik}^{(s)}$, and $\Delta\epsilon_{ik}^{(a)}$ are statistically independent. Scattering due to $\Delta\epsilon_{ik}^{(t)}$ is called depolarized scattering, because in general the degree of polarization in the scattered light is smaller than that of the incident light.

Scattering Coefficient

A quantity that is used to describe the efficiency of the scattering process is the scattering coefficient R, which is defined in terms of the quantities shown in Fig. 8.1.3. Here a beam of light of intensity I_0 illuminates a scattering region of volume V, and the intensity I_s of the scattered light is measured at a distance L from the interaction region. It is reasonable to assume that the intensity of the scattered light increases linearly with the intensity I_0 of the incident light

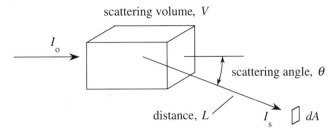

FIGURE 8.1.3   Quantities used to define the scattering coefficient.

and with the volume V of the interaction region, and that it obeys the inverse square law with respect to the distance L to the point of observation. We can hence represent I_s as

$$I_s = \frac{I_0 R V}{L^2}, \tag{8.1.4}$$

where the constant of proportionality R is known as the scattering coefficient.

We now assume that the scattered light falls onto a small detector of projected area dA. The power hitting the detector is given by $dP = I_s dA$. Since the detector subtends a solid angle at the scattering region given by $d\Omega = dA/L^2$, the scattered power per unit solid angle is given by $dP/d\Omega = I_s L^2$, or by

$$\frac{dP}{d\Omega} = I_0 R V. \tag{8.1.5}$$

Either Eq. (8.1.4) or (8.1.5) can be taken as the definition of the scattering coefficient. For scattering of visible light through an angle of 90 degrees, R has the value 2×10^{-8} cm^{-1} for air and 1×10^{-6} cm^{-1} for water.

Scattering Cross Section

It is also useful to define the scattering cross section. We consider a beam of intensity I_0 falling onto an individual molecule, as shown in Fig. 8.1.4. We let P denote the total power of the radiation scattered by this molecule. We assume that P increases linearly with I_0 according to

$$P = \sigma I_0, \tag{8.1.6}$$

where the constant of proportionality σ is known as the total scattering cross section. Since I_0 has the dimensions of power per unit area, we see that σ has the dimensions of an area, which is why it is called a cross section. The cross section can be interpreted as the effective geometrical area of the molecule for removing light from the incident beam.

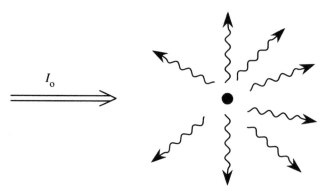

FIGURE 8.1.4   Scattering of light by a molecule.

We also define a differential cross section. Rather than describing the total scattered power, this quantity describes the power dP scattered in some particular direction into the element of solid angle $d\Omega$. We assume that the scattered power per unit solid angle, $dP/d\Omega$, increases linearly with the incident intensity according to

$$\frac{dP}{d\Omega} = I_0 \frac{d\sigma}{d\Omega}, \tag{8.1.7}$$

where $d\sigma/d\Omega$ is known as the differential cross section. Clearly, since P is equal to $\int (dP/d\Omega)d\Omega$, it follows from Eqs. (8.1.6) and (8.1.7) that

$$\sigma = \int_{4\pi} \frac{d\sigma}{d\Omega} \, d\Omega. \tag{8.1.8}$$

Let us next see how to relate the differential scattering cross section to the scattering coefficient. If each of the $\mathscr{N}$ molecules contained in the volume V of Fig. 8.1.3 scatters independently, then the total power per unit solid angle of the scattered light will be $\mathscr{N}$ times larger than the result given in Eq. (8.1.7). Consequently, by comparison with Eq. (8.1.5), we see that the scattering coefficient is given by

$$R = \frac{\mathscr{N}}{V} \frac{d\sigma}{d\Omega}. \tag{8.1.9}$$

One should be wary about taking this equation to be a general result. Recall that a completely homogeneous medium does not scatter light at all, which implies that for such a medium R would be equal to zero and not to $(\mathscr{N}/V)(d\sigma/d\Omega)$. In the next section we examine the conditions under which it is valid to assume that each molecule scatters independently. As a general rule, Eq. (8.1.9) is valid for dilute media and is entirely invalid for condensed matter.

8.2. Microscopic Theory of Light Scattering

Let us now consider light scattering in terms of the field scattered by each molecule contained within the interaction region. Such a treatment is particularly well suited for the case of scattering from a dilute gas, where collective effects due to the interaction of the various molecules are relatively unimportant. (Light scattering from condensed matter is more conveniently treated using the thermodynamic formalism presented in the next section.) As illustrated in Fig. 8.2.1, we assume that the optical field

$$\tilde{\mathbf{E}} = \mathbf{E}_0 e^{-i\omega t} + \text{c.c.} \tag{8.2.1}$$

of intensity $I_0 = (nc/2\pi)|E_0|^2$ is incident on a molecule whose linear dimensions are assumed to be much smaller than the wavelength of light. In response to the applied field, the molecule develops the dipole moment

$$\tilde{\mathbf{p}} = \alpha(\omega)\mathbf{E}_0 e^{-i\omega t} + \text{c.c.}, \tag{8.2.2}$$

where $\alpha(\omega)$ is the polarizability of the particle. Explicit formulas for $\alpha(\omega)$ for certain types of scatterers are given below, but for reasons of generality we leave the form of $\alpha(\omega)$ unspecified for the present.

As a consequence of the time-varying dipole moment given by Eq. (8.2.2), the particle will radiate. The intensity of this radiation at a distance L from the scatterer is given by the magnitude of the Poynting vector (see, for example, Jackson, 1982, Section 9.2) as

$$I_s = \frac{n\langle \ddot{\tilde{p}}^2 \rangle}{4\pi c^3 L^2} \sin^2\phi = \frac{n\omega^4 |\alpha(\omega)|^2 |E_0|^2}{2\pi c^3 L^2} \sin^2\phi. \tag{8.2.3}$$

The angular brackets in the first form imply that the time average of the enclosed quantity is to be taken. As shown in Fig. 8.2.1, ϕ is the angle between the induced dipole moment of the particle and the direction $\mathbf{r}$ to the point of observation.

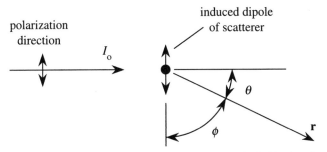

FIGURE 8.2.1 Geometry of light scattering from an individual molecule.

We next use Eq. (8.2.3) to derive an expression for the differential scattering cross section. As in the derivation of Eq. (8.1.5), the scattered power per unit solid angle is given by $dP/d\Omega = I_s L^2$. We introduce the differential cross section of Eq. (8.1.7), $d\sigma/d\Omega = (dP/d\Omega)/I_0 = I_s L^2/I_0$, which through use of Eq. (8.2.3) becomes

$$\frac{d\sigma}{d\Omega} = \frac{\omega^4}{c^4} |\alpha(\omega)|^2 \sin^2 \phi. \tag{8.2.4}$$

We note that this expression for the differential cross section $d\sigma/d\Omega$ predicts a $\sin^2 \phi$ dependence for *any* functional form for $\alpha(\omega)$. This result is a consequence of our assumption that the scattering particle is small compared to an optical wavelength and hence that the scattering is due solely to electric dipole and not to higher-order multipole processes. Since the angular dependence of $d\sigma/d\Omega$ is contained entirely in the $\sin^2 \phi$ term, we can immediately obtain an expression for the total scattering cross section by integrating $d\sigma/d\Omega$ over all solid angles, yielding

$$\sigma = \int_{4\pi} d\Omega \frac{d\sigma}{d\Omega} = \frac{8\pi}{3} \frac{\omega^4}{c^4} |\alpha(\omega)|^2. \tag{8.2.5}$$

In deriving Eq. (8.2.4) for the differential scattering cross section, we assumed that the incident light was linearly polarized, and for convenience we took the direction of polarization to lie in the plane of Fig. 8.2.1. For this direction of polarization, the scattering angle θ and the angle ϕ of Eq. (8.2.3) are related by $\theta + \phi = 90$ degrees, and hence for this direction of polarization Eq. (8.2.4) can be expressed in terms of the scattering angle as

$$\left(\frac{d\sigma}{d\Omega}\right)_p = \frac{\omega^4}{c^4} |\alpha(\omega)|^2 \cos^2 \theta. \tag{8.2.6}$$

Other types of polarization can be treated by allowing the incident field to have a component perpendicular to the plane of Fig. 8.2.1. For this component ϕ is equal to 90 degrees for any value of the scattering angle θ, and hence for this component the differential cross section is given by

$$\left(\frac{d\sigma}{d\Omega}\right)_s = \frac{\omega^4}{c^4} |\alpha(\omega)|^2 \tag{8.2.7}$$

for any value of θ. Since unpolarized light consists of equal intensities in the two orthogonal polarization directions, the differential cross section for unpolarized light is obtained by averaging Eqs. (8.2.6) and (8.2.7), giving

$$\left(\frac{d\sigma}{d\Omega}\right)_{\text{unpolarized}} = \frac{\omega^4}{c^4} |\alpha(\omega)|^2 \frac{1}{2}(1 + \cos^2 \theta). \tag{8.2.8}$$

As an example of the use of these equations, we consider scattering from an atom whose optical properties can be described by the Lorentz model of the atom (that is, we model the atom as a simple harmonic oscillator). According to Eqs. (1.4.17) and (1.4.10) and the relation of $\chi(\omega) = N\alpha(\omega)$, the polarizability of such an atom is given by

$$\alpha(\omega) = \frac{e^2/m}{\omega_0^2 - \omega^2 - 2i\omega\gamma}, \tag{8.2.9}$$

where ω_0 is the resonance frequency and γ is the dipole damping rate. Through use of this expression, the total scattering cross section given by Eq. (8.2.5) becomes

$$\sigma = \frac{8\pi}{3}\left(\frac{e^2}{mc^2}\right)^2 \frac{\omega^4}{\left(\omega_0^2 - \omega^2\right)^2 + 4\omega^2\gamma^2}. \tag{8.2.10}$$

The frequency dependence of the scattering cross section predicted by this equation is illustrated in Fig. 8.2.2. Equation (8.2.10) can be simplified under several different limiting conditions. In particular, we find that

$$\sigma = \frac{8\pi}{3}\left(\frac{e^2}{mc^2}\right)^2 \frac{\omega^4}{\omega_0^4} \qquad \text{for} \quad \omega \ll \omega_0, \tag{8.2.11a}$$

$$\sigma = \frac{2\pi}{3}\left(\frac{e^2}{mc^2}\right)^2 \frac{\omega_0^2}{(\omega_0 - \omega)^2 + \gamma^2} \quad \text{for} \quad \omega \simeq \omega_0, \tag{8.2.11b}$$

$$\sigma = \frac{8\pi}{3}\left(\frac{e^2}{mc^2}\right)^2 \qquad\qquad \text{for} \quad \omega \gg \omega_0. \tag{8.2.11c}$$

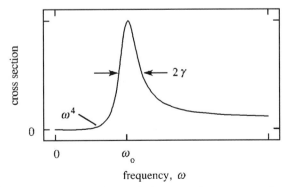

FIGURE 8.2.2   Frequency dependence of the scattering cross section of a Lorentz oscillator.

Equation (8.2.11a) shows that the scattering cross section increases as the fourth power of the optical frequency ω in the limit $\omega \ll \omega_0$. This result leads, for example, to the prediction that the sky is blue, since the shorter wavelengths of sunlight are scattered far more efficiently in the earth's atmosphere than are the longer wavelengths. Scattering in this limit is often known as Rayleigh scattering. Equation (8.2.11b) shows that near the atomic resonance frequency the dependence of the scattering cross section on the optical frequency has a Lorentzian lineshape. Equation (8.2.11c) shows that for very large frequencies the scattering cross section approaches a constant value. This value is of the order of the square of the "classical" electron radius, $r_e = e^2/mc^2$. Scattering in this limit is known as Thompson scattering.

As a second example of the application of Eq. (8.2.5), we consider scattering from a collection of small dielectric spheres. We take ϵ_1 to be the dielectric constant of the material within each sphere and ϵ to be that of the surrounding medium. We assume that each sphere is small in the sense that its radius a is much smaller than the wavelength of the incident radiation. We can then calculate the polarizability of each sphere using the laws of electrostatics. It is straightforward to show (see, for example, Stratton, 1941, or Jackson, 1982) that the polarizability is given by the expression

$$\alpha = \frac{\epsilon_1 - \epsilon}{\epsilon_1 + 2\epsilon} a^3. \qquad (8.2.12)$$

Note that α depends upon frequency only through any possible frequency dependence of ϵ or of ϵ_1. Through use of Eq. (8.2.5), we find that the scattering cross section is given by

$$\sigma = \frac{8\pi}{3} \frac{\omega^4}{c^4} a^6 \left(\frac{\epsilon_1 - \epsilon}{\epsilon_1 + 2\epsilon} \right)^2. \qquad (8.2.13)$$

Note that, as in the low-frequency limit of the Lorentz atom, the cross section scales as the fourth power of the frequency. Note also that the cross section scales as the square of the volume of each particle.

Let us now consider the rather subtle problem of calculating the total intensity of the light scattered from a collection of molecules. We recall from the discussion of Fig. 8.1.2 that only the fluctuations in the optical properties of the medium can lead to light scattering. As shown in Fig. 8.2.3, we divide the total scattering volume V into a large number of identical small regions of volume V'. We assume that V' is sufficiently small that all of the molecules within V' radiate essentially in phase. The intensity of the light emitted by the atoms in V' in some particular direction can thus be represented as

$$I_{V'} = \nu^2 I_{\text{mol}}, \qquad (8.2.14)$$

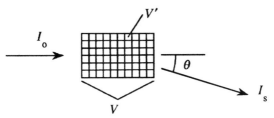

FIGURE 8.2.3   Light scattering from a collection of molecules.

where ν represents the number of molecules in V' and I_{mol} denotes the intensity of the light scattering by a single molecule.

We next calculate the total intensity of the scattered radiation from the entire volume V. We recall from the discussion of Section 8.1 that, for each volume element V', there will be another element whose radiated field tends to interfere destructively with that from V'. Insofar as each volume element contains exactly the same number of molecules, the cancellation will be complete. However, any deviation of ν from its mean value $\bar{\nu}$ can lead to a net intensity of the scattered radiation. The contribution to the net scattered intensity from volume element V' is thus given by $\overline{\Delta \nu^2} I_{\mathrm{mol}}$, where $\overline{\Delta \nu^2} = \overline{\nu^2} - \bar{\nu}^2$. The intensity of the radiation scattered from the total volume V is then given by

$$I_V = I_{\mathrm{mol}} \overline{\Delta \nu^2} \frac{V}{V'}, \qquad (8.2.15)$$

where the last factor V/V' gives the total number of regions of volume V' contained within the interaction volume V. This result shows how the total scattered intensity I_V depends upon the fluctuations in the number density of molecules. We see that the scattered intensity I_V vanishes if the fluctuation $\overline{\Delta \nu^2}$ vanishes.

For the case of a medium sufficiently dilute that the locations of the individual molecules are uncorrelated (that is, for an ideal gas), we can readily calculate the mean fluctuation $\overline{\Delta \nu^2}$ in the number of particles. If N denotes the mean number density of particles, then the mean number of particles in V' is given by

$$\bar{\nu} = NV', \qquad (8.2.16)$$

and the mean fluctuation is given by

$$\overline{\Delta \nu^2} = \overline{\nu^2} - \bar{\nu}^2 = \bar{\nu}, \qquad (8.2.17)$$

where the last equality follows from the properties of the Poisson probability distribution, which are obeyed by uncorrelated particles. We hence find from

Eqs. (8.2.15) through (8.2.17) that

$$I_V = \bar{v}\frac{V}{V'}I_{\text{mol}} = NVI_{\text{mol}} = \mathscr{N}I_{\text{mol}}. \tag{8.2.18}$$

Hence, for an ideal gas the total intensity is simply the intensity of the light scattered by a single molecule multiplied by the total number of molecules, $\mathscr{N} = NV$. Consequently the scattering coefficient R and differential cross section $d\sigma/d\Omega$ introduced in Section 8.1 are related by Eq. (8.1.9), that is, by

$$R = N\frac{d\sigma}{d\Omega}. \tag{8.2.19}$$

By introducing Eq. (8.2.4) into this expression, we find that the scattering coefficient is given by

$$R = N\frac{\omega^4}{c^4}|\alpha(\omega)|^2 \sin^2\phi. \tag{8.2.20}$$

If the scattering medium is sufficiently dilute that its refractive index can be represented as

$$n = 1 + 2\pi N\alpha(\omega), \tag{8.2.21}$$

Equation (8.2.20) can be rewritten as

$$R = \frac{\omega^4}{c^4}\frac{|n-1|^2}{4\pi^2 N}\sin^2\phi. \tag{8.2.22}$$

This result can be used to determine the number density N of molecules in a gaseous sample in terms of two optical constants: the refractive index n and scattering coefficient R at a fixed angle ϕ. In fact, the first accurate measurement of Loschmidt's number (the number density of molecules at standard temperature and pressure, $N_0 = 2.7 \times 10^{19}$ cm^{-3}) was performed through application of Eq. (8.2.22).

8.3. Thermodynamic Theory of Scalar Light Scattering

We next develop a macroscopic description of the light scattering process. We consider the case in which light scattering occurs as the result of fluctuations in the (scalar) dielectric constant and in which these fluctuations are themselves the result of fluctuations in thermodynamic variables, such as the material density and temperature. We assume, as in Fig. 8.2.3 in the preceding section, that the scattering volume V can be divided into a number of smaller volumes V' having the property that all atoms in V' radiate essentially in phase in the θ direction. We let $\Delta\epsilon$ denote the fluctuation of the dielectric constant averaged over the volume V'. Since $\epsilon = 1 + 4\pi\chi$, the fluctuation in the susceptibility is then given by $\Delta\chi = \Delta\epsilon/4\pi$. Because of this change in the susceptibility, the volume V' develops the additional polarization

$$\tilde{\mathbf{P}} = \Delta\chi\tilde{\mathbf{E}}_0 = \frac{\Delta\epsilon}{4\pi}\tilde{\mathbf{E}}_0 \tag{8.3.1}$$

and hence the additional dipole moment

$$\tilde{\mathbf{p}} = V'\tilde{\mathbf{P}} = V'\frac{\Delta\epsilon}{4\pi}\tilde{\mathbf{E}}_0. \tag{8.3.2}$$

The intensity $I_s = (nc/4\pi)\langle\tilde{\mathbf{E}}_s^2\rangle$ of the radiation emitted by this oscillating dipole moment is obtained by introducing Eq. (8.3.2) into Eq. (8.2.3), to obtain

$$I_s = I_0\frac{\omega^4 V'^2\langle\Delta\epsilon^2\rangle\sin^2\phi}{16\pi^2 L^2 c^4}, \tag{8.3.3}$$

where, as before, ϕ is the angle between $\tilde{\mathbf{p}}$ and the direction to the point of observation, and where we have introduced the intensity $I_0 = (nc/4\pi)\langle\tilde{\mathbf{E}}_0^2\rangle$ of the incident light. Equation (8.3.3) gives the intensity of the light scattered from one cell. The total intensity from all the cells is V/V' times as large, since the fluctuations in the dielectric constant for different cells are uncorrelated.

We next need to calculate the mean square fluctuation in the dielectric constant, $\langle\Delta\epsilon^2\rangle$, for any one cell. We take the density ρ and temperature T as the independent thermodynamic variables. We then express the change in the dielectric constant as

$$\Delta\epsilon = \left(\frac{\partial\epsilon}{\partial\rho}\right)_T\Delta\rho + \left(\frac{\partial\epsilon}{\partial T}\right)_\rho\Delta T. \tag{8.3.4}$$

To good accuracy (the error is estimated to be of the order of 2%; see Fabelinskii, 1968), we can usually ignore the second term, since the dielectric constant typically depends much more strongly on density than on temperature.[*] We thus find

$$\langle\Delta\epsilon^2\rangle = \left(\frac{\partial\epsilon}{\partial\rho}\right)^2\langle\Delta\rho^2\rangle,$$

which can be expressed as

$$\langle\Delta\epsilon^2\rangle = \gamma_e^2\frac{\langle\Delta\rho^2\rangle}{\rho_0^2}, \tag{8.3.5}$$

where ρ_0 denotes the mean density of the material and where we have introduced the electrostrictive constant γ_e, which is defined by [†]

$$\gamma_e = \left(\rho\frac{\partial\epsilon}{\partial\rho}\right)_{\rho=\rho_0}. \tag{8.3.6}$$

---

[*] For this reason, it is not crucial that we retain the subscript T on $\partial\epsilon/\partial\rho$.

[†] The reason why γ_e is called the electrostrictive constant will be described in Section 9.1.

The quantity $\langle \Delta \rho^2 \rangle / \rho_0^2$ appearing in Eq. (8.3.5) can be calculated using the laws of statistical mechanics. The result (see, for example, Fabelinskii, 1968, Appendix I, Eq. I.13; or Landau and Lifshitz, 1969) is

$$\frac{\langle \Delta \rho^2 \rangle}{\rho_0^2} = \frac{kT C_T}{V'} \tag{8.3.7}$$

where

$$C_T = -\frac{1}{V} \left(\frac{\partial V}{\partial p} \right)_T \tag{8.3.8}$$

is the isothermal compressibility. Note that the result given by Eq. (8.3.7) (whose proof is outside the subject area of this book) makes sense: fluctuations are driven by thermal excitation; the larger the compressibility, the larger will be the resulting excursion; and the smaller the volume under consideration, the easier it is to change its mean density.

By introducing Eqs. (8.3.5) and (8.3.7) into Eq. (8.3.3) and multiplying the result by the total number of cells, V/V', we find that the total intensity of the scattered radiation is given by

$$I_s = I_0 \frac{\omega^4 V}{16\pi^2 L^2 c^4} \gamma_e^2 C_T kT \sin^2 \phi. \tag{8.3.9a}$$

We can use this result to find that the scattering coefficient R defined by Eq. (8.1.4) is given by

$$R = \frac{\omega^4}{16\pi^2 c^4} \gamma_e^2 C_T kT \sin^2 \phi. \tag{8.3.9b}$$

Ideal Gas

As an example, let us apply the result given by Eq. (8.3.9a) to light scattering from an ideal gas, for which the equation of state is of the form

$$pV = \mathcal{N} kT, \tag{8.3.10}$$

where $\mathcal{N}$ denotes the total number of molecules in the gas. We then find that $(\partial V / \partial p)_T = -\mathcal{N} kT / p^2$ and hence that the isothermal compressibility is given by

$$C_T = \frac{\mathcal{N} kT}{V p^2} = \frac{1}{p} = \frac{V}{\mathcal{N} kT}. \tag{8.3.11}$$

We next assume that $\epsilon - 1$ is linearly proportional to ρ, so that we can represent ϵ as $\epsilon = 1 + A\rho$ for some constant A. We hence find that $\partial \epsilon / \partial \rho = A$, or that

$\partial\epsilon/\partial\rho = (\epsilon - 1)/\rho$, and that the electrostrictive constant is given by

$$\gamma_e = \epsilon - 1. \tag{8.3.12}$$

If we now introduce Eqs. (8.3.11) and (8.3.12) into Eq. (8.3.9a), we find that the scattered intensity can be expressed as

$$I_s = I_0 \frac{\omega^4 V}{16\pi^2 L^2 c^4} \frac{(\epsilon - 1)^2}{N} \sin^2 \phi, \tag{8.3.13}$$

where we have introduced the mean density of particles $N = \mathcal{N}/V$. Through use of Eq. (8.1.4), we can write this result in terms of the scattering coefficient as

$$R = \frac{(\epsilon - 1)^2 \omega^4 \sin^2 \phi}{16\pi^2 c^4 N}. \tag{8.3.14}$$

Note that, since $\epsilon - 1$ is equal to $2(n - 1)$ for a dilute gas (i.e., for $\epsilon - 1 \ll 1$), this result is in agreement with the prediction of the microscopic model of light scattering for an ideal gas, given by Eq. (8.2.22).

Spectrum of the Scattered Light

The analysis just presented has led to an explicit prediction (8.3.9a) for the *total* intensity of the light scattered as the result of the fluctuations in the density (and hence the dielectric constant) of a material system in thermal equilibrium. In order to determine the *spectrum* of the scattered light, we have to examine the dynamical behavior of the density fluctuations that give rise to light scattering. As before (see the discussion associated with Eq. (8.3.4)), we represent the fluctuation in the dielectric constant as

$$\Delta\tilde{\epsilon} = \left(\frac{\partial\epsilon}{\partial\rho}\right)\Delta\tilde{\rho}. \tag{8.3.15}$$

We now choose the entropy s and pressure p to be the independent thermodynamic variables. We can then represent the variation in density, $\Delta\tilde{\rho}$, as

$$\Delta\tilde{\rho} = \left(\frac{\partial\rho}{\partial p}\right)_s \Delta\tilde{p} + \left(\frac{\partial\rho}{\partial s}\right)_p \Delta\tilde{s}. \tag{8.3.16}$$

Here the first term describes adiabatic density fluctuations (that is, acoustic waves) and leads to Brillouin scattering. The second term describes isobaric density fluctuations (that is, entropy or temperature fluctuations) and leads to Rayleigh-center scattering. The two contributions to $\Delta\tilde{\rho}$ are quite different in character and lead to very different spectral distributions of the scattered light, because (as we shall see) the equations of motion for $\Delta\tilde{p}$ and $\Delta\tilde{s}$ are very different.

Brillouin Scattering

The equation of motion for a pressure wave is well known from the field of acoustics and is given by (see, e.g., Fabelinskii, 1968, Section 34.9)

$$\frac{\partial^2 \Delta \tilde{p}}{\partial t^2} - \Gamma' \nabla^2 \frac{\partial \Delta \tilde{p}}{\partial t} - v^2 \nabla^2 \Delta \tilde{p} = 0. \tag{8.3.17}$$

Here v denotes the velocity of sound, which is given in terms of thermodynamic variables by

$$v^2 = \left(\frac{\partial p}{\partial \rho} \right)_s . \tag{8.3.18}$$

The equation for the velocity of sound is conveniently expressed in terms of the compressibility C or in terms of its reciprocal, the bulk modulus K, which are defined by

$$C \equiv \frac{1}{K} = -\frac{1}{V} \frac{\partial V}{\partial p} = \frac{1}{\rho} \frac{\partial \rho}{\partial p}. \tag{8.3.19}$$

The compressibility can be measured either at constant temperature or at constant entropy. The two values of the compressibility, denoted respectively as C_T and C_s, are related by

$$\frac{C_T}{C_s} = \frac{c_p}{c_V} \equiv \gamma, \tag{8.3.20}$$

where c_p is the specific heat (i.e., the heat capacity per unit mass, whose units are erg/g K) at constant pressure, c_V is the specific heat at constant volume, and where their ratio γ is known as the adiabatic index. The velocity of sound as defined by Eq. (8.3.18) can thus be written as

$$v^2 = \frac{K_s}{\rho} = \frac{1}{C_s \rho}. \tag{8.3.21}$$

An important special case of the use of this formula is that of an ideal gas, for which the equation of state is given by Eq. (8.3.10) and the isothermal compressibility is given by Eq. (8.3.11). The adiabatic compressibility is thus given by $C_s = C_T/\gamma = 1/\gamma p$. We hence find from Eq. (8.3.21) that the velocity of sound is given by

$$v = \left(\frac{\gamma p}{\rho} \right)^{1/2} = \left(\frac{\gamma \mathcal{N} kT}{\rho V} \right)^{1/2} = \left(\frac{\gamma kT}{\mu} \right)^{1/2}, \tag{8.3.22}$$

where μ denotes the molecular mass. We thus see that the velocity of sound is of the order of the mean thermal velocity of the molecules of the gas.

TABLE 8.3.1   Typical   sound velocities

Material	v (cm/sec)
Gases	
Dry air	3.31×10^4
He	9.65×10^4
H_2	12.84×10^4
Water vapor	4.94×10^4
Liquids	
CS_2	1.15×10^5
CCl_4	0.93×10^5
Ethanol	1.21×10^5
Water	1.50×10^5
Solids	
Fused silica	5.97×10^5
Lucite	2.68×10^5

The velocity of sound for some common optical materials is listed in Table 8.3.1.

The parameter Γ' appearing in the wave equation (8.3.17) is a damping parameter that can be shown to be expressible as

$$\Gamma' = \frac{1}{\rho}\left[\frac{4}{3}\eta_s + \eta_b + \frac{\kappa}{C_p}(\gamma - 1)\right] \tag{8.3.23}$$

where η_s is the shear viscosity coefficient, η_b is the bulk viscosity coefficient, and κ is the thermal conductivity. For most materials of interest in optics, the last contribution to Γ' is much smaller than the first two. Conventions involving the naming of the viscosity coefficients are discussed briefly in the Appendix to Section 9.6.

As an illustration of the nature of the acoustic wave equation (8.3.17), we consider the propagation of the wave

$$\Delta \tilde{p} = \Delta p e^{i(qz - \Omega t)} + \text{c.c.} \tag{8.3.24}$$

through an acoustic medium. By substituting this form into the acoustic wave equation, we find that q and Ω must be related by a dispersion relation of the form

$$\Omega^2 = q^2(v^2 - i\Omega\Gamma'). \tag{8.3.25}$$

We can rewrite this relation as

$$q^2 = \frac{\Omega^2}{v^2 - i\Omega\Gamma'} = \frac{\Omega^2/v^2}{1 - i\Omega\Gamma'/v^2} \simeq \frac{\Omega^2}{v^2}\left(1 + \frac{i\Omega\Gamma'}{v^2}\right), \tag{8.3.26}$$

which shows that

$$q \simeq \frac{\Omega}{v} + \frac{i\Gamma}{2v}, \tag{8.3.27}$$

where we have introduced the phonon decay rate

$$\Gamma = \Gamma' q^2. \tag{8.3.28}$$

We find by introducing the form for q given by Eq. (8.3.27) into Eq. (8.3.24) that the intensity of the acoustic wave varies spatially as

$$|\Delta p(z)|^2 = |\Delta p(0)|^2 e^{-\alpha_s z}, \tag{8.3.29}$$

where we have introduced the sound absorption coefficient

$$\alpha_s = \frac{q^2 \Gamma'}{v} = \frac{\Gamma}{v}. \tag{8.3.30}$$

It is also useful to define the phonon lifetime as

$$\tau_p = \frac{1}{\Gamma} = \frac{1}{q^2 \Gamma'}. \tag{8.3.31}$$

Next, we calculate the rate at which light is scattered out of a beam of light by these acoustic waves. We assume that the incident optical field is described by

$$\tilde{E}_0(z, t) = E_0 e^{i(\mathbf{k} \cdot \mathbf{r} - \omega t)} + \text{c.c.}, \tag{8.3.32}$$

and that the scattered field obeys the driven wave equation

$$\nabla^2 \tilde{E} - \frac{n^2}{c^2} \frac{\partial^2 \tilde{E}}{\partial t^2} = \frac{4\pi}{c^2} \frac{\partial^2 \tilde{P}}{\partial t^2}. \tag{8.3.33}$$

We take the polarization $\tilde{P}$ of the medium to be given by Eq. (8.3.1) with the variation $\Delta \tilde{\epsilon}$ in dielectric constant given by Eq. (8.3.15), that is, we take $\tilde{P} = (\partial \epsilon / \partial \rho) \Delta \tilde{\rho} \tilde{E}_0 / 4\pi$. We take the variation in density to be given by the first contribution to Eq. (8.3.16), that is, by $\Delta \tilde{\rho} = (\partial \rho / \partial p) \Delta \tilde{p}$, where $\Delta \tilde{p}$ denotes the incremental pressure. We thus find that

$$\begin{aligned}
\tilde{P}(\mathbf{r}, t) &= \frac{1}{4\pi} \left(\frac{\partial \epsilon}{\partial \rho} \right) \left(\frac{\partial \rho}{\partial p} \right)_s \Delta \tilde{p}(\mathbf{r}, t) \tilde{E}_0(z, t) \\
&= \frac{1}{4\pi} \gamma_e C_s \Delta \tilde{p}(\mathbf{r}, t) \tilde{E}_0(z, t),
\end{aligned} \tag{8.3.34}$$

where we have introduced the adiabatic compressibility C_s of Eq. (8.3.19) and the electrostrictive constant of Eq. (8.3.6). We take a typical component of the thermally excited pressure disturbance within the interaction region to be given by

$$\Delta \tilde{p}(\mathbf{r}, t) = \Delta p \, e^{i(\mathbf{q} \cdot \mathbf{r} - \Omega t)} + \text{c.c.} \tag{8.3.35}$$

By combining Eqs. (8.3.33) through (8.3.35), we find that the scattered field must obey the wave equation

$$\nabla^2 \tilde{\mathbf{E}} - \frac{n^2}{c^2}\frac{\partial^2 \tilde{\mathbf{E}}}{\partial t^2} = -\frac{\gamma_e C_s}{c^2}\Big[(\omega - \Omega)^2 E_0 \Delta p^* e^{i(\mathbf{k}-\mathbf{q})\cdot\mathbf{r}-i(\omega-\Omega)t}$$
$$+ (\omega + \Omega)^2 E_0 \Delta p \, e^{i(\mathbf{k}+\mathbf{q})\cdot\mathbf{r}-i(\omega+\Omega)t} + \text{c.c.}\Big] \tag{8.3.36}$$

The first term in this expression leads to Stokes scattering; the second to anti-Stokes scattering. We study these two contributions in turn.

Stokes Scattering (First Term in Eq. (8.3.36))

The polarization is seen to have a component with wavevector

$$\mathbf{k}' \equiv \mathbf{k} - \mathbf{q} \tag{8.3.37}$$

and frequency

$$\omega' \equiv \omega - \Omega, \tag{8.3.38}$$

where the frequency ω and wavevector $\mathbf{k}$ of the incident optical field are related according to

$$\omega = |\mathbf{k}|c/n, \tag{8.3.39}$$

and where the frequency Ω and wavevector $\mathbf{q}$ of the acoustic wave are related according to

$$\Omega = |\mathbf{q}|v. \tag{8.3.40}$$

This component of the polarization can couple efficiently to the scattered optical wave only if its frequency ω' and wavevector $\mathbf{k}'$ are related by the dispersion relation for optical waves, namely

$$\omega' = |\mathbf{k}'|c/n. \tag{8.3.41}$$

In order for Eqs. (8.3.37) through (8.3.41) to be satisfied simultaneously, the sound wave frequency and wavevector must each have a particular value for any scattering direction. For the case of scattering at the angle θ, we must have the situation illustrated in Fig. 8.3.1. Part (a) of this figure shows the relative orientations of the wavevectors of the incident and scattered fields. Part (b) illustrates Eq. (8.3.37) and shows how the wavevector of the acoustic disturbance is related to those of the incident and scattered optical radiation.

Since $|\mathbf{k}|$ is very nearly equal to $|\mathbf{k}'|$ (because Ω is much smaller than ω), diagram (b) shows that

$$|\mathbf{q}| = 2|\mathbf{k}| \sin(\theta/2). \tag{8.3.42}$$

(a)   (b)

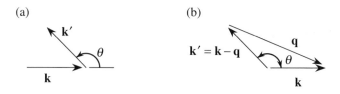

(c)

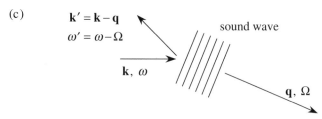

FIGURE 8.3.1   Illustration of Stokes scattering.

The dispersion relation (8.3.40) then shows that the acoustic frequency is given by

$$\Omega = 2|\mathbf{k}|v\sin(\theta/2) = 2n\omega\frac{v}{c}\sin(\theta/2). \tag{8.3.43}$$

We note that the Stokes shift Ω is equal to zero for forward scattering and is maximum for backscattering (i.e., for $\theta = 180$ degrees). The maximum frequency shift is thus given by

$$\Omega_{\text{max}} = 2n\frac{v}{c}\omega. \tag{8.3.44}$$

For $\omega/2\pi = 3 \times 10^{14}$ Hz (i.e., at $\lambda = 1\ \mu$m), $v = 1 \times 10^5$ cm/sec (a typical value), and $n = 1.5$, we obtain $\Omega_{\text{max}}/2\pi = 3 \times 10^9$ Hz.

Stokes scattering can be visualized as the scattering of light from a retreating acoustic wave, as illustrated in part (c) of Fig. 8.3.1.

Anti-Stokes Scattering (Second Term in Eq. (8.3.36))

The analysis here is analogous to that for Stokes scattering. The polarization is seen to have a component with wavevector

$$\mathbf{k}' \equiv \mathbf{k} + \mathbf{q} \tag{8.3.45}$$

and frequency

$$\omega' \equiv \omega + \Omega, \tag{8.3.46}$$

(a)

(b)

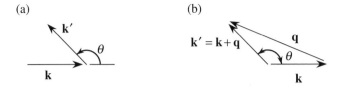

(c)

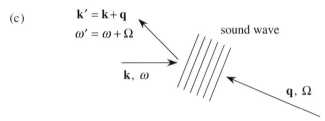

FIGURE 8.3.2   Illustration of anti-Stokes scattering.

where, as before, $\omega = |\mathbf{k}|c/n$ and $\Omega = |\mathbf{q}|v$. This component of the polarization can couple efficiently to an electromagnetic wave only if ω' and $|\mathbf{k}'|$ are related by $\omega' = |\mathbf{k}'|c/n$. We again assume that θ denotes the scattering angle, as illustrated in Fig. 8.3.2. The condition (8.3.45) is illustrated as part (b) of the figure. Since (as before) $|\mathbf{k}|$ is very nearly equal to $|\mathbf{k}'|$, the length of the acoustic wavevector is given by

$$|\mathbf{q}| = 2|\mathbf{k}|\sin(\theta/2). \tag{8.3.47}$$

Hence, by Eq. (8.3.40), the acoustic frequency is given by

$$\Omega = 2n\omega\frac{v}{c}\sin(\theta/2). \tag{8.3.48}$$

Anti-Stokes scattering can be visualized as scattering from an oncoming sound wave, as shown in part (c) of Fig. 8.3.2.

We have thus far ignored attenuation of the acoustic wave in our analysis. If we include this effect, we find that the light scattered into direction θ is not monochromatic but has a spread in angular frequency whose width (FWHM) is given by

$$\delta\omega = 1/\tau_p = \Gamma' q^2 \tag{8.3.49}$$

which becomes, through use of Eq. (8.3.42),

$$\delta\omega = 4\Gamma'|\mathbf{k}|^2\sin^2(\theta/2) = 4n^2\Gamma'\frac{\omega^2}{c^2}\sin^2(\theta/2). \tag{8.3.50}$$

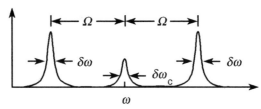

FIGURE 8.3.3   Spectrum showing Brillouin and Rayleigh scattering.

For the case of backscattering ($\theta = 180$ degrees), $\delta\omega/2\pi$ is typically of the order of 100 MHz for organic liquids. Since the acoustic frequency is given by Eq. (8.3.43), we see that the ratio of the linewidth to the Brillouin frequency shift is given by

$$\frac{\delta\omega}{\Omega} = \frac{2n\Gamma'\omega}{vc}\sin(\theta/2). \qquad (8.3.51)$$

The spectrum of the scattered light has the form shown in Fig. 8.3.3.

Rayleigh Center Scattering

We now consider the contribution to $\Delta\tilde{\rho}$ (and hence to $\Delta\tilde{\epsilon}$) resulting from isobaric density fluctuations, which are described by the second term in Eq. (8.3.16) and are proportional to the entropy fluctuation $\Delta\tilde{s}$. Entropy fluctuations are described by the same equation as that describing temperature variations:

$$\rho c_p \frac{\partial \Delta\tilde{s}}{\partial t} - \kappa\nabla^2\Delta\tilde{s} = 0, \qquad (8.3.52)$$

where, as before, c_p denotes the specific heat at constant pressure, and where κ denotes the thermal conductivity. Note that these fluctuations obey a diffusion equation and not a wave equation. A solution to the diffusion equation (8.3.52) is

$$\Delta\tilde{s} = \Delta s_0\, e^{-\delta t} e^{-i\mathbf{q}\cdot\mathbf{r}}, \qquad (8.3.53)$$

where the damping rate of the entropy disturbance is given by

$$\delta = \frac{\kappa}{\rho c_p}q^2. \qquad (8.3.54)$$

We see that, unlike pressure waves, entropy waves do not propagate. As a result, the nonlinear polarization proportional to Δs can give rise only to an unshifted component of the scattered light. The width (FWHM) of this component is given by $\delta\omega_c = \delta$, that is, by

$$\delta\omega_c = \frac{4\kappa}{\rho c_p}|\mathbf{k}|^2\sin^2(\theta/2). \qquad (8.3.55)$$

As a representative case, for liquid water $\kappa = 6$ mW/cm K, $\rho = 1$ g/cm^3, $c_p = 4.2$ J/g K, and the predicted width of the central component for backscattering ($\theta = 180$ degrees) of radiation at 500 nm is $\delta\omega_c/2\pi = 1.4 \times 10^7$ Hz.

It can be shown (Fabelinskii, 1968, Eq. 5.39) that the relative intensities of the Brillouin and Rayleigh center components are given by

$$\frac{I_c}{2I_B} = \frac{c_p - c_v}{c_v} = \gamma - 1. \qquad (8.3.56)$$

Here I_c denotes the integrated intensity of the central component, and I_B that of either of the Brillouin components. This result is known as the Landau–Placzek relation.

8.4. Acoustooptics

The analysis just given of the scattering of light from sound waves can be applied to the situation in which the sound wave is applied to the interaction region externally by means of a transducer. Such acoustooptic devices are useful as intensity or frequency modulators for laser beams or as beam deflectors.

Acoustooptic devices are commonly classified as falling into one of two regimes, each of which will be discussed in greater detail below. These regimes are as follows:

Bragg scattering. This type of scattering occurs for the case of interaction lengths that are sufficiently long that phase-matching considerations become important. Bragg scattering leads to a single diffracted beam. The name is given by analogy to the scattering of X-rays from the atomic planes in a crystal. Bragg scattering can lead to an appreciable scattering efficiency (>50%).

Raman–Nath scattering. This type of scattering occurs in cells with a short interaction length. Phase-matching considerations are not important, and several scattered orders are usually present.

We shall first consider the case of Bragg scattering of light waves; a more precise statement of the conditions under which each type of scattering occurs is given below in connection with the discussion of Raman–Nath scattering.

Bragg Scattering of Light by Sound Waves

The operation of a typical Bragg scattering cell is shown schematically in Fig. 8.4.1. A traveling acoustic wave of frequency Ω and wavelength $\Lambda = 2\pi v/\Omega$ (where v denotes the velocity of sound) is established in the scattering medium. The density variation associated with this acoustic wave produces

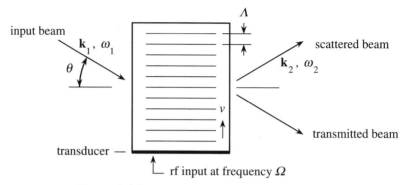

FIGURE 8.4.1   Bragg-type acoustooptic modulator.

a variation in the dielectric constant of the medium, and the incident optical wave scatters from this variation. Although the amplitude of the wave scattered from each acoustic wavefront is typically rather small, the total scattered field can become quite intense if the various contributions add in phase to produce constructive interference. The condition for this to occur is obtained with the help of the construction shown in Fig. 8.4.2 and is given by the relation

$$\lambda = 2\Lambda \sin \theta, \tag{8.4.1}$$

where λ is the wavelength of light in the medium. This condition is known as the Bragg condition. It ensures that the path length difference between rays that reflect from successive acoustic maxima is equal to an optical wavelength. In a typical acoustic–optic device, relevant parameters might be $v = 1.5 \times 10^5$ cm/sec and $\Omega/2\pi = 200$ MHz, which imply that the acoustic wavelength is equal to $\Lambda = 2\pi v/\Omega = 7.5$ μm. If the optical wavelength is 0.5 μm, we see from Eq. (8.4.1) that $\sin \theta = 1/30$ and hence that the deflection angle is given by $2\theta = 4$ degrees.

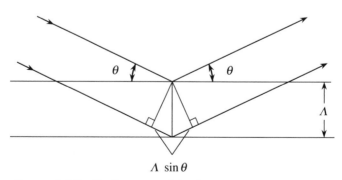

FIGURE 8.4.2   The Bragg condition for acoustooptic scattering.

The Bragg condition given by Eq. (8.4.1) can alternatively be understood as a phase-matching condition. If $\mathbf{k}_1$ denotes the wavevector of the incident optical wave, $\mathbf{k}_2$ that of the diffracted optical wave, and $\mathbf{q}$ that of the acoustic wave, the Bragg condition can be seen with the help of Fig. 8.4.3a to be a statement that

$$\mathbf{k}_2 = \mathbf{k}_1 + \mathbf{q}. \tag{8.4.2}$$

By comparison with the analysis of Section 8.3 for spontaneous Brillouin scattering (and as shown explicitly below), we can see that the frequency of the scattered beam is shifted upward to

$$\omega_2 = \omega_1 + \Omega. \tag{8.4.3}$$

Since Ω is much less than ω_1, we see that ω_2 is approximately equal to ω_1, and hence that $|\mathbf{k}_2| \simeq |\mathbf{k}_1|$. The configuration shown in Fig. 8.4.1 shows the case in which acoustic wave is advancing toward the incident optical wave. For the case of a sound wave propagating in the opposite direction, Eqs. (8.4.2) and (8.4.3) must be replaced by

$$\mathbf{k}_2 = \mathbf{k}_1 - \mathbf{q}, \tag{8.4.4a}$$

$$\omega_2 = \omega_1 - \Omega. \tag{8.4.4b}$$

Figures 8.4.1 and 8.4.2 are unchanged except for the reversal of the direction of the sound velocity vector, although Fig. 8.4.3a must be replaced by Fig. 8.4.3b.

Bragg scattering of light by sound waves can be treated theoretically by considering the time-varying change $\Delta\tilde{\epsilon}$ in the dielectric constant induced by the acoustic density variation $\Delta\tilde{\rho}$. It is usually adequate to assume that $\Delta\tilde{\epsilon}$ scales linearly with $\Delta\tilde{\rho}$, so that

$$\Delta\tilde{\epsilon} = \frac{\partial\epsilon}{\partial\rho}\Delta\tilde{\rho} = \gamma_e\frac{\Delta\tilde{\rho}}{\rho_0}. \tag{8.4.5}$$

Here ρ_0 denotes the mean density of the material, and γ_e denotes the electrostrictive constant defined by Eq. (8.3.6). Equation (8.4.5) applies rigorously

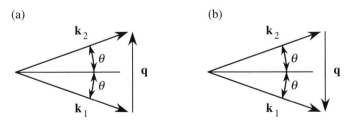

FIGURE 8.4.3   The Bragg condition described as a phase-matching relation.

to the case of liquids, and it predicts the correct qualitative behavior for all materials. For the case of anisotropic materials, the change in the optical properties is described more precisely by means of a tensor relation, which conventionally is given by

$$[\Delta(\epsilon^{-1})]_{ij} = \sum_{kl} p_{ijkl} S_{kl}, \qquad (8.4.6)$$

where the quantity p_{ijkl} is known as the strain-optic tensor and where

$$S_{kl} = \frac{1}{2}\left(\frac{\partial d_k}{\partial x_l} + \frac{\partial d_l}{dx_k}\right) \qquad (8.4.7)$$

is the strain tensor, in which d_k is the k component of the displacement of a particle from its equilibrium position. Whenever the change in the inverse of the dielectric tensor $(\epsilon^{-1})_{ij}$ given by the right-hand side of Eq. (8.4.6) is small, the change in the dielectric tensor ϵ_{ij} is given by

$$(\Delta\epsilon)_{il} = -\sum_{jk} \epsilon_{ij}[\Delta(\epsilon^{-1})]_{jk}\epsilon_{kl}. \qquad (8.4.8)$$

Our theoretical treatment of Bragg scattering assumes the geometry shown in Fig. 8.4.4. The interaction of the incident field

$$\tilde{E}_1 = A_1 e^{i(\mathbf{k}_1 \cdot \mathbf{r} - \omega_1 t)} + \text{c.c.} \qquad (8.4.9)$$

with the acoustic wave of wavevector $\mathbf{q}$ produces the diffracted wave

$$\tilde{E}_2 = A_2 e^{i(\mathbf{k}_2 \cdot \mathbf{r} - \omega_2 t)} + \text{c.c.} \qquad (8.4.10)$$

with $\omega_2 = \omega_1 + \Omega$. The interaction is assumed to be nearly Bragg-matched (i.e., phase-matched) in the sense that

$$\mathbf{k}_2 \simeq \mathbf{k}_1 + \mathbf{q}. \qquad (8.4.11)$$

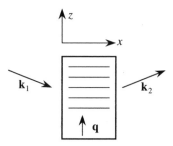

FIGURE 8.4.4   Geometry of a Bragg-type acoustooptic modulator.

The variation of the dielectric constant induced by the acoustic wave is represented as

$$\Delta\tilde{\epsilon} = \Delta\epsilon e^{i(\mathbf{q}\cdot\mathbf{r}-\Omega t)} + \text{c.c.}, \tag{8.4.12}$$

where the complex amplitude $\Delta\epsilon$ is given by $\Delta\epsilon = \gamma_e \Delta\rho/\rho_0$ under those conditions where the change in dielectric constant is accurately predicted by Eq. (8.4.5). More generally, for anisotropic interactions, $\Delta\epsilon$ is the amplitude of the appropriate tensor component of $\Delta\tilde{\epsilon}_{ij}$ given by Eq. (8.4.6). The total optical field $\tilde{E} = \tilde{E}_1 + \tilde{E}_2$ is required to satisfy the wave equation

$$\nabla^2\tilde{E} - \frac{n^2 + \Delta\tilde{\epsilon}}{c^2}\frac{\partial^2\tilde{E}}{\partial t^2} = 0, \tag{8.4.13}$$

where n denotes the refractive index of the material in the absence of the acoustic wave. Since according to Eq. (8.4.12) $\Delta\tilde{\epsilon}$ oscillates at frequency Ω, it couples the optical waves of frequencies ω_1 and $\omega_2 = \omega_1 + \Omega$.

We first consider the portion of Eq. (8.4.13) that oscillates at frequency ω_1. This part is given by

$$\frac{\partial^2 A_1}{\partial x^2} + \frac{\partial^2 A_1}{\partial z^2} + 2ik_{1x}\frac{\partial A_1}{\partial x} + 2ik_{1z}\frac{\partial A_1}{\partial z} - \left(k_{1x}^2 + k_{1z}^2\right)A_1$$

$$+ \frac{n^2\omega_1^2}{c^2}A_1 + \frac{\omega_2^2}{c^2}A_2\Delta\epsilon^* e^{i(\mathbf{k}_2-\mathbf{k}_1-\mathbf{q})\cdot\mathbf{r}} = 0. \tag{8.4.14}$$

This equation can be simplified in the following manner: (1) we introduce the slowly-varying amplitude approximation, which entails ignoring the second-order derivatives, (2) we note that A_1 depends only on x and not on z, since the interaction is invariant to a translation in the z direction, and so we set $\partial A_1/\partial z$ equal to 0, and (3) we note that $k_{1x}^2 + k_{1z}^2 = n^2\omega_1^2/c^2$. Equation (8.4.14) thus becomes

$$2ik_{1x}\frac{dA_1}{dx} = -\frac{\omega_2^2}{c^2}A_2\Delta\epsilon^* e^{i(\mathbf{k}_2-\mathbf{k}_1-\mathbf{q})\cdot\mathbf{r}}. \tag{8.4.15}$$

Next, we note that the propagation vector mismatch $\mathbf{k}_2 - \mathbf{k}_1 - \mathbf{q} \equiv -\Delta\mathbf{k}$ can have a nonzero component only in the x direction, because the geometry we are considering has infinite extent in the z direction, and the z component of the $\mathbf{k}$ wavevector mismatch must therefore vanish. We thus see that

$$(\mathbf{k}_2 - \mathbf{k}_1 - \mathbf{q}) \cdot \mathbf{r} \equiv -\Delta k x, \tag{8.4.16}$$

and hence that Eq. (8.4.15) can be written as

$$\frac{dA_1}{dx} = \frac{i\omega_2^2\Delta\epsilon^*}{2k_{1x}c^2}A_2 e^{-i\Delta kx}. \tag{8.4.17}$$

By a completely analogous derivation, we find that the portion of the wave equation (8.4.13) that describes a wave at frequency ω_2 is given by

$$\frac{dA_2}{dx} = \frac{i\omega_1^2 \Delta\epsilon}{2k_{2x}c^2} A_1 e^{i\Delta kx}. \qquad (8.4.18)$$

Finally, we note that since $\omega_1 \simeq \omega_2 \equiv \omega$ and $k_{1x} \simeq k_{2x} \equiv k_x$, the coupled equations (8.4.17) and (8.4.18) can be written as

$$\frac{dA_1}{dx} = i\kappa A_2 e^{-i\Delta kx}, \qquad (8.4.19a)$$

$$\frac{dA_2}{dx} = i\kappa^* A_1 e^{i\Delta kx}, \qquad (8.4.19b)$$

where we have introduced the coupling constant

$$\kappa = \frac{\omega^2 \Delta\epsilon^*}{2k_x c^2}. \qquad (8.4.20)$$

The solution to these coupled-amplitude equations is particularly simple for the case in which $\tilde{E}_1$ is incident at the Bragg angle. In this case, the interaction is perfectly phase-matched, so that $\Delta k = 0$, and hence Eqs. (8.4.19b) reduce to the set

$$\frac{dA_1}{dx} = i\kappa A_2, \qquad \frac{dA_2}{dx} = i\kappa^* A_1. \qquad (8.4.21)$$

These equations are easily solved using methods similar to those introduced in Chapter 2. The solution appropriate to the boundary conditions illustrated in Fig. 8.4.4 is

$$A_1(x) = A_1(0)\cos(|\kappa|x), \qquad (8.4.22a)$$

$$A_2(x) = \frac{i\kappa^*}{|\kappa|} A_1(0)\sin(|\kappa|x). \qquad (8.4.22b)$$

Note that these solutions obey the relation

$$|A_1(x)|^2 + |A_2(x)|^2 = |A_1(0)|^2, \qquad (8.4.23)$$

which shows that the energy of the optical field is conserved in the Bragg scattering process (since we have assumed that $\Omega \ll \omega$). We define the diffraction efficiency of the Bragg scattering process to be the ratio of the output intensity of the ω_2 wave to the input intensity of the ω_1 wave, and we find that the diffraction efficiency is given by

$$\eta \equiv \frac{|A_2(L)|^2}{|A_1(0)|^2} = \sin^2(|\kappa|L). \qquad (8.4.24)$$

For practical purposes, it is useful to express the coupling constant κ defined by Eq. (8.4.20) in terms of the intensity (i.e., power per unit area) of the acoustic wave. The intensity of a sound wave is given by the relation

$$I = Kv\frac{\langle\Delta\tilde{\rho}^2\rangle}{\rho_0^2} = 2Kv\left|\frac{\Delta\rho}{\rho_0}\right|^2, \qquad (8.4.25)$$

where, as before, $K = 1/C$ is the bulk modulus, v is the sound velocity, and $\Delta\rho$ is the complex amplitude of the density disturbance associated with the acoustic wave. It follows from Eq. (8.4.5) that $\Delta\epsilon$ is equal to $\gamma_e\Delta\rho/\rho$, and hence the acoustic intensity can be written as $I = 2Kv|\Delta\epsilon|^2/\gamma_e^2$. The coupling constant $|\kappa|$ (see Eq. (8.4.20)) can thus be expressed as

$$|\kappa| = \frac{\omega\gamma_e}{2nc\cos\theta}\left(\frac{I}{2Kv}\right)^{1/2}, \qquad (8.4.26)$$

where we have replaced k_x by $n(\omega/c)\cos\theta$.

As an example, we evaluate Eq. (8.4.26) for the case of Bragg scattering in water, which is characterized by the following physical constants: $n = 1.33$, $\gamma_e = 0.82$, $v = 1.5 \times 10^5$ cm/sec, and $K = 2.19 \times 10^{10}$ cm²/dyne. We assume that $\cos\theta \simeq 1$, as is usually the case; that the vacuum optical wavelength is 0.5 μm, so that $\omega = 3.8 \times 10^{15}$ rad/sec; and that the acoustic intensity is 1.0 W/cm² (as might be obtained using 1 W of acoustic power and an acoustic beam diameter of approximately 1 cm), or $I = 10^7$ erg/cm²sec. Under these conditions, Eq. (8.4.26) gives the value $|\kappa| = 1.5$ cm⁻¹. According to Eq. (8.4.24), 100% conversion of the incident beam into the diffracted beam is predicted for $|\kappa|L = \pi/2$, or under the present conditions for a path length through the acoustic beam of $L = 1.0$ cm.

For the case in which the incident beam does not intercept the acoustic wavefronts at the Bragg angle, the theoretical analysis is more complicated because the wavevector mismatch Δk does not vanish. The phase-matching diagrams for the cases of Bragg-angle and non-Bragg-angle incidence are contrasted in Fig. 8.4.5. As discussed in connection with Eq. (8.4.16), the

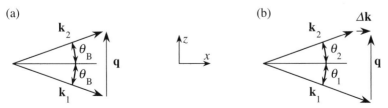

FIGURE 8.4.5 Wavevector diagrams for (a) incidence at the Bragg angle, so that $\Delta k = 0$, and (b) non-Bragg-angle incidence, so that $\Delta k \neq 0$.

wavevector mismatch can have a component only in the x direction, since the medium is assumed to have infinite extent in the z direction.

We first determine the relationship between the wavevector mismatch Δk and the angle of incidence θ_1. We note that the x and z components of the vectors of diagram (b) obey the relations

$$k \cos \theta_1 - k \cos \theta_2 = \Delta k, \tag{8.4.27a}$$

$$k \sin \theta_1 + k \sin \theta_2 = q, \tag{8.4.27b}$$

where we have let $k_1 \simeq k_2 = k$. We note that if the angle of incidence θ_1 is equal to the Bragg angle

$$\theta_B = \sin^{-1} \frac{q}{2k} = \sin^{-1} \frac{\lambda}{2\Lambda}, \tag{8.4.28}$$

then Eqs. (8.4.27b) imply that the diffraction angle θ_2 is also equal to θ_B and that $\Delta k = 0$. For the case in which the light is not incident at the Bragg angle, we set

$$\theta_1 = \theta_B + \Delta\theta, \tag{8.4.29a}$$

where we assume that $\Delta\theta \ll 1$. We note that Eq. (8.4.27b) will be satisfied so long as

$$\theta_2 = \theta_B - \Delta\theta. \tag{8.4.29b}$$

These values of θ_1 and θ_2 are now introduced into Eq. (8.4.27a). The cosine functions are expanded to lowest order in $\Delta\theta$ as

$$\cos(\theta_B \pm \Delta\theta) = \cos\theta_B \mp (\sin\theta_B)\Delta\theta,$$

and we obtain $(2k \sin\theta_B)\Delta\theta = \Delta k$, which through use of Eq. (8.4.28) shows that the wavevector mismatch Δk that occurs as the result of an angular misalignment $\Delta\theta$ is given by

$$\Delta k = -\Delta\theta q. \tag{8.4.30}$$

We next solve Eqs. (8.4.19b) for arbitrary values of Δk. The solution for the case in which no field at frequency ω_2 is applied externally is

$$A_1(x) = e^{-i(1/2)\Delta kx} A_1(0)\left(\cos sx + i\frac{\Delta k}{2s}\sin sx\right), \tag{8.4.31a}$$

$$A_2(x) = i e^{i(1/2)\Delta kx} A_1(0)\frac{\kappa^*}{s}\sin sx, \tag{8.4.31b}$$

where

$$s^2 = |\kappa|^2 + \left(\tfrac{1}{2}\Delta k\right)^2. \tag{8.4.32}$$

The diffraction efficiency for arbitrary Δk is now given by

$$\eta(\Delta k) \equiv \frac{|A_2(L)|^2}{|A_1(0)|^2} = \frac{|\kappa|^2}{|\kappa|^2 + \left(\frac{1}{2}\Delta k\right)^2} \sin^2\left\{\left[|\kappa|^2 + \left(\tfrac{1}{2}\Delta k\right)^2\right]^{1/2}L\right\}.$$

(8.4.33)

We see that for $\Delta k \neq 0$ the maximum efficiency is always less than 100%. Let us examine the rate at which the efficiency decreases as the phase mismatch Δk is increased. We expand $\eta(\Delta k)$ as a power series in Δk as

$$\eta(\Delta k) = \eta(0) + \Delta k \frac{d\eta}{d(\Delta k)}\bigg|_{\Delta k=0} + \frac{1}{2}(\Delta k)^2 \frac{d^2\eta}{d(\Delta k)^2}\bigg|_{\Delta k=0} + \cdots. \quad (8.4.34)$$

By calculating these derivatives, we find that, correct to second order in Δk, the efficiency is given by

$$\eta(\Delta k) = \eta(0)\left[1 - \frac{(\Delta k)^2}{4|\kappa|^2}\left(1 - \frac{|\kappa|K\cos(|\kappa|L)}{\sin(|\kappa|L)}\right)\right], \quad (8.4.35a)$$

where

$$\eta(0) = \sin^2(|\kappa|L). \quad (8.4.35b)$$

One common use of the Bragg acoustooptic effect is to produce an amplitude-modulated laser beam, as illustrated in Fig. 8.4.6. In such a device, the frequency of the electrical signal that is fed to the acoustic transducer is held fixed, but the amplitude of this wave is modulated. As a result, the depth of modulation of the acoustic grating is varied, leading to a modulation of the intensity of the scattered wave.

Another application of Bragg acoustooptic scattering is to produce a beam deflector (Fig. 8.4.7). In such a device, the frequency Ω of the electrical signal that is fed to the acoustic transducer is allowed to vary. As a result, the acoustic wavelength Λ varies, and hence the diffraction angle θ_2 given by Eq. (8.4.29b)

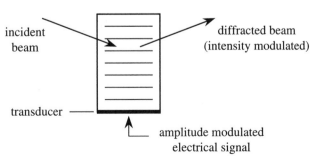

FIGURE 8.4.6   Acoustooptic amplitude modulator.

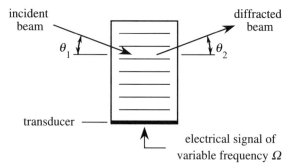

FIGURE 8.4.7   Acoustooptic beam deflector. The angle θ_2 depends on the frequency of Ω of the electrical signal.

can be controlled. It should be noted that the diffraction efficiency given by Eq. (8.4.33) decreases for diffraction at angles different from the Bragg angle, and this effect places limitations on the range of deflection angles that are achievable by means of this technique.

Raman–Nath Effect

The description of Bragg scattering given in the preceding subsection implicitly assumed that the width L of the interaction region was sufficiently large that an incident ray of light would interact with a large number of acoustic wavefronts. As illustrated in Fig. 8.4.8, this condition requires that

$$L \tan \theta_1 \gg \Lambda, \qquad (8.4.36)$$

where Λ is the acoustic wavelength. However, the angle of incidence θ_1 must satisfy the Bragg condition

$$\sin \theta_1 = \frac{\lambda}{2\Lambda} \qquad (8.4.37)$$

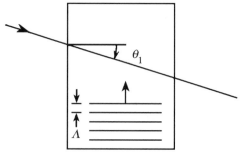

FIGURE 8.4.8   Illustration of the condition under which Bragg scattering occurs.

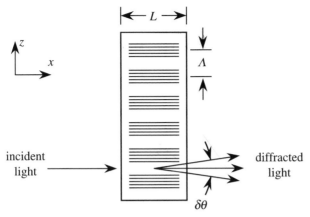

FIGURE 8.4.9   Raman–Nath diffraction.

if efficient scattering is to occur. In most cases of interest, θ_1 is much smaller than unity, and hence $\tan \theta_1 \simeq \theta_1$. Equation (8.4.37) can then be used to eliminate θ_1 from Eq. (8.4.36), which becomes

$$\frac{\lambda L}{\Lambda^2} \gg 1. \tag{8.4.38}$$

If this condition is satisfied, Bragg scattering can occur. Scattering in the opposite limit is known as Raman–Nath scattering.

Raman–Nath scattering can be understood in terms of the diagram shown in Fig. 8.4.9. A beam of light falls onto the scattering cell, typically at near-normal incidence. Because of the presence of the acoustic wave, whose wavelength is denoted Λ, the refractive index of the medium varies spatially with period Λ. The incident light diffracts off this index grating; the characteristic angular spread of the diffracted light is

$$\delta\theta = \frac{\lambda}{\Lambda}. \tag{8.4.39}$$

We now assume that the cell is sufficiently thin that multiple scattering *cannot* occur. This condition can be stated as

$$\delta\theta L < \Lambda. \tag{8.4.40}$$

If $\delta\theta$ is eliminated from this inequality through use of Eq. (8.4.39), we find that

$$\frac{\lambda L}{\Lambda^2} < 1. \tag{8.4.41}$$

We note that this condition is the opposite of the inequality (8.4.38) for the occurrence of Bragg scattering.

We now present a mathematical analysis of Raman–Nath scattering. We assume that the acoustic wave within the scattering cell can be represented as the density variation

$$\Lambda \tilde{\rho} = \Delta \rho e^{i(qz - \Omega t)} + \text{c.c.} \tag{8.4.42}$$

A refractive index variation

$$\Lambda \tilde{n} = \Delta n e^{i(qz - \Omega t)} + \text{c.c.} \tag{8.4.43}$$

is associated with this acoustic wave. We relate the complex amplitude Δn of the refractive index disturbance to the amplitude $\Delta \rho$ of the acoustic wave as follows: We let $\tilde{n} = n_0 + \Delta \tilde{n}$, where $\tilde{n} = \tilde{\epsilon}^{1/2}$ with $\tilde{\epsilon} = \epsilon_0 + \Delta \tilde{\epsilon}$. We thus find that $n_0 = \epsilon_0^{1/2}$ and that $\Delta \tilde{n} = \Delta \tilde{\epsilon}/2n_0$. We now represent $\Delta \tilde{\epsilon}$ as $\Delta \tilde{\epsilon} = (\partial \epsilon / \partial \rho) \Delta \tilde{\rho} = \gamma_e \Delta \tilde{\rho}/\rho_0$ and find that $\Delta \tilde{n} = \gamma_e \Delta \tilde{\rho}/2n_0 \rho_0$, and hence that

$$\Delta n = \frac{\gamma_e \Delta \rho}{2n_0 \rho_0}. \tag{8.4.44}$$

The ensuing analysis is simplified by representing $\Delta \tilde{n}$ using real quantities; we assume that the phase conventions are chosen such that

$$\Delta \tilde{n}(z, t) = 2\Delta n \sin(qz - \Omega t). \tag{8.4.45}$$

The electric field of the incident optical wave is represented as

$$\tilde{E}(\mathbf{r}, t) = A e^{i(kx - \omega t)} + \text{c.c.} \tag{8.4.46}$$

After passing through the acoustic wave, the optical field will have experienced a phase shift

$$\phi = \Delta \tilde{n} \frac{\omega}{c} L = 2\Delta n \frac{\omega}{c} L \sin(qz - \Omega t) \equiv \delta \sin(qz - \Omega t), \tag{8.4.47}$$

where the quantity

$$\delta = 2\Delta n \omega L/c \tag{8.4.48}$$

is known as the modulation index. The transmitted field can hence be represented as $\tilde{E}(\mathbf{r}, t) = A \exp[i(kx - \omega t + \phi)] + \text{c.c.}$, or as

$$\tilde{E}(\mathbf{r}, t) = A e^{i[kx - \omega t + \delta \sin(qz - \Omega t)]} + \text{c.c.} \tag{8.4.49}$$

We see that the transmitted field is phase-modulated in time. To determine the consequences of this form of modulation, we note that Eq. (8.4.49) can be transformed through use of the Bessel function identity

$$e^{i\delta \sin y} = \sum_{l=-\infty}^{\infty} J_l(\delta) e^{ily} \tag{8.4.50}$$

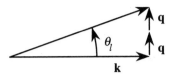

FIGURE 8.4.10   Determination of the diffraction angle.

so that the transmitted field can be expressed as

$$\tilde{E}(\mathbf{r}, t) = A \sum_{l=-\infty}^{\infty} J_l(\delta) e^{i[(kx + lqz) - (\omega + l\Omega)t]} + \text{c.c.} \qquad (8.4.51)$$

We see that the transmitted field is a linear superposition of plane wave components with frequencies $\omega + l\Omega$ and wave vectors $\mathbf{k} + l\mathbf{q}$. As shown in Fig. 8.4.10 (for the case $l = 2$), the lth-order diffracted wave is emitted at angle

$$\theta_l = \tan^{-1} \left(\frac{lq}{k} \right) \simeq \frac{lq}{k} = \frac{l\lambda}{\Lambda}. \qquad (8.4.52)$$

The intensity of the light in this diffraction order is

$$I_l = |A|^2 J_l(\delta)^2, \qquad (8.4.53)$$

where, as before, $\delta \equiv 2\Delta n(\omega/c)L$. Equations (8.4.48) and (8.4.53) constitute the Raman–Nath equations.

It is instructive to repeat this analysis for the case of a standing sound wave. For convenience, we take the resulting modulation of the refractive index to be of the form

$$\tilde{n}(z, t) = 2\Delta n \cos \Omega t \sin qz. \qquad (8.4.54)$$

The phase shift induced in the optical wave is then given by

$$\begin{aligned} \phi &= 2\Delta n \frac{\omega}{c} L \cos \Omega t \sin qz \\ &\equiv \delta \cos \Omega t \sin qz, \end{aligned} \qquad (8.4.55)$$

and the transmitted optical field is given by

$$\tilde{E}(\mathbf{r}, t) = A e^{i(kx - \omega t + \delta \cos \Omega t \sin qz)} + \text{c.c.} \qquad (8.4.56)$$

We now use the Bessel function identity (8.4.50) to transform the factor $\sin qz$ that appears in the exponent of this expression. We find that

$$\tilde{E}(\mathbf{r}, t) = A \sum_{t=-\infty}^{\infty} J_l(\delta \cos \Omega t) \exp[i(kx + lqz) - i\omega t]. \qquad (8.4.57)$$

We see that once again the transmitted field is composed of plane wave components; the lth diffracted order makes an angle

$$\theta_l \simeq \frac{lq}{k} = \frac{l\lambda}{\Lambda} \tag{8.4.58}$$

with the forward direction. The intensity of the lth order is now given by

$$I_l = |A|^2 J_l(\delta \cos \Omega t)^2. \tag{8.4.59}$$

We see that in this case each component is amplitude-modulated.

Problems

1. *Light scattering in air.* Estimate numerically, using the Lorentz model of the atom, the value of the scattering cross section for molecular nitrogen (N_2) for visible light at a wavelength of 500 nm. Use this result to estimate the value of the scattering coefficient R for air at STP. Compare this value with that obtained using Eq. (8.2.22) and the known refractive index of air. (The measured value for 90-degree scattering of unpolarized light, R_{90}^u, is approximately 2×10^{-8} cm^{-1}.) Also, estimate numerically the attenuation distance of light in air, that is, the propagation distance through which the intensity falls to $1/e$ of its initial value due to scattering losses.

[Ans.: $\alpha = (16\pi/3) R_{90}^u = 3 \times 10^{-7}$cm^{-1} = (33 km)$^{-1}$.]

2. *Light scattering in water.* Through use of Eq. (8.3.9b) and handbook values of γ_e and C_T, estimate numerically the value of the scattering coefficient R for liquid water at room temperature for 90-degree scattering of visible light at a wavelength of 500 nm. Use this result to estimate the attenuation distance of light in water.

[Ans: Using the values $C_T = 4.5 \times 10^{-11}$ cm^2/dyne, $n = 1.33$, $\gamma_e = (n^2 - 1)(n^2 + 2)/3 = 0.98$, and $T = 300$ K, we find that $R = 2.8 \times 10^{-6} \sin^2 \phi$ cm^{-1}, and hence that for 90-degree scattering of unpolarized light $R_{90}^u = 1.4 \times 10^{-6}$ cm^{-1}. Thus the attenuation constant is given by $\alpha = (16\pi/3) R_{90}^u = 2.34 \times 10^{-5}$ cm^{-1} = (426 m)$^{-1}$.]

3. *Polarizability of a dielectric sphere.* Verify Eq. (8.2.12).

4. *Acoustic attenuation in water.* Estimate numerically the value of the acoustic absorption coefficient α_s for propagation through water at frequencies of 10^3, 10^6, and 10^9 Hz.

[Ans.: The low-frequency shear viscosity coefficient of water is $\eta_s = 0.01$ dyne sec/cm^2, and the Stokes relation tells us that $\eta_d = -(2/3)\eta_s$. We find that $\Gamma' = 0.66 \times 10^{-2}$ cm^2/sec. Since $\alpha_s = q^2 \Gamma'/v$ and $q = \Omega/v$, where $v = 1.5 \times 10^6$ cm/sec, we find that $\alpha_s = 7.7 \times 10^{-14}$ cm^{-1} at 1 kHz and $\alpha_s = 7.7 \times 10^{-2}$ cm^{-1} at 1 GHz.]

5. *Inverse dielectric tensor.* Verify Eq. (8.4.8).

6. *Solution of the Bragg acoustooptics equations.* Verify Eqs. (8.4.31a) through (8.4.35b).

7. *Acoustooptic beam deflector.* Consider an acoustooptic beam deflector. The incidence angle θ_1 remains fixed while the acoustic frequency Ω is varied to control the deflection angle θ_2. Derive a formula that predicts the maximum useful deflection angle, defined arbitrarily to be that deflection angle for which the diffraction efficiency drops to 50% of its maximum value. Evaluate this formula numerically for the case treated following Eq. (8.4.26), where $|\kappa|L = \pi/2$, $L = 1.1$ cm, and $\Lambda = 30\,\mu$m.

[Ans.: Starting with Eq. (8.4.25), and the readily derived relation $\Delta k = -\frac{1}{2}q\,\delta\theta$, we find that the efficiency drops by 50% when the incidence angle is increased by an amount

$$\delta\theta = \frac{2\sqrt{2}|\kappa|}{q}\left[1 - \frac{|\kappa|L\cos|\kappa|L}{\sin|\kappa|L}\right]^{1/2}.$$

For the case $|\kappa|L = \pi/2$, where the efficiency for $\Delta k = 0$ is 100%, this result simplifies to $\delta\theta = 2\sqrt{2}|\kappa|/q$. For the numerical example, $2\delta\theta = 0.22$ degree.]

References

I. L. Fabelinskii, *Molecular Scattering of Light*, Plenum Press, New York, 1968.

I. L. Fabelinskii, in *Progress in Optics Vol. XXXVII* (E. Wolf, ed.), Elsevier, Amsterdam, 1997.

R. P. Feynman, R. B. Leighton, and M. Sands, *The Feynman Lectures on Physics*, Vol. I, Addison-Wesley, Reading, MA, 1963.

J. D. Jackson, *Classical Electrodynamics*, Wiley, New York, 1982.

L. D. Landau and E. M. Lifshitz, *Electrodynamics of Continuous Media*, Addison-Wesley, Reading, MA, 1960; see especially Chapter 14.

L. D. Landau and E. M. Lifshitz, *Statistical Physics*, Addison-Wesley, Reading, MA, 1969.

J. A. Stratton, *Electromagnetic Theory*, McGraw-Hill, New York, 1941.

A. Yariv, *Quantum Electronics*, Wiley, New York, 1975.

A. Yariv and P. Yeh, *Optical Waves in Crystals*, Wiley, New York, 1984.

Chapter 9

Stimulated Brillouin and Stimulated Rayleigh Scattering

9.1. Stimulated Scattering Processes

We saw in Section 8.1 of the previous chapter that light scattering can occur only as the result of fluctuations in the optical properties of a material system. A light scattering process is said to be *spontaneous* if the fluctuations (typically in the dielectric constant) that cause the light scattering are excited by thermal or by quantum-mechanical zero-point effects. In contrast, a light scattering process is said to be *stimulated* if the fluctuations are induced by the presence of the light field. Stimulated light scattering is typically very much more efficient than spontaneous light scattering. For example, approximately one part in 10^5 of the power contained in a beam of visible light would be scattered out of the beam by spontaneous scattering in passing through 1 cm of liquid water.* In this chapter, we shall see that when the intensity of the incident light is sufficiently large, essentially 100% of a beam of light can be scattered in a 1-cm path as the result of stimulated scattering processes.

In the present chapter we study stimulated light scattering due to induced density variations of a material system. The most important example of such a process is stimulated Brillouin scattering (SBS), which is illustrated schematically in Fig. 9.1.1. This figure shows an incident laser beam of frequency ω_L scattering from the refractive index variation associated with a sound wave of frequency Ω. Since the acoustic wavefronts are moving away from the incident

---

* Recall that the scattering coefficient R is of the order of 10^{-6} cm^{-1} for water.

FIGURE 9.1.1   Stimulated Brillouin scattering.

laser wave, the scattered light is shifted downward in frequency to the Stokes frequency $\omega_S = \omega_L - \Omega$. The reason why this interaction can lead to stimulated light scattering is that the interference of the laser and Stokes fields contains a frequency component at the difference frequency $\omega_L - \omega_S$, which of course is equal to the frequency Ω of the sound wave. The response of the material system to this interference term can act as a source that tends to increase the amplitude of the sound wave. Thus the beating of the laser wave with the sound wave tends to reinforce the Stokes wave, whereas the beating of the laser wave and Stokes waves tends to reinforce the sound wave. Under proper circumstances, the positive feedback described by these two interactions leads to exponential growth of the amplitude of the Stokes wave. SBS was first observed experimentally by Chiao *et al.* (1964).

There are two different physical mechanisms by which the interference of the laser and Stokes waves can drive the acoustic wave. One mechanism is electrostriction, that is, the tendency of materials to become more dense in regions of high optical intensity; this process is described in detail in the next section. The other mechanism is optical absorption. The heat evolved by absorption in regions of high optical intensity tends to cause the material to expand in those regions. The density variation induced by this effect can excite an acoustic disturbance. Absorptive SBS is less commonly used than electrostrictive SBS, since it can occur only in lossy optical media. For this reason we shall treat the electrostrictive case first and return to the case of absorptive coupling in Section 9.6 of this chapter.

There are two conceptually different configurations in which SBS can be studied. One is the SBS generator shown in part (a) of Fig. 9.1.2. In this configuration only the laser beam is applied externally, and both the Stokes and acoustic fields grow from noise within the interaction region. The noise process that initiates SBS is typically the scattering of laser light from thermally generated phonons. For the generator configuration, the Stokes radiation is created at frequencies near that for which the gain of the SBS process is largest. We shall see in Section 9.3 how to calculate this frequency.

Part (b) of Fig. 9.1.2 shows an SBS amplifier. In this configuration both the laser and Stokes fields are applied externally. Strong coupling occurs in this

(a)

ω_L thermal noise
phonons

ω_S

(b)

ω_L ω_S weak Stokes
seed

ω_S

FIGURE 9.1.2 (a) SBS generator; (b) SBS amplifier.

case only if the frequency of the injected Stokes wave is approximately equal to the frequency that would be created by an SBS generator.

In drawing Figs. 9.1.1 and 9.1.2, we have assumed that the laser and Stokes waves are counterpropagating. In fact, the SBS process leads to amplification of a Stokes wave propagating in any direction except for the propagation direction of the laser wave.* However, SBS is usually observed only in the backwards direction, because the spatial overlap of the laser and Stokes beams is largest under these conditions.

9.2. Electrostriction

Electrostriction is the tendency of materials to become compressed in the presence of an electric field. Electrostriction is of interest both as a mechanism leading to a third-order nonlinear optical response and as a coupling mechanism that leads to stimulated Brillouin scattering.

The origin of the effect can be explained in terms of the behavior of a dielectric slab placed in the fringing field of a plane-parallel capacitor. As illustrated in part (a) of Fig. 9.2.1, the slab will experience a force tending to pull it into the region of maximum field strength. The nature of this force can be understood either globally or locally.

We can understand the origin of the electrostrictive force from a global point of view as being a consequence of the maximization of energy. The potential energy per unit volume of a material located in an electric field of field strength

---

* We shall see in Section 9.3 that copropagating laser and Stokes waves could interact only by means of acoustic waves of infinite wavelength, which cannot occur in a medium of finite spatial extent.

(a)

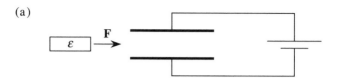

(b)

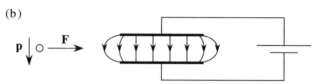

FIGURE 9.2.1   Origin of electrostriction: (a) a dielectric slab near a parallel plate capacitor; (b) a molecule near a parallel plate capacitor.

E is changed with respect to its value in the absence of the field by the amount

$$u = \frac{\epsilon E^2}{8\pi},$$ (9.2.1)

where ϵ is the dielectric constant of the material. Consequently the total energy of the system, $\int u\, dV$, is maximized by allowing the slab to move into the region between the capacitor plates where the field strength is largest.

From a microscopic point of view, we can consider the force acting on an individual molecule placed in the fringing field of the capacitor, as shown in part (b) of Fig. 9.2.1. In the presence of the field $\mathbf{E}$, the molecule develops the dipole moment $\mathbf{p} = \alpha\mathbf{E}$, where α is the molecular polarizability. The energy stored in the polarization of the molecule is given by

$$U = -\int_0^{\mathbf{E}} \mathbf{p} \cdot d\mathbf{E}' = -\int_0^{\mathbf{E}} \alpha\mathbf{E}' \cdot d\mathbf{E}' = -\tfrac{1}{2}\alpha\mathbf{E} \cdot \mathbf{E} \equiv -\tfrac{1}{2}\alpha E^2.$$ (9.2.2)

The force acting on the molecule is then given by

$$\mathbf{F} = -\nabla U = \tfrac{1}{2}\alpha\nabla(E^2).$$ (9.2.3)

We see that each molecule is pulled into the region of increasing field strength.

Next we consider the situation illustrated in Fig. 9.2.2, in which the capacitor is immersed in the dielectric liquid. Molecules are pulled from the surrounding medium into the region between the capacitor plates, thus increasing the density in this region by an amount that we call $\Delta\rho$. We calculate the value of $\Delta\rho$ by means of the following argument: As a result of the increase in density of the material, its dielectric constant changes from its origin value ϵ_0 to the value

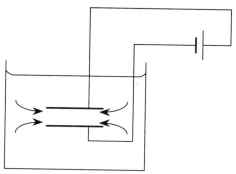

FIGURE 9.2.2   Capacitor immersed in a dielectric liquid.

$\epsilon_0 + \Delta\epsilon$, where

$$\Delta\epsilon = \left(\frac{\partial\epsilon}{\partial\rho}\right)\Delta\rho. \tag{9.2.4}$$

Consequently the field energy density changes by the amount

$$\Delta u = \frac{E^2}{8\pi}\Delta\epsilon = \frac{E^2}{8\pi}\left(\frac{\partial\epsilon}{\partial\rho}\right)\Delta\rho. \tag{9.2.5}$$

However, according to the first law of thermodynamics, this change in energy Δu must be equal to the work Δw performed in compressing the material; the work done per unit volume is given by

$$\Delta w = p_{\mathrm{st}}\frac{\Delta V}{V} = -p_{\mathrm{st}}\frac{\Delta\rho}{\rho}. \tag{9.2.6}$$

Here the strictive pressure p_{st} is the contribution to the pressure of the material that is due to the presence of the electric field. Since $\Delta u = \Delta w$, by equating Eqs. (9.2.5) and (9.2.6) we find that the electrostrictive pressure is given by

$$p_{\mathrm{st}} = -\rho\left(\frac{\partial\epsilon}{\partial\rho}\right)\frac{E^2}{8\pi} \equiv -\gamma_e\frac{E^2}{8\pi}, \tag{9.2.7}$$

where $\gamma_e = \rho(\partial\epsilon/\partial\rho)$ is known as the electrostrictive constant (see also Eq. (8.3.6)). Since p_{st} is negative, the total pressure is reduced in regions of high field strength. The fluid tends to be drawn into these regions, and the density is increased. We calculate the change in density as $\Delta\rho = -(\partial\rho/\partial p)\Delta p$, where we equate Δp with the electrostrictive pressure of Eq. (9.2.7). We write this result as

$$\Delta\rho = -\rho\left(\frac{1}{\rho}\frac{\partial\rho}{\partial p}\right)p_{\mathrm{st}} \equiv -\rho C p_{\mathrm{st}}, \tag{9.2.8}$$

where $C = \rho^{-1}(\partial\rho/\partial p)$ is the compressibility. Combining this result with Eq. (9.2.7), we find that

$$\Delta\rho = \rho C \gamma_e \frac{E^2}{8\pi}. \tag{9.2.9}$$

The derivation of this expression for $\Delta\rho$ has implicitly assumed that the electric field E is a static field. In such a case, the derivatives that appear in the expressions for C and γ_e are to be performed with the temperature T held constant. However, our primary interest is in the case where E represents an optical frequency field; in such a case Eq. (9.2.9) should be replaced by

$$\Delta\rho = \rho C \gamma_e \frac{\langle\tilde{\mathbf{E}}\cdot\tilde{\mathbf{E}}\rangle}{8\pi}, \tag{9.2.10}$$

where the angular brackets denote a time average over an optical period. If $\tilde{\mathbf{E}}(t)$ contains more than one frequency component, so that $\langle\tilde{\mathbf{E}}\cdot\tilde{\mathbf{E}}\rangle$ contains both static components and hypersonic components (as in the case of SBS), C and γ_e should be evaluated at constant entropy to determine the response for the hypersonic components and at constant temperature to determine the response for the static components.

Let us consider the modification of the optical properties of a material system that occurs as a result of electrostriction. We represent the change in the susceptibility in the presence of an optical field as $\Delta\chi = \Delta\epsilon/4\pi$, where $\Delta\epsilon$ is calculated as $(\partial\epsilon/\partial\rho)\Delta\rho$, with $\Delta\rho$ given by Eq. (9.2.10). We thus find that

$$\Delta\chi = \frac{1}{32\pi^2} C \gamma_e^2 \langle\tilde{\mathbf{E}}\cdot\tilde{\mathbf{E}}\rangle. \tag{9.2.11}$$

For simplicity, we consider the case of the monochromatic field

$$\tilde{\mathbf{E}}(t) = \mathbf{E}e^{-i\omega t} + \text{c.c.}; \tag{9.2.12}$$

the case in which $\tilde{\mathbf{E}}(t)$ contains two frequency components that differ by approximately the Brillouin frequency is treated in the following section on SBS. Then, since $\langle\tilde{\mathbf{E}}\cdot\tilde{\mathbf{E}}\rangle = 2\,\mathbf{E}\cdot\mathbf{E}^*$, we see that

$$\Delta\chi = \frac{1}{16\pi^2} C_T \gamma_e^2\, \mathbf{E}\cdot\mathbf{E}^*. \tag{9.2.13}$$

The complex amplitude of the nonlinear polarization that results from this change in the susceptibility can be represented as $\mathbf{P} = \Delta\chi\mathbf{E}$, that is, as

$$\mathbf{P} = \frac{1}{16\pi^2} C_T \gamma_e^2\, |\mathbf{E}|^2\mathbf{E}. \tag{9.2.14}$$

If we write this result in terms of a conventional third-order susceptibility, defined through

$$\mathbf{P} = 3\chi^{(3)}(\omega = \omega + \omega - \omega)|\mathbf{E}|^2\mathbf{E}, \qquad (9.2.15)$$

we find that

$$\chi^{(3)}(\omega = \omega + \omega - \omega) = \frac{1}{48\pi^2}C_T\gamma_e^2. \qquad (9.2.16)$$

For simplicity, we have suppressed the tensor nature of the nonlinear susceptibility in the foregoing discussion. However, we can see from the form of Eq. (9.2.14) that, for an isotropic material, the nonlinear coefficients of Maker and Terhune (see Eq. (4.2.10)) have the form $A = C_T\gamma_e^2/16\pi^2$ and $B = 0$.

Let us estimate numerically the value of $\chi^{(3)}$. We saw in Eq. (8.3.12) that for a dilute gas the electrostrictive constant $\gamma_e \equiv \rho(\partial\epsilon/\partial\rho)$ is given by $\gamma_e = n^2 - 1$. More generally, we can estimate γ_e through use of the Lorentz–Lorenz law (Eq. (3.8.8a)), which leads to the prediction

$$\gamma_e = (n^2 - 1)(n^2 + 2)/3. \qquad (9.2.17)$$

This result shows that γ_e is of the order of unity for condensed matter. The compressibility $C_T = \rho^{-1}(\partial\rho/\partial p)$ is approximately equal to 10^{-10} cm^2 dyne^{-1} for CS$_2$ and is of the same order of magnitude for all condensed matter. We thus find that $\chi^{(3)}(\omega = \omega + \omega - \omega)$ is of the order of 2×10^{-13} esu for condensed matter. For ideal gases, the compressibility C_T is equal to $1/p$, where at 1 atmosphere $p = 10^6$ dyne/cm^2. The electrostrictive constant $\gamma_e = n^2 - 1$ for air at 1 atmosphere is approximately equal to 6×10^{-4}. We thus find that $\chi^{(3)}(\omega = \omega + \omega - \omega)$ is of the order of 1×10^{-15} for gases at 1 atmosphere of pressure.

A very useful, alternative expression for $\chi^{(3)}(\omega = \omega + \omega - \omega)$ can be deduced from expression (9.2.16) by expressing the electrostrictive constant through use of Eq. (9.2.17) and by expressing the compressibility in terms of the material density and velocity of sound through use of Eq. (8.3.21), such that $C_s = 1/v^2\rho$. Similarly, the isothermal compressibility is given by $C_T = \gamma C_s$ where γ is the usual thermodynamic adiabatic index. One thus finds that

$$\chi^{(3)}(\omega = \omega + \omega - \omega) = \frac{1}{48\pi^2}\frac{\gamma}{v^2\rho}\left[\frac{(n^2 - 1)(n^2 + 2)}{3}\right]^2. \qquad (9.2.18)$$

For pulses sufficiently short that heat flow during the pulse is negligible, the factor of γ in the numerator of this expression is to be replaced by unity. As usual, the nonlinear refractive index coefficient n_2 for electrostriction can be deduced from this expression and the result $n_2 = (12\pi^2/n_0^2c)\chi^{(3)}$ obtained earlier (Eq. (4.1.19)).

In comparison with other types of optical nonlinearities, the value of $\chi^{(3)}$ resulting from electrostriction is not usually large. However, it can make an appreciable contribution to total measured nonlinearity for certain optical materials. For the case of optical fibers, Buckland and Boyd (1996, 1997) found that electrostriction can make an approximately 20% contribution to the third-order susceptibility. Moreover, we shall see in the next section that electrostriction provides the nonlinear coupling that leads to stimulated Brillouin scattering, which is often an extremely strong process.

9.3. Stimulated Brillouin Scattering (Induced by Electrostriction)*

Our discussion of spontaneous Brillouin scattering in Chapter 8 presupposed that the applied optical fields are sufficiently weak that they do not alter the acoustic properties of the material system. Spontaneous Brillouin scattering then results from the scattering of the incident radiation off the sound waves that are present in thermal equilibrium.

For an incident laser field of sufficient intensity, even the spontaneously scattered light can become quite intense. The incident and scattered light fields can then beat together, giving rise to density and pressure variations by means of electrostriction. The incident laser field can then scatter off the refractive index variation that accompanies these density variations. The scattered light will be at the Stokes frequency and will add constructively with the Stokes radiation that produced the acoustic disturbance. In this manner, the acoustic and Stokes waves mutually reinforce each other's growth, and each can grow to a large amplitude. This circumstance is depicted in Fig. 9.3.1. Here an incident wave of amplitude E_1, angular frequency ω_1, and wavevector $\mathbf{k}_1$ scatters off a retreating sound wave of amplitude ρ, frequency Ω, and wavevector $\mathbf{q}$ to form a scattered wave of amplitude E_2, frequency ω_2, and wavevector $\mathbf{k}_2$.[†]

Let us next deduce the frequency ω_2 of the Stokes field that is created by the SBS process for the case of an SBS generator (see also part (a) of Fig. 9.1.2). Since the laser field at frequency ω_1 is scattered from a retreating sound wave, the scattered radiation will be shifted downward in frequency to

$$\omega_2 = \omega_1 - \Omega_B. \qquad (9.3.1)$$

---

* Stimulated Brillouin scattering can also be induced by absorptive effects. This less commonly studied case is examined in Section 9.6

[†] We denote the field frequencies as ω_1 and ω_2 rather than ω_L and ω_S so that we can later apply the results of the present treatment to the case of anti-Stokes scattering by identifying ω_1 with ω_{aS} and ω_2 with ω_L. The treatment of the present section assumes only that $\omega_2 < \omega_1$.

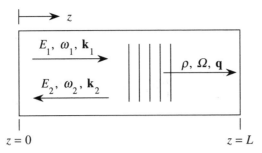

FIGURE 9.3.1 Schematic representation of the stimulated Brillouin scattering process.

Here Ω_B is called the Brillouin frequency, and we shall now see how to determine its value. The Brillouin frequency is related to the acoustic wavevector $\mathbf{q}_B$ by the phonon dispersion relation

$$\Omega_B = |\mathbf{q}_B|v, \qquad (9.3.2)$$

where v is the velocity of sound. By assumption, this sound wave is driven by the beating of the laser and Stokes fields, and its wavevector is therefore given by

$$\mathbf{q}_B = \mathbf{k}_1 - \mathbf{k}_2. \qquad (9.3.3)$$

Since the wavevectors and frequencies of the optical waves are related in the usual manner, that is, by $|\mathbf{k}_i| = n\omega_i/c$, we can use Eq. (9.3.3) and the fact that the laser and Stokes waves are counterpropagating to express the Brillouin frequency of Eq. (9.3.2) as

$$\Omega_B = \frac{v}{c/n}(\omega_1 + \omega_2). \qquad (9.3.4)$$

Equations (9.3.1) and (9.3.4) are now solved simultaneously to obtain an expression for the Brillouin frequency in terms of the frequency ω_1 of the applied field only, that is, we eliminate ω_2 from these equations to obtain

$$\Omega_B = \frac{\frac{2v}{c/n}\omega_1}{1 + \frac{v}{c/n}}. \qquad (9.3.5)$$

However, since v is very much smaller than c/n for all known materials, it is an excellent approximation to take the Brillouin frequency to be

$$\Omega_B = \frac{2v}{c/n}\omega_1. \qquad (9.3.6)$$

At this same level of approximation, the acoustic wavevector is given by

$$\mathbf{q}_B = 2\mathbf{k}_1. \qquad (9.3.7)$$

For the case of the SBS amplifier configuration (see part (b) of Fig. 9.1.2), the Stokes wave is imposed externally and its frequency ω_2 is known *a priori*. The frequency of the driven acoustic wave will then be given by

$$\Omega = \omega_1 - \omega_2, \tag{9.3.8}$$

which in general will be different from the Brillouin frequency of Eq. (9.3.6). As we shall see below, the acoustic wave will be excited efficiently under these circumstances only when ω_2 is chosen such that the frequency difference $|\Omega - \Omega_B|$ is less than or of the order of the Brillouin linewidth Γ_B, which is defined below in Eq. (9.3.14b).

Let us next see how to treat the nonlinear coupling among the three interacting waves. We represent the optical field within the Brillouin medium as $\tilde{E}(z, t) = \tilde{E}_1(z, t) + \tilde{E}_2(z, t)$, where

$$\tilde{E}_1(z, t) = A_1(z, t)e^{i(k_1 z - \omega_1 t)} + \text{c.c.} \tag{9.3.9a}$$

and

$$\tilde{E}_2(z, t) = A_2(z, t)e^{i(-k_2 z - \omega_2 t)} + \text{c.c.} \tag{9.3.9b}$$

Similarly, we describe the acoustic field in terms of the material density distribution

$$\tilde{\rho}(z, t) = \rho_0 + \left[\rho(z, t)e^{i(qz - \Omega t)} + \text{c.c.}\right], \tag{9.3.10}$$

where $\Omega = \omega_1 - \omega_2$, $q = 2k_1$, and ρ_0 denotes the mean density of the medium.

We assume that the material density obeys the acoustic wave equation (see also Eq. (8.3.17))

$$\frac{\partial^2 \tilde{\rho}}{\partial t^2} - \Gamma' \nabla^2 \frac{\partial \tilde{\rho}}{\partial t} - v^2 \nabla^2 \tilde{\rho} = \nabla \cdot \mathbf{f}, \tag{9.3.11}$$

where v is the velocity of sound and Γ' is a damping parameter given by Eq. (8.3.23). The source term on the right-hand side of this equation consists of the divergence of the force per unit volume $\mathbf{f}$, which is given explicitly by

$$\mathbf{f} = \nabla p_{\text{st}}, \qquad p_{\text{st}} = -\gamma_e \frac{\langle \tilde{E}^2 \rangle}{8\pi}. \tag{9.3.12}$$

For the fields given by Eq. (9.3.9), this source term is given by

$$\nabla \cdot \mathbf{f} = \frac{\gamma_e q^2}{4\pi}\left[A_1 A_2^* e^{i(qz - \Omega t)} + \text{c.c.}\right]. \tag{9.3.13}$$

If we now introduce Eqs. (9.3.10) and (9.3.13) into the acoustic wave equation (9.3.11) and assume that the acoustic amplitude varies slowly (if at all) in space and time, we obtain the result

$$-2i\Omega\frac{\partial\rho}{\partial t} + \left(\Omega_B^2 - \Omega^2 - i\Omega\Gamma_B\right)\rho - 2iqv^2\frac{\partial\rho}{\partial z} = \frac{\gamma_e q^2}{4\pi}A_1 A_2^*, \quad (9.3.14a)$$

where we have introduced the Brillouin linewidth

$$\Gamma_B = q^2\Gamma'; \quad (9.3.14b)$$

its reciprocal $\tau_p = \Gamma_B^{-1}$ gives the phonon lifetime.

Equation (9.3.14a) can often be simplified substantially by omitting the last term on its left-hand side. This term describes the propagation of phonons. However, hypersonic phonons are strongly damped and thus propagate only over very short distances before being absorbed.* Since the phonon propagation distance is typically small compared to the distance over which the source term on the right-hand side of Eq. (9.3.14a) varies significantly, it is conventional to drop the term containing $\partial\rho/\partial z$ in describing SBS. This approximation can break down, however, as discussed by Chiao (1965) and by Kroll and Kelley (1971). If we drop the spatial derivative term in Eq. (9.3.14a) and assume steady-state conditions so that $\partial\rho/\partial t$ also vanishes, we find that the acoustic amplitude is given by

$$\rho(z, t) = \frac{\gamma_e q^2}{4\pi}\frac{A_1 A_2^*}{\Omega_B^2 - \Omega^2 - i\Omega\Gamma_B}. \quad (9.3.15)$$

The spatial evolution of the optical fields is described by the wave equation

$$\frac{\partial^2 \tilde{E}_i}{\partial z^2} - \frac{1}{(c/n)^2}\frac{\partial^2 \tilde{E}_i}{\partial t^2} = \frac{4\pi}{c^2}\frac{\partial^2 \tilde{P}_i}{\partial t^2}, \quad i = 1, 2. \quad (9.3.16)$$

The nonlinear polarization, which acts as a source term in this equation, is given by

$$\tilde{P} = \Delta\chi\tilde{E} = \frac{\Delta\epsilon}{4\pi}\tilde{E} = \frac{1}{4\pi\rho_0}\gamma_e\tilde{\rho}\tilde{E}. \quad (9.3.17)$$

We next determine the contributions to $\tilde{P}$ that can act as phase-matched source terms for the laser and Stokes fields. These contributions are given by

$$\tilde{P}_1 = p_1 e^{i(k_1 z - \omega_1 t)} + \text{c.c.}, \quad \tilde{P}_2 = p_2 e^{i(k_2 z - \omega_2 t)} + \text{c.c.}, \quad (9.3.18)$$

where

$$p_1 = \frac{\gamma_e}{4\pi\rho_0}\rho A_2, \quad p_2 = \frac{\gamma_e}{4\pi\rho_0}\rho^* A_1. \quad (9.3.19)$$

---

* We can estimate this distance as follows: According to Eq. (8.3.30), the sound absorption coefficient is given by $\alpha_s = \Gamma_B/v$, where by Eqs. (8.3.23) and (8.3.28) Γ_B is of the order of $\eta_s q^2/\rho_0$. For the typical values $v = 1 \times 10^5$ cm/sec, $\eta_s = 10^{-2}$ dyne cm/sec^2, $q = 4\pi \times 10^4$ cm^{-1}, and $\rho = 1$ cm^{-3}, we find that $\Gamma_B = 1.6 \times 10^8$ sec^{-1} and $\alpha_s^{-1} = 6.3$ μm.

We introduce Eqs. (9.3.9) into the wave equation (9.3.16) along with Eqs. (9.3.18) and (9.3.19), make the slowly-varying amplitude approximation, and obtain the equations

$$\frac{\partial A_1}{\partial z} + \frac{1}{c/n}\frac{\partial A_1}{\partial t} = \frac{i\omega\gamma_e}{2nc\rho_0}\rho A_2, \tag{9.3.20a}$$

$$-\frac{\partial A_2}{\partial z} + \frac{1}{c/n}\frac{\partial A_2}{\partial t} = \frac{i\omega\gamma_e}{2nc\rho_0}\rho^* A_1. \tag{9.3.20b}$$

In these equations ρ is given by the solution to Eq. (9.3.14a). Furthermore, we have dropped the distinction between ω_1 and ω_2 by setting $\omega = \omega_1 \simeq \omega_2$.

Let us now consider steady-state conditions. In this case the time derivatives appearing in Eqs. (9.3.20) can be dropped, and ρ is given by Eq. (9.3.15). The coupled-amplitude equations then become

$$\frac{dA_1}{dz} = \frac{i\omega q^2\gamma_e^2}{8\pi nc\rho_0}\frac{|A_2|^2 A_1}{\Omega_B^2 - \Omega^2 - i\Omega\Gamma_B}, \tag{9.3.21a}$$

$$\frac{dA_2}{dz} = \frac{-i\omega q^2\gamma_e^2}{8\pi nc\rho_0}\frac{|A_1|^2 A_2}{\Omega_B^2 - \Omega^2 + i\Omega\Gamma_B}. \tag{9.3.21b}$$

We see from the form of these equations that SBS is a pure gain process, that is, that the SBS process is automatically phase-matched. For this reason, it is possible to introduce coupled equations for the intensities of the two interacting optical waves. Defining the intensities as $I_i = (nc/2\pi)A_1 A_1^*$, we find from Eqs. (9.3.21) that

$$\frac{dI_1}{dz} = -gI_1 I_2 \tag{9.3.22a}$$

and

$$\frac{dI_2}{dz} = -gI_1 I_2. \tag{9.3.22b}$$

In these equations g is the SBS gain factor, which to good approximation is given by

$$g = g_0\frac{(\Gamma_B/2)^2}{(\Omega_B - \Omega)^2 + (\Gamma_B/2)^2}, \tag{9.3.23}$$

where the line-center gain factor is given by

$$g_0 = \frac{\gamma_e^2\omega^2}{nvc^3\rho_0\Gamma_B}. \tag{9.3.24}$$

The solution to Eqs. (9.3.22) under general conditions will be described below. Note, however, that in the constant-pump limit $I_1 = $ constant, the solution to Eq. (9.3.22b) is

$$I_2(z) = I_2(L)e^{gI_1(L-z)}. \tag{9.3.25}$$

In this limit a Stokes wave injected into the medium at $z = L$ experiences exponential growth as it propagates through the medium. It should be noted that the line-center gain factor g_0 of Eq. (9.3.24) is independent of the laser frequency ω, because the Brillouin linewidth Γ_B is proportional to ω^2 (recall that, according to Eq. (8.3.28), Γ_B is proportional to q^2 and that q is proportional to ω). An estimate of the size of g_0 for the case of CS_2 at a wavelength of 1 μm can be made as follows: $\omega = 2\pi \times 3 \times 10^{14}$ rad/sec, $n = 1.67$, $v = 1.1 \times 10^5$ cm/sec, $\rho_0 = 1.26$ g/cm^3, $\gamma_e = 2.4$, and $\tau_p = \Gamma_B^{-1} = 4 \times 10^{-9}$ sec, giving $g_0 = 1.5 \times 10^{-14}$ cm sec/erg, which in conventional laboratory units becomes $g_0 = 0.15$ cm/MW. The Brillouin gain factors and spontaneous linewidths $\Delta\nu = \Gamma_B/2\pi$ are listed in Table 9.3.1 for a variety of materials.

The theoretical treatment just presented can also be used to describe the propagation of a wave at the anti-Stokes frequency, $\omega_{aS} = \omega_L + \Omega_B$. Equations (9.3.22) were derived for the geometry of Fig. 9.3.1 under the assumption that $\omega_1 > \omega_2$. We can treat anti-Stokes scattering by identifying ω_1 with ω_{aS} and ω_2 with ω_1. We then find that the constant-pump approximation corresponds to the case $I_2(z) = $ constant, and that the solution to Eq. (9.3.22a)

TABLE 9.3.1 Properties of stimulated Brillouin scattering for a variety of materials[a]

Substance	$\Omega_B/2\pi$ (MHz)	$\Gamma_B/2\pi$ (MHz)	g_0 (cm/MW)	$g_B^a(\max)/\alpha$ (cm^2/MW)
CS_2	5850	52.3	0.15	0.14
Acetone	4600	224	0.02	0.022
Toluene	5910	579	0.013	
CCl_4	4390	520	0.006	0.013
Methanol	4250	250	0.013	0.013
Ethanol	4550	353	0.012	0.010
Benzene	6470	289	0.018	0.024
H_2O	5690	317	0.0048	0.0008
Cyclohexane	5550	774	0.0068	
CH_4(1400 atm)	150	10	0.1	
Optical glasses	15,000–26,000	10–106	0.004–0.025	
SiO_2	25,800	78	0.0045	

[a] Values are quoted for a wavelength of 0.694 μm. The quantity $\Gamma_B/2\pi$ is the full width at half maximum in ordinary frequency units of the SBS gain spectrum. The last column gives a parameter used to describe the process of absorptive SBS, which is discussed in Section 9.6. To convert to other laser frequencies ω, recall that Ω_B is proportional to ω, Γ is proportional to ω^2, g_0 is independent of ω, and g_B^a (max) is proportional to ω^{-3}.

is $I_1(z) = I_1(0)e^{-gI_2z}$. Since the anti-Stokes wave at frequency ω_1 propagates in the positive z direction, we see that it experiences attenuation due to the SBS process.

Pump Depletion Effects in SBS

We have seen (Eq. (9.3.25)) that, in the approximation in which the pump intensity is taken to be spatially invariant, the Stokes wave experiences exponential growth as it propagates through the Brillouin medium. Once the Stokes wave has grown to an intensity comparable to that of the pump wave, significant depletion of the pump wave must occur, and under these conditions we must solve the coupled-intensity equations (9.3.22) simultaneously in order to describe the SBS process. We see from these equations that $dI_1/dz = dI_2/dz$ and hence that

$$I_1(z) = I_2(z) + C, \qquad (9.3.26)$$

where the value of the integration constant C depends on the boundary conditions. Using this result, Eq. (9.3.22b) can be expressed as

$$\frac{dI_2}{I_2(I_2 + C)} = -g\, dz. \qquad (9.3.27)$$

This equation can be integrated formally as

$$\int_{I_2(0)}^{I_2(z)} \frac{dI_2}{I_2(I_2 + C)} = -\int_0^z g\, dz', \qquad (9.3.28)$$

which implies that

$$\ln\left\{ \frac{I_2(z)[I_2(0) + C]}{I_2(0)[I_2(z) + C]} \right\} = -gCz. \qquad (9.3.29)$$

Since we have specified the value of I_1 at $z = 0$, it is convenient to express the constant C defined by Eq. (9.3.26) as $C = I_1(0) - I_2(0)$. Equation (9.3.29) is now solved algebraically for $I_2(z)$, yielding

$$I_2(z) = \frac{I_2(0)[I_1(0) - I_2(0)]}{I_1(0)\exp\{gz[I_1(0) - I_2(0)]\} - I_2(0)}. \qquad (9.3.30a)$$

According to Eq. (9.3.26), $I_1(z)$ can be found in terms of this expression as

$$I_1(z) = I_2(z) + I_1(0) - I_2(0). \qquad (9.3.30b)$$

Equations (9.3.30) give the spatial distribution of the field intensities in terms of the boundary values $I_1(0)$ and $I_2(0)$. However, the boundary values that are known physically are $I_1(0)$ and $I_2(L)$; see Fig. 9.3.2. In order to find the

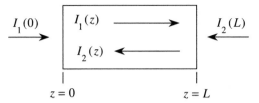

FIGURE 9.3.2   Geometry of an SBS amplifier. The boundary values $I_1(0)$ and $I_2(L)$ are known.

unknown quantity $I_2(0)$ in terms of the known quantities $I_1(0)$ and $I_2(L)$, we set z equal to L in Eq. (9.3.30a) and write the resulting expression as follows:

$$I_2(L) = \frac{I_1(0)[I_2(0)/I_1(0)][1 - I_2(0)/I_1(0)]}{\exp\{g I_1(0)L[1 - I_2(0)/I_1(0)]\} - I_2(0)/I_1(0)}. \quad (9.3.31)$$

This expression is a transcendental equation giving the unknown quantity $I_2(0)/I_1(0)$ in terms of the known quantities $I_1(0)$ and $I_2(L)$.

The results given by Eqs. (9.3.30) and (9.3.31) can be used to analyze the SBS amplifier shown in Fig. 9.3.2. The transfer characteristics of such an amplifier are illustrated in Fig. 9.3.3. Here the vertical axis gives the fraction of the laser intensity that is transferred to the Stokes wave, and the horizontal axis is the quantity $G = g I_1(0)L$ which gives the exponential gain experienced by a *weak* Stokes input. The various curves are labeled according to the ratio of input intensities, $I_2(L)/I_1(0)$. For sufficiently large values of the exponential gain, essentially complete transfer of the pump energy to the Stokes beam is possible.

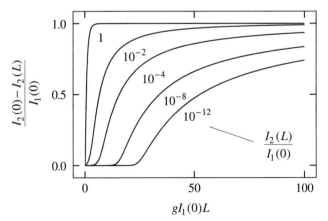

FIGURE 9.3.3   Intensity transfer characteristics of an SBS amplifier.

SBS Generator

For the case of an SBS generator, no Stokes field is injected externally into the interaction region, and hence the value of the Stokes intensity near the Stokes input face $z = L$ is not known *a priori*. In this case, the SBS process is initiated by Stokes photons that are created by spontaneous Brillouin scattering involving the laser beam near its exit plane $z = L$. We therefore expect that the effective Stokes input intensity $I_2(L)$ will be proportional to the local value of the laser intensity $I_1(L)$; we designate the constant of proportionality as f so that

$$I_2(L) = f I_1(L). \tag{9.3.32}$$

We estimate the value of f as follows: We first consider the conditions that apply below the threshold for the occurrence of SBS, such that the SBS reflectivity $R = I_2(0)/I_1(0)$ is much smaller than unity. Under these conditions the laser intensity is essentially constant throughout the medium, and the Stokes output intensity is related to the Stokes input intensity by $I_2(0) = I_2(L)e^G$, where $G = g I_1(0)L$. However, since $I_2(L) = f I_1(0)$ (because $I_1(z)$ is constant), the SBS reflectivity can be expressed as

$$R \equiv \frac{I_2(0)}{I_1(0)} = f e^G. \tag{9.3.33}$$

Laboratory experience has shown that the SBS process displays an apparent threshold. One often defines the SBS threshold as the condition that the reflectivity R reach some prescribed value R_{th}; the value $R_{th} = 0.01$ is a convenient choice. This reflectivity occurs for the specific value G_{th} of the gain parameter $G = g I_1(0)L$. For a wide variety of materials and laser wavelengths, it is found that G_{th} typically lies in the fairly narrow range of 25 to 30. The actual value of G_{th} for a particular situation can be deduced theoretically from a consideration of the thermal fluctuations that initiate the SBS process; see, for instance, Boyd *et al.* (1990) for details. Since G_{th} is approximately 25–30, we see from Eq. (9.3.33) that f is of the order of $\exp(-G_{th})$, or approximately 10^{-12} to 10^{-11}. An order-of-magnitude estimate based on the properties of spontaneous scattering performed by Zel'dovich *et al.* (1985) reaches the same conclusion.

We next calculate the SBS reflectivity R for the general case $G > G_{th}$ (i.e., above threshold) through use of Eq. (9.3.31), which we write as

$$\frac{I_2(L)}{I_1(0)} = \frac{R(1 - R)}{\exp[G(1 - R)] - R}. \tag{9.3.34}$$

To good approximation, the term $-R$ can be dropped from the denominator of the right-hand side of this equation. In order to determine the ratio $I_2(L)/I_1(0)$

that appears on the left-hand side of Eq. (9.3.34), we express Eq. (9.3.30b) as

$$I_1(L) - I_2(L) = I_1(0) - I_2(0).$$

Through use of Eq. (9.3.32) and the smallness of f, we can replace the left-hand side of this equation by $f^{-1}I_2(L)$. We now multiply both sides of the resulting equation by $f/I_1(0)$ to obtain the result $I_2(L)/I_1(0) = f(1 - R)$. This expression is substituted for the left-hand side of Eq. (9.3.34), which is then solved for G, yielding the result

$$\frac{G}{G_{th}} = \frac{G_{th}^{-1}\ln R + 1}{1 - R}, \tag{9.3.35}$$

where we have substituted G_{th} for $-\ln f$.

The nature of this solution is illustrated in Fig. 9.3.4, where the SBS reflectivity $R = I_2(0)/I_1(0)$ is shown plotted as a function of $G = gI_1(0)L$ for the value $G_{th} = 25$. We see that essentially no Stokes light is created for G less than G_{th} and that the reflectivity rises rapidly for laser intensities slightly above this threshold value. In addition, for $G \gg G_{th}$ the reflectivity asymptotically approaches 100%. Well above the threshold for SBS (i.e., for $G \gtrsim 3G_{th}$), Eq. (9.3.35) can be approximated as $G/G_{th} \simeq 1/(1 - R)$, which shows that the SBS reflectivity in this limit can be expressed as

$$R = 1 - \frac{1}{G/G_{th}} \quad \text{(for } G \gg G_{th}\text{)}. \tag{9.3.36}$$

Since the intensity $I_1(L)$ of the transmitted laser beam is given by $I_1(L) = I_1(0)(1 - R)$, in the limit of validity of Eq. (9.3.36) the intensity of the

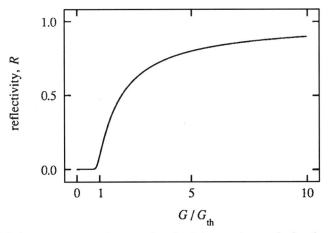

FIGURE 9.3.4 Dependence of the SBS reflectivity on the weak-signal gain $G = gI_1(0)L$.

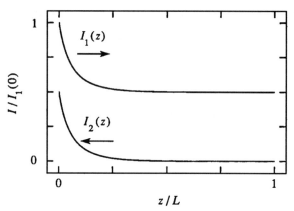

FIGURE 9.3.5   Distribution of the laser and Stokes intensities within the interaction region of an SBS generator.

transmitted beam is given by

$$I_1(L) = \frac{G_{th}}{gL};   \tag{9.3.37}$$

here G_{th}/gL can be interpreted as the input laser intensity at the threshold for SBS. Hence the transmitted intensity is "clamped" at the threshold value for the occurrence of SBS.

Once the value of the Stokes intensity at the plane $z = 0$ is known from Eq. (9.3.35), the distributions of the intensities within the interaction region can be obtained from Eqs. (9.3.20). Figure 9.3.5 shows the distribution of intensities within an SBS generator.*

Let us estimate the minimum laser power P_{th} required to excite SBS under optimum conditions. We assume that a laser beam having a gaussian transverse profile is focused tightly into a cell containing a Brillouin-active medium. The characteristic intensity of such a beam at the beam waist is given by $I = P/\pi w_0^2$, where w_0 is the beam waist radius. The interaction length L is limited to the characteristic diffraction length $b = 2\pi w_0^2/\lambda$ of the beam. The product $G = gIL$ is thus given by $G = 2gP/\lambda$, and by equating this expression with the threshold value G_{th} we find that the minimum laser power required to excite SBS is of the order of

$$P_{th} = \frac{G_{th}\lambda}{2g}.   \tag{9.3.38}$$

---

* Figure 9.3.5 is plotted for the case $G_{th} = 10$. The physically realistic case of $G_{th} = 25$ produces a much less interesting graph because the perceptible variation in intensities occurs in a small region near $z = 0$.

For $\lambda = 1.06$ μm, $G_{th} = 25$, and $g = 0.15$ cm/MW (the value for CS_2) we find that P_{th} is equal to 9 kW. For other organic liquids the minimum power is approximately 10 times larger.

Transient and Dynamical Features of SBS

The phonon lifetime for stimulated Brillouin scattering in liquids is of the order of several nanoseconds. Since Q-switched laser pulses have a duration of the order of several nanoseconds, and mode-locked laser pulses can be much shorter, it is normal for experiments on SBS to be performed in the transient regime. The nature of transient SBS has been treated by Kroll (1965), by Pohl, Maier, and Kaiser (1968), and by Pohl and Kaiser (1970).

The SBS equations can be solved including the transient nature of the phonon field. This was done first by Carman *et al.* (1970) and the results have been summarized by Zel'dovich *et al.* (1985). One finds that

$$I_S(L, T) \simeq \begin{cases} I_N \exp\left(-2\Gamma_B T + 2\sqrt{2(gIL)(\Gamma_B T)}\right) & \Gamma_B T < gIL/2, \\ I_N \exp(gIL) & \Gamma_B T > gIL/2. \end{cases}$$

$$(9.3.39)$$

Here I_N is the effective noise input that initiates the SBS process, gIL is the usual single pass gain, Γ_B is the phonon damping rate, and T is the laser pulse duration.

We can use this result to predict how the SBS threshold intensity I_{th} is increased through use of a short laser pulse. We require that in either limit given above the single pass amplification must equal the threshold value, which we take to be $\exp(25)$. We then find that

$$g I_{th} l = \begin{cases} (12.5 + 2\Gamma_B T)^2/2\Gamma_B T & \Gamma_B T < 12.5, \\ 25 & \Gamma_B T > 12.5. \end{cases}$$

$$(9.3.40)$$

This functional dependence is illustrated in Fig. 9.3.6. Note that even for laser pulses as long as twice the phonon lifetime, the threhold for SBS is raised by a factor of approximately two.

The SBS process is characterized by several different time scales, including the transit time of light through the interaction region, the laser pulse duration, and the phonon lifetime. Consequently, the SBS process can display quite rich dynamical effects. One of these effects is pulse compression, the tendecncy of the SBS Stokes pulse to be shorter (at times very much shorter) than the incident laser pulse. This process is described in Problem 5 at the end of this chapter. When SBS is excited by a multi-longtudinal-mode laser, new types of dynamical behavior can occur. Here the various laser modes beat together

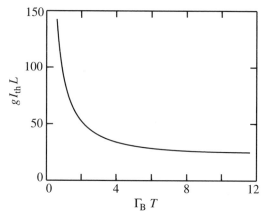

FIGURE 9.3.6   Dependence of the SBS threshold intensity I_{th} on the laser pulse duration T.

leading to modulation in time of the laser intensity within the interaction region. This situation has been analyzed by Narum *et al.* (1986). In addition, the stochastic properties of SBS have been studied in considerable detail. SBS is initiated by noise in the form of thermally excited phonons. Since the SBS process involves nonlinear amplification (nonlinear because of pump depletion effects) in a medium with an effectively nonlocal response (nonlocal because the Stokes and laser fields are counterpropagating), the stochastic properties of the SBS output can be quite different from those of the phonon noise field that initiates SBS. These properties have been studied, for instance, by Gaeta and Boyd (1991). In addition, when SBS is excited by two counterpropagating pump fields, it can display even more complex behavior, including instability and chaos, as studied by Narum *et al.* (1988), Gaeta *et al.* (1989), and Kulagin *et al.* (1991).

9.4. Phase Conjugation by Stimulated Brillouin Scattering

It was noted even in the earliest experiments on stimulated Brillouin scattering (SBS) that the Stokes radiation was emitted in a highly collimated beam in the backward direction. In fact, the Stokes radiation was found to be so well collimated that it was efficiently fed back into the exciting laser, often leading to the generation of new spectral components in the output of the laser (Goldblatt and Hercher, 1968). These effects were initially explained as a purely geometrical effect resulting from the long but thin shape of the interaction region.

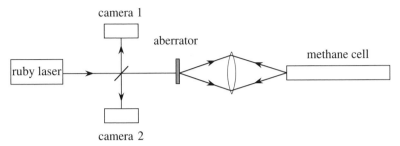

FIGURE 9.4.1 Setup of first experiment on phase conjugation by stimulated Brillouin scattering.

The first indication that the backscattered light was in fact the phase conjugate of the input was provided by an experiment of Zel'dovich *et al.* (1972). The setup used in this experiment is shown in Fig. 9.4.1. The output of a single-mode ruby laser was focused into a cell containing methane gas at a pressure of 125 atmospheres. This cell was constructed in the shape of a cylindrical, multimode waveguide and served to confine the radiation in the transverse dimension. A strong SBS signal was generated from within this cell. A glass plate that had been etched in hydrofluoric acid was placed in the incident beam to serve as an aberrator. Two cameras were used to monitor the transverse intensity distributions of the incident laser beam and of the Stokes return.

The results of this experiment are summarized in the photographs taken by V. V. Ragulsky that are reproduced in Fig. 9.4.2. Part (a) of this figure shows the laser beam shape as recorded by camera 1, and part (b) shows the Stokes

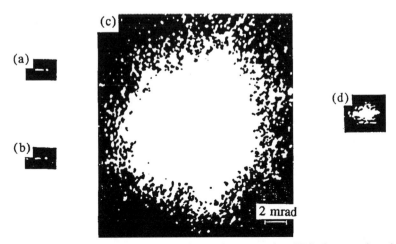

FIGURE 9.4.2 Results of the first experiment demonstrating SBS phase conjugation.

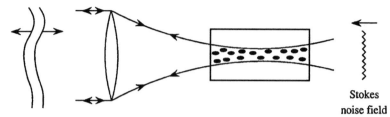

FIGURE 9.4.3   Origin of phase conjugation by SBS. The highly aberrated incident wavefront produces a highly nonuniform intensity distribution (and hence a nonuniform gain distribution) in the focal region of the lens.

beam shape as recorded by camera 2. The similarity of the spot sizes and shapes indicates that the return beam is the phase conjugate of the incident beam. These highly elongated beam shapes are a consequence of the unusual mode pattern of the laser used in these experiments. Part (c) of the figure shows the spot size recorded by camera 2 when the SBS cell had been replaced by a conventional mirror. The spot size in this case is very much larger than that of the incident beam; this result shows the severity of the distortions impressed on the beam by the aberrator. Part (d) of the figure shows the spot size of the return beam when the aberrator was removed from the beam path. This spot size is larger than that shown in part (b). This result shows that SBS forms a more accurate conjugate of the incident light when the beam is highly distorted than when the beam is undistorted.*

   The results of the experiment of Zel'dovich *et al.* are somewhat surprising, because it is not clear from inspection of the coupled-amplitude equations that describe the SBS process why SBS should lead to phase conjugation. We recall that the reason why degenerate four-wave mixing leads to phase conjugation is that the source term driving the output wave A_4 in the coupled-amplitude equations describing four-wave mixing (see, for example, Eq. (7.1.31b)) is proportional to the complex conjugate of the input wave amplitude, that is, to A_3^*. However, for the case of SBS, Eq. (9.3.21b) shows that the output wave amplitude A_2 is driven by a term proportional to $|A_1|^2 A_2$, which contains no information regarding the phase of the input wave A_1.

   The reason why SBS leads to the generation of a phase-conjugate wave is in fact rather subtle (Zel'dovich *et al.*, 1972; Sidorovich, 1976). As illustrated in Fig. 9.4.3, we consider a badly aberrated optical wave that is focused into the SBS interaction region. Since the wave is highly aberrated, a highly nonuniform

---

* The conclusion that SBS forms a better phase conjugate of an aberrated beam than of an unaberrated beam is not true in all cases, and appears to be a consequence of the details of the geometry of the experiment of Zel'dovich *et al.*

intensity distribution (i.e., a volume speckle pattern) is created in the focal region of the wave. Since the gain experienced by the Stokes wave depends upon the local value of the laser intensity (see, for example, Eq. (9.3.22b)), a nonuniform gain distribution for the Stokes wave is therefore present in the focal volume. We recall that SBS is initiated by noise, that is, by spontaneously generated Stokes photons. The noise field that leads to SBS initially contains all possible spatial Fourier components. However, the portion of the noise field that experiences the maximum amplification is the portion whose intensity distribution best matches the nonuniform gain distribution. This portion of the noise field must have wavefronts that match those of the incident laser beam, and hence corresponds to the phase conjugate of the incident laser field.

In order to make this argument more precise, we consider the intensity equation satisfied by the Stokes field (see also Eq. (9.3.22b)),

$$\frac{dI_S}{dz} = -g I_L I_S. \tag{9.4.1}$$

Since we are now considering the case where I_L and I_S possess nonuniform transverse distributions, it is useful to consider the total power in each wave (at fixed z), defined by

$$P_L = \int I_L \, dA, \qquad P_S = \int I_S \, dA, \tag{9.4.2}$$

where the integrals are to be carried out over an area large enough to include essentially all of the power contained in each beam. Equation (9.4.1) can then be rewritten in the form

$$\frac{dP_S}{dz} = -g \frac{P_L P_S}{A} C, \tag{9.4.3}$$

where $A = \int dA$ and where

$$C = \frac{\langle I_L I_S \rangle}{\langle I_L \rangle \langle I_S \rangle} \tag{9.4.4}$$

represents the normalized spatial cross-correlation function of the laser and Stokes field intensity distributions. Here the angular brackets are defined so that $\langle x \rangle = \int x \, dA / A$, where x denotes I_L, I_S, or the product $I_L I_S$.

We see that the power gain experienced by the Stokes wave depends not only on the total power in the laser wave, but also on the degree of correlation between the laser and Stokes wave intensity distributions. If I_L and I_S are completely uncorrelated, so that $\langle I_L I_S \rangle = \langle I_L \rangle \langle I_S \rangle$, the correlation function C takes on the value unity. C is equal to unity also for the case in which both I_L and I_S are spatially uniform. However, if I_L and I_S are correlated, for example because the laser and Stokes fields are phase conjugates of one another, the correlation function can be greater than unity.

A limiting case is that in which the laser field is so badly aberrated that the transverse variations in the complex field amplitude obey gaussian statistics. In such a case, the probability density function for the laser intensity fluctuations is given by (see, for example, Goodman, 1985)

$$P(I) = \frac{1}{I_0} e^{-I/I_0}. \tag{9.4.5}$$

The moments of this distribution are given in general by $\langle I^n \rangle = n \langle I \rangle^n$, and in particular the second moment is given by

$$\langle I^2 \rangle = 2 \langle I \rangle^2. \tag{9.4.6}$$

For that portion of the Stokes field that is the phase conjugate of the laser field, the intensity I_S will be proportional to I_L, and we see from Eqs. (9.4.4) and (9.4.6) that C will be equal to 2. Hence the exponential gain $G \equiv g P_L C L / A$ experienced by the phase-conjugate portion of the noise field will be two times larger than that experienced by any other mode of the noise field. Since the threshold for SBS corresponds to G of the order of 30, the phase-conjugate portion of the SBS signal at threshold will be approximately exp(15) times larger than that of any other component.

On the basis of the argument just presented, we expect that high-quality phase conjugation will occur only if a large number of speckles of the laser intensity distribution are present within the interaction volume. We now determine the conditions under which the number of speckles will be large. We assume that in the focal region the incident laser field has transverse wavefront irregularities on a distance scale as small as a. Each such region will diffract the incident beam into a cone with a characteristic angular spread of $\theta = \lambda/a$. Hence the speckle pattern will look appreciably different after the beam has propagated through the longitudinal distance Δz such that $\theta \Delta z = a$. These considerations show that $\Delta z = a^2/\lambda$. We hence expect that SBS will lead to a high-quality phase-conjugate signal only if the transverse extent of the interaction region is much larger than a and if the longitudinal extent of the interaction region is much longer than Δz. In addition, the quality of the phase-conjugate signal can be degraded if there is poor spatial overlap of the various spatial Fourier components of the laser beam. For example, a highly aberrated beam will spread with a large angular divergence $\theta = \lambda/a$. If those components of the beam with large divergence angle θ fail to overlap the strong central portion of the beam, they will be reflected with low efficiency, leading to a degradation of the quality of the phase-conjugation process. To avoid the possibility of such effects, SBS phase conjugation is often performed using the waveguide geometry shown in Fig. 9.4.1.

One of the applications of SBS phase conjugation is in the design of high-power laser systems. Phase conjugation can be used to correct for aberrations caused, for instance, by thermal stresses induced in the laser gain medium. One example of a high-power laser system that makes use of SBS phase conjugation to maintain control of the polarization properties of the laser output has been described by Bowers *et al.* (1997).

9.5. Stimulated Brillouin Scattering in Gases

We next consider stimulated Brillouin scattering (SBS) in gases. As before (Eq. (9.3.24)), the steady-state line-center gain factor for SBS is given by

$$g_0 = \frac{\gamma_e^2 \omega^2}{\rho_0 n v c^3 \Gamma_B} \tag{9.5.1}$$

with the electrostrictive constant γ_e given by Eq. (8.3.12) and with the Brillouin linewidth given to good approximation by (see also Eqs. (8.3.23) and (9.3.14b))

$$\Gamma_B = (2\eta_s + \eta_d) \, q^2 / \rho_0. \tag{9.5.2}$$

For the case of an ideal gas, we can readily predict the values of the material parameters appearing in these equations (Loeb, 1961). First, we can assume the validity of the Stokes relation (see also the discussion in the Appendix to Section 9.6), which states that the shear and dilation viscosity coefficients are related by $\eta_d = -\frac{2}{3}\eta_s$, and hence we find that

$$\Gamma_B = \frac{4}{3}\eta_s q^2 / \rho_0. \tag{9.5.3}$$

The shear viscosity coefficient η_s can be shown from kinetic theory to be given by

$$\eta_s = \frac{1}{3} N m \bar{v} L, \tag{9.5.4}$$

where N is the atomic number density, m is the molecular mass, $\bar{v}$ is the mean molecular velocity given by $\bar{v} = (8kT/\pi m)^{1/2}$, and L is the mean free path given by $L = (\sqrt{2}\pi d^2 N)^{-1}$ with d denoting the molecular diameter. We hence find that the shear viscosity coefficient is given by

$$\eta_s = \frac{2}{3\pi^{3/2}} \frac{\sqrt{kTm}}{d^2}. \tag{9.5.5}$$

Note that the shear viscosity coefficient is independent of the molecular number density N. The measured (and theoretical) value of the shear viscosity coefficient for N_2 at standard temperature and pressure is $\eta_s = 1.8 \times 10^{-4}$ dyne sec/cm^2.

By introducing expression (9.5.5) for the viscosity into Eq. (9.5.2) and replacing q by $2n\omega/c$, we find that the Brillouin linewidth is given by

$$\Gamma_B = \frac{32}{9\pi^{3/2}} \frac{n^2\omega^2}{c^2} \frac{\sqrt{kT/m}}{d^2N}. \tag{9.5.6}$$

If we assume that the incident optical radiation has a wavelength λ of 1.06 μm, we find that the Brillouin linewidth for N_2 at standard temperature and pressure is equal to $\Gamma_B = 2.77 \times 10^9$ rad/sec and hence that the Brillouin linewidth in ordinary frequency units is given by $\delta\nu$(FWHM) $= \Gamma_B/2\pi = 440$ MHz.

The velocity of sound v, which appears in Eq. (9.5.1), is given for an ideal gas by $v = (\gamma kT/m)^{1/2}$, where γ, the ratio of specific heats, is equal to $5/3$ for a monatomic gas and $7/5$ for a diatomic gas. In addition, the electrostrictive constant γ_e can be estimated as $\gamma_e = \rho(\partial\epsilon/\partial\rho)$ with $(\partial\epsilon/\partial\rho)$ taken as the essentially constant quantity $(\epsilon - 1)/\rho$.

The dependence of g_0 on material parameters can be determined by combining these results with Eq. (9.5.1) to obtain

$$g_0 = \frac{9\pi^{3/2}N^2m^2d^2(\partial\epsilon/\partial\rho)^2}{32\gamma^{1/2}n^3ckT}. \tag{9.5.7}$$

However, in order to obtain a numerical estimate of g_0, it is often more convenient to evaluate the expression (9.5.1) for g_0 directly with the numerical value of Γ_B obtained from Eq. (9.5.6). For N_2 gas at standard temperature and pressure and for a wavelength of 1.06 μm, we take the values $\omega = 1.8 \times 10^{15}$ rad/sec, $n = 1.0003$, $v = 3.3 \times 10^4$ cm/sec, $\gamma_e = n^2 - 1 = 6 \times 10^{-4}$, and we thereby obtain

$$g_0 = 3.8 \times 10^{-19} \frac{\text{cm sec}}{\text{erg}} = 3.8 \frac{\text{cm}}{\text{TW}}. \tag{9.5.8}$$

Note from Eq. (9.5.7) that g_0 scales quadratically with molecular density. Hence, at a pressure of 100 atmospheres the gain factor of N_2 is equal to $g_0 = 0.042$ cm/MW, which is comparable to that of typical organic liquids.

One advantage of the use of gases as the active medium for Brillouin scattering is that the gain for SBS scales with molecular number density as N^2, whereas the gain for stimulated Raman scattering, which is often a competing process, scales as N (see, for example, Eqs. (9.3.19a), (9.3.19b), and (9.3.20)). At pressures greater than 10 atmospheres, the gain for SBS typically exceeds that of stimulated Raman scattering. Moreover, through the use of rare gases (which have no vibrational modes), it is possible to suppress the occurrence of stimulated Raman scattering altogether.

TABLE 9.5.1 Gain factors, phonon lifetimes, and frequency shifts for some compressed Brillouin-active gases at a wavelength of 249 nm[a]

Gas	p (atm)	g_0 (cm/MW)	τ (nsec)	$\Omega_B/2\pi$ (GHz)	g_R (cm/MW)
SF_6	15.5	2.5×10^{-2}	1	0.9	3×10^{-4}
	10	0.9×10^{-2}	0.6		2×10^{-4}
Xe	39	4.4×10^{-2}	2	0.12	0
	10	1.8×10^{-2}	0.4		
Ar	10	1.5×10^{-4}	0.1	3	0
N_2	10	1.7×10^{-4}	0.2	3	3×10^{-5}
CH_4	10	8×10^{-4}	0.1	3	1×10^{-3}

[a] For comparison, the gain factor g_R for forward stimulated Raman scattering is also listed. (After Damzen and Hutchinson, 1983.)

Some parameters relevant to SBS at the 249 nm wavelength of the KrF laser have been compiled by Damzen and Hutchinson (1983) and are presented in Table 9.5.1.

9.6. General Theory of Stimulated Brillouin and Stimulated Rayleigh Scattering

In this section we develop a theoretical model that can treat both stimulated Brillouin and stimulated Rayleigh scattering. These two effects can conveniently be treated together because they both entail the scattering of light from inhomogeneities in thermodynamic quantities. For convenience, we choose the temperature T and density ρ to be the independent thermodynamic variables. The theory that we present incorporates both electrostrictive and absorptive coupling of the radiation to the material system. Our analysis therefore describes the following four scattering processes:

1. *Electrostrictive stimulated Brillouin scattering.* The scattering of light from sound waves that are driven by the interference of the laser and Stokes fields through the process of electrostriction.
2. *Thermal stimulated Brillouin scattering.* The scattering of light from sound waves that are driven by the absorption and subsequent thermalization of the optical energy, leading to temperature and hence to density variations within the medium.

3. *Electrostrictive stimulated Rayleigh scattering.* The scattering of light from isobaric density fluctuations that are driven by the process of electrostriction.

4. *Thermal stimulated Rayleigh scattering.* The scattering of light from isobaric density fluctuations that are driven by the process of optical absorption.

Our analysis is based on the three equations of hydrodynamics (Hunt, 1955; Kaiser and Maier, 1972). The first of these equations is the equation of continuity

$$\frac{\partial \tilde{\rho}_t}{\partial t} + \tilde{\mathbf{u}}_t \cdot \nabla \tilde{\rho}_t + \tilde{\rho}_t \nabla \cdot \tilde{\mathbf{u}}_t = 0, \tag{9.6.1}$$

where $\tilde{\rho}_t$ is the mass density of the fluid and $\tilde{\mathbf{u}}_t$ is the velocity of some small volume element of the fluid.* The second equation is the equation of momentum transfer. It is a generalization of the Navier–Stokes equation and is given by

$$\tilde{\rho}_t \frac{\partial \tilde{\mathbf{u}}_t}{\partial t} + \tilde{\rho}_t (\tilde{\mathbf{u}}_t \cdot \nabla)\tilde{\mathbf{u}}_t = \tilde{\mathbf{f}} - \nabla \tilde{p}_1 + (2\eta_s + \eta_d)\nabla(\nabla \cdot \tilde{\mathbf{u}}_t) - \eta_s \nabla \times (\nabla \times \tilde{\mathbf{u}}_t).$$
$$\tag{9.6.2}$$

Here $\tilde{\mathbf{f}}$ represents the force per unit volume of any externally imposed forces; for the case of electrostriction, $\tilde{\mathbf{f}}$ is given by (see also Eq. (9.3.12))

$$\tilde{\mathbf{f}} = -\frac{\gamma_e}{8\pi} \nabla \langle \tilde{\mathbf{E}} \cdot \tilde{\mathbf{E}} \rangle, \tag{9.6.3}$$

where $\tilde{\mathbf{E}}$ denotes the instantaneous value of the time-varying applied total electric field and γ_e represents the electrostrictive coupling constant

$$\gamma_e = \rho \frac{\partial \epsilon}{\partial \rho}. \tag{9.6.4}$$

The second term on the right-hand side of Eq. (9.6.2) denotes the force due to the gradient of the pressure $\tilde{p}_t$. In the third term, η_s denotes the shear viscosity coefficient and η_d denotes the dilational viscosity coeficient. When the Stokes relation is satisfied, as it is for example for an ideal gas, these coefficients are related by

$$\eta_d = -\tfrac{2}{3}\eta_s. \tag{9.6.5}$$

The coefficients are defined in detail in the Appendix at the end of this section.

---

* The subscript t stands for *total*; we shall later linearize these equations to find the equations satisfied by the linearized quantities, which we shall designate by nonsubscripted symbols.

The last of three principal equations of hydrodynamics is the equation of heat transport, given by

$$\tilde{\rho}_t C_v \frac{\partial \tilde{T}_t}{\partial t} + \tilde{\rho}_t c_v (\tilde{\mathbf{u}} \cdot \nabla \tilde{T}_t) + \tilde{\rho}_t c_v \left(\frac{\gamma - 1}{\beta_p} \right) (\nabla \cdot \tilde{\mathbf{u}}_t) = - \nabla \cdot \tilde{\mathbf{Q}} + \tilde{\phi}_\eta + \tilde{\phi}_{\text{ext}}.$$
(9.6.6)

Here $\tilde{T}_t$ denotes the local value of the temperature, c_v the specific heat at constant volume, $\gamma = c_p / c_v$ the adiabatic index, $\beta_p = -\tilde{\rho}^{-1} (\partial \tilde{\rho} / \partial \tilde{T})_p$ the thermal expansion coefficient, and $\tilde{\mathbf{Q}}$ the heat flux vector. For heat flow due to thermal conduction, $\tilde{\mathbf{Q}}$ satisfies the equation

$$\nabla \cdot \tilde{\mathbf{Q}} = -\kappa \nabla^2 \tilde{T}_t,$$
(9.6.7)

where κ denotes the thermal conductivity. $\tilde{\phi}_\eta$ denotes the viscous energy deposited within the medium per unit volume per unit time and is given by

$$\tilde{\phi}_\eta = \sum_{ij} (2\eta_s d_{ij} d_{ji} + \eta_d d_{ii} d_{jj}),$$
(9.6.8a)

where

$$d_{ij} = \frac{1}{2} \left(\frac{\partial \tilde{u}_i}{\partial x_j} + \frac{\partial \tilde{u}_j}{\partial x_i} \right)$$
(9.6.8b)

is the rate-of-dilation tensor. Finally, $\tilde{\phi}_{\text{ext}}$ gives the energy per unit time per unit volume delivered to the medium from external sources. Absorption of the optical wave provides the contribution

$$\tilde{\phi}_{\text{ext}} = \alpha \frac{nc}{4\pi} \langle \tilde{E}^2 \rangle,$$
(9.6.9)

to this quantity, where α is the optical absorption coefficient.

The acoustic equations are now derived by linearizing the hydrodynamic equations about the nominal conditions of the medium. In particular, we take

$$\tilde{\rho}_t = \rho_0 + \tilde{\rho} \quad \text{with} \quad |\tilde{\rho}| \ll \rho_0,$$
(9.6.10a)
$$\tilde{T}_t = T_0 + \tilde{T} \quad \text{with} \quad |\tilde{T}| \ll T_0,$$
(9.6.10b)
$$\tilde{\mathbf{u}}_t = \tilde{\mathbf{u}} \quad \text{with} \quad |\tilde{\mathbf{u}}| \ll v,$$
(9.6.10c)

where v denotes the velocity of sound. Note that we have assumed that the medium is everywhere motionless in the absence of the acoustic disturbance. We can reliably use the linearized form of the resulting equations so long as the indicated inequalities are satisfied.

We substitute the expansions (9.6.10) into the hydrodynamic equations (9.6.1), (9.6.2), and (9.6.6), drop any term that contains more than one small quantity, and subtract the unperturbed, undriven solution containing only $\tilde{\rho}_0$

and $\tilde{T}_0$. The continuity equation (9.6.1) then becomes

$$\frac{\partial \tilde{\rho}}{\partial t} + \rho_0 \nabla \cdot \tilde{\mathbf{u}} = 0. \tag{9.6.11}$$

In order to linearize the momentum transport equation (9.6.2), we first express the total pressure $\tilde{p}_t$ as

$$\tilde{p}_t = p_0 + \tilde{p} \quad \text{with} \quad |\tilde{p}| \ll p_0. \tag{9.6.12}$$

Since we have taken T and ρ as the independent thermodynamic variables, we can express $\tilde{p}$ as

$$\tilde{p} = \left(\frac{\partial p}{\partial \rho}\right)_T \tilde{\rho} + \left(\frac{\partial p}{\partial T}\right)_\rho \tilde{T}, \tag{9.6.13}$$

or as

$$\tilde{p} = \frac{v^2}{\gamma}(\tilde{\rho} + \beta_p \rho_0 \tilde{T}), \tag{9.6.14}$$

where we have expressed $(\partial p/\partial \rho)_T$ as $\gamma^{-1}(\partial p/\partial \rho)_s = v^2/\gamma$ with $v^2 = (\partial p/\partial \rho)_s$ representing the square of the velocity of sound, and where we have expressed $(\partial p/\partial T)_\rho$ as $\gamma^{-1}(\partial p/\partial \rho)_s(\partial \rho/\partial T)_p = v^2\beta_p\rho_0/\gamma$ with β_p representing the thermal expansion coefficient at constant pressure. Through use of Eq. (9.6.14), the linearized form of Eq. (9.6.2) becomes

$$\rho_0\frac{\partial \tilde{\mathbf{u}}}{\partial t} + \frac{v^2}{\gamma}\nabla\tilde{\rho} + \frac{v^2\beta_p\rho_0}{\gamma}\nabla\tilde{T} - (2\eta_s + \eta_d)\nabla(\nabla \cdot \tilde{\mathbf{u}}) + \eta_s\nabla \times (\nabla \times \tilde{\mathbf{u}}) = \tilde{\mathbf{f}}. \tag{9.6.15}$$

Finally, the linearized form of the energy transport equation, Eq. (9.6.6), becomes

$$\rho_0 c_v\frac{\partial \tilde{T}}{\partial t} + \frac{\rho_0 c_v(\gamma - 1)}{\beta_p}(\nabla \cdot \tilde{\mathbf{u}}) - \kappa\nabla^2\tilde{T} = \tilde{\phi}_{\text{ext}}. \tag{9.6.16}$$

Note that the viscous contribution to the heat input, $\tilde{\phi}_\eta$, does note contribute in the linear approximation.

Equations (9.6.11), (9.6.15), and (9.6.16) constitute the three linearized equations of hydrodynamics for the quantities $\tilde{\mathbf{u}}$, $\tilde{\rho}$, and $\tilde{T}$. The continuity equation in its linearized form (Eq. (9.6.11)) can be used to eliminate the variable $\tilde{\mathbf{u}}$ from the remaining two equations. To do so, we take the divergence of the equation of momentum transfer (9.6.15) and use Eq. (9.6.11) to eliminate the terms containing $\nabla \cdot \tilde{\mathbf{u}}$. We obtain

$$-\frac{\partial^2 \tilde{\rho}}{\partial t^2} + \frac{v^2}{\gamma}\nabla^2\tilde{\rho} + \frac{v^2\beta_p\rho_0}{\gamma}\nabla^2\tilde{T} + \frac{2\eta_s + \eta_d}{\rho_0}\frac{\partial}{\partial t}(\nabla^2\tilde{\rho}) = \frac{\gamma_e}{8\pi}\nabla^2\langle\tilde{E}^2\rangle,$$

(9.6.17)

where we have explicitly introduced the form of $\tilde{\mathbf{f}}$ from Eq. (9.6.3). Also, the energy transport equation (9.6.16) can then be expressed through use of Eqs. (9.6.9) and (9.6.11) as

$$\rho_0 c_v \frac{\partial \tilde{T}}{\partial t} - \frac{c_v(\gamma - 1)}{\beta_p}\frac{\partial \tilde{\rho}}{\partial t} - \kappa\nabla^2\tilde{T} = \frac{nc\alpha}{4\pi}\langle\tilde{E}^2\rangle. \qquad (9.6.18)$$

Equations (9.6.17) and (9.6.18) constitute two coupled equations for the thermodynamic variables $\tilde{\rho}$ and $\tilde{T}$, and they show how these quantities are coupled to one another and are driven by the applied optical field.

In the absence of the driving terms appearing on their right-hand sides, Eqs. (9.6.17) and (9.6.18) allow solutions of the form of damped, freely propagating acoustic waves

$$\tilde{F}(z, t) = Fe^{-\Omega(t - z/v)}e^{-\alpha_s z} + \text{c.c.} \qquad (9.6.19)$$

where F denotes either ρ or T, and where the sound absorption coefficient α_s is given for low frequencies $(\Omega \ll \rho_0 v^2/(2\eta_s + \eta_d))$ by

$$\alpha_s = \frac{\Omega^2}{2\rho_0 v^3}\left[(2\eta_s + \eta_d) + (\gamma - 1)\frac{\kappa}{c_p}\right]. \qquad (9.6.20)$$

For details, see the article by Sette (1961).

We next study the nature of the solution to Eqs. (9.6.17) and (9.6.18) in the presence of their driving terms. We assume that the total optical field can be represented as

$$\tilde{E}(z, t) = A_1 e^{i(k_1 z - \omega_1 t)} + A_2 e^{i(-k_2 z - \omega_2 t)} + \text{c.c.} \qquad (9.6.21)$$

We first determine the response of the medium at the beat frequency between these two applied field frequencies. This disturbance will have frequency

$$\Omega = \omega_1 - \omega_2 \qquad (9.6.22)$$

and wavenumber

$$q = k_1 + k_2 \qquad (9.6.23)$$

and can be taken to be of the form

$$\tilde{\rho}(z, t) = \rho e^{i(qz - \Omega t)} + \text{c.c.}, \qquad (9.6.24)$$

$$\tilde{T}(z, t) = T e^{i(qz - \Omega t)} + \text{c.c.} \qquad (9.6.25)$$

For the present, we are interested only in the steady-state response of the medium, and hence we assume that the amplitudes A_1, A_2, ρ, and T are time-independent. We introduce the fields $\tilde{E}$, $\tilde{\rho}$, and $\tilde{T}$ given by Eqs. (9.6.21) through (9.6.25) into the coupled acoustic equations (9.6.17) and (9.6.18). The parts of these equations that oscillate at frequency Ω are given respectively by

$$-\left(\Omega^2 + i\Omega\Gamma_B - \frac{v^2q^2}{\gamma}\right)\rho + \frac{v^2\beta_p\rho_0q^2}{\gamma}T = \frac{\gamma_e q^2}{4\pi}A_1A_2^* \qquad (9.6.26)$$

and

$$-\left(i\Omega - \frac{1}{2}\gamma\Gamma_R\right)T + \frac{i(\gamma-1)\Omega}{\beta_p\rho_0}\rho = \frac{nc\alpha}{2\pi c_v\rho_0}A_1A_2^*. \qquad (9.6.27)$$

Here we have introduced the Brillouin linewidth

$$\Gamma_B = (2\eta_s + \eta_d)q^2/\rho_0, \qquad (9.6.28)$$

whose reciprocal $\tau_p = \Gamma_B^{-1}$ is the phonon lifetime, and the Rayleigh linewidth

$$\Gamma_R = \frac{2\kappa q^2}{\rho_0 c_p}, \qquad (9.6.29)$$

whose reciprocal $\tau_R = \Gamma_R^{-1}$ is characteristic decay time of the isobaric density disturbances that give rise to Rayleigh scattering.

In deriving Eqs. (9.6.26) and (9.6.27) we have ignored those terms that contain the spatial derivatives of ρ and T. This approximation is equivalent to assuming that the material excitations are strongly damped and hence do not propagate over any appreciable distances. This approximation is valid so long as

$$q \gg \left|\frac{1}{\rho}\frac{\partial\rho}{\partial z}\right|, \left|\frac{1}{T}\frac{\partial T}{\partial z}\right| \quad \text{and} \quad q^2 \gg \left|\frac{1}{\rho}\frac{\partial^2\rho}{\partial z^2}\right|, \left|\frac{1}{T}\frac{\partial^2 T}{\partial z^2}\right|.$$

These inequalities are usually satisfied. Recall that a similar approximation was introduced in Section 9.3 in the derivation of Eq. (9.3.15).

We next solve Eq. (9.6.27) algebraically for T and introduce the resulting expression into Eq. (9.6.26). We obtain the equation

$$\left[-\left(\Omega^2 + i\Omega\Gamma_B - \frac{v^2q^2}{\gamma}\right) + \frac{v^2q^2\Omega(\gamma-1)}{(\Omega + \frac{1}{2}i\gamma\Gamma_R)\gamma}\right]\rho$$

$$= \left[\gamma_e - \frac{i\gamma_a qv}{\Omega + \frac{1}{2}i\gamma\Gamma_R}\right]\frac{q^2}{4\pi}A_1A_2^*, \qquad (9.6.30)$$

where we have introduced the absorptive coupling constant

$$\gamma_a = \frac{2\alpha n v^2 c \beta_p}{c_P \Omega_B} \tag{9.6.31}$$

with $\Omega_B = qv$. Equation (9.6.30) shows how the amplitude ρ of the acoustic disturbance depends on the amplitudes A_1 and A_2 of the two optical fields. Both Brillouin and Rayleigh contributions to ρ are contained in Eq. (9.6.30).

It is an empirical fact (see, for example, Fig. 8.1.1) that the spectrum for Brillouin scattering does not appreciably overlap that for Rayleigh scattering. Equation (9.6.30) can thus be simplified by considering the resonant contributions to the two processes separately. First, we consider the case of stimulated Brillouin scattering (SBS). In this case Ω^2 is approximately equal to $\Omega_B^2 = v^2 q^2$, and hence the denominator $\Omega + \frac{1}{2} i \gamma \Gamma_R$ is nonresonant. We can thus drop the contribution $\frac{1}{2} i \gamma \Gamma_R$ in comparison with Ω in these denominators. Equation (9.6.30) then shows that the Brillouin contribution to ρ is given by

$$\rho_B = \frac{-(\gamma_e - i \gamma_a q v / \Omega) q^2}{4\pi (\Omega^2 + i \Omega \Gamma_B - v^2 q^2)} A_1 A_2^*. \tag{9.6.32}$$

The other resonance in Eq. (9.6.30) occurs at $\Omega = 0$ and leads to stimulated Rayleigh scattering (SRLS). For $|\Omega| \lesssim \Gamma_R$, the Brillouin denominator $\Omega^2 + i \Omega \Gamma_B - v^2 q^2 / \gamma$ is nonresonant and can be approximated by $-v^2 q^2 / \gamma$. Equation (9.6.30) thus becomes

$$\rho_R = \left[\frac{\gamma_e \left(\Omega \frac{1}{2} i \gamma \Gamma_R \right) - i \gamma_a \Omega_B}{\Omega + \frac{1}{2} i \Gamma_R} \right] \frac{1}{4\pi v^2} A_1 A_2^*. \tag{9.6.33}$$

We next calculate the nonlinear polarization as

$$\tilde{p}^{NL} = \Delta \chi \tilde{E} = \frac{\Delta \epsilon}{4\pi} \tilde{E} = \frac{1}{4\pi} \left(\frac{\partial \epsilon}{\partial \rho} \right)_T \tilde{\rho} \tilde{E} = \frac{\gamma_e}{4\pi \rho_0} \tilde{\rho} \tilde{E}, \tag{9.6.34}$$

where $\tilde{p}$ and $\tilde{E}$ are given by Eqs. (9.6.24) and (9.6.21), respectively. We represent the nonlinear polarization in terms of its complex amplitudes as

$$\tilde{P}^{NL} = p_1 e^{i(k_1 z - \omega_1 t)} + p_2 e^{i(-k_2 z - \omega_2 t)} + \text{c.c.} \tag{9.6.35}$$

with

$$p_1 = \frac{\gamma_e}{4\pi \rho_0} \rho A_2, \qquad p_2 = \frac{\gamma_e}{4\pi \rho_0} \rho^* A_1. \tag{9.6.36}$$

This form of the nonlinear polarization is now introduced into the wave equation, which we write in the form (see also Eq. (2.1.22))

$$-\nabla^2 [A_n(\mathbf{r}) e^{i\mathbf{k}_n \cdot \mathbf{r}}] - \frac{\epsilon \omega_n^2}{c^2} A_n(\mathbf{r}) e^{i\mathbf{k}_n \cdot \mathbf{r}} = \frac{4\pi \omega_n^2}{c^2} p_n e^{i\mathbf{k}_n \cdot \mathbf{r}}. \tag{9.6.37}$$

We next make the slowly-varying amplitude approximation and find that the field amplitudes obey the equations

$$\left(\frac{d}{dz} + \tfrac{1}{2}\alpha\right)A_1 = \frac{2\pi i\omega}{nc}p_1, \tag{9.6.38a}$$

$$\left(\frac{d}{dz} - \tfrac{1}{2}\alpha\right)A_2 = \frac{-2\pi i\omega}{nx}p_2, \tag{9.6.38b}$$

where we have introduced the real part of the refractive index $n = Re\sqrt{\epsilon}$ and the optical absorption coefficient $\alpha = (2\omega/c)\,\mathrm{Im}\sqrt{\epsilon}$. Equations (9.6.38) can be used to describe either SBS or SRLS, depending on whether form (9.6.32) or (9.6.33) is used to determine the factor ρ that appears in the expression (9.3.36) for the nonlinear polarization. Since in either case ρ is proportional to the produce $A_1A_2^*$, Eqs. (9.6.38) can be written as

$$\frac{dA_1}{dz} = \kappa|A_2|^2A_1 - \tfrac{1}{2}\alpha A_1, \tag{9.6.39a}$$

$$\frac{dA_2}{dz} = \kappa^*|A_1|^2A_2 + \tfrac{1}{2}\alpha A_2, \tag{9.6.39b}$$

where for SBS κ is given by

$$\kappa_B = -\frac{q^2\omega}{8\pi\rho_0 nc}\frac{i\gamma_e(\gamma_e - i\gamma_a)}{(\Omega^2 + i\Omega\Gamma_B - v^2q^2)}, \tag{9.6.40a}$$

and for SRLS is given by

$$\kappa_R = \frac{i\gamma_e\omega}{8\pi\rho_0 ncv^2}\left[\frac{\gamma_e\left(\Omega + \tfrac{1}{2}i\gamma\Gamma_R\right) - i\gamma_a\Omega_B}{\Omega + \tfrac{1}{2}i\Gamma_R}\right]. \tag{9.6.40b}$$

We now introduce the intensities

$$I_i = \frac{nc}{2\pi}|A_i|^2 \tag{9.6.41}$$

of the two interacting optical waves and use Eqs. (9.6.39) to calculate the spatial rate of change of the intensities as

$$\frac{dI_1}{dz} = -gI_1I_2 - \alpha I_1, \tag{9.6.42a}$$

$$\frac{dI_2}{dz} = -gI_1I_2 + \alpha I_2, \tag{9.6.42b}$$

where we have introduced the gain factor

$$g = -\frac{4\pi}{nc}\mathrm{Re}\,\kappa. \tag{9.6.43}$$

For the case of SBS, we find that the gain factor can be expressed as

$$g_B = g_B^e + g_B^a, \tag{9.6.44a}$$

where

$$g_B^e = \frac{\omega^2 \gamma_e^2}{\rho_0 n v c^2 \Gamma_B} \frac{1}{1 + (2\Delta\Omega/\Gamma_B)^2} \tag{9.6.44b}$$

and

$$g_B^a = \frac{-\omega^2 \gamma_e \gamma_a}{2\rho_0 n v c^2 \Gamma_B} \frac{4\Delta\Omega/\Gamma_B}{1 + (2\Delta\Omega/\Gamma_B)^2} \tag{9.6.44c}$$

denote the electrostrictive and absorptive contributions to the SBS gain factor, respectively. Here we have introduced the detuning from the Brillouin resonance given by $\Delta\Omega = \Omega_B - \Omega$, where $\Omega_B = qv = (k_1 + k_2)v$ and where $\Omega = \omega_1 - \omega_2$. The electrostrictive contribution is maximum for $\Delta\Omega = 0$, where it attains the value

$$g_B^e(\max) = \frac{\omega^2 \gamma_e^2}{\rho_0 n v c^3 \Gamma_B}. \tag{9.6.45}$$

Since (according to Eq. (9.6.28)) Γ_B is proportional to q^2 and hence to ω^2, the gain for electrostrictive SBS is independent of the laser frequency. The absorptive contribution is maximum for $\Delta\Omega = -\Gamma_B/2$, that is, when the Stokes wave (at frequency ω_2) is detuned by one-half the spontaneous Brillouin linewidth Γ_B to the low-frequency side of resonance. The maximum value of the gain for this process is

$$g_B^a(\max) = \frac{\omega^2 \gamma_e \gamma_a}{2\rho_0 n v c^3 \Gamma_B}. \tag{9.6.46}$$

Note that since Γ_B is proportional to q^2 and (according to Eq. (9.6.31)) γ_a is proportional to q^{-1}, the absorptive SBS gain factor is proportional to q^3 and hence depends on the laser frequency as ω^{-3}. Since the gain factor for thermal SBS is linearly proportional to the optical absorption coefficient α (by Eqs. (9.6.31) and (9.6.46)), the gain for thermal SBS can be made to exceed that for electrostrictive SBS by adding an absorber such as a dye to the Brillouin-active medium. As shown in Table 9.3.1, this effect occurs roughly for absorption coefficients greater than 1 cm^{-1}.*

The spectral dependence of the two contributions to the SBS gain is shown schematically in Fig. 9.6.1.

* The quantity g_B^e (max) is designated g_0 in Table 9.3.1.

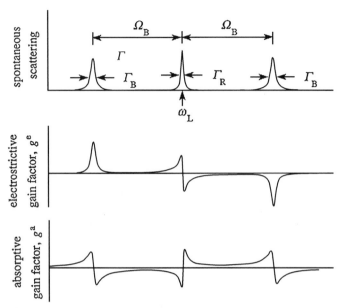

FIGURE 9.6.1   Gain spectra for stimulated Brillouin scattering and stimulated Rayleigh scattering, showing their electrostrictive and absorptive contributions. For comparison, the spectrum of spontaneous Brillouin and Rayleigh scattering is also shown.

For the case of stimulated Rayleigh scattering, we can express the gain factor appearing in Eqs. (9.6.42) through use of Eqs. (9.6.40b) and (9.6.43) as

$$g_R = g_R^e + g_R^a, \tag{9.6.47}$$

where

$$g_R^e = \frac{-\omega \gamma_e^2 (\gamma - 1)}{4 \rho_0 n^2 c^2 v^2} \left[\frac{4\Omega / \Gamma_R}{1 + (2\Omega / \Gamma_R)^2} \right] \tag{9.6.48}$$

and

$$g_R^a = \frac{\omega \gamma_e \gamma_a \Omega_B}{2 \rho_0 n^2 c^2 v^2 \Gamma_R} \left[\frac{4\Omega / \Gamma_R}{1 + (2\Omega / \Gamma_R)^2} \right] \tag{9.6.49}$$

denote the electrostrictive and absorptive contributions to the gain factor, respectively. The contribution g_R^e gives rise to electrostrictive stimulated Rayleigh scattering. The gain factor for this process is maximum for $\Omega = -\Gamma_R/2$ and has the value

$$g_R^e (\text{max}) = \frac{\omega \gamma_e^2 (\gamma - 1)}{4 \rho_0 n^2 c^2 v^2}. \tag{9.6.50}$$

TABLE 9.6.1   Properties of stimulated Rayleigh scattering for a variety of materials at a wavelength of 694 nm[a]

| Substance | Gain factor | | Linewidth |
	g_e(max) (cm/MW)	g_a(max)/α (cm^2/MW)	$\delta\nu_R$ (MHz)
CCl$_4$	2.6×10^{-4}	0.82	17
Methanol	8.4×10^{-4}	0.32	20
CS$_2$	6.0×10^{-4}	0.62	36
Benzene	2.2×10^{-4}	0.57	24
Acetone	2.0×10^{-4}	0.47	21
H$_2$O	0.02×10^{-4}	0.019	27.5
Ethanol		0.38	18

[a] After Kaiser and Maier (1972).

Note that this quantity scales linearly with laser frequency. The absorptive contribution g_R^a gives rise to thermal SRLS. The gain for this process is maximum for $\Omega = \Gamma_R/2$ and has the value

$$g_R^a(\text{max}) = \frac{\omega \gamma_e \gamma_a \Omega_B}{2\rho_0 n^2 c^2 v^2 \Gamma_R}. \tag{9.6.51}$$

Since Γ_R scales with the laser frequency as ω^2, γ_a scales as $1/\omega$, and Ω_B scales as ω, we see that the gain factor for thermal SRLS scales with the laser frequency as $1/\omega$.

As can be seen from Table 9.6.1, Γ_R is often of the order of 10 MHz, which is much narrower than the linewidths of pulsed lasers. In such cases, laser linewidth effects can often be treated in an approximate fashion by convolving the gain predicted by Eqs. (9.6.48) and (9.6.49) with the laser lineshape. If the laser linewidth Γ_L is much broader than Γ_R, the maximum gain for absorptive SRLS is then given by Eq. (9.6.51) with Γ_R replaced by Γ_L. Under these conditions g_R^a (max) is independent of the laser frequency.

We note by inspection of Table 9.6.1 that g_R^a(max) is very much larger than g_R^e(max) except for extremely small values of the absorption coefficient. The two gains become comparable for $\alpha \simeq 10^{-3}$ cm^{-1}, which occurs only for unusually pure materials.

We also see by comparison of Eqs. (9.6.51) and (9.6.46) that the ratio of the two thermal gain factors is given by

$$\frac{g_R^a(\text{max})}{g_B^a(\text{max})} = \frac{2\Gamma_B}{\Gamma_R}. \tag{9.6.52}$$

Comparison of Tables 9.3.1 and 9.6.1 shows that for a given material the ratio Γ_B/Γ_R is typically of the order of 100. Hence, when thermal stimulated scattering occurs, the gain for thermal SRLS is much larger than that for thermal SBS, and most of the energy is emitted by this process.

The frequency dependence of the gain for stimulated Rayleigh scattering is shown in Fig. 9.6.1. Note that electrostrictive SRLS gives rise to gain for Stokes shifted light but that thermal SRLS gives rise to gain for anti-Stokes scattering (Herman and Gray, 1967). This result can be understood from the point of view that n_2 is positive for electrostriction but is negative for the process of heating and subsequent thermal expansion. We saw in the discussion of two-beam coupling presented in Section 7.4 that the lower-frequency wave experiences gain for n_2 positive and loss for n_2 negative.

Appendix: Definition of the Viscosity Coefficients

The viscosity coefficients are defined as follows: The component t_{ij} of the stress tensor gives the i component of the force per unit area on an area element whose normal is in the j direction. We represent the stress tensor as

$$t_{ij} = -p\delta_{ij} + \sigma_{ij},$$

where p is the pressure and σ_{ij} is the contribution to the stress tensor due to viscosity. If we assume that σ_{ij} is linearly proportional to the rate of deformation

$$d_{ij} = \frac{1}{2}\left[\frac{\partial \tilde{u}_i}{\partial x_j} + \frac{\partial \tilde{u}_j}{\partial x_j}\right],$$

we can represent σ_{ij} as

$$\sigma_{ij} = 2\eta_s d_{ij} + \eta_d \delta_{ij} \sum_k d_{kk},$$

where η_s is the shear viscosity coefficient and η_d is the dilational viscosity coefficient. The quantity $\sum_k d_{kk}$ can be interpreted as follows:

$$\sum_k d_{kk} = \sum_k \frac{\partial \tilde{u}_k}{\partial x_k} = \nabla \cdot \tilde{\mathbf{u}}.$$

In general, η_s and η_d are independent parameters. However, for certain physical systems they are related to one another through a relationship first formulated by Stokes. This relationship results from the assumption that the viscous stress tensor σ_{ij} is traceless. In this case the trace of t_{ij} is unaffected by viscous effects; in other words, the mean pressure $-\frac{1}{3}\sum_i t_{ii}$ is unaffected by the effects of viscosity. The condition that σ_{ij} is traceless implies that the combination

$$\sum_i \sigma_{ii} = 2\eta_s \sum_i d_{ii} + 3\eta_d \sum_k d_{kk} = (2\eta_s + 3\eta_d) \sum_k d_{kk}$$

vanishes, or that

$$\eta_d = -\tfrac{2}{3}\eta_s.$$

This result is known as the Stokes relation.

The viscosity coefficients η_s and η_d often appear in the combination $2\eta_s + \eta_d$, as they do in Eq. (9.6.2). When the Stokes relation is satisfied, this combination takes the value

$$2\eta_s + \eta_d = \tfrac{4}{3}\eta_s \qquad \text{(Stokes relation valid)}.$$

Under general conditions, such that the Stokes relation is not satisfied, one often defines the bulk viscosity coefficient η_b by

$$\eta_b = \tfrac{2}{3}\eta_s + \eta_d,$$

in terms of which the quantity $2\eta_s + \eta_d$ can be represented as

$$2\eta_s + \eta_d = \tfrac{4}{3}\eta_s + \eta_b \qquad \text{(in general)}.$$

Note that η_B vanishes identically when the Stokes relation is valid, for example, for the case of an ideal gas.

As an example of the use of these relations, we note that the Brillouin linewidth Γ_B introduced in Eqs. (8.3.23), (9.5.2), and (9.6.28) can be represented (ignoring the contribution due to thermal conduction) either as

$$\Gamma_B = (2\eta_s + \eta_d)q^2/\rho_0$$

or as

$$\Gamma_B = \left(\tfrac{4}{3}\eta_s + \eta_B\right)q^2/\rho_0.$$

Problems

1. *Lorentz–Lorenz prediction of the electrostrictive constant.* Verify Eq. (9.2.17).

2. *Angular dependence of SBS.* Generalize the discussion of Section 9.3 to allow for an the angle θ between the laser and Stokes propagation directions to be arbitrary. In particular, determine how the Brillouin frequency Ω_B, the steady-state line-center gain factor g_0, and the phonon lifetime τ_p depend on the angle θ.

[Ans.:

$$\Omega_B(\theta) = \Omega_B(\theta = 180°) \sin\left(\tfrac{1}{2}\theta\right)$$

$$g_0(\theta) = g_0(\theta = 180°)/\sin\left(\tfrac{1}{2}\theta\right)$$

$$\tau_p(\theta) = \tau_p(\theta = 180°)/\sin^2\left(\tfrac{1}{2}\theta\right).]$$

3. *Transverse SBS.* Consider the possibility of exciting SBS in the transverse direction by a laser beam passing through a fused-silica window at near-normal incidence. Assume conditions appropriate to a high-energy laser. In particular, assume that the window is 70 cm in diameter and is uniformly filled with a laser pulse of 10-nsec duration at a wavelength of 350 nm. What is the minimum value of the laser pulse energy for which SBS can be excited? (In fact, transverse SBS has been observed under such conditions similar to those assumed in this problem; see, for example, J. R. Murray, J. R. Smith, R. B. Ehrlich, D. T. Kyrazis, C. E. Thompson, T. L. Weiland, and R. B. Wilcox, *J. Opt. Soc. Am.* **6**, 2402 (1989).)

[Ans.: ∼2 kJ.]

4. *Optical damage considerations and the study of SBS.* The threshold intensity for optical damage to fused silica is approximately 3 GW/cm^2 and is of the same order of magnitude for most optical materials. (See, for example, W. H. Lowdermilk and D. Milam, *IEEE J. Quantum Electron.* **17**, 1888 (1981).) Use this fact and the value of the SBS gain factor at line center quoted in Table 9.3.1 to determine the minimum length of a cell utilizing fused-silica windows that can be used to excite SBS in acetone with a collimated laser beam. Assume that the laser intensity is restricted to 50% of the threshold intensity as a safety factor to avoid damage to the windows. If the laser pulse length is 20 nsec, what is the minimum value of the laser pulse energy per unit area that can be used to excite SBS? (SBS is often excited by tightly focused laser beams rather than by collimated beams to prevent optical damage to the windows of the cell.)

5. *Pulse compression by SBS.* Explain qualitatively why the Stokes radiation excited by SBS in the backward direction can be considerable shorter in duration than the exciting radiation. How must the physical length of the interaction region be related to the duration of the laser pulse in order to observe this effect? Write down the coupled-amplitude equations that are needed to describe this effect, and, if you wish, solve these equations numerically by computer. What determines the minimum value of the duration of the output pulse?

[Hint: Pulse compression by SBS is described in the scientific literature by D. T. Hon, *Opt. Lett.* **5**, 516 (1980) and by S. S. Gulidov, A. A. Mak, and S. B. Papernyi, *JETP Lett.* **47**, 394 (1988).]

6. *Brillouin-enhanced four-wave mixing.* In addition to SBS, light beams can interact in a Brillouin medium by means of the process known as Brillouin-enhanced four-wave mixing (BEFWM), which is illustrated in the figure shown below.

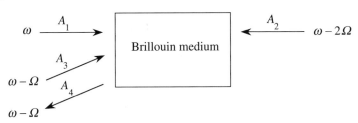

In this process, the incoming signal wave A_3 interferes with the backward-going pump wave A_2 to generate an acoustic wave propagating in the forward direction. The forward-going pump wave scatters from the acoustic wave to generate the phase-conjugate wave A_4. Since A_4 is at the Stokes sideband of A_1, it also undergoes amplification by the usual SBS process. Phase-conjugate reflectivities much larger than 100% have been observed in the BEFWM process. Using the general formalism outlined in Section 9.3, derive the form of the four coupled-amplitude equations that describe BEFWM under steady-state conditions. Solve these equations analytically in the constant-pump approximation.

[Hint: BEFWM has been discussed in the scientific literature. See, for example, M. D. Skeldon, P. Narum, and R. W. Boyd, *Opt. Lett.* **12**, 1211 (1987).]

References

Reviews of Stimulated Light Scattering

R. W. Boyd and G. Grynberg, "Optical Phase Conjugation," in *Contemporary Nonlinear Optics* (G. P. Agrawal and R. W. Boyd, eds.), Academic Press, Boston, 1992.

I. L. Fabelinskii, *Molecular Scattering of Light*, Plenum Press, New York, 1968.

I. L. Fabelinskii, "Stimulated Mandlestam-Brillouin Process," in *Quantum Electronics: A Treatise* (H. Rabin and C. L. Tang, eds.), Vol. I, Part A, Academic Press, New York, 1975.

R. A. Fisher, ed., *Optical Phase Conjugation,* Academic Press, New York, 1983, especially Chapters 6 and 7.

W. Kaiser and M. Maier, in *Laser Handbook* (F. T. Arecchi and E. O. Schulz-DuBois, eds.), North-Holland, 1972.

Y. R. Shen, *Principles of Nonlinear Optics*, Wiley, New York, 1984.

B. Ya. Zel'dovich, N. F. Pilipetsky, and V. V. Shkunov, *Principles of Phase Conjugation*, Springer-Verlag, Berlin, 1985.

Stimulated Brillouin Scattering

I. P. Batra, R. H. Enns, and D. Pohl, *Phys. Stat. Sol.* (6) **48**, 11 (1971).

M. W. Bowers, R. W. Boyd, and A. K. Hankla, *Opt. Lett.* **22**, 360 (1997).

R. W. Boyd, K. Rzążewski, and P. Narum, *Phys. Rev. A* **42**, 5514 (1990).

E. L. Buckland and R. W. Boyd, *Opt. Lett.* **21**, 1117 (1996).

E. L. Buckland and R. W. Boyd, *Opt. Lett.* **22**, 676 (1997).

R. L. Carman, F. Shimizu, C. S. Wang, and N. Bloembergen, *Phys. Rev. A* **2**, 60 (1970).

R. Y. Chiao, Ph.D. dissertation, Massachusetts Institute of Technology, 1965.

R. Y. Chiao, C. H. Townes, and B. P. Stoicheff, *Phys. Rev. Lett.* **12**, 592 (1964).

E. U. Condon, in *Handbook of Physics* (E. U. Condon and H. Odishaw, eds.), McGraw-Hill, New York, 1967, Chapter 1.

M. J. Damzen and H. Hutchinson, *IEEE J. Quantum Electron.* **QE-19**, 7 (1983).

A. L. Gaeta and R. W. Boyd, *Phys. Rev. A* **44**, 3205 (1991).

A. L. Gaeta, M. D. Skeldon, R. W. Boyd, and P. Narum, *J. Opt. Soc. Am. B* **6**, 1709 (1989).

N. Goldblatt and M. Hercher, *Phys. Rev. Lett.* **20**, 310 (1968).

J. W. Goodman, *Statistical Optics*, Wiley, New York, 1985.

E. E. Hagenlocker, R. W. Minck, and W. G. Rado, *Phys. Rev.* **154**, 226 (1967).

R. M. Herman and M. A. Gray, *Phys. Rev. Lett.* **19**, 824 (1967).

F. V. Hunt, *J. Acoust. Soc. Am.* **27**, 1019 (1955); see also *American Institute of Physics Handbook*, McGraw-Hill, New York, 1972, pp. 3–37 ff.

N. M. Kroll, *Appl. Phys.* **36**, 34 (1965).

N. M. Kroll and P. L. Kelly, *Phys. Rev.* **A4**, 763 (1971).

O. Kulagin, G. A. Pasmanik, A. L. Gaeta, T. R. Moore, G. J. Benecke, and R. W. Boyd, *J. Opt. Soc. B* **8**, 2155 (1991).

H. Lamb, *Hydrodynamics*, Dover, New York, 1945.

L. D. Landau and E. M. Lifshitz, *Fluid Mechanics*, Pergamon, London, 1959.

L. B. Loeb, *The Kinetic Theory of Gases,* Dover, New York, 1961.

P. Narum, M. D. Skeldon, and R. W. Boyd, *IEEE J. Quantum Electron.* **QE-22**, 2161 (1986).

P. Narum, M. D. Skeldon, A. L. Gaeta, and R. W. Boyd, *J. Opt. Soc. Am. B* **5**, 623 (1988).

D. Pohl and W. Kaiser, *Phys. Rev.* **B1**, 31 (1970).

D. Pohl, M. Maier, and W. Kaiser, *Phys. Rev. Lett.* **20**, 366 (1968).

D. Sette, in *Handbuch der Physik, XI/1, Acoustics I* (S. Flügge, ed.), Springer-Verlag, Berlin, 1961.

V. G. Sidorovich, *Sov. Phys. Tech. Phys.* **21**, 1270 (1976).

B. Ya. Zel'dovich, V. I. Popovichev, V. V. Ragulsky, and F. S. Faizullov, *JETP Lett.* **15**, 109 (1972).

Chapter 10

Stimulated Raman Scattering and Stimulated Rayleigh-Wing Scattering

10.1. The Spontaneous Raman Effect

The spontaneous Raman effect was discovered by C. V. Raman in 1928. To observe this effect, a beam of light illuminates a material sample (which can be a solid, liquid, or gas), and the scattered light is observed spectroscopically, as illustrated in Fig. 10.1.1. In general, the scattered light contains frequencies different from those of the excitation source. Those new components shifted to lower frequencies are called Stokes components, and those shifted to higher frequencies are called anti-Stokes components. The Stokes components are typically orders of magnitude more intense than the anti-Stokes components.

These properties of Raman scattering can be understood through use of the energy level diagrams shown in Fig. 10.1.2. Raman Stokes scattering consists of a transition from the ground state g to the final state n by means of an intermediate transition to a virtual level associated with excited state n'. Raman anti-Stokes scattering entails a transition from level n to level g with n' serving as the intermediate level. The anti-Stokes lines are typically much weaker than the Stokes lines because, in thermal equilibrium, the population of level n is smaller than the population in level g by the Boltzmann factor $\exp(-\hbar\omega_{ng}/kT)$.

The Raman effect has important spectroscopic applications because transitions that are one-photon forbidden can often be studied using Raman scattering. For example, the Raman transitions illustrated in Fig. 10.1.2 can occur only if the matrix elements $\langle g|\hat{\mathbf{r}}|n'\rangle$ and $\langle n'|\hat{\mathbf{r}}|n\rangle$ are both nonzero, and this fact

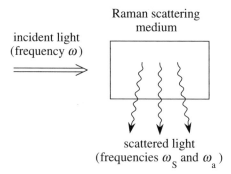

FIGURE 10.1.1   Spontaneous Raman scattering.

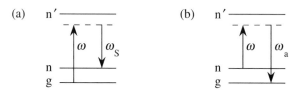

FIGURE 10.1.2   Energy level diagrams describing (a) Raman Stokes scattering and (b) Raman anti-Stokes scattering.

implies (for a material system that possesses inversion symmetry, so that the energy eigenstates possess definite parity) that the states g and n must possess the same parity. But under these conditions the $g \rightarrow n$ transition is forbidden for single-photon electric dipole transitions because the matrix element $\langle g | \hat{\mathbf{r}} | n \rangle$ must necessarily vanish.

10.2.   Spontaneous versus Stimulated Raman Scattering

The spontaneous Raman scattering process described in the previous section is typically a rather weak process. Even for condensed matter, the scattering cross section per unit volume for Raman Stokes scattering is only approximately 10^{-6} cm^{-1}. Hence, in propagating through 1 cm of the scattering medium, only approximately 1 part in 10^6 of the incident radiation will be scattered into the Stokes frequency.

However, under excitation by an intense laser beam, highly efficient scattering can occur as a result of the stimulated version of the Raman scattering process. Stimulated Raman scattering is typically a very strong scattering process: 10% or more of the energy of the incident laser beam is often converted into the Stokes frequency. Another difference between spontaneous and stimulated Raman scattering is that the spontaneous process leads to nearly isotropic emission, whereas the stimulated process leads to emission in a narrow cone in the forward and backward directions. Stimulated Raman scattering was discovered by Woodbury and Ng (1962) and was described more fully by Eckardt *et al.* (1962). The properties of stimulated Raman scattering have been reviewed by Bloembergen (1967), Kaiser and Maier (1972), Penzkofer *et al.* (1979), and Raymer and Walmsley (1990).

The relation between spontaneous and stimulated Raman scattering can be understood in terms of an argument (Hellwarth, 1963) that considers the process from the point of view of the photon occupation numbers of the various field modes. One postulates that the probability per unit time that a photon will be emitted into Stokes mode S is given by

$$P_S = Dm_L(m_S + 1). \qquad (10.2.1)$$

Here m_L is the mean number of photons per mode in the laser radiation, m_S is the mean number of photons in Stokes mode S, and D is a proportionality constant whose value depends on the physical properties of the material medium. This functional form is assumed because the factor m_L leads to the expected linear dependence of the transition rate on the laser intensity, and the factor $m_S + 1$ leads to stimulated scattering through the contribution m_S and to spontaneous scattering through the contribution of unity. This dependence on the factor $m_S + 1$ is reminiscent of the stimulated and spontaneous contributions to the total emission rate for a single-photon transition of an atomic system as treated by the Einstein A and B coefficients. Equation (10.2.1) can be justified by more rigorous treatments; note, for example, that the results of the present analysis are consistent with those of the fully quantum-mechanical treatment of Raymer and Mostowski (1981).

By the definition of P_S as a probability per unit time for emitting a photon into mode S, the time rate of change of the mean photon occupation number for the Stokes mode is given by $dm_S/dt = P_S$ or, through use of Eq. (10.2.1), by

$$\frac{dm_S}{dt} = Dm_L(m_S + 1). \qquad (10.2.2)$$

If we now assume that the Stokes mode corresponds to a wave traveling in the positive z direction at the velocity c/n, as illustrated in Fig. 10.2.1, we see that the time rate of change given by Eq. (10.2.2) corresponds to a spatial growth rate given by

$$\frac{dm_S}{dz} = \frac{1}{c/n}\frac{dm_S}{dt} = \frac{1}{c/n}Dm_L(m_S + 1). \qquad (10.2.3)$$

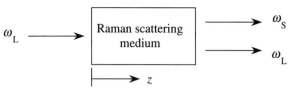

FIGURE 10.2.1   Geometry describing stimulated Raman scattering.

For definiteness, Fig. 10.2.1 shows the laser and Stokes beams propagating in the same direction; in fact, Eq. (10.2.3) applies even if the angle between the propagation directions of the laser and Stokes waves is arbitrary, as long as z is measured along the propagation direction of the Stokes wave.

It is instructive to consider Eq. (10.2.3) in the two opposite limits of $m_S \ll 1$ and $m_S \gg 1$. In the first limit, where the occupation number of the Stokes mode is much less than unity, Eq. (10.2.3) becomes simply

$$\frac{dm_S}{dz} = \frac{1}{c/n} D m_L \qquad \text{(for } m_S \ll 1\text{)}. \qquad (10.2.4)$$

The solution to this equation for the geometry of Fig. 10.2.1 under the assumption that the laser field is unaffected by the interaction (and hence that m_L is independent of z) is

$$m_S(z) = m_S(0) + \frac{1}{c/n} D m_L z \qquad \text{(for } m_S \ll 1\text{)}, \qquad (10.2.5)$$

where $m_S(0)$ denotes the photon occupation number associated with the Stokes field at the input to the Raman medium. This limit corresponds to spontaneous Raman scattering; the Stokes intensity increases in proportion to the length of the Raman medium and hence to the total number of molecules contained in the interaction region.

The opposite limiting case is that in which there are many photons in the Stokes mode. In this case Eq. (10.2.3) becomes

$$\frac{dm_S}{dz} = \frac{1}{c/n} D m_L m_S \qquad \text{(for } m_S \gg 1\text{)}, \qquad (10.2.6)$$

whose solution (again under the assumption of an undepleted input field) is

$$m_S(z) = m_S(0)e^{Gz} \qquad \text{(for } m_S \gg 1\text{)}, \qquad (10.2.7)$$

where we have introduced the Raman gain coefficient

$$G = \frac{D m_L}{c/n}. \qquad (10.2.8)$$

Again $m_S(0)$ denotes the photon occupation number associated with the Stokes field at the input to the Raman medium. If no field is injected into the Raman medium, $m_S(0)$ represents the quantum noise associated with the vacuum state, which is equivalent to one photon per mode. Emission of the sort described by Eq. (10.2.7) is called stimulated Raman scattering. The Stokes intensity is seen to grow exponentially with propagation distance through the medium, and large values of the Stokes intensity are routinely observed at the output of the interaction region.

We see from Eq. (10.2.8) that the Raman gain coefficient can be related simply to the phenomenological constant D introduced earlier. However, we see from Eq. (10.2.5) that the strength of spontaneous Raman scattering is also proportional to D. Since the strength of spontaneous Raman scattering is often described in terms of a scattering cross section, it is thus possible to determine a relationship between the gain coefficient G for stimulated Raman scattering and the cross section for spontaneous Raman scattering. This relationship is derived as follows:

Since one laser photon is lost for each Stokes photon that is created, the occupation number of the laser field changes as the result of spontaneous scattering into one particular Stokes mode in accordance with the relation $dm_L/dz = -dm_S/dz$, with dm_S/dz given by Eq. (10.2.4). However, since the system can radiate into a large number of Stokes modes, the total rate of loss of laser photons is given by

$$\frac{dm_L}{dz} = -Mb\frac{dm_S}{dz} = \frac{-Dm_LMb}{c/n}, \tag{10.2.9}$$

where M is the total number of modes into which the system can radiate and where b is a geometrical factor that accounts for the fact that the angular distribution of scattered radiation may be nonuniform and hence that the scattering rate into different Stokes modes may be different. Explicitly, b is the ratio of the angularly averaged Stokes emission rate to the rate in the direction of the particular Stokes mode S for which D (and hence the Raman gain coefficient) is to be determined. If $|f(\theta, \phi)|^2$ denotes the angular distribution of the Stokes radiation, b is then given by

$$b = \frac{\int |f(\theta, \phi)|^2 d\Omega/4\pi}{|f(\theta_S, \phi_S)|^2}, \tag{10.2.10}$$

where (θ_S, ϕ_S) gives the direction of the particular Stokes mode for which D is to be determined.

The total number of Stokes modes into which the system can radiate is given by the expression (see, for example, Boyd, 1983, Eq. (3.4.4))

$$M = \frac{V\omega_S^2\Delta\omega}{\pi^2(c/n)^3}, \tag{10.2.11}$$

where V denotes the volume of the region in which the modes are defined and where $\Delta\omega$ denotes the linewidth of the scattered Stokes radiation. The rate of loss of laser photons is conventionally described by the cross section σ for Raman scattering, which is defined by the relation

$$\frac{dm_L}{dz} = -N\sigma m_L, \qquad (10.2.12)$$

where N is the number density of molecules. By comparison of Eqs. (10.2.9) and (10.2.12), we see that we can express the parameter D in terms of the cross section σ by

$$D = \frac{N\sigma(c/n)}{Mb}. \qquad (10.2.13)$$

This expression for D, with M given by Eq. (10.2.11), is now substituted into expression (10.2.8) for the Raman gain coefficient to give the result

$$G = \frac{N\sigma\pi^2c^3m_L}{V\omega_S^2\Delta\omega bn^3} \equiv \frac{N\pi^2c^3m_L}{V\omega_S^2bn^3}\left(\frac{\partial\sigma}{\partial\omega}\right)_0, \qquad (10.2.14)$$

where in obtaining the second form we have used the definition of the spectral density of the scattering cross section to express σ in terms of its line-center value $(\partial\sigma/\partial\omega)_0$ as

$$\sigma = \left(\frac{\partial\sigma}{\partial\omega}\right)_0\Delta\omega. \qquad (10.2.15)$$

Equation (10.2.14) gives the Raman gain coefficient in terms of the number of laser photons per mode, m_L. In order to express the gain coefficient in terms of the laser intensity, which can be measured directly, we assume the geometry shown in Fig. 10.2.2. The laser intensity I_L is equal to the number of photons contained in this region multiplied by the energy per photon and divided by the cross-sectional area of the region and by the transit time through the region,

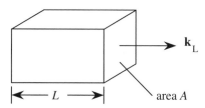

FIGURE 10.2.2   Geometry of the region within which the laser and Stokes modes are defined.

that is,

$$I_L = \frac{m_L \hbar \omega_L}{A(nL/c)} = \frac{m_L \hbar \omega_L c}{Vn}, \tag{10.2.16}$$

where $V = AL$. Through use of this result, the Raman gain coefficient of Eq. (10.2.14) can be expressed as

$$G = \frac{N\pi^2 c^2}{\omega_S^2 bn^2 \hbar \omega_L} \left(\frac{\partial \sigma}{\partial \omega} \right)_0 I_L. \tag{10.2.17}$$

It is sometimes convenient to express the Raman gain coefficient not in terms of the spectral cross section $(\partial \sigma / \partial \omega)_0$ but in terms of the differential spectral cross section $(\partial^2 \sigma / \partial \omega \, \partial \Omega)_0$, where $d\Omega$ is an element of solid angle. These quantities are related by

$$\left(\frac{\partial \sigma}{\partial \omega} \right)_0 = 4\pi b \left(\frac{\partial^2 \sigma}{\partial \omega \partial \Omega} \right)_0, \tag{10.2.18}$$

where b is the factor defined in Eq. (10.2.10) that accounts for the possible nonuniform angular distribution of the scattered Stokes radiation. Through use of this relation, Eq. (10.2.17) becomes

$$G = \frac{4\pi^3 N c^2}{\omega_S^2 \hbar \omega_L n_S^2} \left(\frac{\partial^2 \sigma}{\partial \omega \partial \Omega} \right)_0 I_L \tag{10.2.19}$$

Some of the parameters describing stimulated Raman scattering are listed in Table 10.2.1 for a number of materials.

10.3. Stimulated Raman Scattering Described by the Nonlinear Polarization

Next, we develop a classical (that is, non-quantum-mechanical) model that describes stimulated Raman scattering (Garmire *et al.*, 1963). We assume that the optical radiation interacts with a vibrational mode of a molecule, as illustrated in Fig. 10.3.1. We assume that the vibrational mode can be described as a simple harmonic oscillator of resonance frequency ω_v and damping constant γ, and we denote by $\tilde{q}$ the deviation of the internuclear distance from its equilibrium value q_0. The equation of motion describing the molecule vibration is thus

$$\frac{d^2 \tilde{q}}{dt^2} + 2\gamma \frac{d\tilde{q}}{dt} + \omega_v^2 \tilde{q} = \frac{\tilde{F}(t)}{m}, \tag{10.3.1}$$

where $\tilde{F}(t)$ denotes any force that acts on the vibrational degree of freedom and where m represents the reduced nuclear mass.

TABLE 10.2.1   Properties of stimulated Raman scattering for several materials[a]

Substance	Frequency shift v_0 (cm^{-1})	Linewidth Δv (cm^{-1})	Cross section $N(d\sigma/d\Omega)_0$ (10^{-8} cm^{-1} sec^{-1})	Gain factor[b] G/I_L (cm/GW)
Liquid O_2	1552	0.117	0.48 ± 0.14	14.5 ± 4
Liquid N_2	2326.5	0.067	0.29 ± 0.09	16 ± 5
Benzene	992	2.15	3.06	2.8
CS_2	655.6	0.50	7.55	24
Nitrobenzene	1345	6.6	6.4	2.1
Bromobenzene	1000	1.9	1.5	1.5
Chlorobenzene	1002	1.6	1.5	1.9
Toluene	1003	1.94	1.1	1.2
NiNbO$_3$	256	23	381	8.9
	637	20	231	9.4
Ba$_2$NaNb$_5$O$_{15}$	650			6.7
LiTaO$_3$	201	22	238	4.4
SiO$_2$	467			0.8
Methane gas	2916		(10 atm)[c]	0.66
H$_2$ gas	4155		(>10 atm)	1.5
H$_2$ gas (rotat.)	450		(>0.5 atm)	0.5
Deuterium gas	2991		(>10 atm)	1.1
N$_2$ gas	2326		(10 atm)[c]	0.071
O$_2$ gas	1555		(10 atm)[c]	0.016

[a] After Kaiser and Maier (1972) and Simon and Tittel (1994). All transitions are vibrational except for the 450 cm^{-1} hydrogen transition which is rotational.
[b] Measured at 694 nm unless stated otherwise.
[c] Measured at 500 nm.

The key assumption of the theory is that the optical polarizability of the molecule (which is typically predominantly electronic in origin) is not constant, but depends on the internuclear distance according to the equation

$$\tilde{\alpha}(t) = \alpha_0 + \left(\frac{\partial\alpha}{\partial q}\right)_0 \tilde{q}(t). \tag{10.3.2}$$

Here α_0 is the polarizability of a molecule in which the internuclear distance is held fixed at its equilibrium value. According to Eq. (10.3.2), when the

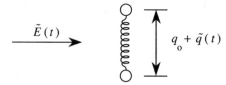

FIGURE 10.3.1   Molecular description of stimulated Raman scattering.

molecule is set into oscillation its polarizability will be modulated in time, and hence the refractive index of a collection of coherently oscillating molecules will be modulated in time in accordance with the relations

$$\tilde{n}(t) = \sqrt{\tilde{\epsilon}(t)} = [1 + 4\pi N\tilde{\alpha}(t)]^{1/2}. \tag{10.3.3}$$

The temporal modulation of the refractive index will modify a beam of light in passing through the medium. In particular, frequency sidebands separated from the laser frequency by $\pm\omega_v$ will be impressed upon the transmitted laser beam.

Next, we examine how the molecular vibrations can be driven coherently by an applied optical field. In the presence of the optical field $\tilde{E}(z, t)$, each molecule will become polarized, and the induced dipole moment of a molecule located at coordinate z will be given by

$$\tilde{\mathbf{p}}(z, t) = \alpha \, \tilde{\mathbf{E}}(z, t). \tag{10.3.4}$$

The energy required to establish this oscillating dipole moment is given by

$$W = \tfrac{1}{2}\langle \tilde{\mathbf{p}}(z, t) \cdot \tilde{\mathbf{E}}(z, t) \rangle = \tfrac{1}{2}\alpha \langle \tilde{E}^2(z, t) \rangle, \tag{10.3.5}$$

where the angular brackets denote a time average over an optical period. The applied optical field hence exerts a force given by

$$\tilde{F} = \frac{dW}{dq} = \frac{1}{2}\left(\frac{d\alpha}{dq}\right)_0 \langle \tilde{E}^2(z, t) \rangle \tag{10.3.6}$$

on the vibrational degree of freedom. In particular, if the applied field contains two frequency components, Eq. (10.3.6) shows that the molecular coordinate will experience a time-varying force at the beat frequency between the two field components.

The origin of stimulated Raman scattering can be understood schematically in terms of the interactions shown in Fig. 10.3.2. Part (a) of the figure shows how

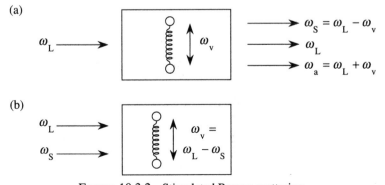

FIGURE 10.3.2 Stimulated Raman scattering.

molecular vibrations modulate the refractive index of the medium at frequency ω_v and thereby impress frequency sidebands onto the laser field. Part (b) shows how the Stokes field at frequency $\omega_S = \omega_L - \omega_v$ can beat with the laser field to produce a modulation of the total intensity of the form

$$\tilde{I}(t) = I_0 + I_1 \cos(\omega_L - \omega_S)t. \qquad (10.3.7)$$

This modulated intensity coherently excites the molecular oscillation at frequency $\omega_v = \omega_L - \omega_S$. The two processes shown in parts (a) and (b) of the figure reinforce one another in the sense that the interaction shown in part (b) leads to a stronger molecular vibration, which by the interaction shown in part (a) leads to a stronger Stokes field, which in turn leads to a stronger molecular vibration.

To make these ideas quantitative, let us assume that the total optical field can be represented as

$$\tilde{E}(z, t) = A_L e^{i(k_L z - \omega_L t)} + A_S e^{i(k_S z - \omega_S t)} + \text{c.c.} \qquad (10.3.8)$$

According to Eq. (10.3.6) the time-varying part of the applied force is then given by

$$\tilde{F}(z, t) = \left(\frac{\partial \alpha}{\partial q}\right)_0 \left[A_L A_S^* e^{i(Kz - \Omega t)} + \text{c.c.}\right], \qquad (10.3.9)$$

where we have introduced the notation

$$K = k_L - k_S \quad \text{and} \quad \Omega = \omega_L - \omega_S. \qquad (10.3.10)$$

We next find the solution to Eq. (10.3.1) with a force term of the form of Eq. (10.3.9). We adopt a trial solution of the form

$$\tilde{q} = q(\Omega) e^{i(Kz - \Omega t)} + \text{c.c.} \qquad (10.3.11)$$

We insert Eqs. (10.3.9) and (10.3.11) into Eq. (10.3.1), which becomes

$$-\omega^2 q(\Omega) - 2i\omega\gamma q(\Omega) + \omega_v^2 q(\Omega) = \frac{1}{m}\left(\frac{\partial \alpha}{\partial q}\right)_0 A_L A_S^*,$$

and hence we find that the amplitude of the molecular vibration is given by

$$q(\Omega) = \frac{(1/m)[\partial\alpha/\partial q]_0 A_L A_S^*}{\omega_v^2 - \Omega^2 - 2i\Omega\gamma}. \qquad (10.3.12)$$

Since the polarization of the medium is given according to Eqs. (10.3.2) and (10.3.4) by

$$\tilde{P}(z,t) = N\tilde{p}(z,t) = N\tilde{\alpha}(z,t)\tilde{E}(z,t)$$
$$= N\left[\alpha_0 + \left(\frac{\partial\alpha}{\partial q}\right)_0 \tilde{q}(z.t)\right]\tilde{E}(z,t), \quad (10.3.13)$$

the nonlinear part of the polarization is given by

$$\tilde{P}^{NL}(z,t) = N\left(\frac{\partial\alpha}{\partial a}\right)_0 \left[q(\omega)e^{i(Kz-\Omega t)} + \text{c.c.}\right]$$
$$\times \left[A_L e^{i(k_L z - \omega_L t)} + A_S e^{i(K_S z - \omega_S t)} + \text{c.c.}\right]. \quad (10.3.14)$$

The nonlinear polarization is seen to contain several different frequency components. The part of this expression that oscillates at frequency ω_S is known as the Stokes polarization and is given by

$$\tilde{P}_S^{NL}(z,t) = P(\omega_S)e^{i\omega_S t} + \text{c.c.} \quad (10.3.15)$$

with a complex amplitude given by

$$P(\omega_S) = N\left(\frac{\partial\alpha}{\partial q}\right)_0 q^*(\Omega)A_L e^{ik_S z}. \quad (10.3.16)$$

By introducing the expression (10.3.12) for $q(\Omega)$ into this equation, we find that the complex amplitude of the Stokes polarization is given by

$$P(\omega_S) = \frac{(N/m)(\partial\alpha/\partial q)_0^2 |A_L|^2 A_S}{\omega_v^2 - \Omega^2 + 2i\Omega\gamma}e^{ik_S z}. \quad (10.3.17)$$

We now define the Raman susceptibility through the expression

$$P(\omega_S) = 6\chi_R(\omega_S)|A_L|^2 A_S e^{ik_S z}, \quad (10.3.18)$$

where for notational convenience we have introduced $\chi_R(\omega_S)$ as a shortened form of $\chi^{(3)}(\omega_S = \omega_S + \omega_L - \omega_L)$. By comparison of Eqs. (10.3.17) and (10.3.18), we find that the Raman susceptibility is given by

$$\chi_R(\omega_S) = \frac{(N/6m)(\partial\alpha/\partial q)_0^2}{\omega_v^2 - (\omega_L - \omega_S)^2 + 2i(\omega_L - \omega_S)\gamma}. \quad (10.3.19a)$$

The real and imaginary parts of $\chi_R(\omega_S) \equiv \chi_R'(\omega_S) + i\chi_R''(\omega_S)$ are illustrated in Fig. 10.3.3.

Near the Raman resonance, the Raman susceptibility can be approximated as

$$\chi_R(\omega_S) = \frac{(N/12m\omega_v)(\partial\alpha/\partial q)_0^2}{[\omega_S - (\omega_L - \omega_v)] + i\gamma}. \quad (10.3.19b)$$

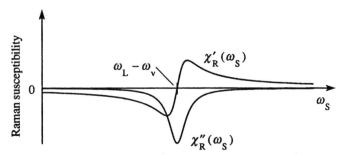

FIGURE 10.3.3   Resonance structure of the Raman susceptibility.

Note that, at the exact Raman resonance (that is, for $\omega_S = \omega_L - \omega_v$), the Raman susceptibility is negative imaginary. (We shall see below that consequently the Stokes wave experiences amplification.)

In order to describe explicitly the spatial evolution of the Stokes wave, we use Eqs. (10.3.8), (10.3.15), (10.3.18), and (9.3.19) for the nonlinear polarization for the driven wave equation (2.1.16). We then find that the evolution of the field amplitude A_S is given in the slowly-varying amplitude approximation by

$$\frac{dA_S}{dz} = -\alpha_S A_S, \qquad (10.3.20)$$

where

$$\alpha_S = -12\pi i \frac{\omega_S}{n_S c} \chi_R(\omega_S) |A_L|^2 \qquad (10.3.21)$$

is the Stokes wave "absorption" coefficient. Since the imaginary part of $\chi_R(\omega_S)$ is negative, the real part of the absorption coefficient is negative, implying that the Stokes wave actually experiences exponential growth. Note that α_S depends only in the modulus of the complex amplitude of the laser field. Raman Stokes amplification is thus a process for which the phase-matching condition is automatically satisfied. Alternatively, Raman Stokes amplification can be said to be a pure gain process.

We can also predict the spatial evolution of a wave at the anti-Stokes frequency through use of the results of the calculation just completed. In the derivation of Eq. (10.3.19a), no assumptions were made regarding the sign of $\omega_L - \omega_S$. We can thus deduce the form of the anti-Stokes susceptibility by formally replacing ω_S by ω_a in Eq. (10.3.19a) to obtain the result

$$\chi_R(\omega_a) = \frac{(N/6m)(\partial\alpha/\partial q)_0^2}{\omega_v^2 - (\omega_L - \omega_a)^2 + 2i(\omega_L - \omega_a)\gamma}. \qquad (10.3.22)$$

Since ω_S and ω_a are related through

$$\omega_L - \omega_S = -(\omega_L - \omega_a), \qquad (10.3.23)$$

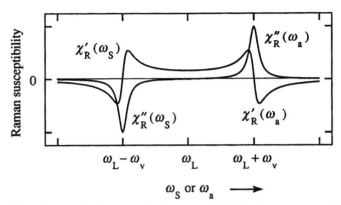

FIGURE 10.3.4   Relation between Stokes and anti-Stokes Raman susceptibilities.

we see that

$$\chi_R(\omega_a) = \chi_R(\omega_S)^*. \tag{10.3.24}$$

The relation between the Stokes and anti-Stokes Raman susceptibilities is illustrated in Fig. 10.3.4. Near the Raman resonance, Eq. (10.3.22) can be approximated by

$$\chi_R(\omega_a) = \frac{(N/12m\omega_v)(\partial\alpha/\partial q)_0^2}{[\omega_a - (\omega_L + \omega_v)] + i\gamma}, \tag{10.3.25}$$

and at the exact resonance the Raman susceptibility is positive imaginary. The amplitude of the anti-Stokes wave hence obeys the propagation equation

$$\frac{dA_a}{dz} = -\alpha_a A_a, \tag{10.3.26}$$

where

$$\alpha_a = -12\pi \frac{\omega_a}{n_a c} \chi_R(\omega_a)|A_L|^2. \tag{10.3.27}$$

For a positive imaginary $\chi_R(\omega_a)$, α_a is positive real, implying that the anti-Stokes wave experiences attenuation.

However, it is found experimentally (Terhune, 1963) that the anti-Stokes wave is generated with appreciable efficiency, at least in certain directions. The origin of anti-Stokes generation is an additional contribution to the nonlinear polarization beyond that described by the Raman susceptibility of Eq. (10.3.25). Inspection of Eq. (10.3.14) shows that there is a contribution to the anti-Stokes polarization

$$\tilde{P}_a^{NL}(z, t) = P(\omega_a)e^{-i\omega_a t} + \text{c.c.} \tag{10.3.28}$$

that depends on the Stokes amplitude and which is given by

$$P(\omega_a) = N\left(\frac{\partial \alpha}{\partial q}\right)_0 q(\Omega)A_L = \frac{(N/m)(\partial \alpha/\partial q)_0^2 A_L^2 A_S^*}{\omega_v^2 - \Omega^2 - 2i\Omega\gamma}e^{i(2k_L - k_S)z}. \quad (10.3.29)$$

(Recall that $\Omega \equiv \omega_L - \omega_S = \omega_a - \omega_L$.) This contribution to the nonlinear polarization can be described in terms of a four-wave mixing susceptibility $\chi_F(\omega_a) \equiv \chi^{(3)}(\omega_a = \omega_L + \omega_L - \omega_S)$, which is defined by the relation

$$P(\omega_a) = 3\chi_F(\omega_a)A_L^2 A_S^* e^{i(2k_L - k_S)z} \quad (10.3.30)$$

and which is hence equal to

$$\chi_F(\omega_a) = \frac{(N/3m)(\partial \alpha/\partial q)_0^2}{\omega_v^2 - (\omega_L - \omega_a)^2 + 2i(\omega_L - \omega_a)\gamma}. \quad (10.3.31)$$

We can see by comparison with Eq. (10.3.22) that

$$\chi_F(\omega_a) = 2\chi_R(\omega_a). \quad (10.3.32)$$

The total polarization at the anti-Stokes frequency is the sum of the contributions described by Eqs. (10.3.22) and (10.3.31) and is hence given by

$$P(\omega_a) = 6\chi_R(\omega_a)|A_L|^2 A_a e^{ik_a z} + 3\chi_F(\omega_a)A_L^2 A_S^* e^{i(2k_L - k_s)z}. \quad (10.3.33)$$

Similarly, there is a four-wave mixing contribution to the Stokes polarization described by

$$\chi_F(\omega_S) = \frac{(N/3m)(\partial \alpha/\partial q)_0^2}{\omega_v^2 - (\omega_L - \omega_a)^2 + 2i(\omega_L - \omega_S)\gamma}, \quad (10.3.34)$$

so that the total polarization at the Stokes frequency is given by

$$P(\omega_S) = 6\chi_R(\omega_S)|A_L|^2 A_S e^{ik_s z} + 3\chi_F(\omega_S)A_L^2 A_a^* e^{i(2k_L - k_a)z}. \quad (10.3.35)$$

The Stokes four-wave mixing susceptibility is related to the Raman Stokes susceptibility by

$$\chi_F(\omega_S) = 2\chi_R(\omega_S) \quad (10.3.36)$$

and to the anti-Stokes susceptibility through

$$\chi_F(\omega_S) = \chi_F(\omega_a)^*. \quad (10.3.37)$$

The spatial evolution of the Stokes and anti-Stokes fields is now obtained by introducing Eqs. (10.3.33) and (10.3.35) into the driven wave Eq. (2.1.16). We assume that the medium is optically isotropic and that the slowly-varying amplitude and constant-pump approximations are valid. We find that the field amplitudes obey the set of coupled equations

$$\frac{dA_S}{dz} = -\alpha_S A_S + \kappa_S A_a^* e^{i\Delta kz}, \tag{10.3.38a}$$

$$\frac{dA_a}{dz} = -\alpha_a A_a + \kappa_a A_S^* e^{i\Delta kz}, \tag{10.3.38b}$$

where we have introduced nonlinear absorption and coupling coefficients

$$\alpha_j = \frac{-12\pi i \omega_j}{n_j c} \chi_R(\omega_j)|A_L|^2, \qquad j = S, a, \tag{10.3.39a}$$

$$\kappa_j = \frac{6\pi i \omega_j}{n_j c} \chi_F(\omega_j) A_L^2, \qquad j = S, a \tag{10.3.39b}$$

and have defined the wavevector mismatch

$$\Delta k = \Delta \mathbf{k} \cdot \hat{\mathbf{z}} = (2\mathbf{k}_L - \mathbf{k}_S - \mathbf{k}_a) \cdot \hat{\mathbf{z}}. \tag{10.3.40}$$

The form of Eqs. (10.3.38) shows that each of the Stokes and anti-Stokes amplitudes is driven by a Raman gain or loss term (the first term on the right-hand side) and by a phase-matched four-wave mixing term (the second). The four-wave mixing term is an effective driving term only when the wavevector mismatch Δk is small. For a material with normal dispersion, the refractive index experienced by the laser wave is always less than the mean of those experienced by the Stokes and anti-Stokes waves, as illustrated in part (a) of Fig. 10.3.5. For this reason, perfect phase matching ($\Delta k = 0$) can always be

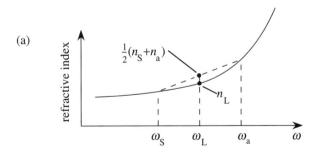

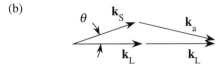

FIGURE 10.3.5 Phase-matching relations for Stokes and anti-Stokes coupling in stimulated Raman scattering.

achieved if the Stokes wave propagates at some nonzero angle with respect to the laser wave, as illustrated in part (b) of the figure. For angles appreciably different from this phase-matching angle, Δk is large, and only the first term on the right-hand side of each of Eqs. (10.3.38) is important. For these directions, the two equations decouple, and the Stokes sideband experiences gain and the anti-Stokes sideband experiences loss. However, for directions such that Δk is small, both driving terms on the right-hand sides of Eqs. (10.3.38) are important, and the two equations must be solved simultaneously. In the next section, we shall see how to solve these equations and shall see that both Stokes and anti-Stokes radiation can be generated in directions for which Δk is small.

10.4. Stokes–Anti-Stokes Coupling in Stimulated Raman Scattering

In this section, we study the nature of the solution to the equations describing the propagation of the Stokes and anti-Stokes waves. We have just seen that these equations are of the form

$$\frac{dA_1}{dz} = -\alpha_1 A_1 + \kappa_1 A_2^* e^{i\Delta kz}, \tag{10.4.1a}$$

$$\frac{dA_2^*}{dz} = -\alpha_2^* A_2^* + \kappa_2^* A_1 e^{-i\Delta kz}. \tag{10.4.1b}$$

In fact, equations of this form are commonly encountered in nonlinear optics and also describe, for example, any forward four-wave mixing process in the constant-pump approximation. The ensuing discussion of the solution to these equations is simplified by first rewriting Eqs. (10.4.1) as

$$e^{-i\Delta kz/2}\left(\frac{dA_1}{dz} + \alpha_1 A_1\right) = \kappa_1 A_2^* e^{i\Delta kz/2}, \tag{10.4.2a}$$

$$e^{i\Delta kz/2}\left(\frac{dA_2^*}{dz} + \alpha_2^* A_2^*\right) = \kappa_2^* A_1 e^{-i\Delta kz/2}, \tag{10.4.2b}$$

from which it follows that the equations can be expressed as

$$\left(\frac{d}{dz} + \alpha_1 + \frac{i\Delta k}{2}\right) A_1 e^{-i\Delta kz/2} = \kappa_1 A_2^* e^{i\Delta kz/2}, \tag{10.4.3a}$$

$$\left(\frac{d}{dz} + \alpha_2^* - \frac{i\Delta k}{2}\right) A_2^* e^{i\Delta kz/2} = \kappa_2^* A_1 e^{-i\Delta kz/2}. \tag{10.4.3b}$$

The form of these equations suggests that we introduce the new variables F_1 and F_2 defined by

$$F_1 = A_1 e^{-i\Delta kz/2} \quad \text{and} \quad F_2^* = A_2^* e^{i\Delta kz/2}, \tag{10.4.4}$$

so that Eqs. (10.4.3) become

$$\left(\frac{d}{dz} + \alpha_1 + i\frac{\Delta k}{2}\right)F_1 = \kappa_1 F_2^*, \tag{10.4.5a}$$

$$\left(\frac{d}{dz} + \alpha_2^* - i\frac{\Delta k}{2}\right)F_2^* = \kappa_2^* F_1. \tag{10.4.5b}$$

We now eliminate F_2^* algebraically from this set of equations to obtain the single equation

$$\left(\frac{d}{dz} + \alpha_2^* - i\frac{\Delta k}{2}\right)\left(\frac{d}{dz} + \alpha_1 + i\frac{\Delta k}{2}\right)F_1 = \kappa_1\kappa_2^* F_1. \tag{10.4.6}$$

We solve this equation by adopting a trial solution of the form

$$F_1(z) = F_1(0)e^{gz}, \tag{10.4.7}$$

where g represents an unknown spatial growth rate. We substitute this form into Eq. (10.4.6) and find that this equation is satisfied by the trial solution if g satisfies the algebraic equation

$$\left(g + \alpha_2^* - \frac{i\Delta k}{2}\right)\left(g + \alpha_1 + \frac{i\Delta k}{2}\right) = \kappa_1\kappa_2^*. \tag{10.4.8}$$

In general, this equation possesses two solutions, which are given by

$$g_\pm = -\tfrac{1}{2}(\alpha_1 + \alpha_2^*) \pm \tfrac{1}{2}[(\alpha_1 - \alpha_2^* + i\Delta k)^2 + 4\kappa_1\kappa_2^*]^{1/2}. \tag{10.4.9}$$

Except for special values of α_1, α_2, κ_1, κ_2, and Δk, the two values of g given by Eq. (10.4.9) are distinct. Whenever the two values of g are distinct, the general solution for F is given by

$$F_1 = F_1^+(0)e^{g_+z} + F_1^-(0)e^{g_-z}, \tag{10.4.10}$$

and hence through use of Eq. (10.4.4) we see that the general solution for A_1 is of the form

$$A_1(z) = (A_1^+ e^{g_+z} + A_1^- e^{g_-z})e^{i\Delta kz/2}. \tag{10.4.11}$$

Here A_1^+ and A_1^- are constants of integration whose values must be determined from the relevant boundary conditions. The general form of the solution for $A_2^*(z)$ is readily found by substituting Eq. (10.4.11) into Eq. (10.4.3a), which becomes

$$\left(g_+ + \alpha_1 + i\frac{\Delta k}{2}\right)A_1^+ e^{g+z} + \left(g_- + \alpha_1 + i\frac{\Delta k}{2}\right)A_1^- e^{g-z} = \kappa_1 A_2^* e^{i\Delta kz/2}.$$

This equation is now solved for $A_2^*(z)$ to obtain

$$A_2^*(z)$$
$$= \left[\left(\frac{g_+ + \alpha_1 + i\Delta k/2}{\kappa_1}\right)A_1^+ e^{g+z} + \left(\frac{g_- + \alpha_1 + i\Delta k/2}{\kappa_1}\right)A_1^- e^{g-z}\right]e^{-i\Delta kz/2}.$$
$$\text{(10.4.12)}$$

If we define constants A_2^+ and A_2^- by means of the equation

$$A_2^*(z) = (A_2^{+*}e^{g+z} + A_2^{-*}e^{g-z})e^{-i\Delta kz/2}, \qquad (10.4.13)$$

we see that the amplitudes $A_1^{\pm}$ and $A_2^{\pm}$ are related by

$$\frac{A_2^{\pm*}}{A_1^{\pm}} = \frac{g_{\pm} + \alpha_1 + i\Delta k/2}{\kappa_1}. \qquad (10.4.14)$$

This equation shows how the amplitudes A_2^+ and A_1^+ are related in the part of the solution that grows as $\exp(g_+ z)$, and similarly how the amplitudes A_2^- and A_1^- are related in the part of the solution that grows as $\exp(g_- z)$. We can think of Eq. (10.4.14) as specifying the eigenmodes of propagation of the Stokes and anti-Stokes waves.

As written, Eq. (10.4.14) appears to be asymmetric with respect to the roles of the ω_1 and ω_2 fields. However, this asymmetry occurs in appearance only. Since $g_{\pm}$ depends upon $\alpha_1, \alpha_2, \kappa_1, \kappa_2$, and Δk, the right-hand side of Eq. (10.4.14) can be written in a variety of equivalent ways, some of which display the symmetry of the interaction more explicitly. We next rewrite Eq. (10.4.14) in such a manner.

One can show by explicit calculation using Eq. (10.4.9) that the quantities g_+ and g_- are related by

$$\left(g_+ + \alpha_1 + \frac{i\Delta k}{2}\right)\left(g_- + \alpha_1 + \frac{i\Delta k}{2}\right) = -\kappa_1\kappa_2^*. \qquad (10.4.15)$$

In addition, one can see by inspection of Eq. (10.4.9) that their difference is given by

$$g_+ - g_- = [(\alpha_1 - \alpha_2^* + i\Delta k)^2 + 4\kappa_1\kappa_2^*]^{1/2}. \qquad (10.4.16a)$$

By substitution of Eq. (10.4.9) into this last equation, it follows that

$$g_+ - g_- = \pm[2g_{\pm} + (\alpha_1 + \alpha_2^*)], \qquad (10.4.16b)$$

where on the right-hand side either both pluses or both minuses must be used.

Furthermore, one can see from Eq. (10.4.9) that

$$g_+ + g_- = -(\alpha_1 + \alpha_2^*). \tag{10.4.16c}$$

By rearranging this equation and adding $i\,\Delta k/2$ to each side, it follows that

$$\left(g_\pm + \alpha_1 + \frac{i\,\Delta k}{2}\right) = -\left(g_\mp + \alpha_2^* - \frac{i\,\Delta k}{2}\right), \tag{10.4.17a}$$

$$\left(g_\pm + \alpha_2^* + \frac{i\,\Delta k}{2}\right) = -\left(g_\mp + \alpha_1 + \frac{i\,\Delta k}{2}\right). \tag{10.4.17b}$$

Through use of Eqs. (10.4.15) and (10.4.17a), Eq. (10.4.14) can be expressed as

$$\frac{A_2^{\pm*}}{A_1^\pm} = \frac{g_\pm + \alpha_1 + i\,\Delta k/2}{\kappa_1} = \frac{-\kappa_2^*}{g_\mp + \alpha_1 + i\,\Delta k/2} = \frac{\kappa_2^*}{g_\pm + \alpha_2^* - i\,\Delta k/2}. \tag{10.4.18}$$

By taking the geometric mean of the last and third-last forms of this expression, we find that the ratio $A_2^{\pm*}/A_1^\pm$ can be written as

$$\frac{A_2^{\pm*}}{A_1^\pm} = \left[\frac{\kappa_2^*(g_\pm + \alpha_1 + i\,\Delta k/2)}{\kappa_1(g_\pm + \alpha_2^* - i\,\Delta k/2)}\right]^{1/2}; \tag{10.4.19}$$

this form shows explicitly the symmetry between the roles of the ω_1 and ω_2 fields.

Next, we find the form of the solution when the boundary conditions are such that the input fields are known at the plane $z = 0$, that is, when $A_1(0)$ and $A_2^*(0)$ are given. We proceed by finding the values of the constants of integration A_1^+ and A_1^-. Equation (10.4.11) is evaluated at $z = 0$ to give the result

$$A_1(0) = A_1^+ + A_1^-, \tag{10.4.20a}$$

and Eq. (10.4.12) is evaluated at $z = 0$ to give the result

$$A_2^*(0) = \left(\frac{g_+ + \alpha_1 + i\,\Delta k/2}{\kappa_1}\right)A_1^+ + \left(\frac{g_- + \alpha_1 + i\,\Delta k/2}{\kappa_1}\right)A_1^-. \tag{10.4.20b}$$

We rearrange Eq. (10.4.20a) to find that $A_1^- = A_1(0) - A_1^+$, and we substitute this form into Eq. (10.4.20b) to obtain

$$A_2^*(0) = \left(\frac{g_+ - g_-}{\kappa_1}\right)A_1^+ + \left(\frac{g_- + \alpha_1 + i\,\Delta k/2}{\kappa_1}\right)A_1(0).$$

We solve this equation for A_1^+ to obtain

$$A_1^+ = \left(\frac{\kappa_1}{g_+ - g_-}\right)A_2^*(0) - \left(\frac{g_- + \alpha_1 + i\,\Delta k/2}{g_+ - g_-}\right)A_1(0). \quad (10.4.21a)$$

If instead we solve Eq. (10.4.20a) for A_1^+ and substitute the result $A_1^+ = A_1(0) - A_1^-$ into Eq. (10.4.20b), we find that

$$A_2^*(0) = \left(\frac{g_- - g_+}{\kappa_1}\right)A_1^- - \left(\frac{g_+ + \alpha_1 + i\,\Delta k/2}{\kappa_1}\right)A_1(0),$$

which can be solved for A_1^- to obtain

$$A_1^- = -\left(\frac{\kappa_1}{g_+ - g_-}\right)A_2^*(0) + \left(\frac{g_+ + \alpha_1 + i\,\Delta k/2}{g_+ - g_-}\right)A_1(0). \quad (10.4.21b)$$

The expressions (10.4.21) for the constants A_1^+ and A_1^- are now substituted into Eqs. (10.4.11) and (10.4.12) to give the solution for the spatial evolution of the two interacting fields in terms of their boundary values as

$$A_1(z) = \frac{1}{g_+ - g_-}\left\{\left[\kappa_1 A_2^*(0) - \left(g_- + \alpha_1 + \frac{i\,\Delta k}{2}\right)A_1(0)\right]e^{g_+ z}\right.$$

$$\left. - \left[\kappa_1 A_2^*(0) - \left(g_+ + \alpha_1 + \frac{i\,\Delta k}{2}\right)A_1(0)\right]e^{g_- z}\right\}e^{i\,\Delta k z/2} \quad (10.4.22)$$

and

$$A_2^*(z) = \frac{1}{g_+ - g_-}\left\{\left[\left(g_+ + \alpha_1 + \frac{i\,\Delta k}{2}\right)A_2^*(0) + \kappa_2^* A_1(0)\right]e^{g_+ z}\right.$$

$$\left. - \left[\left(g_- + \alpha_1 + \frac{i\,\Delta k}{2}\right)A_2^*(0) + \kappa_2^* A_1(0)\right]e^{g_- z}\right\}e^{-i\,\Delta k z/2}. \quad (10.4.23)$$

Through use of Eqs. (10.4.17), the second form can be written in terms of α_2 instead of α_1 as

$$A_2^*(z) = \frac{1}{g_+ - g_-}\left\{\left[-\left(g_- + \alpha_2^* - \frac{i\,\Delta k}{2}\right)A_2^*(0) + \kappa_2^* A_1(0)\right]e^{g_+ z}\right.$$

$$\left. + \left[\left(g_+ + \alpha_2^* - \frac{i\,\Delta k}{2}\right)A_2^*(0) - \kappa_2^* A_1(0)\right]e^{g_- z}\right\}e^{-i\,\Delta k z/2}. \quad (10.4.24)$$

Before applying the results of the derivation just performed to the case of stimulated Raman scattering, let us make sure that the solution makes sense by applying it to several specific limiting cases.

Dispersionless, Nonlinear Medium without Gain or Loss

For a medium without gain (or loss), we set $\alpha_1 = \alpha_2 = 0$. Also, since the medium is lossless and dispersionless, $\chi^{(3)}(\omega_1 = 2\omega_0 - \omega_2)$ must equal $\chi^{(3)}(\omega_2 = 2\omega_0 - \omega_1)$, and hence the product $\kappa_1 \kappa_2^*$ that appears in the solution is equal to

$$\kappa_1 \kappa_2^* = \frac{36\pi^2 \omega_1 \omega_2}{c^2} |\chi^{(3)}(\omega_1 = 2\omega_0 - \omega_2)|^2 |A_0|^4, \qquad (10.4.25)$$

which is a real, positive quantity. We allow Δk to be arbitrary, to allow the possibility that $\mathbf{k}_1$ and $\mathbf{k}_2$ are not parallel to $\mathbf{k}_0$. Under these conditions, the coupled gain coefficient of Eq. (10.4.9) reduces to

$$g_\pm = \pm[\kappa_1 \kappa_2^* - (\Delta k/2)^2]^{1/2}. \qquad (10.4.26)$$

We see that, so long as Δk is not too large, the root g_+ will be a positive real number corresponding to amplification, whereas the root g_- will be a negative real number corresponding to attenuation. However, if the wavevector mismatch becomes so large that Δk^2 exceeds $4\kappa_1 \kappa_2^*$, both roots will become pure imaginary, indicating that each eigensolution shows oscillatory spatial behavior. According to Eq. (10.4.14), the ratio of amplitudes corresponding to each eigensolution is given by

$$\frac{A_2^{\pm*}}{A_1^{\pm}} = \frac{g_\pm + i\Delta k/2}{\kappa_1}. \qquad (10.4.27)$$

The right-hand side of this expression simplifies considerably for the case of perfect phase matching ($\Delta k = 0$) and becomes $\pm(\kappa_2^*/\kappa_1)^{1/2}$. If we also choose our phase conventions so that A_0 is purely real, we find that expression reduces to

$$\frac{A_2^{\pm*}}{A_1^{\pm}} = \pm i\left(\frac{\omega_2}{\omega_1}\right)^{1/2} \simeq \pm i, \qquad (10.4.28)$$

which shows that the two frequency sidebands are phased by $\pm\pi/2$ radians in each of the eigensolutions.

Medium without a Nonlinearity

One would expect on physical grounds that, for a medium in which $\chi^{(3)}$ vanishes, the solution would reduce to the usual case of the free propagation of the ω_1 and ω_2 waves. By setting $\kappa_1 = \kappa_2 = 0$ in Eq. (10.4.9), and assuming for simplicity that Δk vanishes, we find that

$$g_+ = -\alpha_2^* \quad \text{and} \quad g_- = -\alpha_1^*. \qquad (10.4.29)$$

The eigenamplitudes are found most readily from Eq. (10.4.19). If we assume that κ_1 and κ_2 approach zero in such a manner that κ_2^*/κ_1 remains finite, we find from Eq. (10.4.19) that

$$\frac{A_2^{+*}}{A_1^+} = \infty, \qquad \frac{A_2^{-*}}{A_1^-} = 0. \qquad (10.4.30)$$

Thus the positive root corresponds to a wave at frequency ω_2, which propagates according to

$$A_2^*(z) = A_2^*(0)e^{g+z} = A_2^*(0)e^{-\alpha_2^*}(z), \qquad (10.4.31a)$$

whereas the negative root corresponds to a wave at frequency ω_1, which propagates according to

$$A_1(z) = A_1(0)e^{g-z} = A_1(0)e^{-\alpha_1 z}. \qquad (10.4.31b)$$

Stokes–Anti-Stokes Coupling in Stimulated Raman Scattering

Let us now apply this analysis to the case of stimulated Raman scattering (see also Bloembergen and Shen, 1964). For definiteness, we associate ω_1 with the Stokes frequency ω_S and ω_2 with the anti-Stokes frequency ω_a. The nonlinear absorption coefficients α_S and α_a and coupling coefficients κ_S and κ_a are given by Eqs. (10.3.39) with the nonlinear susceptibilities given by Eqs. (10.3.19b), (10.3.25), (10.3.31), and (10.3.34). In light of the relations

$$\chi_F(\omega_S) = \chi_F(\omega_a)^* = 2\chi_R(\omega_S) = 2\chi_R(\omega_a)^* \qquad (10.4.32)$$

among the various elements of the susceptibility, we find that the absorption and coupling coefficients can be related to each other as follows:

$$\alpha_a = -\alpha_S^* \left(\frac{n_S \omega_a}{n_a \omega_S}\right), \qquad (10.4.33a)$$

$$\kappa_S = -\alpha_S e^{2i\phi_L}, \qquad (10.4.33b)$$

$$\kappa_a = \alpha_S^* \left(\frac{n_S \omega_a}{n_a \omega_S}\right) e^{2i\phi_L}, \qquad (10.4.33c)$$

where ϕ_L is the phase of the pump laser defined through

$$A_L = |A_L|e^{i\phi_L}, \qquad (10.4.34)$$

and where the Stokes amplitude absorption coefficient is given explicitly by

$$\alpha_S = \frac{-i\pi \omega_S N (\partial\alpha/\partial q)_0^2 |A_L|^2}{2n_S c \omega_v [\omega_S - (\omega_L - \omega_v) + i\gamma]}. \qquad (10.4.35)$$

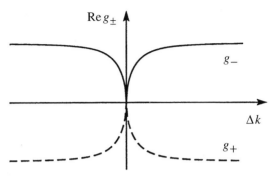

FIGURE 10.4.1   Dependence of the coupled gain on the wavevector mismatch.

If we now introduce the relations (10.4.33) into the expression (10.4.9) for the coupled gain coefficient, we find the gain eigenvalues are given by

$$g_\pm = -\tfrac{1}{2}\alpha_S\left(1 - \frac{n_S\omega_a}{n_a\omega_S}\right)$$

$$\pm\frac{1}{2}\left\{\left[\alpha_S\left(1 + \frac{n_S\omega_a}{n_a\omega_S}\right) + i\,\Delta k\right]^2 - 4\alpha_S^2\frac{n_S\omega_a}{n_a\omega_S}\right\}^{1/2}.$$

(10.4.36)

It is usually an extremely good approximation to set the factor $n_S\omega_L/n_a\omega_S$ equal to unity. In this case Eq. (10.4.36) simplifies to

$$g_\pm = \pm[i\alpha_S\Delta k - (\Delta k/2)^2]^{1/2}.$$

(10.4.37)

The dependence of $g_\pm$ on the phase mismatch is shown graphically in Fig. 10.4.1.* Equation (10.4.37) leads to the perhaps surprising result that the coupled gain $g_\pm$ vanishes in the limit of perfect phase matching. The reason for this behavior is that, for sufficiently small Δk, the anti-Stokes wave (which normally experiences loss) is so strongly coupled to the Stokes wave (which normally experiences gain) that it prevents the Stokes wave from growing exponentially.

It is also instructive to study the expression (10.4.37) for the coupled gain in the limit in which $|\Delta k|$ is very large. For $|\Delta k| \gg |\alpha_S|$, Eq. (10.4.37) becomes

$$g_\pm = \pm i\frac{\Delta k}{2}\left(1 - \frac{4i\alpha_S}{\Delta k}\right) \simeq \pm\left(\alpha_S + \tfrac{1}{2}i\,\Delta k\right).$$

(10.4.38)

---

* The graph has the same visual appearance whether the approximate form (10.4.37) or the exact form (10.4.36) is plotted.

Through use of Eq. (10.4.14), we find that the ratio of sidemode amplitudes associated with each of these gain eigenvalues is given by

$$\frac{A_a^{+*}}{A_S^+} = -2 - i\frac{\Delta k}{\alpha_S} \simeq i\frac{\Delta k}{\alpha_S}, \qquad (10.4.39a)$$

$$\frac{A_a^{-*}}{A_S^-} = 0. \qquad (10.4.39b)$$

Since we have assumed that $|\Delta k|$ is much larger than $|\alpha_S|$, we see that the + mode is primarily anti-Stokes, whereas the − mode is primarily Stokes.*

Let us now examine more carefully the nature of the decreased gain that occurs near $\Delta k = 0$. By setting $\Delta k = 0$ in the exact expression (10.4.36) for the coupled gain, we find that the gain eigenvalues become

$$g_+ = 0, \qquad g_- = -\alpha_S\left(1 - \frac{n_S\omega_a}{n_a\omega_S}\right). \qquad (10.4.40)$$

Note that $|g_-|$ is much smaller than $|\alpha_S|$ but does not vanish identically. Note also that with the sign convention used here at resonance g_- is a negative quantity. We find from Eq. (10.4.14) that to good approximation

$$\frac{A_a^{\pm*}}{A_S^{\pm}} = -1; \qquad (10.4.41)$$

thus each eigensolution is seen to be an equal combination of Stokes and anti-Stokes components, as mentioned in our discussion of Fig. 10.4.1.

Next, let us consider the spatial evolution of the field amplitudes under the assumptions that $\Delta k = 0$ and that their values are known at $z = 0$. We find from Eqs. (10.4.22) and (10.4.23) that

$$A_S(z) = \frac{-1}{1 - n_S\omega_a/n_a\omega_S}\left\{\left[A_a^*(0)e^{2i\phi_L} + \frac{n_S\omega_a}{n_a\omega_S}A_S(0)\right]\right.$$
$$\left. -[A_a^*(0)e^{2i\phi_L} + A_S(0)]e^{g_-z}\right\}, \qquad (10.4.42a)$$

$$A_a^*(z) = \frac{1}{1 - n_S\omega_a/n_a\omega_S}\left\{\left[A_a^*(0) + \frac{n_S\omega_a}{n_a\omega_S}A_S(0)e^{-2i\phi_L}\right]\right.$$
$$\left. -\frac{n_S\omega_a}{n_a\omega_S}[A_a^*(0) + A_S(0)e^{-2i\phi_L}]e^{g_-z}\right\}. \qquad (10.4.42b)$$

---

* Recall that at resonance α_S is real and *negative*; hence $g_- = -\alpha_S - \frac{1}{2}i\,\Delta k$ has a positive real part and leads to amplification.

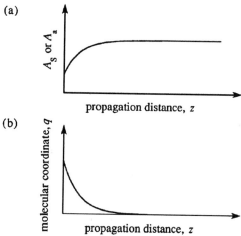

FIGURE 10.4.2 Nature of Raman amplification for the case of perfect phase matching $(\Delta k = 0)$.

Note that, since g_- is negative, the second term in each expression experiences exponential decay, and as $z \rightarrow \infty$ the field amplitudes approach the asymptotic values

$$A_S(z \rightarrow \infty) = \frac{-1}{1 - n_S\omega_a/n_a\omega_S} \left[A_a^*(0)e^{2i\phi_L} + \frac{n_S\omega_a}{n_a\omega_S} A_S(0) \right], \quad (10.4.43\text{a})$$

$$A_a^*(z \rightarrow \infty) = \frac{1}{1 - n_S\omega_a/n_a\omega_S} \left[A_a^*(0) + \frac{n_S\omega_a}{n_a\omega_S} A_S(0)e^{-2i\phi_L} \right]. \quad (10.4.43\text{b})$$

Note that each field is amplified by the factor $(1 - n_S\omega_a/n_a\omega_S)^{-1}$. The nature of this amplification is illustrated in part (a) of Fig. 10.4.2. We see that after propagating through a distance of several times $1/g_-$, the field amplitudes attain constant values and no longer change with propagation distance.

To see why field amplitudes remain constant, it is instructive to consider the nature of the molecule vibration in the simultaneous presence of the laser, Stokes, and anti-Stokes fields, that is, in the field

$$\tilde{E}(z, t) = A_L e^{i(k_L z - \omega_L t)} + A_S e^{i(k_S z - \omega_S t)} + A_a e^{i(k_a z - \omega_a t)} + \text{c.c.}, \quad (10.4.44)$$

where $k_L - k_S = k_a - k_L \equiv K$ and $\omega_L - \omega_S = \omega_a - \omega_L \equiv \Omega$. The solution to the equation of motion (10.3.1) for the molecular vibration with the force term given by Eqs. (10.3.6) and (10.4.44) is given by

$$\tilde{q}(z, t) = q(\Omega)e^{i(Kz - \Omega t)} + \text{c.c.},$$

where

$$g(\Omega) = \frac{(1/m)(\partial\alpha/\partial q)_0(A_L A_S^* + A_a A_L^*)}{\omega_v^2 - \Omega^2 - 2i\Omega\gamma}.$$ (10.4.45)

We can see from Eqs. (10.4.43) that, once the field amplitudes have attained their asymptotic values, the combination $A_L A_S^* + A_a A_L^*$ vanishes, implying that the amplitude $q(\omega)$ of the molecular vibration also vanishes asymptotically, as illustrated in part (b) of Fig. 10.4.2.

10.5. Stimulated Rayleigh-Wing Scattering

Stimulated Rayleigh-wing scattering is the light-scattering process that results from the tendency of anisotropic molecules to become aligned along the electric field vector of an optical wave. Stimulated Rayleigh-wing scattering was described theoretically by Bloembergen and Lallemand (1966) and by Chiao *et al.* (1966), and was observed experimentally by Mash *et al.* (1965) and Cho *et al.* (1967). Other early studies were conducted by Denariez and Bret (1968) and by Foltz *et al.* (1968).

The molecular orientation effect was described in Section 4.4 for the case in which the applied optical field $\tilde{E}(t)$ contains a single frequency component, and it was found that the average molecular polarizability is modified by the presence of the applied field. The molecular polarizability can be expressed as

$$\langle\alpha\rangle = \alpha_0 + \alpha_{\mathrm{NL}},$$ (10.5.1)

where the usual, weak-field polarizability is given by

$$\alpha_0 = \tfrac{1}{3}\alpha_\| + \tfrac{2}{3}\alpha_\perp,$$ (10.5.2)

where $\alpha_\|$ and $\alpha_\perp$ denote the polarizabilities measured parallel to the perpendicular to the symmetry axis of the molecule, respectively (see Fig. 10.5.1). In addition, the lowest-order nonlinear contribution to the polarizability is given by

$$\alpha_{\mathrm{NL}} = \bar{\alpha}_2 \langle \tilde{E}^2 \rangle,$$ (10.5.3)

where

$$\bar{\alpha}_2 = \frac{8\pi}{45} \frac{(\alpha_\| - \alpha_\perp)^2}{kT}.$$ (10.5.4)

In order to describe stimulated Rayleigh-wing scattering, we need to determine the response of the molecular system to an optical field that contains both

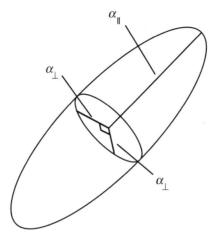

FIGURE 10.5.1 Illustration of the polarizabilities of an anisotropic molecule for the case $\alpha_\parallel > \alpha_\perp$.

laser and Stokes components, which we describe by the equation

$$\tilde{E}(\mathbf{r}, t) = A_L e^{i(k_L z - \omega_L t)} + A_S e^{i(-k_S z - \omega_S t)} + \text{c.c.} \qquad (10.5.5)$$

For the present, we assume that the laser and Stokes waves are linearly polarized in the same direction and are counterpropagating. The analysis for the case in which the waves have arbitrary polarization and/or are copropagating is somewhat more involved and is discussed briefly below.

Since the intensity, which is proportional to $\langle \tilde{E}^2 \rangle$, now contains a component at the beat frequency $\omega_L - \omega_S$, the nonlinear contribution to the mean polarizability $\langle \alpha \rangle$ is no longer given by Eq. (10.5.3), which was derived for the case of a monochromatic field. We assume that, in general, α_{NL} is described by the equation

$$\tau \frac{d\alpha_{\mathrm{NL}}}{dt} + \alpha_{\mathrm{NL}} = \bar{\alpha}_2 \overline{\tilde{E}^2}. \qquad (10.5.6)$$

In this equation τ represents the molecular orientation relaxation time and is the characteristic response time of the SRWS process; see Table 10.5.1 for typical values of τ. Equation (10.5.6) has the form of a Debye relaxation equation; recall that we have studied equations of this sort in our general discussion of two-beam coupling in Section 7.4.

If Eq. (10.5.6) is solved in steady state with $\tilde{E}(t)$ given by Eq. (10.5.5), we find that the nonlinear contribution to the polarizability of a molecule located

TABLE 10.5.1   Properties of SRWS for several materials

Substance	G (cm/GW)	τ (psec)	$\Delta\nu = 1/2\pi\tau$ (GHz)
CS_2	3	2	80
Nitrobenzene	3	48	3.3
Bromobenzene	1.4	15	10
Chlorobenzene	1.0	8	20
Toluene	1.0	2	80
Benzene	0.6	3	53

at position z is given by

$$\alpha_{NL}(z, t) = 2\bar{\alpha}_2(A_L A_L^* + A_S A_S^*) + \left(\frac{2\bar{\alpha}_2 A_L A_S^* e^{iqz - \Omega t}}{1 - i\Omega\tau} \right) + \text{c.c.}, \quad (10.5.7)$$

where we have introduced the wavevector magnitude q and frequency Ω associated with the material excitation, which are given by

$$q = k_L + k_S, \qquad \Omega = \omega_L - \omega_S. \quad (10.5.8)$$

Note that because the denominator of the second term in the expression for $\alpha_{NL}(z, t)$ is a complex quantity, the nonlinear response will in general be shifted in phase with respect to the intensity distribution associated with the interference of the laser and Stokes fields. We shall see below that this phase shift is the origin of the gain of the stimulated Rayleigh-wing scattering process.

We next derive the equation describing the propagation of the Stokes field. This derivation is formally identical to that presented in Section 7.4 in our general discussion of two-beam coupling. To apply that treatment to the present case, we need to determine the values of the refractive indices n_0 and n_2 that are relevant to the problem at hand. We find that n_0 is obtained from the usual Lorentz–Lorenz law (see also Eq. (3.8.8a)) as

$$\frac{n_0^2 - 1}{n_0^2 + 2} = \frac{4\pi}{3} n\alpha_0 \quad (10.5.9a)$$

and that the nonlinear refractive index is given (see also Eqs. (4.1.18) and (4.4.26)) by

$$n_2 = \left(\frac{n_0^2 + 2}{3} \right)^4 \frac{2\pi}{n_0^2 c} \bar{\alpha}_2 N. \quad (10.5.9b)$$

Then, as in Eq. (7.4.15), we find that the spatial evolution of the Stokes wave is described by

$$\frac{dA_S}{dz} = \frac{i n_0 n_2 \omega_S}{2\pi}(A_L A_L^* + A_S A_S^*)A_S + \frac{i n_0 n_2 \omega_S}{2\pi}\frac{A_L A_L^* A_S}{1 + i\Omega\tau}. \qquad (10.5.10)$$

Here the first term on the right-hand side leads to a spatial variation of the phase of the Stokes, whereas the second term leads to both a phase variation and to amplification of the Stokes wave. The gain associated with stimulated Rayleigh-wing scattering can be seen more clearly in terms of the equation relating the intensities of the two waves, which are defined by

$$I_j = \frac{n_0 c}{2\pi}|A_j|^2, \qquad j = L, S. \qquad (10.5.11)$$

The spatial variation of the intensity of the Stokes wave is therefore described by

$$\frac{dI_S}{dz} = \frac{n_0 c}{2\pi}\left[A_S \frac{dA_S^*}{dz} + A_S^* \frac{dA_S}{dz}\right]. \qquad (10.5.12)$$

Through use of Eq. (10.5.10), we can write this result as

$$\frac{dI_S}{dz} = g_{RW} I_L I_S, \qquad (10.5.13)$$

where we have introduced the gain factor g_{RW} for stimated Raleigh-wing scattering, which is given by

$$g_{RW} = g_{RW}^{(max)}\left(\frac{2\Omega\tau}{1 + \Omega^2\tau^2}\right), \qquad (10.5.14a)$$

where $g_{RW}^{(max)}$ denotes the maximum value of the gain factor, which is given by

$$g_{RW}^{(max)} = \frac{n_2 \omega_S}{c} = \left(\frac{n_0^2 + 2}{3}\right)^4 \frac{16\pi^2 \omega_S N(\alpha_\parallel - \alpha_\perp)^2}{45 k T n_0^2 c^2}. \qquad (10.5.14b)$$

We have made use of Eqs. (10.5.4) and (10.5.9b) in obtaining the second form of the expression for $g_{RW}^{(max)}$.

The frequency dependence of the gain factor for stimulated Rayleigh-wing scattering as predicted by Eq. (10.5.14a) is illustrated in Fig. 10.5.2. We see that amplification of the ω_S wave occurs for $\omega_S < \omega_L$ and that attenuation occurs for $\omega_S > \omega_L$. The maximum gain occurs when $\Omega \equiv \omega_L - \omega_S$ is equal to $1/\tau$.

The nature of the stimulated Rayleigh-wing scattering process is illustrated schematically in Fig. 10.5.3. The interference of the forward-going wave of frequency ω_L and wavevector magnitude k_L and the backward-going wave

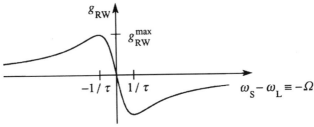

FIGURE 10.5.2   Frequency dependence of the gain factor for stimulated Rayleigh-wing scattering.

of frequency ω_S and wavevector magnitude k_S produces a fringe pattern that moves slowly through the medium in the forward direction with phase velocity $v = \Omega/q$. The tendency of the molecules to become aligned along the electric field vector of the total optical wave leads to planes of maximum molecular alignment alternating with planes of minimum molecular alignment. As mentioned above, these planes are shifted in phase with respect to the maxima and minima of the intensity distributions. The scattering of the laser field from this periodic array of aligned molecules leads to the generation of the Stokes wave. The scattered radiation is shifted to lower frequencies because the material disturbance causing the scattering is moving in the forward direction. The scattering process shows gain because the generation of Stokes radiation tends to reinforce the modulated portion of the interference pattern, which leads to increased molecular alignment and thus to increased scattering of Stokes radiation.

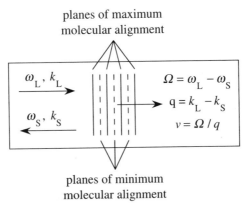

FIGURE 10.5.3   Nature of stimulated Rayleigh-wing scattering.

Polarization Properties of Stimulated Rayleigh-Wing Scattering

A theoretical analysis of the polarization properties of stimulated Rayleigh-wing scattering has been conducted by Chiao and Godine (1969). The details of their analysis are quite complicated; here we shall simply quote some of their principal results.

In order to treat the polarization properties of stimulated Rayleigh-wing scattering, one must consider the tensor properties of the material response. The analysis of Chiao and Godine presupposes that the nonlinear contribution to the susceptibility obeys the equation of motion

$$\tau \frac{d}{dt} \Delta \chi_{ik} + \Delta \chi_{ik} = C \left(\langle \tilde{E}_i \tilde{E}_k \rangle - \tfrac{1}{3} \delta_{ik} \langle \tilde{\mathbf{E}} \cdot \tilde{\mathbf{E}} \rangle \right), \qquad (10.5.15)$$

where, ignoring for the present local-field corrections, the proportionality constant C is given by

$$C = \frac{N(\alpha_\parallel - \alpha_\perp)^2}{15kT}. \qquad (10.5.16)$$

Note that the trace of the right-hand side of Eq. (10.5.15) vanishes, as required by the fact that Rayleigh-wing scattering is described by a traceless, symmetric permittivity tensor.

By requiring that the Stokes wave obey the wave equation with a susceptibility given by the solution to Eq. (10.5.15), and taking account of rotation of the pump laser polarization (see, for example, the discussion in Section 4.2), Chiao and Godine calculate the gain factor for stimulated Rayleigh-wing scattering for arbitrary polarization of the laser and Stokes fields. Some of their results for special polarization cases are summarized in Table 10.5.2.

For any state of polarization of the pump wave, some particular polarization of the Stokes wave will experience maximum gain. As a consequence of the large value of the gain required to observe stimulated light scattering ($g_{RW} I_L L \simeq 25$), the light generated by stimulated Rayleigh-wing scattering will have a polarization that is nearly equal to that for which the gain is maximum. The relation between the laser polarization and the Stokes polarization for which the gain is maximum is illustrated in Table 10.5.3. Note that the generated wave will be nearly, but not exactly, the polarization conjugate (in the sense of vector phase conjugation, as discussed in Section 7.2) of the incident laser wave. In particular, the polarization ellipse of the generated wave will be rounder and tilted with respect to that of the laser wave.

Zel'dovich and Yakovleva (1980) have studied theoretically the polarization properties of stimulated Rayleigh-wing scattering for the case in which

TABLE 10.5.2 Dependence of the gain factor for stimulated Rayleigh-wing scattering in the backward direction on the polarization of the laser and Stokes waves for the cases of linear and circular polarization[a]

Laser polarization	$\updownarrow$	$\updownarrow$	◯	◯
Stokes polarization	$\updownarrow$	↔	◯	◯
Gain factor	1	3/4	3/2	1/6

[a] The arrows on the circles denote the direction in which the electric field vector rotates in time at a fixed position in space. The gain factors are given relative to that given by Eqs. (10.5.14) for the case of linear and parallel polarization.

the pump radiation is partially polarized. They predict that essentially perfect vector phase conjugation can be obtained by stimulated Rayleigh-wing scattering for the case in which the pump radiation is completely depolarized in the sense that the state of polarization varies randomly over the transverse dimensions of the laser beam. The wavefront-reconstructing properties of stimulated Rayleigh-wing scattering have been studied experimentally by Kudriavtseva *et al.* (1978), and the vector phase conjugation properties have been studied experimentally by Miller *et al.* (1990).

The analysis of stimulated Rayleigh-wing scattering in the forward and near-forward direction is much more complicated than that of backward stimulated Rayleigh-wing scattering because the possibility of Stokes–anti-Stokes coupling (as described in Section 10.4 for stimulated Raman scattering) must be included in the analysis. This situation has been described by Chiao *et al.* (1966) and by Chiao and Godine (1969).

TABLE 10.5.3 Relation between laser polarization and the Stokes polarization experiencing maximum gain in backward stimulated Rayleigh-wing scattering

Laser	$\updownarrow$	◯	◯
Stokes	$\updownarrow$	◯	◯

Problems

1. *Estimation of the properties of stimulated Raman scattering.* By making reasonable assumptions regarding the value of the parameter $(d\alpha/dq)$, perform an order-of-magnitude estimate of the gain factor for stimulated Raman scattering for condensed matter, and compare this value with the measured values given in Table 10.2.1.

2. *Polarization properties of stimulated Rayleigh-wing scattering.* By carrying out the prescription described in the first full paragraph following Eq. (10.5.16), verify that the entries in Table 10.5.2 are correct.

References

Stimulated Raman Scattering

N. Bloembergen and Y. R. Shen, *Phys. Rev. Lett.* **12**, 504 (1964).

N. Bloembergen, *Am. J. Phys.* **35**, 989 (1967).

R. W. Boyd, *Radiometry and the Detection of Optical Radiation*, Wiley, New York, 1983.

G. Eckhardt, R. W. Hellwarth, F. J. McClung, S. E. Schwarz, D. Weiner, and E. J. Woodbury, *Phys. Rev. Lett.* **9** (1962).

E. Garmire, F. Pandarese, and C. H. Townes, *Phys. Rev. Lett.* **11**, 160 (1963).

R. W. Hellwarth, *Phys. Rev.* **130**, 1850 (1963).

W. Kaiser and M. Maier, in *Laser Handbook* (F. T. Arecchi and E. O. Schulz-DuBois, eds.), North-Holland, 1972.

A. Penzkofer, A. Laubereau, and W. Kaiser, *Prog. Quantum Electron.* **6**, 55 (1979).

M. G. Raymer and J. Mostowski, *Phys. Rev.* **A24**, 1980 (1981).

M. G. Raymer and I. A. Walmsley, in *Progress in Optics*, Vol. 28 (E. Wolf, ed.), North-Holland, Amsterdam, 1990.

Y. R. Shen and N. Bloembergen, *Phys. Rev.* **137**, 1787 (1965).

U. Simon and F. K. Tittel, in *Methods of Experimental Physics,* Vol. III (R. G. Hulet and F. B. Dunning, eds.), Academic Press, San Diego, 1994.

R. W. Terhune, *Bull. Am. Phys. Soc.* **8**, 359 (1963).

E. J. Woodbury and W. K. Ng, *Proc. I.R.E.* **50**, 2367 (1962).

Stimulated Rayleigh-Wing Scattering

N. Bloembergen and P. Lallemand, *Phys. Rev. Lett.* **16**, 81 (1966).

R. Y. Chiao and J. Godine, *Phys. Rev.* **185**, 430 (1969).

R. Y. Chiao, P. L. Kelley, and E. Garmire, *Phys. Rev. Lett.* **17**, 1158 (1966).

C. W. Cho, N. D. Foltz, D. H. Rank, and T. A. Wiggins, *Phys. Rev. Lett.* **18**, 107 (1967).

M. Denariez and G. Bret, *Phys. Rev.* **177**, 171 (1968).

N. D. Foltz, C. W. Cho, D. H. Rank, and T. A. Wiggins, *Phys. Rev.* 165, 396 (1968).

A. D. Kudriavtseva, A. I. Sokolovskaia, J. Gazengel, N. Phu Xuan, and G. Rivore, *Opt. Commun.* 26, 446 (1978).

D. I. Mash, V. V. Morozov, V. S. Starunov, and I. L. Fabelinskii, *JETP Lett.* 2, 25 (1965).

E. J. Miller, M. S. Malcuit, and R. W. Boyd, *Opt. Lett.* 15, 1189 (1990).

B. Ya. Zel'dovich and T. V. Yakovleva, *Sov. J. Quantum Electron.* 10, 501 (1980).

Chapter 11

The Electrooptic and Photorefractive Effects

11.1. Introduction to the Electrooptic Effect

The electrooptic effect is the change in refractive index of a material induced by the presence of a dc (or low-frequency) electric field.

In some materials, the change in refractive index depends linearly on the strength of the applied electric field. This change is known as the linear electrooptic effect or Pockels effect. The linear electrooptic effect can be described in terms of a nonlinear polarization given by

$$P_i(\omega) = 2 \sum_{jk} \chi_{ijk}^{(2)}(\omega = \omega + 0) E_j(\omega) E_k(0). \qquad (11.1.1)$$

Since the linear electrooptic effect can be described by a second-order nonlinear susceptibility, it follows from the general discussion of Section 1.5 that a linear electrooptic effect can occur only for materials that are noncentrosymmetric. Although the linear electrooptic effect can be described in terms of a second-order nonlinear susceptibility, a very different mathematical formalism has historically been used to describe the electrooptic effect; this formalism is described in Section 11.2 of this chapter.

In centrosymmetric materials (such as liquids and glasses), the lowest-order change in the refractive index depends quadratically on the strength of the applied dc (or low-frequency) field. This effect is known as the Kerr electrooptic effect* or as the quadratic electrooptic effect. It can be described in terms of

---

* The quadratic electrooptic effect is often referred to simply as the Kerr effect. More precisely, it is called the Kerr electrooptic effect to distinguish it from the Kerr magnetooptic effect.

a nonlinear polarization given by

$$P_i(\omega) = 3 \sum_{jkl} \chi_{ijkl}^{(3)}(\omega = \omega + 0 + 0) E_j(\omega) E_k(0) E_l(0). \quad (11.1.2)$$

11.2. Linear Electrooptic Effect

In this section we develop a mathematical formalism that describes the linear electrooptic effect. In an anisotropic material, the constitutive relation between the field vectors $\mathbf{D}$ and $\mathbf{E}$ has the form

$$D_i = \sum_j \epsilon_{ij} E_j \quad (11.2.1a)$$

or explicitly,

$$\begin{bmatrix} D_x \\ D_y \\ D_z \end{bmatrix} = \begin{bmatrix} \epsilon_{xx} & \epsilon_{xy} & \epsilon_{xz} \\ \epsilon_{yx} & \epsilon_{yy} & \epsilon_{yz} \\ \epsilon_{zx} & \epsilon_{zy} & \epsilon_{zz} \end{bmatrix} \begin{bmatrix} E_x \\ E_y \\ E_z \end{bmatrix}. \quad (11.2.1b)$$

For a lossless, non–optically active material, the dielectric permeability tensor ϵ_{ij} is represented by a real symmetric matrix, which therefore has six independent elements, i.e., $\epsilon_{xx}, \epsilon_{yy}, \epsilon_{zz}, \epsilon_{xy} = \epsilon_{yx}, \epsilon_{xz} = \epsilon_{zx}$, and $\epsilon_{yz} = \epsilon_{zy}$. A general mathematical result states that any real, symmetric matrix can be expressed in diagonal form by means of an orthogonal transformation. Physically, this result implies that there exists some new coordinate system (X, Y, Z), related to the coordinate system x, y, z of Eq. (11.2.1b) by rotation of the coordinate axes, in which Eq. (11.2.1b) has the much simpler form

$$\begin{bmatrix} D_X \\ D_Y \\ D_Z \end{bmatrix} = \begin{bmatrix} \epsilon_{XX} & 0 & 0 \\ 0 & \epsilon_{YY} & 0 \\ 0 & 0 & \epsilon_{ZZ} \end{bmatrix} \begin{bmatrix} E_X \\ E_Y \\ E_Z \end{bmatrix}. \quad (11.2.2)$$

This new coordinate system is known as the principal-axis system, because in it the dielectric tensor is represented as a diagonal matrix.

We next consider the energy density per unit volume,

$$U = \frac{1}{8\pi} \mathbf{D} \cdot \mathbf{E} = \frac{1}{8\pi} \sum_{ij} \epsilon_{ij} E_i E_j, \quad (11.2.3)$$

associated with a wave propagating through the anisotropic medium. In the principal-axis coordinate system, the energy density can be represented as

$$U = \frac{1}{8\pi} \left[\frac{D_X^2}{\epsilon_{XX}} + \frac{D_Y^2}{\epsilon_{YY}} + \frac{D_Z^2}{\epsilon_{ZZ}} \right]. \quad (11.2.4)$$

This result shows that the surfaces of constant energy density in $\mathbf{D}$ space are ellipsoids. The shape of these ellipsoids can be described in terms of the coordinates (X, Y, Z) themselves. If we let

$$X = \left(\frac{1}{8\pi U}\right)^{1/2} D_X, \qquad Y = \left(\frac{1}{8\pi U}\right)^{1/2} D_Y, \qquad Z = \left(\frac{1}{8\pi U}\right)^{1/2} D_Z,$$

$$(11.2.5)$$

Eq. (11.2.4) becomes

$$\frac{X^2}{\epsilon_{XX}} + \frac{Y^2}{\epsilon_{YY}} + \frac{Z^2}{\epsilon_{ZZ}} = 1. \tag{11.2.6}$$

The surface described by this equation is known as the *optical indicatrix* or as the *index ellipsoid*. The equation describing the index ellipsoid takes on its simplest form in the principal-axis system; in other coordinate systems it is given by the general expression for an ellipsoid, which we write in the form

$$\left(\frac{1}{n^2}\right)_1 x^2 + \left(\frac{1}{n^2}\right)_2 y^2 + \left(\frac{1}{n^2}\right)_3 z^2 + 2\left(\frac{1}{n^2}\right)_4 yz$$

$$+ 2\left(\frac{1}{n^2}\right)_5 xz + 2\left(\frac{1}{n^2}\right)_6 xy = 1.$$

$$(11.2.7)$$

The coefficients $(1/n^2)_i$ are optical constants that describe the optical indicatrix in the new coordinate system; they can be expressed in terms of the coefficients $\epsilon_{XX}, \epsilon_{YY}, \epsilon_{ZZ}$ by means of the standard transformation laws for coordinate transformations, but the exact nature of the relationship is of no interest for our present purposes.

The index ellipsoid can be used to describe the optical properties of an anisotropic material by means of the following procedure (Born and Wolf, 1975). For any given direction of propagation within the crystal, a plane perpendicular to the propagation vector and passing through the center of the ellipsoid is constructed. The curve formed by the intersection of this plane with the index ellipsoid forms an ellipse. The semimajor and semiminor axes of this ellipse give the two allowed values of the refractive index for this particular direction of propagation; the orientations of these axes give the polarization directions of the $\mathbf{D}$ vector associated with these refractive indices.

We next study how the optical indicatrix is modified when the material system is subjected to a constant or low-frequency electric field. This modification is conveniently described in terms of the *impermeability tensor* η_{ij}, which is defined by the relation

$$E_i = \sum_j \eta_{ij} D_j. \tag{11.2.8}$$

Note that this relation is the inverse of that given by Eq. (11.2.1a), and hence that η_{ij} is the matrix inverse of ϵ_{ij}, that is, that $\eta_{ij} = (\epsilon^{-1})_{ij}$. We can express the optical indicatrix in terms of the elements of the impermeability tensor by noting that the energy density is equal to $U = (1/8\pi) \sum_{ij} \eta_{ij} D_i D_j$. If we now define coordinates x, y, z by means of relations $x = D_x/(8\pi U)^{1/2}$, etc., we find that the expression for U as a function of $\mathbf{D}$ becomes

$$1 = \eta_{11}x^2 + \eta_{22}y^2 + \eta_{33}z^2 + 2\eta_{12}xy + 2\eta_{23}yz + 2\eta_{13}xz. \quad (11.2.9)$$

By comparison of this expression for the optical indicatrix with that given by Eq. (11.2.7), we find that

$$\left(\frac{1}{n^2}\right)_1 = \eta_{11}, \qquad \left(\frac{1}{n^2}\right)_2 = \eta_{22}, \qquad \left(\frac{1}{n^2}\right)_3 = \eta_{33},$$

$$\left(\frac{1}{n^2}\right)_4 = \eta_{23} = \eta_{32}, \qquad \left(\frac{1}{n^2}\right)_5 = \eta_{13} = \eta_{31}, \qquad \left(\frac{1}{n^2}\right)_6 = \eta_{12} = \eta_{21}. \quad (11.2.10)$$

We next assume that η_{ij} can be expressed as a power series in the strength of the components E_k of the applied electric field as

$$\eta_{ij} = \eta_{ij}^{(0)} + \sum_k r_{ijk} E_k + \sum_{kl} s_{ijkl} E_k E_l + \cdots. \quad (11.2.11)$$

Here r_{ijk} is the tensor that describes the linear electrooptic effect, s_{ijkl} is the tensor that describes the quadratic electrooptic effect, etc. Since the dielectric permeability tensor ϵ_{ij} is real and symmetric, its inverse η_{ij} must also be real and symmetric, and consequently the electrooptic tensor r_{ijk} must be symmetric in its first two indices. For this reason, it is often convenient to represent the third-rank tensor r_{ijk} as a two-dimensional matrix r_{hk} using contracted notation according to the prescription

$$h = \begin{cases} 1 & \text{for} & ij = 11, \\ 2 & \text{for} & ij = 22, \\ 3 & \text{for} & ij = 33, \\ 4 & \text{for} & ij = 23 \text{ or } 32, \\ 5 & \text{for} & ij = 13 \text{ or } 31, \\ 6 & \text{for} & ij = 12 \text{ or } 21. \end{cases} \quad (11.2.12)$$

In terms of this contracted notation, we can express the lowest-order modification of the optical constants $(1/n^2)_i$ that appears in expression (11.2.7) for the optical indicatrix as

$$\Delta\left(\frac{1}{n^2}\right)_i = \sum_j r_{ij} E_j, \quad (11.2.13a)$$

where we have made use of Eqs. (11.2.10) and (11.2.11). This relationship can be written explicitly as

$$
\begin{bmatrix}
\Delta(1/n^2)_1 \\
\Delta(1/n^2)_2 \\
\Delta(1/n^2)_3 \\
\Delta(1/n^2)_4 \\
\Delta(1/n^2)_5 \\
\Delta(1/n^2)_6
\end{bmatrix}
=
\begin{bmatrix}
r_{11} & r_{12} & r_{13} \\
r_{21} & r_{22} & r_{23} \\
r_{31} & r_{32} & r_{33} \\
r_{41} & r_{42} & r_{43} \\
r_{51} & r_{52} & r_{53} \\
r_{61} & r_{62} & r_{63}
\end{bmatrix}
\begin{bmatrix}
E_x \\
E_y \\
E_z
\end{bmatrix}.
\tag{11.2.13b}
$$

The quantities r_{ij} are known as the electrooptic coefficients and give the rate at which the coefficients $(1/n^2)_i$ change with increasing electric field strength.

We remarked earlier that the linear electrooptic effect vanishes for materials possessing inversion symmetry. Even for materials lacking inversion symmetry, where the coefficients do not necessarily vanish, the form of r_{ij} is restricted by any rotational symmetry properties that the material may possess. For example, for any material (such as ADP and KDP) possessing the point group symmetry $\bar{4}2m$, the electrooptic coefficients must be of the form

$$
r_{ij} =
\begin{bmatrix}
0 & 0 & 0 \\
0 & 0 & 0 \\
0 & 0 & 0 \\
r_{41} & 0 & 0 \\
0 & r_{41} & 0 \\
0 & 0 & r_{63}
\end{bmatrix}
\quad \text{(for class } \bar{4}2m\text{),}
\tag{11.2.14}
$$

where we have expressed r_{ij} in the standard crystallographic coordinate system, in which the Z direction represents the optic axis of the crystal. We see from Eq. (11.2.14) that the form of the symmetry properties of the point group $\bar{4}2m$ requires 15 of the electrooptic coefficients to vanish and two of the remaining coefficients to be equal. Hence r_{ij} possesses only two independent elements in this case.

Similarly, the electrooptic coefficients of crystals of class $3m$ (such as lithium niobate) must be of the form

$$
r_{ij} =
\begin{bmatrix}
0 & -r_{22} & r_{13} \\
0 & r_{22} & r_{13} \\
0 & 0 & r_{33} \\
0 & r_{42} & 0 \\
r_{42} & 0 & 0 \\
r_{22} & 0 & 0
\end{bmatrix}
\quad \text{(for class } 3m\text{),}
\tag{11.2.15}
$$

and the electrooptic coefficients of crystals of the class 4mm (such as barium titanate) must be of the form

$$r_{ij} = \begin{bmatrix} 0 & 0 & r_{13} \\ 0 & 0 & r_{13} \\ 0 & 0 & r_{33} \\ 0 & r_{42} & 0 \\ r_{42} & 0 & 0 \\ 0 & 0 & 0 \end{bmatrix} \qquad \text{(for class } 4mm\text{)}. \qquad (11.2.16)$$

The properties of several electrooptic materials are summarized in Table 11.2.1.

TABLE 11.2.1 Properties of several electrooptic materials[a]

Material	Point group	Electrooptic coefficients $(10^{-12}$ m/V$)$	Refractive index
Potassium dihydrogen phosphate, KH$_2$PO$_4$ (KDP)	$\bar{4}2m$	$r_{41} = 8.77$ $r_{63} = 10.5$	$n_0 = 1.514$ $n_e = 1.472$ (at 0.5461 μm)
Potassium dideuterium phosphate, KD$_2$PO$_4$ (KD*P)	$\bar{4}2m$	$r_{41} = 8.8$ $r_{63} = 26.4$	$n_0 = 1.508$ $n_e = 1.468$ (at 0.5461 μm)
Lithium niobate, LiNbO$_3$	$3m$	$r_{13} = 9.6$ $r_{22} = 6.8$ $r_{33} = 30.9$ $r_{42} = 32.6$	$n_0 = 2.3410$ $n_e = 2.2457$ (at 0.5 μm)
Lithium tantalate, LiTaO$_3$	$3m$	$r_{13} = 8.4$ $r_{22} = -0.2$ $r_{33} = 30.5$ $r_{51} = 20$	$n_0 = 2.176$ $n_e = 2.180$ (at 0.633 nm)
Barium titanate, BaTiO$_3$[b]	$4mm$	$r_{13} = 19.5$ $r_{33} = 97$ $r_{42} = 1640$	$n_0 = 2.488$ $n_e = 2.424$ (at 514 nm)
Strontium barium niobate, Sr$_{0.6}$Ba$_{0.4}$NbO$_6$ (SBN:60)	$4mm$	$r_{13} = 55$ $r_{33} = 224$ $r_{42} = 80$	$n_0 = 2.367$ $n_e = 2.337$ (at 514 nm)
Zinc telluride, ZnTe	$\bar{4}3m$	$r_{41} = 4.0$	$n_0 = 2.99$ (at 0.633 μm)

[a] From a variety of sources. See, for example, B. J. Thompson and E. Hartfield in *The Handbook of Optics* (W. G. Driscoll and W. Vaughan, eds.), McGraw-Hill, New York, 1978, and W. R. Cook, Jr. and H. Jaffe, "Electrooptic Coefficients," in *Landolt-Bornstein, New Series*, Vol. II (K.-H. Hellwege, ed.), Springer-Verlag, 1979, pp. 552–651. The electrooptic coefficients are given in the MKS units of m/V. To convert to the cgs units of cm/statvolt each entry should be multiplied by 3×10^4.

[b] $\epsilon_{dc}^{\parallel} = 135$, $\epsilon_{dc}^{\perp} = 3700$.

11.3. Electrooptic Modulators

As an example of the application of the formalism developed in the last section, we now consider how to construct an electrooptic modulator using the material potassium dihydrogen phosphate (KDP). Of course, the analysis is formally identical for any electrooptic material of point group $\bar{4}2m$.

KDP is a uniaxial crystal, and hence in the absence of an applied electric field the index ellipsoid is given in the standard crystallographic coordinate system by the equation

$$\frac{X^2}{n_0^2} + \frac{Y^2}{n_0^2} + \frac{Z^2}{n_e^2} = 1. \tag{11.3.1}$$

Note that this (X, Y, Z) coordinate system is the principal-axis coordinate system in the absence of an applied electric field. If an electric field is applied to crystal, the index ellipsoid becomes modified according to Eqs. (11.2.13b) and (11.2.14) and takes the form

$$\frac{X^2}{n_0^2} + \frac{Y^2}{n_0^2} + \frac{Z^2}{n_e^2} + 2r_{41}E_X YZ + 2r_{41}E_Y XZ + 2r_{63}E_Z XY = 1. \tag{11.3.2}$$

Note that (since cross terms containing YZ, XZ, and XY appear in this equation) the (X, Y, Z) coordinate system is not the principal-axis coordinate system when an electric field is applied to the crystal. Note also that the crystal will no longer necessarily be uniaxial in the presence of a dc electric field.

Let us now assume that the applied electric field has only a Z component, so that Eq. (11.3.2) reduces to

$$\frac{X^2}{n_0^2} + \frac{Y^2}{n_0^2} + \frac{Z^2}{n_e^2} + 2r_{63}E_Z XY = 1. \tag{11.3.3}$$

This special case is often encountered in device applications. The new principal-axis coordinate system can now be found by inspection. If we let

$$X = \frac{x - y}{\sqrt{2}}, \qquad Y = \frac{x + y}{\sqrt{2}}, \qquad Z = z, \tag{11.3.4}$$

we find that Eq. (11.3.3) becomes

$$\left(\frac{1}{n_0^2} + r_{63}E_z\right)x^2 + \left(\frac{1}{n_0^2} - r_{63}E_z\right)y^2 + \frac{z^2}{n_e^2} = 1, \tag{11.3.5}$$

which describes an ellipsoid in its principal-axis system. This ellipsoid can alternatively be written as

$$\frac{x^2}{n_x^2} + \frac{y^2}{n_y^2} + \frac{z^2}{n_e^2} = 1,$$ (11.3.6)

where, in the physically realistic limit $r_{63}E_z \ll 1$, the new principal values of the refractive index are given by

$$n_x = n_0 - \tfrac{1}{2}n_0^3 r_{63} E_z,$$ (11.3.7a)

$$n_y = n_0 + \tfrac{1}{2}n_0^3 r_{63} E_z.$$ (11.3.7b)

Figure 11.3.1 shows how to construct a modulator based on the electrooptic effect in KDP. Part (a) shows a crystal that has been cut so that the optic axis (Z axis) is perpendicular to the plane of the entrance face, which contains the X and Y crystalline axes. Part (b) of the figure shows the same crystal in the presence of a longitudinal (z-directed) electric field $E_z = V/L$, which is established by applying a voltage V between the front and rear faces. The

(a)   (b)

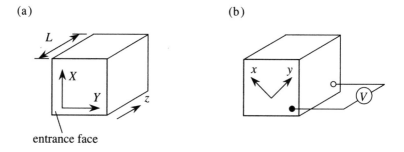

entrance face

(c)

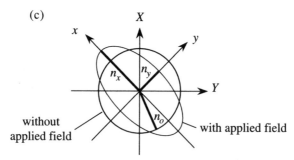

FIGURE 11.3.1 The electrooptic effect in KDP. (a) Principal axes in the absence of an applied field. (b) Principal axes in the presence of an applied field. (c) The intersection of the index ellipsoid with the plane $z = Z = 0$.

principal axes (x, y, z) of the index ellipsoid in the presence of this field are also indicated. In practice, the potential difference is applied by coating the front and rear faces with a thin film of a conductive coating. Historically, thin layers of gold have been used, although more recently the transparent conducting material indium tin oxide has successfully been used.

Part (c) of Fig. 11.3.1 shows the curve formed by the intersection of the plane perpendicular to the direction of propagation (i.e., the plane $z = Z = 0$) with the index ellipsoid. For the case in which no static field is applied, the curve has the form of a circle, showing that the refractive index has the value n_0 for any direction of polarization. For the case in which a field is applied, this curve has the form of an ellipse. In drawing the figure, we have arbitrarily assumed that the factor $r_{63} E_z$ is negative; consequently the semimajor and semiminor axes of this ellipse are along the x and y directions and have lengths n_x and n_y respectively.

Let us next consider a beam of light propagating in the $z = Z$ direction through the modulator crystal shown in Fig. 11.3.1. A wave polarized in the x direction propagates with a different phase velocity than a wave polarized in the y direction. In propagating through the length L of the modulator crystal, the x and y polarization components will thus acquire the phase difference

$$\Gamma = (n_y - n_x)\frac{\omega L}{c}, \tag{11.3.8}$$

which is known as the retardation. By introducing Eqs. (11.3.7) into this expression we find that

$$\Gamma = \frac{n_0^3 r_{63} E_z \omega L}{c}.$$

Since $E_z = V/L$, this result shows that the retardation introduced by a longitudinal electrooptic modulator depends only on the voltage V applied to the modulator and is independent of the length of the modulator. In particular, the retardation can be represented as

$$\Gamma = \frac{n_0^3 r_{63} \omega V}{c}. \tag{11.3.9}$$

It is convenient to express this result in terms of the quantity

$$V_{\lambda/2} = \frac{\pi c}{\omega n_0^3 r_{63}}, \tag{11.3.10}$$

which is known as the half-wave voltage. Eq. (11.3.9) then becomes

$$\Gamma = \pi \frac{V}{V_{\lambda/2}}. \tag{11.3.11}$$

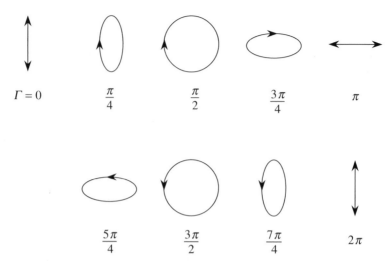

FIGURE 11.3.2   Polarization ellipses describing the light leaving the modulator of Fig. 11.3.1 for various values of the retardation. In all cases, the input light is linearly polarized in the vertical (X) direction.

Note that a half-wave (π radians) of retardation is introduced when the applied voltage is equal to the half-wave voltage. Half-wave voltages of typical electrooptic materials are of the order of 10 kV for visible light.

Since the x and y polarization components of a beam of light generally experience different phase shifts in propagating through an electrooptic crystal, the state of polarization of the light leaving the modulator will generally be different from that of the incident light. Figure 11.3.2 shows how the state of polarization of the light leaving the modular depends upon the value of the retardation Γ for the case in which vertically (X) polarized light is incident on the modulator. Note that light of any ellipticity can be produced by controlling the voltage V applied to the modulator.

Figure 11.3.3 shows one way of constructing an intensity modulator based on the configuration shown in Fig. 11.3.1. The incident light is passed through a

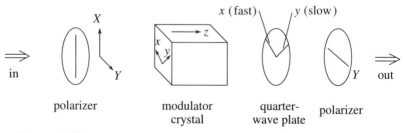

FIGURE 11.3.3   Construction of a voltage-controllable intensity modulator.

linear polarizer whose transmission axis is oriented in the X direction. The light then enters the modulator crystal, where its x and y polarization components propagate with different velocities and acquire a phase difference, whose value is given by Eq. (11.3.11). The light leaving the modulator then passes through a quarter-wave plate oriented so that its fast and slow axes coincide with the x and y axes of the modulator crystal, respectively. The beam of light thereby acquires the additional retardation $\Gamma_B = \pi/2$. For reasons that will become apparent later, Γ_B is called the bias retardation. The total retardation is then given by

$$\Gamma = \pi \frac{V}{V_{\lambda/2}} + \frac{\pi}{2}. \tag{11.3.12}$$

In order to analyze the operation of this modulator, let us represent the electric field of the incident radiation after passing through the initial polarizer as

$$\tilde{\mathbf{E}} = \mathbf{E}_{\text{in}} e^{-i\omega t} + \text{c.c.}, \tag{11.3.13a}$$

where

$$\mathbf{E}_{\text{in}} = E_{\text{in}} \hat{\mathbf{X}} = \frac{E_{\text{in}}}{\sqrt{2}} (\hat{\mathbf{x}} + \hat{\mathbf{y}}). \tag{11.3.13b}$$

After the beam passes through the modulator crystal and quarter-wave plate, the phase of the y polarization component will be shifted with respect to that of the x polarization component by an amount Γ, so that (to within an unimportant overall phase factor) the complex field amplitude becomes

$$\mathbf{E} = \frac{E_{\text{in}}}{\sqrt{2}} (\hat{\mathbf{x}} + e^{i\Gamma} \hat{\mathbf{y}}). \tag{11.3.14}$$

Only the $\hat{\mathbf{Y}} = (-\hat{\mathbf{x}} + \hat{\mathbf{y}})/\sqrt{2}$ component of this field will be transmitted by the final polarizer. The field amplitude measured after this polarizer is hence given by $\mathbf{E}_{\text{out}} = (\mathbf{E} \cdot \hat{\mathbf{Y}})\hat{\mathbf{Y}}$, or as

$$\mathbf{E}_{\text{out}} = \frac{E_{\text{in}}}{2} (-1 + e^{i\Gamma}) \hat{\mathbf{Y}}. \tag{11.3.15}$$

If we now define the transmission T of the modulator of Fig. 11.3.3 as

$$T = \frac{|\mathbf{E}_{\text{out}}|^2}{|\mathbf{E}_{\text{in}}|^2}, \tag{11.3.16}$$

we find through use of Eq. (11.3.15) that the transmission is given by

$$T = \sin^2(\Gamma/2). \tag{11.3.17}$$

The functional form of these transfer characteristics is shown in Fig. 11.3.4. We see that the transmission can be made to vary from zero to one by varying the total retardation between zero and π radians. We can also see the motivation for inserting the quarter-wave plate into the setup of Fig. 11.3.3 in order

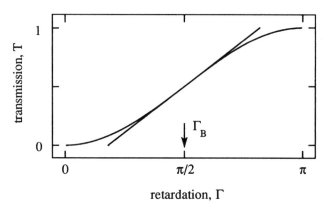

FIGURE 11.3.4   Transmission characteristics of the electrooptic modulator shown in Fig. 11.3.3.

to establish the bias retardation $\Gamma_B = \pi/2$. For the case in which the applied voltage V vanishes, the total retardation will be equal to the bias retardation, and the transmission of the modulator will be 50%. Since the transmission T varies approximately linearly with the retardation Γ for retardations near $\Gamma = \pi/2$, the transmission will vary nearly linearly with the value V of the applied voltage. For example, if the applied voltage is given by

$$V(t) = V_m \sin \omega_m t, \qquad (11.3.18)$$

the retardation will be given by

$$\Gamma = \frac{\pi}{2} + \frac{\pi V_m}{V_{\lambda/2}} \sin \omega_m t. \qquad (11.3.19)$$

The transmission predicted by Eq. (11.3.17) is hence given by

$$
\begin{aligned}
T &= \sin^2\left(\frac{\pi}{4} + \frac{\pi V_m}{2V_{\lambda/2}} \sin \omega_m t\right) \\
&= \frac{1}{2}\left[1 + \sin\left(\frac{\pi V_m}{V_{\lambda/2}} \sin \omega_m t\right)\right],
\end{aligned}
$$

which, for $\pi V_m / V_{\lambda/2} \ll 1$, becomes

$$T = \frac{1}{2}\left(1 + \frac{\pi V_m}{V_{\lambda/2}} \sin \omega_m t\right). \qquad (11.3.20)$$

The electrooptic effect can also be used to construct a phase modulator for light. For example, if the light incident on the electrooptic crystal of Fig. 11.3.1 is linearly polarized along the x (or the y) axis of the crystal, the light will propagate with its state of polarization unchanged but with its phase shifted by an amount that depends on the value of the applied voltage. The voltage-dependent part of the phase shift is hence given by

$$\phi = (n_x - n_0)\frac{\omega L}{c} = -\frac{n_0^3 r_{63} E_z \omega L}{2c} = \frac{n_0^3 r_{63} V \omega}{2c}. \qquad (11.3.21)$$

11.4.   Introduction to the Photorefractive Effect

The photorefractive effect is the change in refractive index of an optical material that results from the optically induced redistribution of electrons and holes. The photorefractive effect is quite different from most of the other nonlinear-optical effects described in this book in that it cannot be described by a nonlinear susceptibility $\chi^{(n)}$ for any value of n. The reason is that, under a wide range of conditions, the change in refractive index in steady state is independent of the intensity of the light that induces the change. Because the photorefractive effect cannot be described by means of a nonlinear susceptibility, special methods must be employed to describe it; these methods are described in the next several sections. The photorefractive effect tends to give rise to a strong optical nonlinearity; experiments are routinely performed using milliwatts of laser power. However, the effect tends to be rather slow, with response times of 0.1 sec being typical.

The origin of the photorefractive effect is illustrated schematically in Fig. 11.4.1. We imagine that a photorefractive crystal is illuminated by two intersecting beams of light of the same frequency. These beams interfere to produce the spatially modulated intensity distribution $I(x)$ shown in the upper graph. Free charge carriers, which we assume to be electrons, are generated through photoionization at a rate that is proportional to the local value of the optical intensity. These carriers can diffuse through the crystal or can drift in response to a static electric field. Both processes are observed experimentally. In drawing the figure we have assumed that diffusion is the dominant process, in which case the electron density is smallest in the regions of maximum optical intensity, because electrons have preferentially diffused away from these regions. The spatially varying charge distribution $\rho(x)$ gives rise to a spatially varying electric field distribution, whose form is shown in the third graph. Note that the maxima of the field $E(x)$ are shifted by 90 degrees with respect to those of the charge density distribution $\rho(x)$. The reason for this behavior is that the Maxwell equation $\nabla \cdot \mathbf{D} = 4\pi \rho$ when applied to the present situation implies that $dE/dx = 4\pi\rho/\epsilon$, and the spatial derivative that appears in this equation leads to a 90-degree phase shift between $E(x)$ and $\rho(x)$. The last graph in the figure shows the refractive index variation $\Delta n(x)$ that is produced through the linear electrooptic effect (Pockels effect) by the field $E(x)$.* Note that $\Delta n(x)$

---

* In drawing the figure, we have assumed that the electrooptic coefficient is positive. Note that the relation $\Delta(1/n^2) = r_{\text{eff}}E$ implies that $\Delta n = -\frac{1}{2}n^3 r_{\text{eff}}E$.

TABLE 11.4.1   Properties of some photorefractive crystals[a]

Material	Useful wavelength range (μm)	Carrier drift length $\mu\tau E$ at $E = 2$ kV/cm(μm)	τ_d (sec)	$n^3 r_{\text{eff}}$ (pm/V)	$n^3 r_{\text{eff}}/\epsilon_{dc}$ (pm/V)
InP:Fe	0.85–1.3	3	10^{-4}	52	4.1
GaAs:Cr	0.8–1.8	3	10^{-4}	43	3.3
LiNbO$_3$:Fe^{3+}	0.4–0.7	$<10^{-4}$	300	320	11
Bi$_{12}$SiO$_{20}$	0.4–0.7	3	10^5	82	1.8
Sr$_{0.4}$Ba$_{0.6}$Nb$_2$O$_6$	0.4–0.6	—	10^2	2460	4.0
BaTiO$_3$	0.4–0.9	0.1	10^2	11,300	4.9
KNbO$_3$	0.4–0.7	0.3	10^{-3}	690	14

[a] τ is the carrier recombination time; τ_d is the dielectric relaxation time in the dark. Adapted from Glass *et al.* (1984).

is shifted by 90 degrees with respect to the intensity distribution $I(x)$ that produces it. This phase shift has the important consequence that it can lead to the transfer of energy between the two incident beams. This transfer of energy is described in Section 11.6

The properties of some photorefractive crystals are summarized in Table 11.4.1.

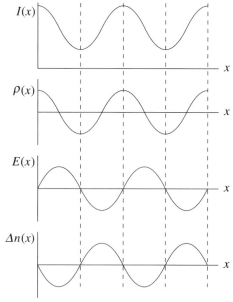

FIGURE 11.4.1   Origin of the photorefractive effect.

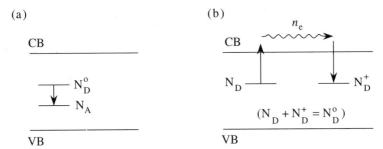

FIGURE 11.5.1 Energy levels and populations of the model of the photorefractive effect due to Kukhtarev *et al.*

11.5. Photorefractive Equations of Kukhtarev *et al.*

In this section we see how to describe the photorefractive effect by means of a model (see Fig. 11.5.1) due to Kukhtarev and co-workers.* This model presupposes that the photorefractive effect is due solely to one type of charge carrier, which for definiteness we assume to be the electron. As illustrated in part (a) of the figure, we assume that the crystal contains N_A acceptors and N_D^0 donors per unit volume, with $N_A \ll N_D^0$. We assume that the acceptor levels are completely filled with electrons that have fallen from the donor levels, and that these filled acceptor levels cannot be ionized by thermal or optical effects. Thus, at temperature $T = 0$ and in the absence of an optical field, each unit volume of the crystal contains N_A ionized donors, N_A electrons bound to acceptor impurities, and $N_D^0 - N_A$ neutral donor levels that can participate in the photorefractive effect. We further assume that electrons can be excited thermally or optically from the donor levels into the conduction band, as illustrated in part (b) of the figure. We let n_e, N_D^+, and N_D denote the number densities of conduction band electrons, ionized donors, and un-ionized donors, respectively. Note that $N_D + N_D^+$ must equal N_D^0, but that N_D^+ is not necessarily equal to n_e, because some donors lose their electrons to the acceptors and because electrons can migrate within the crystal, leading to regions that are not electrically neutral.

We next assume that the variation in level populations can be described by the rate equations

$$\frac{\partial N_D^+}{\partial t} = (sI + \beta)\left(N_D^0 - N_D^+\right) - \gamma n_e N_D^+, \tag{11.5.1}$$

$$\frac{\partial n_e}{\partial t} = \frac{\partial N_D^+}{\partial t} + \frac{1}{e}(\nabla \cdot \mathbf{j}), \tag{11.5.2}$$

---

* See Kukhtarev *et al.* (1977, 1979). This model is also described in several of the chapters of the book edited by Günter and Huignard (1988).

where s is a constant proportional to the photoionization cross section of a donor, β is the thermal generation rate, γ is the recombination coefficient, $-e$ is the charge of the electron, and $\mathbf{j}$ is the electrical current density. Equation (11.5.1) states that the ionized donor concentration can increase by thermal ionization or photoionization of un-ionized donors and can decrease by recombination. Equation (11.5.2) states that the mobile electron concentration can increase in any small region either because of the ionization of donor atoms or because of the flow of electrons into the local region. The flow of current is described by the equation

$$\mathbf{j} = n_e e \mu \mathbf{E} + eD\nabla n_e + \mathbf{j}_{ph}, \tag{11.5.3}$$

where μ is the electron mobility, D is the diffusion constant (which by the Einstein relation is equal to $k_B T \mu / e$), and $\mathbf{j}_{ph}$ is the photovoltaic (also known as the photogalvanic) contribution to the current. The last contribution results from the tendency of the photoionization process to eject the electron in a preferred direction in anisotropic crystals. For some materials (such as barium titanate and bismuth silicon oxide) this contribution to $\mathbf{j}$ is negligible, although for others (such as lithium niobate) it is very important. For lithium niobate $\mathbf{j}_{ph}$ has the form $\mathbf{j}_{ph} = pI\hat{\mathbf{c}}$, where $\hat{\mathbf{c}}$ is a unit vector in the direction of the optic axis of the crystal and p is a constant. The importance of the photovoltaic current has been discussed by Glass (1978).

The field $\mathbf{E}$ appearing in Eq. (11.5.3) is the static (or possibly low-frequency) electric field appearing within the crystal due to any applied voltage or to any charge separation within the crystal. It must satisfy the Maxwell equation

$$\epsilon_{dc}\nabla \cdot \mathbf{E} = -4\pi e(n_e + N_A - N_D^+), \tag{11.5.4}$$

where ϵ_{dc} is the static dielectric constant of the crystal. The modification of the optical properties is described by assuming that the optical-frequency dielectric constant is changed by an amount

$$\Delta\epsilon = -\epsilon^2 r_{\text{eff}} |E|. \tag{11.5.5}$$

For simplicity, here we are treating the dielectric properties in the scalar approximation; the tensor properties can be treated explicitly using the formalism developed in Section 11.2.* Note that the scalar form of Eq. (11.2.13a) is $\Delta(1/\epsilon) = r_{\text{eff}} |\mathbf{E}|$, from which Eq. (11.5.5) follows directly. The optical field $\tilde{E}_{\text{opt}}$ is assumed to obey the wave equation

$$\nabla^2 \tilde{E}_{\text{opt}} + \frac{1}{c^2}\frac{\partial^2}{\partial t^2}(\epsilon + \Delta\epsilon)\tilde{E}_{\text{opt}} = 0. \tag{11.5.6}$$

---

* See also the calculation of r_{eff} for one particular case in Eqs. (11.6.14b) in the next section.

Equations (11.5.1) through (11.5.6) constitute the photorefractive equations of Kukhtarev *et al.* They have been solved in a variety of special cases and have been found to provide an adequate description of most photorefractive phenomena. We shall consider their solution in special cases in the next two sections.

11.6. Two-Beam Coupling in Photorefractive Materials

Under certain circumstances, two beams of light can interact in a photorefractive crystal in such a manner that energy is transferred from one beam to the other. This process, which is often known as two-beam coupling, can be used, for example, to amplify a weak, image-bearing signal beam by means of an intense pump beam. Exponential gains of 10 per centimeter are routinely observed.

A typical geometry for studying two-beam coupling is shown in Fig. 11.6.1. Signal and pump waves, of amplitudes A_s and A_p respectively, interfere to form a nonuniform intensity distribution within the crystal. Because of the nonlinear response of the crystal, this nonuniform intensity distribution produces a refractive index grating within the material. However, this grating is displaced from the intensity distribution in the direction of the positive (or negative, depending on the sign of the dominant charge carrier and the sign of the effective electrooptic coefficient) crystalline c axis. As a result of this phase shift, the light scattered from A_p and A_s interferes constructively with A_s, whereas the light scattered from A_s into A_p interferes destructively with A_p, and consequently the signal wave is amplified whereas the pump wave is attenuated.

In order to describe this process mathematically, we assume that the optical field within the crystal can be represented as

$$\tilde{E}_{\mathrm{opt}}(\mathbf{r}, t) = [A_p(z)e^{i\mathbf{k}_p \cdot \mathbf{r}} + A_s(z)e^{i\mathbf{k}_s \cdot \mathbf{r}}]e^{-i\omega t} + \text{c.c.} \qquad (11.6.1)$$

We assume that $A_p(z)$ and $A_s(z)$ are slowly varying functions of the coordinate z. The intensity distribution of the light within the crystal can be expressed as

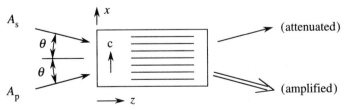

FIGURE 11.6.1 Typical geometry for studying two-beam coupling in a photorefractive crystal.

$I = (n_0 c/4\pi)\langle \tilde{E}^2_{\text{opt}} \rangle$ or as

$$I = I_0 + (I_1 e^{iqx} + \text{c.c.}), \tag{11.6.2a}$$

where

$$I_0 = \frac{n_0 c}{2\pi}(|A_p|^2 + |A_s|^2),$$

$$I_1 = \frac{n_0 c}{2\pi}(A_p A_s^*)(\hat{\mathbf{e}}_\mathbf{p} \cdot \hat{\mathbf{e}}_\mathbf{s}), \quad \text{and} \quad \mathbf{q} \equiv q\hat{\mathbf{x}} = \mathbf{k_p} - \mathbf{k_s}. \tag{11.6.2b}$$

Here $\hat{e}_p$ and $\hat{e}_s$ are the polarization unit vectors of the pump and signal waves, which are assumed to be linearly polarized. The quantity $\mathbf{q}$ is known as the grating wavevector. Note that the intensity distribution can also be described by the expression

$$I = I_0[1 + m\cos(qx + \phi)], \tag{11.6.3}$$

where $m = 2|I_1|/I_0$ is known as the modulation index and where $\phi = \tan^{-1}(\text{Im}I_1/\text{Re}I_1)$.

In order to determine how the optical properties of the photorefractive material are modified by the presence of the pump and signal fields, we first solve Eqs. (11.5.1) through (11.5.4) of Kukhtarev *et al.* to find the static electric field $\mathbf{E}$ induced by the intensity distribution of Eqs. (11.6.2). This static electric field can then be used to calculate the change in the optical-frequency dielectric constant through use of Eq. (11.5.5). Since Eqs. (11.5.1) through (11.5.4) are nonlinear (i.e., they contain products of the unknown quantities n_e, N_D^+, $\mathbf{j}$, and $\mathbf{E}$), they cannot easily be solved exactly. For this reason, we assume that the depth of modulation m is small (i.e., $|I_1| \ll I_0$) and seek an approximate steady-state solution of Eqs. (11.5.1) through (11.5.4) in the form

$$E = E_0 + (E_1 e^{iqx} + \text{c.c.}), \qquad j = j_0 + (j_1 e^{iqx} + \text{c.c.}),$$

$$n_e = n_{e0} + (n_{e1} e^{iqx} + \text{c.c.}), \quad N_D^+ = N_{D0}^+ + (N_{D1}^+ d^{iqx} + \text{c.c.}), \tag{11.6.4}$$

where $E = E\hat{\mathbf{x}}$ and $\mathbf{j} = j\hat{\mathbf{x}}$. We assume that the quantities E_1, j_1, n_{e1}, and N_{D1} are small in the sense that the product of any two of them can be neglected.

We next introduce Eqs. (11.6.4) into Eqs. (11.5.1) through (11.5.4) and equate terms with common x dependences. We thereby find several sets of equations. The set that is independent of the x coordinate depends only on the large quantities (subscript zero) and is given (in the same order as Eqs. (11.5.1) through (11.5.4)) by

$$(sI_0 + \beta)\left(N_D^0 - N_{D0}^+\right) = \gamma n_{e0} N_{D0}^+, \tag{11.6.5a}$$

$$j_0 = \text{constant}, \tag{11.6.5b}$$

$$j_0 = n_{e0} e \mu E_0 + j_{\text{ph},0}, \tag{11.6.5c}$$

$$N_{D0}^+ = n_{e0} + N_A. \tag{11.6.5d}$$

Equations (11.6.5a) and (11.6.5d) can be solved directly to determine the mean electron density n_{e0} and mean ionized donor density N_{D0}^+. Since in most realistic cases the inequality $n_{e0} \ll N_A$ is satisfied, the densities are given simply by

$$N_{Do}^+ = N_A, \tag{11.6.6a}$$

$$n_{e0} = \frac{(sI_0 + \beta)\left(N_D^0 - N_A\right)}{\gamma N_A}. \tag{11.6.6b}$$

The two remaining equations (11.6.5b) and (11.6.5c) determine the mean current density j_0 and mean field E_0. Let us assume for simplicity that the photovoltaic contribution j_{ph} is negligible for the material under consideration. The value of E_0 then depends on the properties of any external electric circuit to which the crystal is connected. In the common situation in which no voltage is externally applied to the crystal, E_0 and hence j_0 vanish.

We next consider the equation for the first-order quantities (quantities with the subscript 1) by considering the portions of Eqs. (11.5.1) through (11.5.4) with the spatial dependence e^{iqx}. The resulting equations are (we assume that $E_0 = 0$)

$$sI_1\left(N_D^0 - N_A\right) - (sI_0 + \beta)N_{D1}^+ = \gamma n_{e0} N_{D1}^+ + \gamma n_{e1} N_A, \tag{11.6.7a}$$

$$j_1 = 0, \tag{11.6.7b}$$

$$-n_{e0} e E_1 = iq k_B T n_{e1}, \tag{11.6.7c}$$

$$iq \epsilon_{dc} E_1 = -4\pi e (n_{e1} - N_{D1}^+). \tag{11.6.7d}$$

We solve these equations algebraically (again assuming that $n_{e0} \ll N_A$) to find that the amplitude of the spatially varying part of the static electric field is given by

$$E_1 = -i\left(\frac{sI_1}{sI_0 + \beta}\right)\left(\frac{E_D}{1 + E_D/E_q}\right), \tag{11.6.8}$$

where we have introduced the characteristic field strengths

$$E_D = \frac{qk_B T}{e}, \qquad E_q = \frac{4\pi e}{\epsilon_{de} q} N_{\text{eff}}, \tag{11.6.9}$$

where $N_{\text{eff}} = N_A(N_D^0 - N_A)/N_D^0$ can be interpreted as an effective trap density. Note that in the common circumstance where $N_A \ll N_D^0$, N_{eff} is given approximately by $N_{\text{eff}} \simeq N_A$. The quantity E_D is called the diffusion field strength and is a measure of the field strength required to inhibit the separation of charge due to thermal agitation. The quantity E_q is called the maximum space charge field and is a measure of the maximum electric field that can be created by redistributing charge of mean density $e N_{\text{eff}}$ over the characteristic distance $2\pi/q$. Note from Eq. (11.6.8) that E_1 is shifted in phase with respect to the intensity distribution I_1 and that E_1 is proportional to the depth of modulation m in the common case of $\beta \ll s I_0$.

Recall that the change in the optical-frequency dielectric constant is proportional to the amplitude E_1 of the spatially modulated component of the static electric field. For this reason, it is often of practical interest to maximize the value of E_1. We see from Eq. (11.6.8) that E_1 is proportional to the product of the factor $s I_1/(s I_0 + \beta)$, which can be maximized by increasing the depth of modulation $m = 2|I_1|/I_0$,[*] with the factor $E_D/(1 + E_D/E_q)$. Since each of the characteristic field strengths E_D and E_q depends on the grating wavevector, this second factor can be maximized by using the optimum value of q. To show the dependence of E_1 on q, we can rewrite Eq. (11.6.8) as

$$E_1 = -i\left(\frac{s I_1}{s I_0 + \beta}\right) E_{\text{opt}} \frac{2(q/q_{\text{opt}})}{1 + (q/q_{\text{opt}})^2}, \qquad (11.6.10a)$$

where

$$q_{\text{opt}} = \left(\frac{4\pi N_{\text{eff}} e^2}{k_B T \epsilon_{dc}}\right)^{1/2}, \qquad E_{\text{opt}} = \left(\frac{\pi N_{\text{eff}} k_B T}{\epsilon_{dc}}\right)^{1/2}. \qquad (11.6.10b)$$

Note that q_{opt} is of the order of magnitude of the inverse of the Debye screening distance.

The dependence of E_1 on q is shown in Fig. 11.6.2. Note that the grating wavevector q can be varied experimentally by controlling the angle between the pump and signal beams, since (see Fig. 11.6.1) q is given by the formula

$$q = 2n\frac{\omega}{c}\sin\theta. \qquad (11.6.11)$$

Through an experimental determination of the optimum value of the magnitude of the grating wavevector, the value of the effective trap density N_{eff} can be obtained through use of Eq. (11.6.10b).

Let us next calculate the spatial growth rate that the signal wave experiences as the result of two-beam coupling in photorefractive materials. For simplicity,

---

[*] Recall, however, that the present derivation is valid only if $m \ll 1$.

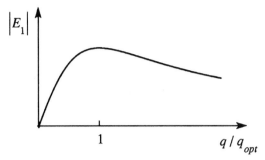

FIGURE 11.6.2 Dependence of the modulated component of the space charge field on the magnitude q of the grating wavevector.

we assume that the photoionization rate sI_0 is much greater than the thermal ionization rate β (which is the usual case in practice), so that the field amplitude E_1 of Eq. (11.6.8) can be expressed through use of Eqs. (11.6.2) as

$$E_1 = -i\frac{A_p A_s^*}{|A_s|^2 + |A_p|^2}(\hat{e}_p \cdot \hat{e}_s)E_m,\qquad(11.6.12a)$$

where

$$E_m = \frac{E_D}{1 + E_D/E_q}.\qquad(11.6.12b)$$

According to Eq. (11.5.5), this field produces a change in the dielectric constant of amplitude $\Delta\epsilon = -\epsilon^2 r_{\text{eff}}E_1$. For the particular geometry of Fig. 11.6.1, the product $\epsilon^2 r_{\text{eff}}$ has the form (Feinberg $et\ al.$, 1980; see also Feinberg and MacDonald, 1989)

$$\epsilon^2 r_{\text{eff}} = \sum_{ijklm} r_{ijk}\left(\epsilon_{il}\hat{e}_l^s\right)\left(\epsilon_{jm}\hat{e}_m^p\right)\hat{q}_k,\qquad(11.6.13)$$

where $\hat{e}_l^s$ and $\hat{e}_m^p$ denote the l and m cartesian components of the polarization unit vectors of the signal and pump waves, respectively, and $\hat{q}_k$ denotes the k cartesian component of a unit vector in the direction of the grating vector. For crystals of point group $4mm$ (such as barium titanate), one finds that for ordinary waves

$$r_{\text{eff}} = r_{13}\sin\left(\frac{\alpha_s + \alpha_p}{2}\right)\qquad(11.6.14a)$$

and that for extraordinary waves

$$r_{\text{eff}} = n^{-4}\big[n_0^4 r_{13}\cos\alpha_s\cos\alpha_p + 2n_e^2 n_0^2 r_{42}\cos\tfrac{1}{2}(\alpha_s + \alpha_p)$$
$$+ n_e^4 r_{33}\sin\alpha_s\sin\alpha_p\big]\sin\tfrac{1}{2}(\alpha_s + \alpha_p).\qquad(11.6.14b)$$

Here α_s and α_p denote the angles between the propagation vectors of the signal and pump waves and the positive crystalline c axis, respectively, and n is the refractive index experienced by the beam that scatters off the grating.

Note from Table 11.2.1 that for barium titanate the electrooptic coefficient r_{42} is much larger than either r_{13} or r_{33}. We see from Eqs. (11.6.14) that only through the use of light of extraordinary polarization can one utilize this large component of the electrooptic tensor.

The change in the dielectric constant $\Delta\epsilon = -\epsilon^2 r_{\mathrm{eff}} E_1$ produces a nonlinear polarization given by

$$P^{\mathrm{NL}} = \left(\frac{\Delta\epsilon}{4\pi}e^{i\mathbf{q}\cdot\mathbf{r}} + \mathrm{c.c.}\right)(A_s e^{i\mathbf{k_s}\cdot\mathbf{r}} + A_p e^{i\mathbf{k_p}\cdot\mathbf{r}}). \qquad (11.6.15)$$

Recall that $\mathbf{q} = \mathbf{k_p} - \mathbf{k_s}$. The part of the nonlinear polarization having the spatial variation $\exp(i\mathbf{k_s}\cdot\mathbf{r})$ can act as a phase-matched source term for the signal wave and is given by

$$P_s^{\mathrm{NL}} = \frac{\Delta\epsilon^*}{4\pi}A_p e^{i\mathbf{k_s}\cdot\mathbf{r}} = \frac{-i\epsilon^2 r_{\mathrm{eff}} E_m}{4\pi}\frac{|A_p|^2 A_s}{|A_p|^2 + |A_s|^2}e^{i\mathbf{k_s}\cdot\mathbf{r}}. \qquad (11.6.16a)$$

Likewise, the portion of P^{NL} that can act as a phase-matched source term for the pump wave is given by

$$P_p^{\mathrm{NL}} = \frac{\Delta\epsilon}{4\pi}A_s e^{i\mathbf{k_p}\cdot\mathbf{r}} = \frac{i\epsilon^2 r_{\mathrm{eff}} E_m}{4\pi}\frac{|A_s|^2 A_p}{|A_p|^2 + |A_s|^2}e^{i\mathbf{k_p}\cdot\mathbf{r}}. \qquad (11.6.16b)$$

We next derive coupled-amplitude equations for the pump and signal fields using the formalism described in Section 2.1. We define z_s and z_p to be distances measured along the signal and pump propagation directions. We find that in the slowly-varying amplitude approximation the signal amplitude varies as

$$2ik\frac{dA_s}{dz_s}e^{i\mathbf{k_s}\cdot\mathbf{r}} = -4\pi\frac{\omega^2}{c^2}P_s^{\mathrm{NL}}, \qquad (11.6.17a)$$

which through use of Eq. (11.6.16a) becomes

$$\frac{dA_s}{dz_s} = \frac{\omega}{2c}n^3 r_{\mathrm{eff}} E_m\frac{|A_p|^2 A_s}{|A_p|^2 + |A_s|^2}. \qquad (11.6.17b)$$

We find that the intensity $I_s = (nc/2\pi)|A_s|^2$ of the signal wave varies spatially as $dI_s/dz_s = A_s^* dA_s/dz_s + \mathrm{c.c.}$, or as

$$\frac{dI_s}{dz_s} = \Gamma\frac{I_s I_p}{I_s + I_p}, \qquad (11.6.18a)$$

where*

$$\Gamma = \frac{\omega}{c} n^3 r_{\text{eff}} E_m.$$ (11.6.18b)

A similar derivation shows that the pump intensity varies spatially as

$$\frac{dI_p}{dz_p} = -\Gamma \frac{I_s I_p}{I_s + I_p}.$$ (11.6.18c)

Note that Eq. (11.6.18a) predicts that the signal intensity grows exponentially with propagation distance in the common limit $I_s \ll I_p$.[†] The strong amplification available from photorefractive two-beam coupling allows this process to be used for various practical applications. For the application of photorefractive two-beam coupling to the design of efficient polarizers, see Heebner *et al.* (2000).

The treatment of two-beam coupling given above has assumed that the system is in steady state. Two-beam coupling under transient conditions can also be treated using the material equations of Kukhtarev *et al.* It has been shown (Kukhtarev *et al.*, 1977; Refrégier *et al.*, 1985; Valley, 1987) that, under the assumption that $n_e \ll N_D^+$, $N_D^+ \ll N_D^0$, and $\beta \ll sI_0$, the electric field amplitude E_1 obeys the equation

$$\tau \frac{\partial E_1}{\partial t} + E_1 = -i E_m \frac{A_p A_s^*}{|A_p|^2 + |A_s|^2} (\hat{e}_p \cdot \hat{e}_s)$$ (11.6.19)

with E_m given by Eq. (11.6.12b) and with the response time τ given by

$$\tau = \tau_d \frac{1 + E_D / E_M}{1 + E_D / E_q}$$ (11.6.20a)

where

$$\tau_d = \frac{\epsilon_{dc}}{4\pi e \mu n_{e0}}, \qquad E_M = \frac{\gamma N_A}{q\mu}.$$ (11.6.20b)

---

* Following convention, we use the same symbol Γ to denote the photofrefractive gain coefficient and the retardation of Section 11.3.

[†] Here we implicitly assuming that Γ is a positive quantity. If Γ is negative, the wave that we have been calling the pump wave will be amplified and the wave that we have been calling the signal wave will be attenuated. The sign of Γ depends on the sign of r_{eff}, which can be either positive or negative, and on the sign of E_m. Note that, according to Eqs. (11.6.9) and (11.6.12b), the sign of E_m depends upon the sign of the dominant charge carrier (our derivation has assumed the case of an electron) and upon the sign of q, which is the x component of $\mathbf{k_p} - \mathbf{k_s}$. For the case of barium titanate, the dominant charge carriers are usually holes, and the wave whose wavevector has a positive component along the crystalline c axis is amplified.

Note that the photorefractive response time τ scales linearly with the dielectric relaxation time τ_d.* Since the mean electron density n_{e0} increases linearly with optical intensity (see Eq. (11.6.6b)), we see that the photorefractive response time becomes faster when the crystal is excited using high optical intensities.

We next write the coupled-amplitude equations for the pump and signal fields in terms of the field amplitude E_1 as

$$\frac{\partial A_p}{\partial x_p} = \frac{-\omega}{2n_p c} r_{\text{eff}} A_s E_1, \tag{11.6.21a}$$

$$\frac{\partial A_s}{\partial x_s} = \frac{-\omega}{2n_s c} r_{\text{eff}} A_p E_1. \tag{11.6.21b}$$

Equations (11.6.19) through (11.6.21) describe the transient behavior of two-beam coupling.

11.7. Four-Wave Mixing in Photorefractive Materials

Next we consider the mutual interaction of four beams of light in a photorefractive crystal. We assume the geometry of Fig. 11.7.1. Note that the pump beams 1 and 2 are counterpropagating, as are beams 3 and 4. Thus the interaction shown in the figure can be used to generate beam 4 as the phase conjugate of beam 3.

The general problem of the interaction of four beams of light in a photorefractive material is very complicated, because the material response consists of four distinct gratings, namely, one grating due to the interference of beams 1 and 3 and of 2 and 4, one grating due to the interference of beams 1 and 4 and of 2 and 3, one grating due to the inference of beams 1 and 2, and one grating due to the interference of beams 3 and 4. However, under certain experimental situations, only one of these gratings leads to appreciable nonlinear coupling among the beams. If one assumes that the polarizations, propagation directions, and coherence properties of the input beams are selected so that only the grating due to the interference of beams 1 and 3 and beams 2 and 4 is important, the coupled-amplitude equations describing the propagation of the four beams become (Cronin-Golomb et al., 1984; see also

---

* The dielectric relaxation time is the characteristic time in which charge imbalances neutralize in a conducting material. The expression for the dielectric relaxation time is derived by combining the equation of continuity $\partial\rho/\partial t = -\nabla\cdot\mathbf{j}$ with Ohm's law in the form $\mathbf{j} = \sigma\mathbf{E}$ to find that $\partial\rho/\partial t = -\sigma\nabla\cdot\mathbf{E} = -(\sigma/\epsilon_{dc})\nabla\cdot\mathbf{D} = -(4\pi\sigma/\epsilon_{dc})\rho \equiv -\rho/\tau_d$. By equating the electrical conductivity σ with the product $n_{e0}e\mu$, we obtain the expression for τ_d quoted in the text.

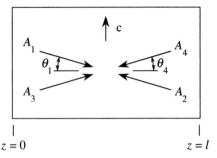

FIGURE 11.7.1 Geometry of four-wave mixing in a photorefractive material.

Fischer *et al.*, 1981)

$$\frac{dA_1}{dz} = -\frac{\gamma}{S_0}(A_1 A_3^* + A_2^* A_4)A_3 - \alpha A_1, \qquad (11.7.1\text{a})$$

$$\frac{dA_2}{dz} = -\frac{\gamma}{S_0}(A_1^* A_3 + A_2 A_4^*)A_4 + \alpha A_2, \qquad (11.7.1\text{b})$$

$$\frac{dA_3}{dz} = \frac{\gamma}{S_0}(A_1^* A_3 + A_2 A_4^*)A_1 - \alpha A_3, \qquad (11.7.1\text{c})$$

$$\frac{dA_4}{dz} = \frac{\gamma}{S_0}(A_1 A_3^* + A_2^* A_4)A_2 + \alpha A_4. \qquad (11.7.1\text{d})$$

In these equations, we have introduced the following quantities:

$$\gamma = \frac{\omega r_{\mathrm{eff}} n_0^3 E_m}{2c \cos \theta} \qquad (11.7.2\text{a})$$

with E_m given by Eq. (11.6.12b),

$$S_0 = \sum_{i=1}^{4} |A_i|^2, \qquad (11.7.2\text{b})$$

and $\alpha = \frac{1}{2}\alpha_0 / \cos \theta$, where α_0 is the intensity absorption coefficient of the material and where for simplicity we have assumed that $\theta = \theta_1 = \theta_4$.

Cronin-Golomb *et al.* (1984) have shown that Eqs. (11.7.1) can be solved for a large number of cases of interest. The solutions show a variety of interesting features, including amplified reflection, self-oscillation, and bistability.

Externally Self-Pumped Phase-Conjugate Mirror

One interesting feature of four-wave mixing in photorefractive materials is that it can be used to construct a self-pumped phase-conjugate mirror of the sort illustrated in Fig. 11.7.2. In such a device, only the signal wave A_3 is

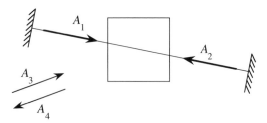

FIGURE 11.7.2   Geometry of the externally self-pumped phase-conjugate mirror. Only the A_3 wave is applied externally; this wave excites the oscillation of the waves A_1 and A_2, which act as pump waves for the four-wave mixing process that generates the conjugate wave A_4.

applied externally. Waves A_1 and A_2 grow from noise within the resonator that surrounds the photorefractive crystal. Oscillation occurs because wave A_1 is amplified at the expense of wave A_3 by the process of two-beam coupling. The output wave A_4 is generated by four-wave mixing involving waves A_1, A_2, and A_3. Such a device was constructed by White *et al.* (1982) and is described further by Cronin-Golomb *et al.* (1984).

Internally Self-Pumped Phase-Conjugate Mirror

Even more remarkable than the the device just described is the internally self-pumped phase conjugate mirror, which is illustrated in Fig. 11.7.3. Once again, only the signal wave A_3 is applied externally. By means of a complicated nonlinear process analogous to self-focusing, beams A_1 and A_2 are created. Reflection of these waves at the corner of the crystal feeds these waves back into the path of the applied wave A_3. Four-wave mixing processes involving waves A_1, A_2, and A_3 then create the output wave A_4 as the phase conjugate of A_3. This device was first demonstrated by Feinberg (1982) and analyzed

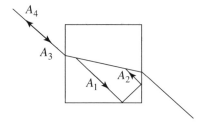

FIGURE 11.7.3   Geometry of the internally self-pumped phase conjugate mirror. Only the A_3 wave is applied externally; this wave excites the oscillation of the waves A_1 and A_2, which act as pump waves for the four-wave mixing process that generates the conjugate wave A_4.

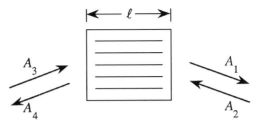

FIGURE 11.7.4   Geometry of the double phase-conjugate mirror. Waves A_2 and A_3 are applied externally and need not be phase-coherent. The generated wave A_1 is the phase conjugate of A_2, and the generated wave A_4 is the phase conjugate of A_3.

theoretically by MacDonald and Feinberg (1983). Because of the complicated nature of the coupling that occurs in this device, it can produce complicated dynamical behavior including deterministic chaos, as demonstrated by Gauthier *et al.* (1987). Because of the ease wth which a phase-conjugate signal can be produced in such a device, it lends itself to practical applications, such as new types of interferometers (Gauthier *et al.*, 1989).

Double Phase-Conjugate Mirror

Another application of four-wave mixing in photorefractive crystals is the double phase-conjugate mirror of Fig. 11.7.4. In such a device the waves A_2 and A_3 are applied externally; these waves are assumed to be mutually incoherent, so that no gratings are formed by their interference. The nonlinear interaction leads to the generation of the output wave A_1, which is the phase conjugate of A_2, and to the output wave A_4, which is the phase conjugate of A_3. However, A_1 is phase-coherent with A_3, whereas A_4 is phase-coherent with A_2. The double phase-conjugate mirror possesses the remarkable property that one of the output waves can be an amplified phase-conjugate wave, even though the two input waves are mutually incoherent.

The nature of the nonlinear coupling that produces the double phase-conjugate mirror can be understood from the coupled-amplitude equations (11.7.1). For simplicity, we consider the limit in which α is negligible and in which the input waves A_2 and A_3 are not modified by the nonlinear interaction, so that only Eqs. (11.7.1a) and (11.7.1d) need to be considered. We see that each output wave is driven by two terms, one of which is a two-beam-coupling term that tends to amplify the output wave, and the other of which is a four-wave-mixing term that causes each output to be the phase conjugate of its input wave. It has been shown by Cronin-Golomb *et al.* (1984) and by Weiss *et al.* (1987) that the requirement for the generation of the two output waves is that $|\gamma|l$ be greater than 2. Operation of the double phase-conjugate mirror has been demonstrated experimentally by Weiss *et al.* (1987).

Other Applications of Photorefractive Nonlinear Optics

Because the photorefractive effect leads to a large nonlinear response, it lends itself to a variety of applications. Many of these have applications have been reviewed by Günter and Huignard (1988, 1989) and by Boyd and Grynberg (1992).

Problems

1. *Numerical evaluation of photorefractive quantities.* Consider the process of two-beam coupling in barium titanate in the geometry of Fig. 11.6.1. Estimate the numerical values of the physical quantities E_D, E_q, E_{opt}, E_1, r_{eff}, $\Delta\epsilon$, and Γ. Assume that the effective trap density N_{eff} is equal to 10^{12} cm^{-3}, that the thermal generation rate is negligible, that the modulation index m is 10^{-3}, and that $\theta_s = \theta_p = 5$ degrees.

2. *Transient two-beam coupling.* Verify Eq. (11.6.19).

3. *Relation between electrooptic and nonlinear optics tensors.* Determine the mathematical relationship between the second-order susceptibility $\chi^{(2)}_{ijk}$ and the linear electrooptic coefficient r_{ijk}. Similarly, determine the mathematical relationship between the third-order susceptibility $\chi^{(3)}_{ijkl}$ and the quadratic electrooptic coefficient s_{ijkl}.

References

Electrooptic Effect

M. Born and E. Wolf, *Principles of Optics*, Pergamon, London, 1975.

W. R. Cook, Jr., and J. Jaffe, "Electrooptic Coefficients," in *Landolt-Bornstein, New Series,* Vol. II (K.-H. Hellwege, ed.), Springer-Verlag, Berlin, 1979, pp. 552–651.

I. P. Kaminow, *An Introduction to Electrooptic Devices*, Academic Press, New York, 1974.

B. J. Thompson and E. Hartfield, in *Handbook of Optics* (W. G. Driscoll and W. Vaughan, eds.), McGraw-Hill, New York, 1978.

A. Yariv and P. Yeh, *Optical Waves in Crystals*, Wiley, New York, 1984.

Photorefractive Effect

R. W. Boyd and G. Grynberg, "Optical Phase Conjugation," in *Contemporary Nonlinear Optics* (G. P. Agrawal and R. W. Boyd, eds.), Academic Press, Boston, 1992.

M. Cronin-Golomb, B. Fischer, J. O. White, and A. Yariv, *IEEE J. Quantum Electron.* **20**, 12 (1984).

J. Feinberg, *Opt. Lett.* **7**, 486 (1982).

J. Feinberg and K. R. MacDonald, in *Photorefractive Materials and Their Applications*, Vol. II (P. Günter and J.-P. Huignard, eds.), Springer-Verlag, Berlin, 1989.

J. Feinberg, D. Heiman, A. R. Tanguay, Jr. and R. W. Hellwarth, *J. Appl. Phys.* **5**, 1297 (1980).

B. Fischer, M. Cronin-Golomb, J. O. White, and A. Yariv, *Opt. Lett.* **6**, 519 (1981).

D. J. Gauthier, R. W. Boyd, R. K. Jungquist, J. B. Lisson, and L. L. Voci, *Opt. Lett.* **14**, 325 (1989).

D. J. Gauthier, P. Narum, and R. W. Boyd, *Phys. Rev. Lett.* **58**, 16 (1987).

A. M. Glass, *Opt. Eng.* **17**, 470 (1978).

A. M. Glass, D. von der Linde, and T. J. Negran, *Appl. Phys. Lett.* **25**, 233 (1974).

A. M. Glass, A. M. Johnson, D. H. Olson, W. Simpson, and A. A. Ballman, *Appl. Phys. Lett.* **44**, 948 (1984).

P. Günter and J.-P. Huignard, eds., *Photorefractive Materials and Their Applications*, Springer-Verlag, Berlin, Part I, 1988; Part II, 1989.

J. E. Heebner, R. S. Bennink, R. W. Boyd, and R. A. Fisher, *Opt. Lett.* **25**, 257 (2000).

N. Kukhtarev, V. B. Markov, and S. G. Odulov, *Opt. Commun.* **23**, 338 (1977).

N. Kukhtarev, V. B. Markov, S. G. Odulov, M. S. Soskin, and V. L. Vinetskii, *Ferroelectrics* **22**, 949–960, 961–964 (1979).

K. R. MacDonald and J. Feinberg, *J. Opt. Soc. Am.* **73**, 548 (1983).

Ph. Refrégier, L. Solymar, H. Rabjenbach, and J. P. Huignard, *J. Appl. Phys* **58**, 45 (1985).

G. C. Valley, *J. Opt. Soc. Am. B* **4**, 14, 934 (1987).

S. Weiss, S. Sternklar, and B. Fischer, *Opt. Lett.* **12**, 114 (1987).

J. O. White, M. Cronin-Golomb, B. Fischer, and A. Yariv, *Appl. Phys. Lett.* **40**, 450 (1982).

Chapter 12

Optically Induced Damage and Multiphoton Absorption

12.1. Introduction to Optical Damage

A topic of great practical importance is optically induced damage of optical components. Optical damage is important because it ultimately limits the maximum amount of power that can be transmitted through a particular optical material. Optical damage thus imposes a constraint on the efficiency of many nonlinear optical processes, by limiting the maximum field strength E that can be used to excite the nonlinear response without the occurrence of optical damage. In this context, it is worth pointing out that present laser technology can produce laser beams of sufficient intensity to exceed the damage thresholds of all known materials.

There are several different physical mechanisms that can lead to optically induced damage. These mechanisms, and an approximate statement of the conditions under which each might be observed, are as follows:

- Linear absorption, leading to localized heating and cracking of the optical material. This is the dominant damage mechanism for continuous-wave and long-pulse ($\gtrsim 1$ μsec) laser beams.
- Avalanche breakdown, which is the dominant mechanism for pulsed lasers (shorter than $\lesssim 1$ μsec) for intensities in the range of 10^9 W/cm^2 to 10^{12}/cm^2.

- Multiphoton ionization or multiphoton dissociation of the optical material, which is the dominant mechanism for intensities in the range 10^{12} to 10^{16} W/cm^2.

- Direct (single cycle) field ionization, which is the dominant mechanism for intensities $>10^{10}$ W/cm^2.

We next present a more detailed description of several of these mechanisms.

Let us begin by briefly summarizing some of the basic empirical observations regarding optical damage. When a collimated laser beam interacts with an optical material, optical damage usually occurs at a lower threshold on the surfaces than in the interior. This observation suggests that cracks and other imperfections on an optical surface can serve to initiate the process of optical damage, either by enhancing the local field strength in regions near the cracks or by providing a source of nearly free electrons needed to initiate the avalanche breakdown process. It is also observed (Lowdermilk and Milam, 1981) that surface damage occurs with a lower threshold at the exiting surface than at the entering surface of an optical material. One mechanism leading to this behavior results from the nature of the electromagnetic boundary conditions at a dielectric/air interface, which lead to a deenhancement in field strength at the entering surface and an enhancement at the exiting surface. This process is illustrated pictorially in Fig. 12.1.1. Other physical mechanisms leading to the same sort of front/back asymmetry include the facts that (1) plasma formation which accompanies localized damage tends to shield that region from further damage for the entering but not the exiting surface and (2) diffraction from

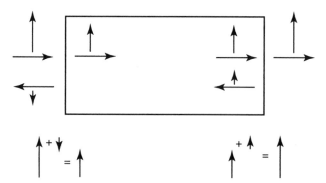

FIGURE 12.1.1   For a collimated laser beam, optical damage tends to occur at the exiting surface of an optical material, because the boundary conditions on the electric field vector lead to an deenhancement at the entering surface and an enhancement at the exiting surface.

defects at the front surface can lead to significant intensity variation (hot spots) at the exiting surface. This effect has been described, for instance, by Genin *et al.* (2000).

12.2. Avalanche-Breakdown Model

The avalanche-breakdown mechanism is believed to be the dominant damage mechanism for most pulsed lasers. The nature of this mechanism is that a small number N_0 of free electrons initially present within the optical material are accelerated to high energies through their interaction with the laser field. These electrons can then impact-ionize other atoms within the material, thereby producing additional electrons which are subsequently accelerated by the laser field and which eventually produce still more electrons. Some fraction of the energy imparted to each electron will lead to a localized heating of the material, which can eventually lead to damage of the material due to cracking or melting. The few electrons initially present within the material are created by one of several processes, including thermal excitation, quantum mechanical tunnelling by means of the Keldysh mechanism (Ammosov *et al.*, 1986), multiphoton excitation, or free electrons resulting from crystal defects.

Let us next describe the avalanche-breakdown model in a more quantitative manner. We note that the energy Q imparted to an electron initially at rest and subjected to an electric field $\tilde{E}$ (assumed quasistatic for present) for a time duration t is given by

$$Q = e\tilde{E}d \quad \text{where} \quad d = \tfrac{1}{2}at^2 = \tfrac{1}{2}(e\tilde{E}/m)t^2 \qquad (12.2.1)$$

or

$$Q = e^2\tilde{E}^2t^2/2m \quad \text{for} \quad t \leq \tau. \qquad (12.2.2)$$

This result holds for times $t \lesssim \tau$, where τ is the mean time between collisions. For longer time durations, the total energy imparted to the electron will be given approximately by the energy imparted to the electron in time interval τ (that is, by $e^2E^2\tau^2/2m$) multiplied by the number of such time intervals (that is, by t/τ), giving

$$Q = e^2\tilde{E}^2t\tau/2m \quad t > \tau. \qquad (12.2.3)$$

The rate at which the electron gains energy is given in this limit as*

$$P = \frac{dQ}{dt} = e^2 \tilde{E}^2 \tau / 2m. \tag{12.2.4}$$

We next assume that the density of free electrons $N(t)$ changes in time according to

$$\frac{dN}{dt} = \frac{fNP}{W} \tag{12.2.5}$$

where f is the fraction of the absorbed power P that leads to further ionization, so that $1 - f$ represents the fraction that leads to heating, where W represents the ionization threshold of the material under consideration, and where P is given by Eq. (12.2.4). The solution to Eq. (12.2.5) is thus

$$N(t) = N_0 e^{gt} \quad \text{where} \quad g = \frac{f e^2 \tilde{E}^2 \tau}{2m W}. \tag{12.2.6}$$

We next introduce the assumption that optical damage will occur if the electron density $N(T_p)$ at the end of the laser pulse of duration T_p exceeds some damage threshold value N_{th}, which is often assumed to be of the order of 10^{18} cm^{-3}. The condition for the occurrence of laser damage can thus be expressed as

$$\frac{f e^2 \tilde{E}^2 \tau T_p}{2m W} > \ln(N_{th}/N_0). \tag{12.2.7}$$

The right-hand side of this equality depends only weakly on the assumed values of N_{th} and N_0 and can be taken to have a value of the order of 30. This result can be used to find that the threshold intensity for producing laser damage is given by

$$I_{th} = \frac{nc}{4\pi} \langle \tilde{E}^2 \rangle = \frac{nc}{2\pi} \frac{m W}{f e^2 \tau T_p} \ln(N_{th}/N_0). \tag{12.2.8}$$

If we evaluate this expression under the assumption that $n \approx 1$, $W = 5$ eV, $\tau \approx 10^{-15}$ sec, $T_p \approx 10^{-9}$ sec, and $f \approx 0.01$, we find that $I_{th} \simeq 40$ GW/cm^2, in reasonable agreement with measured values.

---

* This result can also be deduced by noting that the rate of Joule heating of a conducting material is given by

$$NP = \tfrac{1}{2} \sigma \tilde{E}^2$$

where N is the number density of electrons and σ is the electrical conductivity, which according to the standard Drude formula is given by

$$\sigma = \frac{(Ne^2/m)\tau}{1 + \omega^2 \tau^2}.$$

This result constitutes a generalization of that of Eq. (12.2.4) and reduces to it in the limit $\omega\tau \ll 1$.

12.3. Influence of Laser Pulse Duration

There is a well-established scaling law that relates the laser damage threshold to the laser pulse duration T_p for pulse durations in the approximate range of 10 psec to 10 nsec. In particular, this scaling law states that the fluence (energy per unit area) required to produce damage increases with pulse duration as $T_p^{1/2}$, and correspondingly the intensity required to produce laser damage decreases with pulse duration as $T_p^{-1/2}$. This scaling law can be interpreted as a statement that (for this range of pulse durations) optical damage depends not solely on laser fluence or on laser intensity but rather upon their geometrical mean. Some data illustrating this scaling law are shown in Fig. 12.3.1, and more information regarding this law can be found in Lowdermilk and Milam (1981).

The $T_p^{1/2}$ scaling law can be understood, at least in general terms, by noting that the avalanche-breakdown model ascribes the actual damage mechanism to rapid localized heating of the optical material. The local temperature distribution $T(\mathbf{r}, t)$ obeys the heat transport equation (see also Eq. (4.5.2))

$$\rho C \frac{\partial T}{\partial t} - \kappa \nabla^2 T = N(1 - f)P, \qquad (12.3.1)$$

where f, N, and P have the same meanings as in the previous section, κ is the thermal conductivity, and ρC is the heat capacity per unit volume. Let us

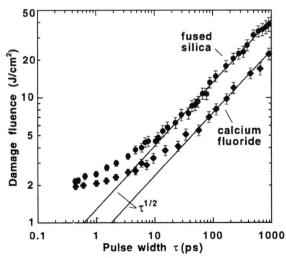

FIGURE 12.3.1   Measured dependence of laser damage threshold on laser pulse duration (Stuart *et al.*, 1995).

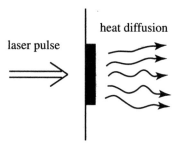

FIGURE 12.3.2   Illustration of the diffusion of heat following absorption of an intense laser pulse.

temporarily ignore the source term on the right-hand side of this equation, and estimate the distance L over which a temperature rise ΔT will diffuse in a time interval T_p. Replacing derivatives with ratios and assuming diffusion in only one dimension, as indicated symbolically in Fig. 12.3.2, we find that

$$\rho C \frac{\Delta T}{T_p} = \kappa \frac{\Delta T}{L^2}, \qquad (12.3.2)$$

or that

$$L = (DT_p)^{1/2} \quad \text{where} \quad D = \kappa/\rho c \text{ is the diffusion constant.} \quad (12.3.3)$$

The heat deposited by the laser pulse is thus spread out over a region of dimension L that is proportional to $T_p^{1/2}$, and the threshold for optical damage will be raised by this same factor. Although this explanation for the $T_p^{1/2}$ dependence is widely quoted, and although it leads to the observed dependence on the pulse duration T_p, some doubt has been expressed (Bloembergen, 1997) regarding whether values of D for typical materials are sufficiently large for thermal diffusion to be important. Nontheless, detailed numerical calculations (Stuart *et al.*, 1995, 1996) that include the effects of multiphoton ioization, Joule heating, and avalanche ionization are in good agreement with experimental results.

12.4.   Direct Photoionization

In this process the laser field strength is large enough to rip electrons away from the atomic nucleus. This process is expected to become dominant if the peak laser field strength exceeds the atomic field strength $E_{at} = e/a_0^2 = 6 \times 10^9$ V/cm. Fields this large are obtained at intensities of

$$I_{at} = \frac{nc}{8\pi} E_{at}^2 = 4 \times 10^{16} \text{ W/cm}^2.$$

For laser pulses of duration 100 fsec or longer, laser damage can occur at much lower intensities by means of the other processes described above. Direct photoionization is described in more detail in Chapter 13.

12.5. Multiphoton Absorption and Multiphoton Ionization

In this section we calculate the rate at which multiphoton absorption processes occur. Some examples of multiphoton absorption processes are shown schematically in Fig. 12.5.1. Two-photon absorption was first reported experimentally by Kaiser and Garrett (1961).

Some of the reasons for current interest in the field of multiphoton absorption include the following:

1. Multiphoton spectroscopy can be used to study high-lying electronic states and states not accessible from the ground state because of selection rules.
2. Two-photon microscopy has been used to eliminate much of the background associated with imaging through highly scattering materials, both because most materials scatter less strongly at longer wavelengths and because two-photon excitation provides sensitivity only in the focal volume of the incident laser beam. Such behavior is shown in Fig. 12.5.2.
3. Multiphoton absorption and multiphoton ionization can lead to laser damage of optical materials and be used to write permanent refractive index structures into the interior of optical materials. See for instances the articles listed at the end of this chapter under the topic Optical Damage with Femtosecond Laser Pulses.

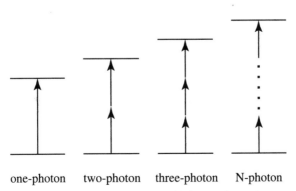

     one-photon   two-photon   three-photon   N-photon

FIGURE 12.5.1   Several examples of multiphoton absorption processes.

FIGURE 12.5.2   Fluorescence from a dye solution (20 μM solution of fluorescein in water) under (a) one-photon excitation and (b) two-photon excitation. Note that under two-photon excitation, fluorescence is excited only at the focal spot of the incident laser beam. Photographs courtesy of W. Webb.

4. Multiphoton absorption constitutes a nonlinear loss mechanism that can limit the efficiency of nonlinear optical devices such as optical switches (see also the discussion in Section 7.3).

In principle, we already know how to calculate multiphoton absorption rates by means of the formulas presented earlier in Chapter 3. For instance, the linear absorption rate is proportional to Im $\chi^{(1)}(\omega)$. Similarly, the two-photon absorption rate is proportional to Im $\chi^{(3)}(\omega = \omega + \omega - \omega)$. We have already seen how to calculate these quantities. However, the method we used to calculate $\chi^{(3)}$ becomes tedious to apply to higher-order processes (e.g., $\chi^{(5)}$ for three-photon absorption, etc.). For this reason, we now develop a simpler approach that generalizes more easily to N-photon absorption for arbitrary N.

Theory of Single- and Multiphoton Absorption and Fermi's Golden Rule

Let us next see how to use the laws of quantum mechanics to calculate single- and multi-photon absorption rates. We begin by deriving the standard result for the single-photon absorption rate, and we then generalize this result to higher-order processes.

The calculation uses procedures similar to those used in Section 3.2 to calculate the nonlinear optical susceptibility. We assume that the atomic wavefunction $\psi(\mathbf{r}, t)$ obeys the time-dependent Schrödinger equation

$$i\hbar \frac{\partial \psi(\mathbf{r}, t)}{\partial t} = \hat{H} \psi(\mathbf{r}, t), \tag{12.5.1}$$

where the Hamiltonian $\hat{H}$ is represented as

$$\hat{H} = \hat{H}_0 + \hat{V}(t). \tag{12.5.2}$$

Here $\hat{H}_0$ is the Hamiltonian for a free atom and

$$\hat{V}(t) = -\hat{\mu}\tilde{E}(t) \quad \text{where} \quad \hat{\mu} = -e\hat{r} \tag{12.5.3}$$

is the interaction energy with the applied optical field. For simplicity we take this field as a monochromatic wave of the form

$$\tilde{E}(t) = E e^{-i\omega t} + \text{c.c.} \tag{12.5.4}$$

which is switched on suddenly at time $t = 0$.

We assume that the solutions to Schrödinger's equation for a free atom are known, and that the wavefunctions associated with the energy eigenstates can be represented as

$$\psi_n(\mathbf{r}, t) = u_n(\mathbf{r}) e^{-i\omega_n t} \quad \text{where} \quad \omega_n = E_n/\hbar. \tag{12.5.5}$$

We see that expression (12.5.5) will satisfy Schrödinger's equation (12.5.1) (with $\hat{H}$ set equal to $\hat{H}_0$) if $u_n(\mathbf{r})$ satisfies the eigenvalue equation

$$\hat{H}_0 u_n(\mathbf{r}) = E_n u_n(\mathbf{r}). \tag{12.5.6}$$

We now return to the general problem of solving Schrödinger's equation in the presence of a time-dependent interaction potential $\hat{V}(t)$:

$$i\hbar \frac{\partial \psi(\mathbf{r}, t)}{\partial t} = (\hat{H}_0 + \hat{V}(t))\psi(\mathbf{r}, t). \tag{12.5.7}$$

Since the energy eigenstates of $\hat{H}_0$ form a complete set, we can express the solution to Eq. (12.5.7) as a linear combination of these eigenstates, that is, as

$$\psi(\mathbf{r}, t) = \sum_l a_l(t) u_l(\mathbf{r}) e^{-i\omega_l t}. \tag{12.5.8}$$

We introduce Eq. (12.5.8) into Eq. (12.5.7) and find that

$$i\hbar \sum_l \frac{da_l}{dt} u_l(\mathbf{r}) e^{-i\omega_l t} + i\hbar \sum_l (-i\omega_l) a_l(t) u_l(\mathbf{r}) e^{-i\omega_l t}$$

$$= \sum_l a_l(t) E_l u_l(\mathbf{r}) e^{-i\omega_l t} + \sum_l a_l(t) \hat{V} u_l(\mathbf{r}) e^{-i\omega_l t} \tag{12.5.9}$$

where (since $E_l = \hbar\omega_l$) clearly the second and third terms cancel. To simplify this expression further, we multiply both sides (from the left) by $u_m^*(\mathbf{r})$ and integrate over all space. Making use of the orthonormality condition

$$\int u_m^*(r)u_l(r)d^3r = \delta_{ml},\qquad(12.5.10)$$

we obtain

$$i\hbar \frac{da_m}{dt} = \sum_l a_l(t)V_{ml}e^{-i\omega_{lm}t},\qquad(12.5.11)$$

where $\omega_{lm} = \omega_l - \omega_m$ and where

$$V_{ml} = \int u_m^*(\mathbf{r})\hat{V}u_l(\mathbf{r})d^3r\qquad(12.5.12)$$

are the matrix elements of the interaction Hamiltonian $\hat{V}$. Equation (12.5.11) is a matrix form of the Schrödinger equation.

Oftentimes, as in the case at hand, Eq. (12.5.11) cannot be solved exactly and must be solved using perturbation techniques. To this end, we introduce an expansion parameter λ which is assumed to vary continuously between zero and one; the value $\lambda = 1$ is taken to correspond to the physical situation at hand. We replace V_{ml} by λV_{ml} in Eq. (12.5.11) and expand $a_m(t)$ in powers of the interaction as

$$a_m(t) = a_m^{(0)}(t) + \lambda a_m^{(1)}(t) + \lambda^2 a_m^{(2)}(t) + \cdots.\qquad(12.5.13)$$

By equating powers of λ on each side of the resulting form of Eq. (12.5.11) we obtain the set of equations

$$\frac{da_m^{(N)}}{dt} = (i\hbar)^{-1}\sum_l a_l^{(N-1)}V_{ml}e^{-i\omega_{lm}t}, \quad N = 1, 2, 3\ldots.\qquad(12.5.14)$$

Linear Absorption

Let us first see how to use Eq. (12.5.14) to describe linear absorption. We set $N = 1$ to correspond to an interaction first-order in the field. We also assume that in the absence of the applied laser field the atom is in the state g (typically the ground state) so that

$$a_g^{(0)}(t) = 1, \qquad a_l^{(0)}(t) = 0 \quad \text{for} \quad l \neq g\qquad(12.5.15)$$

for all times t. Through use of Eqs. (12.5.3) and (12.5.4), we represent V_{mg} as

$$V_{mg} = -\mu_{mg}(Ee^{-i\omega t} + E^*e^{i\omega t}).\qquad(12.5.16)$$

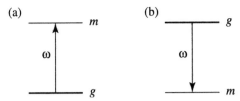

FIGURE 12.5.3   (a) The first term in Eq. (12.5.17) describes the process of one-photon absorption, whereas (b) the second term describes the process of stimulated emission.

Equation (12.5.14) then becomes

$$\frac{da_m^{(1)}}{dt} = -(i\hbar)^{-1}\mu_{mg}\left[Ee^{i(\omega_{mg}-\omega)t} + E^*e^{i(\omega_{mg}+\omega)t}\right].$$

This equation can be integrated to give

$$a_m^{(1)}(t) = -(i\hbar)^{-1}\mu_{mg}\int_0^t dt'\left[Ee^{i(\omega_{mg}-\omega)t'} + E^*e^{i(\omega_{mg}+\omega)t'}\right]$$

$$\tag{12.5.17}$$

$$= \frac{\mu_{mg}E}{\hbar(\omega_{mg}-\omega)}\left[e^{i(\omega_{mg}-\omega)t} - 1\right] + \frac{\mu_{mg}E^*}{\hbar(\omega_{mg}+\omega)}\left[e^{i(\omega_{mg}+\omega)t} - 1\right].$$

The resonance structure of this expression is illustrated schematically in Fig. 12.5.3. Note that the first term in this expression can become resonant for the process of one-photon absorption, and that (if state m lies below state g) the second term can become resonant for the process of stimulated emission. As our present interest is in the process of one-photon absorption, we drop the second term from consideration. The neglect of the second term is known as the rotating wave approximation. Since $a_m^{(1)}(t)$ is a probability amplitude, the probability $p_m^{(1)}(t)$ that the atom is in state m at time t is given by

$$p_m^{(1)}(t) = \left|a_m^{(1)}(t)\right|^2 = \frac{|\mu_{mg}E|^2}{\hbar^2}\left|\frac{e^{i(\omega_{mg}-\omega)t} - 1}{\omega_{mg} - \omega}\right|^2$$

$$\tag{12.5.18}$$

$$= \frac{|\mu_{mg}E|^2}{\hbar^2}\frac{4\sin^2[(\omega_{mg}-\omega)t/2]}{(\omega_{mg}-\omega)^2} \equiv \frac{|\mu_{mg}E|^2}{\hbar^2}f(t),$$

where

$$f(t) = \frac{4\sin^2[(\omega_{mg}-\omega)t/2]}{(\omega_{mg}-\omega)^2}.$$

$$\tag{12.5.19}$$

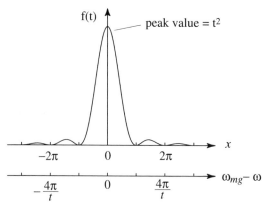

FIGURE 12.5.4   Approximation of $f(t)$ of Eq. (12.5.20) as a Dirac delta function.

Let us examine the time dependence of this expression for large values of the interaction time t. Note that we can express $f(t)$ as

$$f(t) = t^2 \left(\frac{\sin^2 x}{x^2} \right) \quad \text{where} \quad x \equiv (\omega_{mg} - \omega)t/2. \quad (12.5.20)$$

Note further (see also Fig. 12.5.4) that the peak value of $f(t)$ is t^2, but that the width of the central peak is of the order of $2\pi/t$. Thus the area under the central peak is of the order of $2\pi t$, and for large t the function becomes highly peaked. These facts suggest that, for large t, $f(t)$ is proportional to t times a Dirac delta function. In fact, it can be shown that

$$\lim_{t \to \infty} f(t) = 2\pi t \delta(\omega_{mg} - \omega). \quad (12.5.21)$$

Thus for large t the probability to be in the upper level m can be represented, at least formally, by

$$p_m^{(1)}(t) = \frac{2\pi |\mu_{mg} E|^2 t}{\hbar^2} \delta(\omega_{mg} - \omega). \quad (12.5.22)$$

This result is somewhat unphysical because of the presence of the delta function on the right-hand side. In physically realistic situations, the transition frequency ω_{mg} is not perfectly well defined, but is spread into a continuous distribution by various line-broadening mechanisms (Fig. 12.5.5). One often expresses this thought by saying that the final state m is spread into a density of final states $\rho_f(\omega_{mg})$. In the context of atomic physics, $\rho_f(\omega_{mg})$ is often known as the atomic lineshape function. It is defined so

$$\rho_f(\omega_{mg}) d\omega_{mg} = \text{probability that the transition frequency}$$
$$\text{lies between } \omega_{mg} \text{ and } \omega_{mg} + d\omega_{mg}. \quad (12.5.23)$$

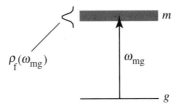

FIGURE 12.5.5   Level m is spread into a density of states described by the function $\rho_f(\omega_{mg})$.

The density of final states is normalized such that

$$\int_0^\infty \rho_f(\omega_{mg})\, d\omega_{mg} = 1. \tag{12.5.24}$$

A well-known example of a density of final states is the Lorentzian lineshape function

$$\rho_f(\omega_{mg}) = \frac{1}{\pi} \frac{\Gamma/2}{(\bar{\omega}_{mg} - \omega_{mg})^2 + (\Gamma/2)^2} \tag{12.5.25}$$

where $\bar{\omega}_{mg}$ is the line-center transition frequency and Γ is the full-width at half maximum of the distribution in angular frequency units. For a transition broadened by the finite lifetime of its upper level, Γ is the population decay rate of the upper level.

For a transition characterized by a density of final states, the probability $p_m^{(1)}(t)$ to be in the upper level given by Eq. (12.5.22) must be averaged over all possible values of the transition frequency. One obtains

$$p_m^{(1)}(t) = \frac{2\pi |\mu_{mg} E|^2 t}{\hbar^2} \int_0^\infty \rho_f(\omega_{mg}) \delta(\omega_{mg} - \omega) d\omega_{mg}$$

$$\tag{12.5.26}$$

$$= \frac{2\pi |\mu_{mg} E|^2 t}{\hbar^2} \rho_f(\omega_{mg} = \omega).$$

The notation $\rho_f(\omega_{mg} = \omega)$ means that the density of final states is to be evaluated at the frequency ω of the incident laser light. Since the probability for the atom to be in the upper level is seen to increase linearly with time, we can define a transition rate for linear absorption by

$$R_{mg}^{(1)} = \frac{p_m^{(1)}(t)}{t} = \frac{2\pi |\mu_{mg} E|^2}{\hbar^2} \rho_f(\omega_{mg} = \omega). \tag{12.5.27}$$

This result is a special case of Fermi's golden rule. Linear absorption is often described in terms of an absorption cross section $\sigma_{mg}^{(1)}(\omega)$, defined such that

$$R_{mg}^{(1)} = \sigma_{mg}^{(1)}(\omega) I, \tag{12.5.28}$$

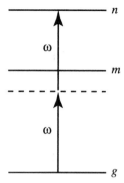

FIGURE 12.5.6   Definition of energy levels used in the calculation of the two-photon transition rate.

where $I = (nc/2\pi)|E|^2$. By comparison with Eq. (12.5.27) we find that

$$\sigma_{mg}^{(1)}(\omega) = \frac{4\pi^2}{nc} \frac{|\mu_{mg}|^2}{\hbar^2} \rho_f(\omega_{mg} = \omega). \qquad (12.5.29)$$

Two-Photon Absorption

Let us next treat the case of two-photon absorption. To do so, we need to solve the set of equations (12.5.14) for $N = 1$ and $N = 2$ to obtain the probability amplitude $a_n^{(2)}(t)$ for the atom to be in level n at time t. The conventions for labeling the various levels are shown in Fig. 12.5.6. Our strategy is to solve Eq. (12.5.14) first for $N = 1$ to obtain $a_m^{(1)}(t)$, which is then used on the right-hand side of Eq. (12.5.14) with $N = 2$. In fact, the expression we obtain for $a_m^{(1)}(t)$ is identical to that of Eq. (12.5.17), obtained in our treatment of linear absorption. We again drop the second term (which does not lead to two-photon absorption). In addition, we express V_{nm} as follows:

$$V_{nm} = -\mu_{nm}(Ee^{-i\omega t} + E^*e^{i\omega t})$$
$$\simeq -\mu_{nm}Ee^{-i\omega t}. \qquad (12.5.30)$$

Here we have dropped the negative-frequency contribution to V_{nm} for reasons analogous to those described above in connection with Eq. (12.5.17). We thus obtain

$$\frac{d}{dt}a_n^{(2)}(t) = (i\hbar)^{-1}\sum_m a_m^{(1)}(t)V_{nm}e^{-i\omega_{mn}t}$$
$$= -(i\hbar)^{-1}\sum_m \frac{\mu_{nm}\mu_{mg}E^2}{\hbar(\omega_{mg} - \omega)}\left[e^{i(\omega_{ng}-2\omega)t} - e^{i(\omega_{nm}-\omega)t}\right]. \qquad (12.5.31)$$

We next drop the second term in square brackets, which describes the transient response of the system but does not lead to two-photon absorption. The resulting equation can be integrated directly to obtain

$$a_n^{(2)}(t) = \sum_m \frac{\mu_{nm}\mu_{mg}E^2}{\hbar^2(\omega_{mg} - \omega)} \left[\frac{e^{i(\omega_{mg} - 2\omega)t} - 1}{\omega_{ng} - 2\omega} \right].$$ (12.5.32)

The calculation now proceeds analogously to that for the linear absorption. The probability to be in level n is given by

$$p_n^{(2)}(t) = \left| a_n^{(2)}(t) \right|^2 = \left| \sum_m \frac{\mu_{nm}\mu_{mg}E^2}{\hbar^2(\omega_{mg} - \omega)} \right|^2 \left| \frac{e^{i(\omega_{mg} - 2\omega)t} - 1}{\omega_{ng} - 2\omega} \right|^2.$$ (12.5.33)

For large t, the expression becomes (see Eqs. (12.5.18)–(12.5.22))

$$p_n^{(2)}(t) = \left| \sum_m \frac{\mu_{nm}\mu_{mg}E^2}{\hbar^2(\omega_{mg} - \omega)} \right|^2 2\pi t \delta(\omega_{ng} - 2\omega),$$ (12.5.34)

and if we assume that level n is smeared into a density of states we obtain

$$p_n^{(2)}(t) = \left| \sum_m \frac{\mu_{nm}\mu_{mg}E^2}{\hbar^2(\omega_{mg} - \omega)} \right|^2 2\pi t \rho_f(\omega_{ng} = 2\omega).$$ (12.5.35)

Since the probability for the atom to be in the upper level is seen to increase linearly with time, we can define a transition rate for two-photon absorption given by

$$R_{ng}^{(2)} = \frac{p_n^{(2)}(t)}{t}.$$ (12.5.36)

It is convenient to recast this result in terms of a two-photon cross section defined by

$$R_{ng}^{(2)} = \sigma_{ng}^{(2)}(\omega)I^2.$$ (12.5.37)

We obtain

$$\sigma_{ng}^{(2)}(\omega) = \frac{8\pi^3}{n^2 c^2} \left| \sum_m \frac{\mu_{nm}\mu_{mg}}{\hbar^2(\omega_{mg} - \omega)} \right|^2 \rho_f(\omega_{ng} = 2\omega).$$ (12.5.38)

Experimentally, two-photon cross sections are often quoted with intensities measured in photons cm^{-2} sec^{-1}. With this convention, Eqs. (12.5.37) and (12.5.38) must be replaced by

$$R_{ng}^{(2)} = \bar{\sigma}_{ng}^{(2)}(\omega)\bar{I}^2 \quad \text{where} \quad \bar{I} = \frac{nc}{2\pi\hbar\omega}|E|^2$$ (12.5.39)

and where

$$\bar{\sigma}_{ng}^{(2)}(\omega) = \frac{8\pi^3\omega^2}{n^2c^2} \left| \sum_m \frac{\mu_{nm}\mu_{mg}}{\hbar(\omega_{mg} - \omega)} \right|^2 \rho_f(\omega_{ng} = 2\omega). \quad (12.5.40)$$

We can perform a numerical estimate of $\bar{\sigma}^{(2)}$ by assuming that a single level dominates the sum in Eq. (12.5.40) and assuming that the one-photon transition is highly nonresonant so that $\omega_{mg} - \omega \approx \omega$. We also assume that the laser frequency is tuned to the peak of the two photon resonance, so that $\rho_f(\omega_{ng} = 2\omega) \approx (2\pi\Gamma_n)^{-1}$, where Γ_n is the width of level n. We then obtain

$$\bar{\sigma}_{ng}^{(2)} \approx \frac{4\pi^2|\mu_{nm}\mu_{mg}|^2}{\hbar^2c^2\Gamma_n}. \quad (12.5.41)$$

To evaluate this expression, we assume that both μ_{nm} and μ_{mg} are of the order of $ea_0 = 2.5 \times 10^{-18}$ esu and that $\Gamma_n = 2\pi(1 \times 10^{13})$ rad/sec. We then obtain

$$\bar{\sigma}_{ng}^{(2)} \approx 10^{-49} \frac{\text{cm}^4 \text{ s}}{\text{photon}^2}. \quad (12.5.42)$$

This value is in good order-of-magnitude agreement with those measured by Xu and Webb (1996) for a variety of molecular fluorophores. There can be considerable variation in the values of molecular two-photon cross sections. Drobizhev *et al.* (2001) report a two-photon cross section as large as 1.1×10^{-46} cm^4 sec/photon^2 in a dendrimer molecule.

Multiphoton Absorption

The results of this section are readily generalized to higher-order processes. One obtains, for instance, the following set of relations:

$$R_{mg}^{(1)} = \left| \frac{\mu_{mg}E}{\hbar} \right|^2 2\pi\rho_f(\omega_{mg} - \omega),$$

$$R_{ng}^{(2)} = \left| \sum_m \frac{\mu_{nm}\mu_{mg}E^2}{\hbar^2(\omega_{mg} - \omega)} \right|^2 2\pi\rho_f(\omega_{ng} - 2\omega),$$

$$R_{og}^{(3)} = \left| \sum_{mn} \frac{\mu_{on}\mu_{nm}\mu_{mg}E^3}{\hbar^3(\omega_{ng} - 2\omega)(\omega_{mg} - \omega)} \right|^2 2\pi\rho_f(\omega_{og} - 3\omega),$$

$$R_{pg}^{(4)} = \left| \sum_{omn} \frac{\mu_{po}\mu_{on}\mu_{nm}\mu_{mg}E^4}{\hbar^4(\omega_{og} - 3\omega)(\omega_{ng} - 2\omega)(\omega_{mg} - \omega)} \right|^2 2\pi\rho_f(\omega_{pg} - 4\omega),$$

etc.

Problems

1. *Relation between the two-photon absorption cross section and* $\chi^{(3)}$. Derive an expression relating the two-photon absorption cross section $\sigma^{(2)}$ to the third-order susceptibility $\chi^{(3)}$. Be sure to indicate the frequency dependence of $\chi^{(3)}$.

2. *Multiphoton absorption coefficients.* Starting with the expressions for the rates of one-, two-, and three-photon absorption quoted above, deduce expressions for the one-, two-, and three-photon absorption coefficients α, β, and γ defined by the equation

$$\frac{dI}{dz} = -\alpha I - \beta I^2 - \gamma I^3.$$

Make order-of-magnitude estimates of β and γ for condensed matter, and compare them to typical measured values as tabulated in the scientific literature.

References

Optical Damage

M. V. Ammosov, N. B. Delone, and V. P. Krainov, *Sov. Phys. JETP* **64**, 1191 (1986).

N. Bloembergen, *J. Nonlinear Opt. Phys. Materials* **6**, 377 (1997).

D. Du, X. Liu, G. Korn, J. Squier, and G. Mourou, *Appl. Phys. Lett.* **64**, 3071 (1994).

F. Y. Genin, M. D. Feit, M. R. Kozlowski, A. M. Rubenchik, A. Salleo, and J. Yoshiyama, *Appl. Opt.* **39**, 3654 (2000).

B. C. Stuart, M. D. Feit, A. M. Rubenchik, B. W. Shore, and M. D. Perry, *Phys. Rev. Lett.* **74**, 2248 (1995).

B. C. Stuart, M. D. Feit, S. Herman, A. M. Rubenchik, B. W. Shore, and M. D. Perry, *Phys. Rev. B* **53**, 1749 (1996).

Reviews of Optical Damage

N. Bloembergen, *IEEE J. Quantum Electron.* **10**, 375 (1974).

W. H. Lowdermilk and D. Milam, *IEEE J. Quantum Electron.* **17**, 1888 (1981).

A. A. Manenkov and A. M. Prokhorov, *Sov. Phys. Usp.* **29**, 104 (1986).

Y. P. Raizer, *Sov. Phys. JETP* **21**, 1009 (1965).

R. M. Wood, *Laser Damage in Optical Materials*, Adam Hilger, Bristol, 1986.

Optical Damage with Femtosecond Laser Pulses

K. M. Davis, K. Miura, N. Sugimoto, and K. Hirao, *Opt. Lett.* **21**, 1729 (1996).

E. N. Glezer, M. Milosavljevic, L. Huang, R. J. Finlay, T.-H. Her, J. P. Callan, and E. Mazur, *Opt. Lett.* **21**, 2023 (1996).

P. P. Pronko, S. K. Dutta, J. Squier, J. V. Rudd, D. Du, and G. Mourou, *Opt. Commun.* **114**, 106 (1995).

Multiphoton Absorption

W. Denk, J. H. Strickler, and W. W. Webb, *Science* **248**, 73 (1990).

M. Drobizhev, A. Karotki, A. Rebane, and C. W. Spangler, *Opt. Lett.* **26**, 1081 (2001).

W. Kaiser and C. G. B. Garrett, *Phys. Rev. Lett.* **7**, 229 (1961).

C. Xu and W. W. Webb, *J. Opt. Soc. Am. B* **13**, 481 (1996).

C. Xu and W. W. Webb, Chapter 11 of *Topics in Fluorescence Spectroscopy*, Vol. 5: *Nonlinear and Two-Photon-Induced Fluorescence* (J. Lakowicz, ed.), Plenum Press, New York, 1997.

Chapter 13

Ultrafast and Intense-Field Nonlinear Optics

13.1. Introduction

There is currently great interest in the physics of ultrashort laser pulses. Recent advances have led to the generation of laser pulses with durations of the order of 1 attosecond. Ultrashort pulses can be used to probe the properties of matter on extremely short time scales. Within the context of nonlinear optics, ultrashort laser pulses are of interest for two separate reasons. The first reason is that the nature of nonlinear optical interactions is often profoundly modified through the use of ultrashort laser pulses. The next two sections of this chapter treat various aspects of the resulting modifications of the nature of nonlinear optical interactions. The second reason is that ultrashort laser pulses tend to possess extremely high peak intensities, because laser pulse energies tend to be established by the energy-storage capabilities of laser gain media, and thus short laser pulses tend to have much higher peak powers than longer pulses. The second half of this chapter is devoted to a survey of the sorts of nonlinear optical processes that can be excited by extremely intense laser fields.

13.2. Ultrashort Pulse Propagation Equation

In this and the following section we treat aspects of the propagation of ultrashort laser pulses through optical systems. Some physical processes that will be included in this analysis include self-steepening leading to optical shock-wave formation, the influence of higher-order dispersion, and space-time coupling effects. In the present section we derive a form of the pulse propagation equation relevant to the propagation of an ultrashort laser pulse through a nonlinear, dispersive nonlinear optical medium. In many ways, this equation can be considered to be a generalization of the pulse propagation equation (the so-called nonlinear Schrödinger equation) of Section 7.5. We begin with the wave equation in the time domain (see, for instance, Eq. (2.1.14)) which we express as

$$-\nabla^2 \tilde{E} + \frac{1}{c^2}\frac{\partial^2 \tilde{D}^{(1)}}{\partial t^2} = -\frac{4\pi}{c^2}\frac{\partial^2 \tilde{P}}{\partial t^2}. \tag{13.2.1}$$

We express the field quantities is terms of their Fourier transforms as

$$\tilde{E}(\mathbf{r}, t) = \int E(\mathbf{r}, \omega)e^{-i\omega t}d\omega/2\pi, \tag{13.2.2a}$$

$$\tilde{D}^{(1)}(\mathbf{r}, t) = \int D^{(1)}(\mathbf{r}, \omega)e^{-i\omega t}d\omega/2\pi, \tag{13.2.2b}$$

$$\tilde{P}(\mathbf{r}, t) = \int P(\mathbf{r}, \omega)e^{-i\omega t}d\omega/2\pi, \tag{13.2.2c}$$

where all of the integrals are to be performed over the range $-\infty$ to ∞. We assume that $D^{(1)}(\mathbf{r}, \omega)$ and $E(\mathbf{r}, \omega)$ are related by the usual linear dispersion relation as

$$D^{(1)}(\mathbf{r}, \omega) = \epsilon^{(1)}(\omega)E(\mathbf{r}, \omega) \tag{13.2.3}$$

and that $\tilde{P}$ represents the nonlinear part of the material response. By introducing these forms into Eq. (13.2.1), we obtain a relation that can be regarded as the wave equation in the frequency domain and which is given by

$$\nabla^2 E(\mathbf{r}, \omega) + \epsilon^{(1)}(\omega)(\omega^2/c^2)E(\mathbf{r}, \omega) = -(4\pi\omega^2/c^2)P(\mathbf{r}, \omega). \tag{13.2.4}$$

Our goal is to derive a wave equation for the slowly varying field amplitude $\tilde{A}(\mathbf{r}, t)$ defined by

$$\tilde{E}(\mathbf{r}, t) = \tilde{A}(\mathbf{r}, t)\, e^{i(k_0 z - \omega_0 t)} + \text{c.c.}, \tag{13.2.5}$$

where ω_0 is the carrier frequency and k_0 is the linear part of the wavevector at the carrier frequency. We represent $\tilde{A}(r, t)$ in terms of its spectral content as

$$\tilde{A}(\mathbf{r}, t) = \int A(\mathbf{r}, \omega)\, e^{-i\omega t}d\omega/2\pi. \tag{13.2.6}$$

Note that $E(\mathbf{r}, \omega)$ and $A(\mathbf{r}, \omega)$ are related by

$$E(\mathbf{r}, \omega) \simeq A(\mathbf{r}, \omega - \omega_0)e^{ik_0 z}. \tag{13.2.7}$$

In terms of the quantity $A(\mathbf{r}, \omega)$ (the slowly varying field amplitude in the frequency domain) the wave equation (13.2.4) becomes

$$\nabla_\perp^2 A + \frac{\partial^2 A}{\partial z^2} + 2ik_0 \frac{\partial A}{\partial z} + \left[k^2(\omega) - k_0^2\right]A = -\frac{4\pi\omega^2}{c^2}P(\mathbf{r}, \omega)e^{-ik_0 z} \quad (13.2.8)$$

where

$$k^2(\omega) = \epsilon(\omega)(\omega^2/c^2). \quad (13.2.9)$$

We next approximate $k(\omega)$ as a power series in the frequency difference $\omega - \omega_0$ as

$$k(\omega) = k_0 + k_1(\omega - \omega_0) + D \quad \text{where} \quad D = \sum_{n=2}^{\infty} \frac{1}{n!}k_n(\omega - \omega_0)^n \quad (13.2.10)$$

so that $k^2(\omega)$ can be expressed as

$$k^2(\omega) = k_0^2 + 2k_0 k_1(\omega - \omega_0) + 2k_0 D + 2k_1 D(\omega - \omega_0) + k_1^2(\omega - \omega_0)^2 + D^2. \quad (13.2.11)$$

Here D represents high-order dispersion. We have displayed explicitly the linear term $k_1(\omega - \omega_0)$ in the power series expansion because k_1 has a direct physical interpretation as the inverse of the group velocity. We now introduce this expression into the wave equation in the form of Eq. (13.2.8), which then becomes

$$\nabla_\perp^2 A + \frac{\partial^2 A}{\partial z^2} + 2ik_0 \frac{\partial A}{\partial z} + 2k_0 k_1(\omega - \omega_0)A + 2k_0 D A + 2k_1 D(\omega - \omega_0)A$$

$$+ k_1^2(\omega - \omega_0)^2 A = (-4\pi\omega^2/c^2)P(z, \omega)e^{-ik_0 z}, \quad (13.2.12)$$

where we have dropped the contribution D^2 because it is invariably small. We now convert this equation back to the time domain. To do so, we multiply this equation by $\exp\left[-i(\omega - \omega_0)t\right]$ and integrate over all values of $\omega - \omega_0$. We obtain

$$\left[\nabla_\perp^2 + \frac{\partial^2}{\partial z^2} + 2ik_0\left(\frac{\partial}{\partial z} + k_1\frac{\partial}{\partial t}\right) + 2ik_1\tilde{D}\frac{\partial}{\partial t} + 2k_0\tilde{D} - k_1\frac{\partial^2}{\partial t^2}\right]\tilde{A}(\mathbf{r}, t)$$

$$= \frac{4\pi}{c^2}\frac{\partial^2\tilde{P}}{\partial t^2}e^{-i(k_0 z - \omega_0 t)}, \quad (13.2.13)$$

where $\tilde{D}$ represents the differential operator

$$\tilde{D} = \sum_{n=2}^{\infty} \frac{1}{n}k_n\left(i\frac{\partial}{\partial t}\right)^n = -\frac{1}{2}k_2\frac{\partial^2}{\partial t^2} + \cdots. \quad (13.2.14)$$

We now represent the polarization in terms of its slowly varying amplitude $\tilde{p}(\mathbf{r}, t)$ as

$$\tilde{P}(\mathbf{r}, t) = \tilde{p}(\mathbf{r}, t)e^{i(k_0 z - \omega_0 t)} + \text{c.c.} \qquad (13.2.15)$$

For example, for the case of a material with an instantaneous third-order response, the polarization is given by

$$\tilde{p}(\mathbf{r}, t) = 3\chi^{(3)}|\tilde{A}(\mathbf{r}, t)|^2 \tilde{A}(\mathbf{r}, t). \qquad (13.2.16)$$

We thus find that

$$\frac{\partial \tilde{P}}{\partial t} = \left(-i\omega_0 \tilde{p} + \frac{\partial \tilde{p}}{\partial t}\right)e^{i(k_0 z - \omega_0 t)} + \text{c.c.}$$

$$= -i\omega_0 \left[\left(1 + \frac{i}{\omega_0}\frac{\partial}{\partial t}\right)\tilde{p}\right]e^{i(k_0 z - \omega_0 t)} + \text{c.c.} \qquad (13.2.17a)$$

and

$$\frac{\partial^2 \tilde{P}}{\partial t^2} = -\omega_0^2 \left[\left(1 + \frac{i}{\omega_0}\frac{\partial}{\partial t}\right)^2 \tilde{p}\right]e^{i(k_0 z - \omega_0 t)} + \text{c.c.} \qquad (13.2.17b)$$

By introducing this expression into the wave equation in the form (13.2.13), we obtain

$$\left[\nabla_\perp^2 + \frac{\partial^2}{\partial z^2} + 2ik_0\left(\frac{\partial}{\partial z} + k_1 \frac{\partial}{\partial t}\right) + 2k_0 \tilde{D} + 2ik_1 \tilde{D}\frac{\partial}{\partial t} - k_1^2 \frac{\partial^2}{\partial t^2}\right]\tilde{A}(\mathbf{r}, t)$$

$$= -\frac{4\pi\omega_0^2}{c^2}\left(1 + \frac{i}{\omega_0}\frac{\partial}{\partial t}\right)^2 \tilde{p}. \qquad (13.2.18)$$

Next we convert this equation to a retarded time frame specified by the coordinates z' and τ defined by

$$z' = z \qquad \text{and} \qquad \tau = t - \frac{1}{v_g}z = t - k_1 z. \qquad (13.2.19)$$

so that

$$\frac{\partial}{\partial z} = \frac{\partial}{\partial z'} - k_1 \frac{\partial}{\partial \tau} \qquad \text{and} \qquad \frac{\partial}{\partial t} = \frac{\partial}{\partial \tau}. \qquad (13.2.20)$$

The wave equation then becomes

$$\left[\nabla_\perp^2 + \frac{\partial^2}{\partial z'^2} - 2k_1 \frac{\partial}{\partial z'}\frac{\partial}{\partial \tau} + k_1^2 \frac{\partial^2}{\partial \tau^2} + 2ik_0\left(\frac{\partial}{\partial z'} - k_1 \frac{\partial}{\partial \tau} + k_1 \frac{\partial}{\partial \tau}\right)\right.$$

$$\left. + 2k_0 \tilde{D} + 2ik_1 \tilde{D}\frac{\partial}{\partial \tau} - k_1^2 \frac{\partial^2}{\partial \tau^2}\right]\tilde{A}(\mathbf{r}, t) = -\frac{4\pi\omega_0^2}{c^2}\left(1 + \frac{i}{\omega_0}\frac{\partial}{\partial \tau}\right)^2 \tilde{p}.$$

$$(13.2.21)$$

We now make the slowly varying amplitude approximation (that is, we drop the term $\partial^2/\partial z'^2$) and simplify this expression to obtain

$$\left[\nabla_{\perp}^2 - 2k_1 \frac{\partial}{\partial z'} \frac{\partial}{\partial \tau} + 2ik_0 \frac{\partial}{\partial z'} + 2k_0 \tilde{D} + 2ik_1 \tilde{D} \frac{\partial}{\partial \tau} \right] \tilde{A}(\mathbf{r}, t)$$

$$= -\frac{4\pi \omega_0^2}{c^2} \left(1 + \frac{i}{\omega_0} \frac{\partial}{\partial \tau} \right)^2 \tilde{p}. \tag{13.2.22}$$

This equation can alternatively be writen as

$$\left[\nabla_{\perp}^2 + 2ik_0 \frac{\partial}{\partial z'} \left(1 + \frac{ik_1}{k_0} \frac{\partial}{\partial \tau} \right) + 2k_0 \tilde{D} \left(1 + \frac{ik_1}{k_0} \frac{\partial}{\partial \tau} \right) \right] \tilde{A}(\mathbf{r}, t)$$

$$= -\frac{4\pi \omega_0^2}{c^2} \left(1 + \frac{1}{\omega_0} \frac{\partial}{\partial \tau} \right)^2 \tilde{p}. \tag{13.2.23}$$

Note that several of the terms in this equation depend upon the ratio k_1/k_0. This ratio can be approximated as follows: $k_1/k_0 = v_g^{-1}/(n\omega_0/c) = n_g/(n\omega_0)$. Ignoring dispersion, $n_g = n$, so that $k_1/k_0 = 1/\omega_0$. In this approximation the wave equation becomes

$$\left[\nabla_{\perp}^2 + 2ik_0 \frac{\partial}{\partial z'} \left(1 + \frac{i}{\omega_0} \frac{\partial}{\partial \tau} \right) + 2k_0 \tilde{D} \left(1 + \frac{i}{\omega_0} \frac{\partial}{\partial \tau} \right) \right] \tilde{A}(\mathbf{r}, t)$$

$$= -\frac{4\pi \omega_0^2}{c^2} \left(1 + \frac{i}{\omega_0} \frac{\partial}{\partial \tau} \right)^2 \tilde{p}, \tag{13.2.24}$$

which can also be expressed as

$$\left[\left(1 + \frac{i}{\omega_0} \frac{\partial}{\partial \tau} \right)^{-1} \nabla_{\perp}^2 + 2ik_0 \frac{\partial}{\partial z'} + 2k_0 \tilde{D} \right] \tilde{A}(\mathbf{r}, t)$$

$$= -\frac{4\pi \omega_0^2}{c^2} \left(1 + \frac{i}{\omega_0} \frac{\partial}{\partial \tau} \right) \tilde{p}. \tag{13.2.25}$$

This equation can be considered to be a generalization of the nonlinear Schrödinger equation. It includes the effects of higher-order dispersion (through the term that includes $\tilde{D}$), space-time coupling (through the presence of the differential operator on the left-hand side of the equation), and self-steepening (through the presence of the differential operator on the right-hand side). This form of the pulse propagation equation has been described by Brabec and Krausz (1997). It can be used to treat many types of nonlinear response. For instance, for a material displaying an instantaneous third- and fifth-order nonlinearity, $\tilde{p}$ is given by $\tilde{p} = 3\chi^{(3)}|\tilde{A}|^2 \tilde{A} + 10\chi^{(5)}|\tilde{A}|^4 \tilde{A}$.

This equation can also be used to treat a dispersive nonlinear material. For ultrashort laser pulses, the value of $\chi^{(3)}$ can vary appreciably for different frequency components of the pulse. The effects of the dispersion of $\chi^{(3)}$ can be modeled in first approximation (see for instance Diels and Rudolph, 1996, p. 139) by representing $\chi^{(3)}(\omega) \equiv \chi^{(3)}(\omega = \omega + \omega - \omega)$ as

$$\chi^{(3)}(\omega) = \chi^{(3)}(\omega_0) + (\omega - \omega_0)\frac{d\chi^{(3)}}{d\omega}. \qquad (13.2.26)$$

where the derivative is to be evaluated at frequency ω_0. Thus $p(\omega)$ can be represented as

$$p(\omega) = 3\left[\chi^{(3)}(\omega_0) + (\omega - \omega_0)\frac{d\chi^{(3)}}{d\omega}\right]|A(\omega)|^2 A(\omega). \qquad (13.2.27)$$

This relation can be converted to the time domain using the same procedure as that used in going from Eq. (13.2.12) to Eq. (13.2.13). One finds that

$$\tilde{p}(\tau) = 3\left[\chi^{(3)}(\omega_0) + \frac{d\chi^{(3)}}{d\omega}i\frac{\partial}{\partial\tau}\right]|\tilde{A}|^2\tilde{A}. \qquad (13.2.28)$$

This expression for $\tilde{p}$ can be used directly in Eq. (13.2.24) or (13.2.25). However, since Eq. (13.2.26) contains only a linear correction term in $(\omega - \omega_0)$, and consequently Eq. (13.2.28) contains only a contribution first-order in $\partial/\partial\tau$, for reasons of consistency one wants to include in the resulting pulse propagation equation only contributions first-order in $\partial/\partial\tau$. Noting that

$$\left(1 + \frac{i}{\omega_0}\frac{\partial}{\partial\tau}\right)^2 = \left(1 + \frac{2i}{\omega_0}\frac{\partial}{\partial\tau} - \frac{1}{\omega_0^2}\frac{\partial^2}{\partial\tau^2}\right) \approx \left(1 + \frac{2i}{\omega_0}\frac{\partial}{\partial\tau}\right), \qquad (13.2.29)$$

one finds that in this approximation the pulse propagation equation is given by

$$\nabla_\perp^2 + 2ik_0\frac{\partial}{\partial z'}\left[\left(1 + \frac{i}{\omega_0}\frac{\partial}{\partial\tau}\right) + 2k_0\tilde{D}\left(1 + \frac{i}{\omega_0}\frac{\partial}{\partial\tau}\right)\right]\tilde{A}(\mathbf{r},t)$$

$$= (-12\pi/c^2)\,\omega_0^2\,\chi^{(3)}(\omega_0)\left[1 + \left(2 + \frac{\omega_0}{\chi^{(3)}(\omega_o)}\frac{d\chi^{(3)}}{d\omega}\right)\frac{i}{\omega_0}\frac{\partial}{\partial\tau}\right]|\tilde{A}|^2\tilde{A}. \qquad (13.2.30)$$

Procedures for incorporating other sorts of nonlinearities into the present formalism have been described by Gaeta (2000).

13.3. Interpretation of the Ultrashort Pulse Propagation Equation

Let us next attempt to obtain some level of intuitive understanding of the various physical processes described in Eq. (13.2.24). As a first step, let us study a simplified version of this equation obtained by ignoring the correction terms $(i/\omega_0)\partial/\partial\tau$ by replacing the factors $[1 + (i/\omega_0)(\partial/\partial\tau)]$ by unity and by including only the lowest-order contribution (known as second-order dispersion) to $\tilde{D}$. One obtains

$$\frac{\partial A(\mathbf{r}, t)}{\partial z'} = \left[\frac{i}{2k_0}\nabla_\perp^2 - \frac{i}{2}k_2\frac{\partial^2}{\partial\tau^2} + \frac{6\pi i\omega_0}{n_0 c}\chi^{(3)}(\omega_0)\,|\tilde{A}|^2 \right]\tilde{A}. \quad (13.3.1)$$

Written in this form, the equation leads to the interpretation that the field amplitude A varies with propagation distance z' (the left-hand side) because of three physical effects (the three terms on the right-hand side). The term involving the transverse laplacian describes the spreading of the beam due to diffraction, the term involving the second time derivative describes the temporal spreading of the pulse due to group velocity dispersion, and the third term describes the nonlinear acquisition of phase. It is useful to introduce distance scales over which each of the terms becomes appreciable. We define these scales as follows:

$$L_{\text{dif}} = \tfrac{1}{2}k_0 w_0^2 \quad \text{(diffraction length)}, \qquad\qquad (13.3.2a)$$

$$L_{\text{dis}} = T^2/|k_2| \quad \text{(dispersion length)}, \qquad\qquad (13.3.2b)$$

$$L_{\text{NL}} = \frac{n_0 c}{6\pi\omega_0\chi^{(3)}|A|^2} = \frac{1}{(\omega/c)n_2 I} \quad \text{(nonlinear length)}. \quad (13.3.2c)$$

In these equations w_0 is a measure of the characteristic beam radius, and T is a measure of the characteristic pulse duration. The significance of these distance scales is that for a given physical situation the process with the shortest distance scales is expected to be dominant. For reference, note that for fused silica at 800 nm $n_2 = 3.5 \times 10^{-16}$ cm^2/W and $k_2 = 446$ fsec^2/cm. Through use of Eq. (13.3.2b) we see that, for a 20-fsec pulse propagating through fused silica, L_{dis} is approximately 0.9 cm. Thus in propagating through 0.9 cm of fused silica a 20-fsec pulse approximately doubles in pulse duration because of group velocity dispersion.

Self-Steepening

Let us next examine the influence of the correction factor $[1 + (i/\omega_0)(\partial/\partial\tau)]$ on the nonlinear source term of Eq. (13.2.25). To isolate this influence, we drop the correction factor in other places in the equation. Also, for generality, we use the propagation equation in the form given by (13.2.30), which allows the nonlinear response to be dispersive. We also transform back to the laboratory reference frame z, t (not the z', τ frame in which the pulse is nearly stationary) so that the factor $k_1 \partial \tilde{A}/\partial t = (1/v_g)\partial \tilde{A}/\partial t = (n_0^{(g)}/c)\partial \tilde{A}/\partial t$ appears explicitly in the wave equation, which takes the form

$$\frac{\partial \tilde{A}}{\partial z} - \frac{n_0^{(g)}}{c}\frac{\partial \tilde{A}}{\partial t} = \frac{i}{2k_0}\nabla_\perp^2 \tilde{A} - \frac{i}{2}k_2\frac{\partial^2 \tilde{A}}{\partial t^2} + \frac{i6\pi\omega_0}{n_0c}\chi^{(3)}(\omega_0)|\tilde{A}|^2\tilde{A}$$

$$+ \frac{i6\pi\omega_0}{n_0c}\chi^{(3)}(\omega_0)\left(2 + \frac{\omega_0}{\chi^{(3)}(\omega_0)}\frac{d\chi^{(3)}}{d\omega}\right)\frac{i}{\omega_0}\frac{\partial}{\partial t}|\tilde{A}|^2 A. \tag{13.3.3}$$

We now introduce nonlinear coefficients γ_1 and γ_2 defined by

$$\gamma_1 = \frac{6\pi\omega_0}{n_0c}\chi^{(3)}(\omega_0) \quad \text{and} \quad \gamma_2 = \frac{6\pi\omega_0}{n_0c}\chi^{(3)}(\omega_0)\left(1 + \frac{1}{2}\frac{\omega_0}{\chi^{(3)}}\frac{d\chi^{(3)}}{d\omega}\right). \tag{13.3.4}$$

Note that in the absence of dispersion $\gamma_1 = \gamma_2$. In terms of these quantities, Eq. (13.3.3) can be expressed more concisely as

$$\frac{\partial \tilde{A}}{\partial z} - \frac{n_0^{(g)}}{c}\frac{\partial \tilde{A}}{\partial t} = \frac{i}{2k_0}\nabla_\perp^2\tilde{A} - \frac{i}{2}k_2\frac{\partial^2\tilde{A}}{\partial t^2} + i\gamma_1|A|^2A - 2\gamma_2\frac{1}{\omega_0}\frac{\partial}{\partial t}(|\tilde{A}|^2 A). \tag{13.3.5}$$

Next note that the time derivative in the last term can be written as

$$\frac{\partial}{\partial t}(|\tilde{A}|^2\tilde{A}) = \frac{\partial}{\partial t}(\tilde{A}^2\tilde{A}^*) = \tilde{A}^2\frac{\partial \tilde{A}^*}{\partial t} + 2\tilde{A}^*\tilde{A}\frac{\partial \tilde{A}}{\partial t}$$

$$= 2|\tilde{A}|^2\frac{\partial \tilde{A}}{\partial t} + \tilde{A}^2\frac{\partial \tilde{A}^*}{\partial t}. \tag{13.3.6}$$

The first contribution to the last form can be identified as an intensity-dependent contribution to the group velocity. The second contribution does not have a simple physical interpretation, but can be considered to represent a dispersive four-wave mixing term. To proceed we make use of Eq. (13.3.6) to express Eq. (13.3.5) as

$$\frac{\partial \tilde{A}}{\partial z} - \frac{n_{\text{eff}}^{(g)}}{c}\frac{\partial \tilde{A}}{\partial t} = \frac{i}{2k_0}\nabla_\perp^2\tilde{A} - \frac{i}{2}k_2\frac{\partial^2\tilde{A}}{\partial t^2} + i\gamma_1|A|^2A - \frac{2\gamma_2}{\omega_0}\tilde{A}^2\frac{\partial \tilde{A}^*}{\partial t} \tag{13.3.7}$$

where

$$n_{\text{eff}}^{(g)} = n_0^{(g)} + \frac{4\gamma_2c}{\omega_0}|\tilde{A}|^2 \equiv n_0^{(g)} + n_2^{(g)}I. \tag{13.3.8}$$

(a)   (b)   (c)

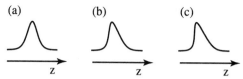

FIGURE 13.3.1   Self-steepening and optical shock formation. (a) The incident optical pulse is assumed to have a Gaussian time evolution. (b) After propagation through a nonlinear medium, the pulse displays self-steepening, typically of the trailing edge. (c) If the self-steepening becomes sufficiently pronounced that the intensity changes instantaneously, an optical shock wave is formed.

In the last form of this relation, we have introduced the coefficient of the intensity dependence of the group index as

$$n_2^{(g)} = \frac{48\pi^2}{n_0^2 c}\chi^{(3)}(\omega_0)\left[1 + \frac{1}{2}\frac{\omega_0}{\chi^{(3)}(\omega_0)}\frac{d\chi^{(3)}}{d\omega}\right]. \tag{13.3.9}$$

We thus see that the last term in Eq. (13.3.5) leads an intensity dependence of the group index n_g as well as to the last term of Eq. (13.3.7), which as mentioned above is a dispersive four-wave mixing contribution. We also see from Eq. (13.3.9) that the intensity dependence of the group index depends both on the susceptibility and on its dispersion.

The intensity dependence of the group velocity leads to the phenomena of self-steepening and optical shock wave formation. These phenomena are illustrated in Fig. 13.3.1. Note that for the usual situation in which $n_2^{(g)}$ is positive, the peak of the pulse is slowed down more than the edges of the pulse, leading to steepening of the trailing edge of the pulse. If this edge becomes infinitely steep, it is said to form an optical shock wave. Self-steepening has been described by DeMartini *et al.* (1967), by Yang and Shen (1984), and by Gaeta (2000). Note also that we can define a self-steepening distance scale analogous to these of Eqs. (13.3.2) by

$$L_{ss} = \frac{cT}{n_2^{(g)}I}. \tag{13.3.10}$$

For the usual situation in which $n_2^{(g)} \approx n_2$, L_{ss} is much larger than L_{NL} (because, except for extremely short pulses, $cT \gg 1/k_0$), and thus self-steepening tends to be difficult to observe.

Space-Time Coupling

Let us now examine the influence of space-time coupling, that is, the influence of the differential operator $[1 + (i/\omega_0)\,\partial/\partial\tau]^{-1}$ on the left-hand side of Eq. (13.2.25). We can see the significance of this effect most simply by considering propagation through a dispersionless, linear material so that the wave equation becomes

$$\left(1 + \frac{i}{\omega_0}\frac{\partial}{\partial \tau}\right)^{-1}\nabla_\perp^2 \tilde{A}(\mathbf{r}, t) + 2ik_0\frac{\partial}{\partial z'}\tilde{A}(\mathbf{r}, t) = 0. \quad (13.3.11)$$

The first term is said to represent space-time coupling because it involves both temporal and spatial derivatives of the field amplitude. To examine the significance of this mathematical form, it is convenient to rewrite this equation as

$$\nabla_\perp^2 \tilde{A}(\mathbf{r}, t) + \left(1 + \frac{i}{\omega_0}\frac{\partial}{\partial \tau}\right)2ik_0\frac{\partial}{\partial z'}\tilde{A}(\mathbf{r}, t) = 0. \quad (13.3.12)$$

Let us first consider the somewhat artificial example of a field of the form $\tilde{A}(\mathbf{r}, t) = a(\mathbf{r})e^{-i\delta\omega t}$; such a field is a monochromatic field at frequency $\omega_0 + \delta\omega$. We substitute this form into Eq. (13.3.12) and obtain

$$\nabla_\perp^2 a(\mathbf{r}) + \left(1 + \frac{\delta\omega}{\omega_0}\right)2ik_0\frac{\partial}{\partial z'}a(\mathbf{r}) = 0, \quad (13.3.13)$$

which can alternatively be expressed as

$$\nabla_\perp^2 a(\mathbf{r}) + 2i\left(k_0 + \delta k\right)\frac{\partial}{\partial z'}a(\mathbf{r}) = 0 \quad (13.3.14)$$

where $\delta k = k_0 (\delta\omega/\omega_0)$. This wave thus diffracts as a wave of frequency $\omega_0 + \delta\omega$ rather than a wave of frequency ω_0. More generally, for the case of an ultrashort pulse, the operator $[1 + (i/\omega_0)\partial/\partial \tau]$ describes the fact that different frequency components of the pulse diffract into different cone angles. Thus, after propagation different frequency components will have different radial dependences. These effects and their implications for self-focusing have been described by Rothenberg (1992).

Supercontinuum Generation

When a short intense pulse propagates through a nonlinear optical medium, it often undergoes significant spectral broadening. This effect was first reported by Alfano and Shapiro (1970). The amount of broadening can be very significant. For instance, using an 80-fsec pulse of peak intensity $\sim 10^{14}$ W/cm^2 propagating through 0.5 mm of ethylene glycol, Fork et al. (1983) observed a broadened spectrum extending from 0.4 ω_0 to 3.3 ω_0, where ω_0 is the central frequency of the input laser pulse. Supercontinuum generation has also been observed in gases (Corkum et al., 1986). Many models have been introduced over the years in attempts to explain supercontinuum generation. At present, it appears that pulse self-steepening (Yang and Shen, 1984) leading to optical shock-wave formation (Gaeta, 2000) is the physical mechanism leading to supercontinuum generation.

13.4.   Intense-Field Nonlinear Optics

Most nonlinear optical phenomena* can be described by assuming that the material polarization can be expanded as a power series in the applied electric field amplitude. This relation in its simplest form is given by

$$\tilde{P}(t) = \chi^{(1)} \tilde{E}(t) + \chi^{(2)} \tilde{E}(t)^2 + \chi^{(3)} \tilde{E}(t)^3 + \cdots. \tag{13.4.1}$$

However, for sufficiently large field strengths, this power series expansion need not converge. We saw in Chapter 6 that under resonant conditions this power-series description breaks down if the Rabi frequency $\Omega = \mu_{ba} E/\hbar$ associated with the interaction of the laser field with the atom becomes comparable to $1/T_1$, where T_1 is the atomic excited-state lifetime. Even under highly nonresonant conditions, Eq. (13.6.1) can become invalid. This breakdown will certainly occur if the laser field amplitude E becomes comparable to or larger than the atomic field strength

$$E_{\text{at}} = \frac{e}{a_0^2} = \frac{e}{(\hbar^2/me^2)^2} = 2 \times 10^7 \text{ statvolt/cm} = 6 \times 10^9 \text{ V/cm}, \tag{13.4.2}$$

which corresponds to an intensity of[†]

$$I_{\text{at}} = \frac{c}{8\pi} E_{\text{at}}^2 = 4 \times 10^{16} \text{ W/cm}^2. \tag{13.4.3}$$

In fact, lasers that can produce intensities larger than 10^{20} W/cm^2 are presently available (Mourou *et al.*, 1998). In this chapter we explore some of the physical phenomena that can occur through use of fields this intense.

Let us consider briefly the conceptual framework one might use to describe intense-field nonlinear optics. Recall that the quantum-mechanical calculation of the nonlinear optical susceptibility presented in Chapter 3 presupposes that the Hamiltonian of an atom in the presence of a laser field is of the form

$$\hat{H} = \hat{H}_0 + \hat{V}(t), \tag{13.4.4}$$

where $\hat{H}_0$ is the Hamiltonian of an isolated atom and $\hat{V}(t) = -\mu \tilde{E}(t)$ represents the interaction energy of the atom with the laser field. Schrödinger's equation

---

* The photorefractive effect of Chapter 11 being an obvious exception.

[†] Here we take the *peak* field strength of the optical wave, which we assume to be linearly polarized, to be E_{at}.

is then solved for this Hamiltonian through use of perturbation theory under the assumption $V(t) \ll H_0$. For the case of intense-field nonlinear optics, the nature of this inequality is the reverse, that is, the interaction energy $V(t)$ is much larger than H_0. This observation suggests that it should prove useful to begin our study of intense-field nonlinear optics by considering the motion of a free electron in an intense laser field.

13.5. Motion of a Free Electron in a Laser Field

Let us initially ignore both relativistic effects and the influence of the magnetic field associated with the laser beam. We assume the laser beam to be linearly polarized and of the form $\tilde{\mathbf{E}}(t) = \tilde{E}(t)\hat{x}$, where $\tilde{E}(t) = Ee^{-i\omega t} +$ c.c. The equation of motion of the electron is then given by

$$m\ddot{\tilde{x}} = -e\tilde{E}(t) \quad \text{or} \quad m\ddot{\tilde{x}} = -eEe^{-i\omega t} + \text{c.c.}, \qquad (13.5.1)$$

which leads to the solution

$$\tilde{x}(t) = xe^{-i\omega t} + \text{c.c.} \qquad (13.5.2)$$

where

$$x = eE/m\omega^2. \qquad (13.5.3)$$

The time-averaged kinetic energy associated with this motion is given by $K = \frac{1}{2}m\langle \dot{\tilde{x}}(t)^2 \rangle$ or, since

$$\dot{\tilde{x}}(t) = (-i\omega x)e^{i\omega t} + \text{c.c.}, \qquad (13.5.4)$$

by

$$K = \frac{e^2 E^2}{m\omega^2} = \frac{e^2 E_{\text{peak}}^2}{4m\omega^2}. \qquad (13.5.5)$$

This energy is known as the jitter energy (as it is associated with the oscillation of the electron about its equilibrium position) or as the ponderomotive energy (Kibble, 1966). This energy can be appreciable. By way of example, consider a laser field of wavelength 1.06 μm. One finds by numerical evaluation that the ponderomotive energy is equal to 13.6 eV (a typical atomic energy) for $I = 1.3 \times 10^{14}$ W/cm^2, is equal to 4.2 keV for $I = I_{\text{at}}$ (which is given by Eq. (13.4.3)), and is equal to $mc^2 = 500$ keV for $I = 4.8 \times 10^{18}$ W/cm^2.

The equation of motion (13.5.1) and its solution (13.5.2) are linear in the laser field amplitude. Both magnetic and relativistic effects can induce nonlinearity in the electronic response. Let us first consider briefly the influence of magnetic effects; see also Problem 1 at the end of this chapter for a more detailed analysis. The electric field of Eq. (13.5.1) has a magnetic field associated with it. Assuming propagation in the z direction, this magnetic field is of the form $\tilde{\mathbf{B}}(t) = \tilde{B}(t)\hat{y}$, where $\tilde{B}(t) = Be^{i\omega t} + $ c.c. and where, assuming propagation in vacuum, $B = E$. Since according to Eq. (13.5.4) the electron has a velocity in the x direction, it will experience a magnetic force $\mathbf{F} = (\mathbf{v}/c) \times \mathbf{B}$ in the z direction. The equation of motion for the z component of the velocity is thus

$$m\ddot{z} = \left[\left(-\frac{ieE}{m\omega}\right)e^{-i\omega t} + \text{c.c.}\right][Be^{-i\omega t} + \text{c.c.}]. \qquad (13.5.6)$$

The right-hand side of this equation consists of terms at zero frequency and at frequencies $\pm 2\omega$. When Eq. (13.5.6) is solved, one find that the z-component of the electron motion consists of oscillations at frequency 2ω and amplitude $eEB/m^2\omega^3$ superposed on a uniform drift velocity. The velocity associated with this motion leads to a magnetic force in the x direction at frequency 3ω. In similar manner, all harmonics of the laser frequency appear in the atomic motion.*

As noted above, relativistic effects also lead to nonlinearities in the atomic response. The origin of this effect is the relativistic change in electron mass that occurs when the electron velocity becomes comparable to the speed of light c. The resulting motion can be described in a relatively straightforward manner. Landau and Lifshitz (1960) show that for a linearly polarized laser beam of peak field strength E_0, i.e., $\tilde{E} = E_0 \cos(\omega t - \omega z/c)$, the electron moves in a figure-8 pattern superposed on a uniform translational motion in the z-direction. In the reference frame moving with the uniform translational velocity, the electron motion can be described the equations

$$x = \frac{\beta c}{\omega} \cos \eta \qquad y = 0 \qquad z = \frac{\beta^2 c}{8\omega} \sin 2\eta, \qquad (13.5.7)$$

where

$$\eta = \omega(t - z/c), \qquad (13.5.8a)$$

$$\beta = eE_0/\gamma'\omega, \qquad (13.5.8b)$$

$$\gamma'^2 = m^2c^2 + e^2E_0^2/2\omega^2. \qquad (13.5.8c)$$

For circularly polarized radiation described by $E_y = E_0 \cos(\omega t - \omega z/c)$, $E_x = E_0 \sin(\omega t - \omega z/c)$, the electron moves with uniform angular velocity in

---

* This conclusion arises, for instance, as a generalization of the results of Problem 7 of Chapter 4.

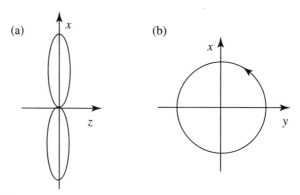

FIGURE 13.5.1   Motion of a free electron in (a) a linearly polarized laser field and (b) a circularly polarized field. Note that for linearly polarized light the motion is in the xz plane and that for circularly polarized light is in the xy plane.

a circle of radius $ecE_0/\gamma\omega^2$; this motion can be described by the equations

$$x = \frac{\beta c}{\omega}\cos\omega t, \qquad y = \frac{\beta c}{\omega}\sin\omega t, \qquad z = 0, \qquad (13.5.9)$$

where β has the same definition as above. These conclusions are summarized in Fig. 13.5.1. More detailed treatments of the motion of a free electron in a laser field can be found in Sarachik and Schappert (1970) and in Castillo-Herrara and Johnston (1993).

It is convenient to introduce a dimensionless parameter a to quantify the strength of the applied laser field. This parameter can be interpreted as the Lorentz-invariant, dimensionless vector potential and is defined by the relation

$$a^2 = \frac{K}{mc^2} = \frac{e^2 E^2}{m^2 c^2 \omega^2}. \qquad (13.5.10)$$

This relation can also be expressed as

$$a^2 = \frac{2}{\pi}\frac{I r_0 \lambda^2}{mc^3}, \qquad (13.5.11)$$

where $r_0 = e^2/mc^2$ is the classical electron radius, $\lambda = 2\pi c/\omega$ is the vacuum wavelength of the laser radiation, and $I = (c/8\pi)E^2$ is the laser intensity. The interpretation of the parameter a is that $a^2 \ll 1$ is the nonrelativistic regime, $a^2 \gtrsim 1$ is the relativisitic regime, and $a^2 \gg 1$ is the ultrarelativistic regime.

13.6. High-Harmonic Generation

High-harmonic generation is a dramatic process in which an intense laser beam* illuminates an atomic medium and all odd harmonics $q\omega$ of the laser frequency ω up to some cutoff order q_{max} are emitted in the forward direction. It is found that most of the harmonics are emitted with comparable efficiency. This observation demonstrates that high-harmonic generation is not a perturbative (i.e., is not a $\chi^{(q)}$) process. For a perturbative process each successively higher order would be expected to be emitted with a smaller efficiency. Harmonic orders as large as $q = 221$ have been observed (Chang *et al.*, 1999). High-harmonic generation is typically observed using laser intensities in the range 10^{14}–10^{16} W/cm^2.

Many of the features of high-harmonic generation can be understood in terms of a model due to Corkum (1993). One imagines an atom in the presence of a linearly polarized laser field sufficiently intense to ionize the atom. Even though the electron kinetic energy K might greatly exceed the ionization potential I_P of the atom, because of the oscillatory nature of the optical field the electron will follow an oscillatory trajectory that returns it to the atomic nucleus once each optical period, as illustrated in Fig. 13.6.1. Because of the $1/r^2$ nature of the nuclear Coulomb potential, the electron will feel an appreciable force and thus an acceleration only when it is very close to the nucleus. The radiated field is proportional to the instantaneous acceleration, and the field radiated by any individual electron will thus consist of a sequence of pulses separated by the optical period of the fundamental laser field. However, in a collection of atoms, roughly half of the ejected electrons will be emitted near the positive maximum of the oscillating laser field and half near the negative maximum, and consequently the emitted radiation will consist of a sequence of pulses separated by half the optical period of the fundamental laser field. These pulses are mutually coherent, and thus the spectrum of the emitted radiation is the Fourier transform of this pulse train, which is a series of components separated by twice the laser frequency. Thus only odd harmonics are emitted, in consistency with the general symmetry properties of centrosymmetric material media, as described in Section 1.5.

Arguments based on energetics can be used to estimate the maximum harmonic order q_{max}. The process of high-harmonic generation is illustrated

---

* Intense in the sense that the ponderomotive energy K is much larger than the ionization potential I_P.

(a)

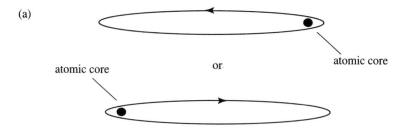

(b)

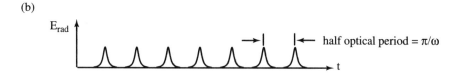

(c)

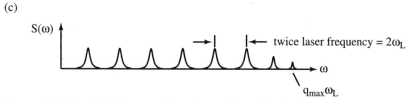

FIGURE 13.6.1   (a) Trajectory of an electron immediately following ionization. The electron experiences the intense laser field, and thus oscillates at frequency ω. It emits a brief pulse of radiation each time it passes near the atomic core. The radiation from a collection of such electrons thus has the form shown in (b). The spectrum of the emitted radiation is determined by the square of the Fourier transform of the pulse train, and thus has the form shown in (c).

symbolically in Fig. 13.6.2. The energy available to the emitted photon is the sum of the available kinetic energy of the electron less the (negative) ionization energy of the atom. This line of reasoning might suggest that $q_{max} \hbar \omega = K + I_P$, but detailed calculations show that the coefficient of the kinetic energy term is in fact 3.17, so that

$$q_{max} \hbar \omega = 3.17K + I_P. \qquad (13.6.1)$$

This prediction is in good agreement with laboratory data.

   We conclude this section with a brief historical summary of progress in the field of intense-field nonlinear optics and high-harmonic generation. In 1979, Agostini *et al.* reported the observation of a phenomenon that has come to be called above-threshold ionization (ATI). This group measured the energy spectra of electrons produced by photoionization and observed multiple peaks

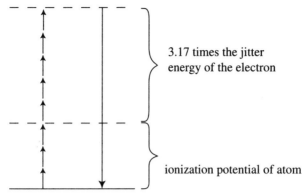

3.17 times the jitter
energy of the electron

ionization potential of atom

FIGURE 13.6.2 Schematic representation of the empirical relation $\hbar\omega q_{max} = 3.17K + I_P$. The numerical factor of 3.17 is a consequence of detailed analysis of the dynamics of an electron interacting simultaneously with an external laser field and the atomic core.

separated by the photon energy $\hbar\omega$. This observation attracted great theoretical interest because according to current theoretical models based on lowest-order perturbation theory only one peak associated with the minimum number of photons needed to produce ionization was expected to be present. More recent work has included the possibility of double ionization in which two electons are ejected as part of the photoionization process (Walker *et al.*, 1994). One of the earliest observations of high-harmonic generation was that of Ferray *et al.* (1988), who observed up to the 33rd harmonic with laser intensities as large as 10^{13} W/cm^2 using Ar, Kr, and Xe gases (Fig. 13.6.3). Kulander and Shore (1989) presented one of the first successful computer models of high-harmonic generation. L'Huillier and Balcou (1993) observed HHG using pulses of 1 psec duration and intensities as large as 10^{15} W/cm^2, and observed harmonics up to the 135th order in Ne. Corkum (1993) presented the theoretical model of HHG described in the previous two paragraphs. Nearly simultaneously, Schafer *et al.* (1993) presented similar ideas along with experimental data. Lewenstein *et al.* (1994) presented a fully quantum-mechanical theory of HHG that clarified the underlying physics and produced quantitative predictions. Chang *et al.* (1997) reported HHG in He excited by 26-fsec laser pulses from a Ti:sapphire laser system operating at 800 nm. They observed harmonic peaks up to a maximum of the 221st order and unresolved structure up to an energy (460 eV or 2.7 nm wavelength) corresponding to the 297th order. Slightly shorter wavelengths ($\lambda = 2.5$ nm, $h\nu = 500$ eV) have been observed by Schnurer *et al.* (1998). Durfee *et al.* (1999) have shown how to phase match the process of HHG by propagating the laser beam through a gas-filled capillary waveguide.

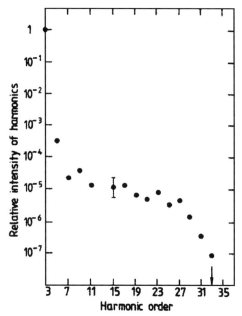

FIGURE 13.6.3   Experimental data of Ferray *et al.* (1988) illustrating high-harmonic generation.

13.7. Nonlinear Optics of Plasmas and Relativistic Nonlinear Optics

A plasma is a partially or fully ionized gas. Plasmas play an important role in nonlinear optics in two different ways: (1) Nonlinear optical processes such as multiphoton ionization can create a plasma. The optical properties of the material system are thereby modified even by the linear response of the plasma. (2) A plasma (no matter how it is generated) can respond in an intrinsically nonlinear manner to an applied optical field. In the present section we briefly survey both sorts of nonlinear optical response.

Let us first consider the process of plasma formation. We let N_e denote the number of free electrons per unit volume and N_i the corresponding number of positive ions. We also let N_T denote the total number of atoms present, both ionized and un-ionized. We assume that these quantities obey the rate equation

$$\frac{dN_e}{dt} = \frac{dN_i}{dt} = (N_T - N_i)\sigma^{(N)}I^N - rN_eN_i. \qquad (13.7.1)$$

Here $\sigma^{(N)}$ is the N-photon absorption cross section (see also Section 12.5) and r is the electron–ion recombination rate. For short laser pulses of the sort often used to study plasma nonlinearities, recombination is an unlikely event and the last term in this equation can usually be ignored. In this case, the electron density increases monotonically during the laser pulse.

Let us next consider the (linear) optical properties of a plasma. We found above (Eqs. (13.5.2) and (13.5.3)) that the position of an electron in the field $\tilde{E}(t) = E e^{-i\omega t} + \text{c.c.}$ will vary according to $\tilde{x}(t) = x e^{-i\omega t} + \text{c.c.}$ where $x = eE/m\omega^2$. The dipole moment associated with this response is $\tilde{p}(t) \equiv p e^{-i\omega t} + \text{c.c.} = -e\tilde{x}(t)$. The polarizability $\alpha(\omega)$ defined by $p = \alpha(\omega)E$ is thus given by

$$\alpha(\omega) = -\frac{e^2}{m\omega^2}. \qquad (13.7.2)$$

The dielectric constant of a collection of such electrons is thus given by

$$\epsilon = 1 + 4\pi N\alpha(\omega) = 1 - \frac{4\pi Ne^2}{m\omega^2}, \qquad (13.7.3)$$

which is often expressed as

$$\epsilon = 1 - \frac{\omega_p^2}{\omega^2}, \qquad \text{where} \qquad \omega_p^2 = \frac{4\pi Ne^2}{m} \qquad (13.7.4)$$

and where ω_p is known as the plasma frequency. For N sufficiently small that $\omega_p^2 < \omega^2$ (an underdense plasma), the dielectric constant is positive, $n = \sqrt{\epsilon}$ is real, and light waves can propagate. Conversely, for N sufficiently large that $\omega_p^2 > \omega^2$ (an overdense plasma), the dielectric constant is negative, $n = \sqrt{\epsilon}$ is imaginary, and light waves cannot propagate.

By way of comparison we recall that for a bound electron the linear polarizability is given (see Eq. (1.4.17) and note that $\chi^{(1)}(\omega) = N\alpha(\omega)$) by

$$\alpha(\omega) = \frac{e^2/m}{\omega_0^2 - \omega^2 - 2i\omega\gamma}, \qquad (13.7.5)$$

which in the highly nonresonant limit $\omega \ll \omega_0$ reduces to

$$\alpha_{\text{bound}} = \frac{e^2}{m\omega_0^2}. \qquad (13.7.6)$$

Note that the polarizability of a free electron is opposite in sign and (for the common situation $\omega \ll \omega_0$) much larger in magnitude than that of a bound electron. Thus the process of plasma formation makes a large *negative* contribution to the refractive index. Note also that we have ignored the contribution of the ionic core to the polarizability because it is very much smaller than the electronic contribution because the mass of the ion is much larger than that of the electron.

Let us next consider nonlinear optical effects that occur within a plasma. There are two primary mechanisms of nonlinearity: (a) ponderomotive effects and (2) relativistic effects.

Ponderomotive effects result from the tendency of charged particles such as electrons to be expelled from regions of high field strength. These effects are important only for laser pulses sufficiently long in duration for particle motion to be important. Ponderomotive effects share an identical origin with the electrostrictive effects described in Section 9.2; the effect is simply given a different name in the context of plasma nonlinearities. Despite the fact that $\alpha(\omega)$ is negative for a free electron (the ponderomotive case) but positive for bulk matter (the electrostrictive case), both effects lead to an increase in refractive index. In the ponderomotive case, the electron, which makes a negative contribution to the refractive index, is expelled from the laser beam, leading to an increase in refractive index.

Another mechanism of nonlinearity in plasmas is relativistic effects. In a sufficiently intense laser beam ($I \gtrsim 10^{18}$ W/cm^2) a free electron can be accelerated to relativistic velocities in a half optical period. This conclusion can be reached by equating the ponderomotive energy K of Section 13.5 with the value mc^2, or can be reached in a more intuitive fashion by noting that a field of strength

$$E_{\text{rel}} = 2mc^2/\lambda e \qquad (13.7.7)$$

will accelerate an electron to relativistic velocities in a half optical period (because $\lambda/2$ is the distance traveled by a relativistic particle in a time $1/2\nu$), and further noting that the corresponding intensity $I = (c/8\pi)E_{\text{rel}}^2$ is of the order of 10^{18} W/cm^2.

Even when the electron velocity is considerably less than velocity of light in vacuum c, appreciable nonlinear effects can be induced by the relativistic change in the electron mass, that is, the change in electron mass from m to γm where

$$\gamma = \frac{1}{\sqrt{1 - v^2/c^2}}. \qquad (13.7.8)$$

The value of the plasma frequency and consequently the refractive index of the plasma is thereby modified such that

$$n^2 = 1 - \frac{\omega_p^2}{\gamma \omega^2}, \qquad (13.7.9)$$

where as before $\omega_p^2 = 4\pi Ne^2/m$. Detailed analysis (Max et al., 1974; Sprangle et al., 1987) shows that the value of the relativistic factor γ to be used in

Eq. (13.7.9) is given in general (that is, even in the strongly relativistic limit) by the expression

$$\gamma^2 = 1 + \frac{e^2 E_0^2}{m^2 \omega^2 c^2} \qquad (13.7.10)$$

where E_0 is the peak field amplitude of the incident laser field. In writing this result in the form shown, we have assumed that the transverse contribution to the velocity is much larger than the longitudinal component. Note that (by comparison of Eq. (13.7.10) with Eqs. (13.5.4), (13.5.3), and (13.7.8)) the correct expression for the dielectric constant is obtained if the *peak* electron velocity is used in conjunction with Eq. (13.7.8) to determine the value of γ to be used in Eq. (13.7.9).

We next calculate the nonlinear coefficient n_2 by determining the lowest-order change in refractive index. The relativistic factor γ is given by the square root of expression (13.7.10), which to lowest order becomes

$$\gamma = 1 + \frac{1}{2} \frac{e^2 E_0^2}{m^2 \omega^2 c^2} \equiv 1 + x, \qquad (13.7.11)$$

where the parameter x has been introduced for future convenience. We can thus write Eq. (13.7.9) as

$$n^2 = 1 - \frac{\omega_p^2}{\omega^2(1+x)} \simeq 1 - \frac{\omega_p^2}{\omega^2}(1-x)$$

$$= n_0^2 + \frac{\omega_p^2}{\omega^2} x, \qquad (13.7.12)$$

where $n_0^2 = 1 - \omega_p^2/\omega^2$. We can thus express n as

$$n \simeq n_0 + \frac{1}{2 n_0} \frac{\omega_p^2}{\omega^2} x \equiv n_0 + n_2 I. \qquad (13.7.13)$$

Setting $I = (n_0 c/8\pi) E_0^2$, we find that

$$n_2 = \frac{2\pi \omega_p^2 e^2}{n_0^2 m^2 c^3 \omega^4}. \qquad (13.7.14)$$

This expression gives the relativistic contribution to the nonlinear refractive index. Note that this process is purely relativistic: Planck's constant does not appear in this expression. Note further that expression (13.7.14) can be rewritten in the intuitively revealing form

$$n_2 = \frac{1}{2\pi n_0^2} \left(\frac{\omega_p}{\omega}\right)^2 \left[\frac{\lambda^2}{(mc^2)/(r_0/c)}\right], \qquad (13.7.15)$$

where $\lambda = 2\pi c/\omega$ and $r_0 = e^2/mc^2 = 2.6 \times 10^{-13}$ is the classical electron radius. The term in square brackets can be intepreted as the fundamental relativistic unit of nonlinear refractive index, that is, an area (λ^2) divided by the fundamental unit of power P_{rel} (the electron rest mass mc^2 divided by the transit time of light across the classical radius of the electron). Numerically we find (in mks units) that

$$P_{\text{rel}} = \frac{mc^2}{(r_0/c)} = \frac{(0.5 \times 10^6)(1.6 \times 10^{-19})}{(2.6 \times 10^{-15}/3 \times 10^8)} = 9.2 \times 10^9 \text{ W.} \quad (13.7.16)$$

At a wavelength of $\lambda = 1 \ \mu\text{m}$, one thus finds that

$$n_2 = \frac{1}{2\pi n_0^2}\left(\frac{\omega_p}{\omega}\right)^2\left(\frac{10^{-12}}{9.2 \times 10^9}\right) = \frac{1}{2\pi n_0^2}\left(\frac{\omega_0}{\omega}\right)^2 1.1 \times 10^{-22} \frac{\text{m}^2}{\text{W}}$$
$$\tag{13.7.17}$$
$$= \frac{1}{2\pi n_0^2}\left(\frac{\omega_p}{\omega}\right)^2 1.1 \times 10^{-26} \frac{\text{cm}^2}{\text{W}}.$$

On the basis of the expression for n_2 just derived, one can calculate the critical power for self-focusing in a plasma. Since in general the expression for the critical power is given by Eq. (7.1.10) as

$$P_{\text{cr}} = \frac{(0.61)^2\pi\lambda^2}{8n_0 n_2} \simeq \frac{\lambda^2}{8n_0 n_2}, \quad (13.7.18)$$

we find through use of Eq. (13.7.14) that

$$P_{\text{cr}} = \frac{\pi}{4}n_0 c\left(\frac{mc^2}{e}\right)^2\left(\frac{\omega}{\omega_p}\right)^2 = 6.7\left(\frac{\omega}{\omega_p}\right)^2 \text{GW.} \quad (13.7.19)$$

An expression for the critical power was first derived by Sprangle et al. (1987) using slightly different assumptions, yielding a similar expression but with the factor $(\pi/4)$ in the first form replaced by 2 and thus the numerical factor 6.7 in the second form replaced by 17. Relativistic self-focusing has been observed experimentally by Monot et al. (1995). Note further that Eq. (13.7.19) can be reexpressed in the suggestive form

$$P_{\text{cr}} = \frac{\pi}{4}n_0\left(\frac{\omega}{\omega_p}\right)^2\left(\frac{mc^2}{r_0/c}\right). \quad (13.7.20)$$

Here, as in Eq. (13.7.16), the last factor denotes the relativistic unit of laser power.

13.8. Nonlinear Quantum Electrodynamics

We saw in the last section that there is a characteristic field strength $E_{\text{rel}} = 2mc^2/e\lambda$ at which relativistic effects become important. There is another field strength E_{QED} at which effects associated with the quantum vacuum become important. This field strength is defined by the relations

$$E_{\text{QED}} = \frac{mc^2}{e\lambdabar_c}, \qquad \text{where} \qquad \lambdabar_c = \frac{\hbar}{mc}. \qquad (13.8.1)$$

Here $\lambdabar_c = 3.6 \times 10^{-11}$ cm is the (reduced) Compton wavelength of the electron. The Compton wavelength is one measure of the size of the electron in the sense that the position of the electron cannot be localized to an accuracy better than $\lambdabar_c$.* The QED field strength is thus a measure of the field strength required to accelerate an electron to relativistic velocities in a distance of the order of the size of the electron. Consequently a field of this magnitude is large enough to lead to the spontaneous creation of electron–positron pairs.

The QED field strength is numerically given by

$$E_{\text{QED}} = 1.32 \times 10^{16} \text{ V/cm} = 4.4 \times 10^{13} \text{ statvolt/cm}. \qquad (13.8.2)$$

The intensity of a wave whose peak field amplitude is equal to E_{QED} is consequently

$$I_{\text{QED}} = \frac{c}{8\pi} E_{\text{QED}}^2 = 4 \times 10^{29} \text{ W/cm}^2. \qquad (13.8.3)$$

This value exceeds the intensity that can be produced by the most powerful lasers currently available. From a different perspective, I_{QED} designates the largest laser intensity that could possibly be produced, in that any larger laser intensity would essentially be instantaneously absorbed as the result of electron–positron creation. Nonetheless, in the rest frame of a relativisitic electron of energy E, produced for example by a particle accelerator, the intensity of a laser beam is greatly increased by relativistic effects. This increase occurs because the electric field strength in the electron rest frame is larger than the field in the laboratory frame by the factor $\gamma = E/mc^2$. In fact, electron–positron pair creation resulting from the interaction of a relativistic

---

* This conclusion follows as a consequence of the time–energy uncertainty relation which we take in the form $\Delta E \, \Delta t \gtrsim \hbar$. We take the energy uncertainty to be the rest energy of an electron $\Delta E = mc^2$ and we set the time uncertainty to be $\Delta t = \Delta x/c$. We thus find that the minimum uncertainty in position is $\Delta x = \hbar/mc$; we take this length as the definition of the reduced Compton wavelength $\lambdabar_c$. This result can be understood more intuitively by noting that a photon of wavelength shorter than $\sim \lambdabar_c$ would have an energy sufficiently large to create an electron–positron pair, thus rendering moot the question of the location of the original electron.

electron with a laser beam has been observed experimentally (Burke *et al.*, 1997).

Nonlinear quantum electrodynamic effects have been predicted even for field strengths considerably smaller than E_{QED}. Euler and Kockel (1935) have shown that there is an intrinsic nonlinearity to the electromagnetic vacuum which leads to a field-dependent dielectric tensor of the form

$$\epsilon_{ik} = \delta_{ik} + \frac{e^4\hbar}{45\pi m^4 c^7}[2(E^2 - B^2)\delta_{ik} + 7B_iB_k]. \tag{13.8.4}$$

Note that the term containing $(E^2 - B^2)$ vanishes for electromagnetic plane waves in vacuum, because of the relation $|\mathbf{E}| = |\mathbf{B}|$. The dielectric response relevant to plane-wave laser beams is thus

$$\epsilon_{ik} = \delta_{ik} + \frac{7e^4\hbar}{45\pi m^4 c^7} B_iB_k, \tag{13.8.5}$$

which can be expressed as

$$\epsilon_{ik} = \delta_{ik} + \frac{7}{45\pi} \frac{e^2}{\hbar c} \frac{B_iB_k}{E_{QED}^2}. \tag{13.8.6}$$

This form of the expression shows that indeed the nonlinear response of the vacuum becomes appreciable only for $B \gtrsim E_{QED}$.

Since the magnetic field (rather than the electric field) appears in the expression for the dielectric response, the tensor properties of the nonlinearity of the vacuum are different from those of most other optical nonlinearities. Nonetheless, by suppressing the tensor nature of the response, one can describe the nonlinearity in terms of a standard third-order susceptibility, and such a description is useful for comparing the size of this effect with that of other nonlinear optical processes. Since $D = \epsilon E = E + 4\pi P$, we see that $D^{NL} = 4\pi P^{NL}$ or that (working in the time domain, making use of Eq. (13.8.6), and now explicitly indicating time-dependent quantities by a tilde)

$$\frac{7}{45\pi} \frac{e^2}{\hbar c} \frac{\tilde{B}^2 \tilde{E}}{E_{QED}^2} = 4\pi \chi^{(3)} \tilde{E}^3. \tag{13.8.7}$$

Since $\tilde{B} = \tilde{E}$, we find that

$$\chi^{(3)} = \frac{7}{180\pi^2} \frac{e^2}{\hbar c} \frac{1}{E_{QED}^2} = 1.44 \times 10^{-32} \text{ esu}. \tag{13.8.8}$$

Recall for comparison that, for CS_2, $\chi^{(3)} = 1.9 \times 10^{-12}$ esu. Alternatively, the nonlinear refractive index coefficient is given by $n_2 = (12\pi^2/c)\chi^{(3)}$ or by

$$n_2 = 5.6 \times 10^{-34} \text{ cm}^2/\text{W}. \tag{13.8.9}$$

We saw earlier in Chapter 7 that strong self-action effects are expected only if the power of a laser beam exceeds the critical power for self-focusing

$$P_{cr} = \frac{\lambda^2}{8 n_0 n_2}. \tag{13.8.10}$$

We find by combining Eqs. (13.8.9) and (13.8.10) that at a wavelength of 1 μm,

$$P_{cr} = 4.4 \times 10^{24} \text{ W}, \tag{13.8.11}$$

which is considerably larger than the power of any laser source currently contemplated.

Problem

1. Consider the nonrelativistic motion of a free electron in the laser field $\tilde{\mathbf{E}}(z, t) = E_0 \cos \omega t \, \hat{\mathbf{x}}$, $\tilde{\mathbf{B}}(z, t) = B_0 \cos \omega t \, \hat{\mathbf{y}}$ with $B_0 = E_0$. Assume that the electron is injected into the field at rest at position $x = 0$, $y = 0$ at time t_0.
(a) Solve the equation of motion for the electron and thereby determine $x(t)$, $y(t)$, $z(t)$, $v_x(t)$, $v_y(t)$, and $v_z(t)$ for all $t > t_0$. Plot the trajectory of the electron of the electron motion, both in the laboratory frame and in a reference frame in which the electron is on average at rest (specify what frame this is). Note that some of these results depend on the value of t_0; for those that do, show plots for several different values of t.
(b) Ignoring the magnetic contribution, calculate the peak and time-averaged kinetic energy of the electron.
(c) Repeat for circular polarization.

References

Sections 13.1 through 13.3: Ultrafast Nonlinear Optics

R. R. Alfano and S. L. Shapiro, *Phys. Rev. Lett.* **24**, 592 (1970).
T. Brabec and F. Krausz, *Phys. Rev. Lett.* **78**, 3282 (1997).
P. B. Corkum, C. Rolland, and T. Srinivasan-Rao, *Phys. Rev. Lett.* **57**, 2268 (1986).
F. DeMartini, C. H. Townes, T. K. Gustafson, and P. L. Kelley, *Phys. Rev.* **164**, 312 (1967).
J.-C. Diels and W. Rudolph, *Ultrashort Laser Pulse Phenomena*, Academic Press, San Diego, 1996.
R. L. Fork, C. V. Shank, C. Hirlimann, R. Yen, and W. J. Tomlinson, *Opt. Lett.* **8**, 1 (1983).
A. L. Gaeta, *Phys. Rev. Lett.* **84**, 3582 (2000).
J. E. Rothenberg, *Opt. Lett.* **17**, 1340 (1992).
G. Yang and Y. R. Shen, *Opt. Lett.* **9**, 510 (1984).

Sections 13.4 and 13.5: Intense-Field Nonlinear Optics and Motion of a Free Electron

C. I. Castillo-Herrara and T. W. Johnston, *IEEE Trans. Plasma Sci.* **21**, 125 (1993).

T. W. B. Kibble, *Phys. Rev. Lett.* **16**, 1054 (1966).

L. D. Landau and E. M. Lifshitz, *Classical Theory of Fields*, Section 48, Pergamon Press, Oxford, 1960.

G. A. Mourou, C. P. J. Barty, and M. D. Perry, *Phys. Today*, January, 22 (1998).

E. S. Sarachik and G. T. Schappert, *Phys. Rev. D* **1**, 2738 (1970).

Section 13.6: High-Harmonic Generation

P. Agostini, F. Fabre, G. Mainfray, G. Petite, and N. K. Rahman, *Phys. Rev. Lett.* **42**, 1127 (1979).

Z. Chang, A. Rundquist, H. Wang, M. M. Murnane, and H. C. Kapteyn, *Phys. Rev. Lett.* **79**, 2967 (1997) and **82**, 2006 (1999).

P. B. Corkum, *Phys. Rev. Lett.* **71**, 1994 (1993).

C. G. Durfee III, A. R. Rundquist, S. Backus, C.e Herne, M. M. Murnane, and H. C. Kapteyn, *Phys. Rev. Lett.* **83**, 2187 (1999).

M. Ferray, A. L'Huillier, X. F. Li, L. A. Lompre, G. Mainfray, and C. Manus, *J. Phys. B*, **21**, L31 (1988).

K. C. Kulander and B. W. Shore, *Phys. Rev. Lett.* **62**, 524 (1989).

M. Lewenstein, Ph. Balcou, M. Yu. Ivanov, A. L'Huillier, and P. B. Corkum, *Phys. Rev. A* **49**, 2117 (1994).

A. L'Huillier and Ph. Balcou, *Phys. Rev. Lett.* **70**, 774 (1993).

X. Li, A. L'Huillier, M. Ferrey, L. Lompre, G. Mainfrey, and C. Magnus, *J. Phys. B* **21**, L31 (1989).

M. Schnürer, Ch. Spielmann, P. Wobrauschek, C. Streli, N. H. Burnett, C. Kan, K. Ferencz, R. Koppitsch, Z. Cheng, T. Brabec, and F. Krausz, *Phys. Rev. Lett.* **80**, 3236 (1998).

K. J. Schafer, B. Yang, L. F. DiMauro, and K. C. Kulander, *Phys. Rev. Lett.* **70**, 1599 (1993).

B. Walker, B. Sheely, L. F. DiMauro, P. Agostini, K. J. Schafer, and K. C. Kulander, *Phys. Rev. Lett.* **73**, 1127 (1994).

Section 13.7: Nonlinear Optics of Plasmas and Relativistic Nonlinear Optics

C. E. Max, J. Arons, and A. B. Langdon, *Phys. Rev. Lett.* **33**, 209 (1974).

P. Monot, T. Auguste, P. Gibbon, F. Jakober, and G. Mainfray, A. Dulieu, M. Louis-Jacquet, G. Malka, and J. L. Miquel, *Phys. Rev. Lett.* **74**, 2953 (1995).

P. Sprangle, C.-M. Tang, and E. Esarez, *IEEE Trans. Plasma Sci.* **15**, 145 (1987).

R. Wagner, S.-Y. Chen, A. Maksemchak, and D. Umstadter, *Phys. Rev. Lett.* **78**, 3125 (1997).

Section 13.8: Nonlinear Quantum Electrodynamics

D. L. Burke, R. C. Field, G. Horton-Smith, J. E. Spencer, D. Walz, S. C. Berridge, W. M. Bugg, K. Shmakov, A. W. Weidemann, C. Bula, K. T. McDonald, E. J. Prebys, C. Bamber, S. J. Boege, T. Koffas, T. Kotseroglou, A. C. Melissinos, D. D. Meyerhofer, D. A. Reis, and W. Ragg, *Phys. Rev. Lett.* **79**, 1626 (1997).

H. Euler and K. Kockel, *Naturwiss.* **23**, 246 (1935).

Appendices

Appendix A.   The Gaussian System of Units

In this appendix we review briefly the basic equations of electromagnetism when written in the gaussian system of units. Conversion between the gaussian and SI (i.e., rationalized MKSA) systems of units is summarized in an additional appendix. The intent of this appendix is to establish notation and not to present a rigorous exposition of electromagnetic theory.

In the gaussian system, mechanical properties are measured in cgs units, that is, distance is measured in centimeters (cm), mass in grams (g), and time in seconds (s). The unit of force is thus the g cm/sec^2, known as a dyne, and the unit of energy is the g cm^2/sec^2, known as the erg. The fundamental electrical unit is a unit of charge, known either as the statcoulomb or simply as the electrostatic unit of charge. It is defined such that the force between two charged point particles, each containing 1 statcoulomb of charge and separated by 1 centimeter, is 1 dyne. More generally, the force between two charged particles of charges q_1 and q_2 separated by the directed distance $\mathbf{r} = r\hat{\mathbf{r}}$ where $\hat{\mathbf{r}}$ is a unit vector in the $\mathbf{r}$ direction is given by

$$\mathbf{F} = \frac{q_1 q_2}{r^2}\hat{\mathbf{r}}. \tag{A.1}$$

The unit of current is thus the statcoulomb/sec, which is known as the statampere, or simply as the electrostatic unit of current. The unit of electrical potential (i.e., potential energy per unit charge) is the erg/statcoulomb, also known as the statvolt.

561

In the gaussian system, Maxwell's equations have the form

$$\nabla \times \mathbf{E} = -\frac{1}{c}\frac{\partial \mathbf{B}}{\partial t}, \tag{A.2a}$$

$$\nabla \times \mathbf{H} = \frac{1}{c}\frac{\partial \mathbf{D}}{\partial t} + \frac{4\pi}{c}\mathbf{J}, \tag{A.2b}$$

$$\nabla \cdot \mathbf{B} = 0, \tag{A.2c}$$

$$\nabla \cdot \mathbf{D} = 4\pi\rho. \tag{A.2d}$$

A remarkable feature of the gaussian system is that the four primary field vectors (i.e., the electric field $\mathbf{E}$, the electric displacement field $\mathbf{D}$, the magnetic induction $\mathbf{B}$, and the magnetic intensity $\mathbf{H}$, as well as the polarization vector $\mathbf{P}$ and the magnetization vector $\mathbf{M}$, which will be introduced shortly) all have the same dimensions, i.e.,

$$[\mathbf{E}] = [\mathbf{D}] = [\mathbf{B}] = [\mathbf{H}] = [\mathbf{D}] = [\mathbf{M}]$$

$$= \frac{\text{statvolt}}{\text{cm}} = \frac{\text{statcoulomb}}{\text{cm}^2} = \text{gauss} = \text{oersted} = \left(\frac{\text{erg}}{\text{cm}^3}\right)^{1/2}. \tag{A.3}$$

By convention the name gauss is used only in reference to the field $\mathbf{B}$ and oersted only with the field $\mathbf{H}$. The two additional quantities appearing in Maxwell's equations are the free charge density ρ, measured in units of statcoulomb/cm^3, and the free current density $\mathbf{J}$, measured in units of statampere/cm^2. Under many circumstances $\mathbf{J}$ is given by the expression

$$\mathbf{J} = \sigma\mathbf{E}, \tag{A.4}$$

which can be considered to be a microscopic form of Ohm's law, where σ is the electrical conductivity, whose units are inverse seconds.

The relationships among the four electromagnetic field vectors are known as the constitutive relations. These relations, even in the presence of nonlinearities, have the form

$$\mathbf{D} = \mathbf{E} + 4\pi\mathbf{P}, \tag{A.5a}$$

$$\mathbf{H} = \mathbf{B} - 4\pi\mathbf{M}. \tag{A.5b}$$

The manner in which the response of a material medium can lead to a nonlinear dependence of $\mathbf{P}$ upon $\mathbf{E}$ is of course the subject of this book. For the limiting case of a purely linear response, the relationships can be expressed (assuming an isotropic medium for notational simplicity) as

$$\mathbf{P} = \chi^{(1)}\mathbf{E}, \tag{A.6a}$$

$$\mathbf{M} = \chi_m^{(1)}\mathbf{H}. \tag{A.6b}$$

Note that the linear electric susceptibility $\chi^{(1)}$ and the linear magnetic suscepti-bility $\chi_m^{(1)}$ are dimensionless quantities. If we now introduce the linear dielectric constant $\epsilon^{(1)}$ (also known as the dielectric permittivity) and the linear magnetic permeability $\mu^{(1)}$, both of which are dimensionless and are defined by

$$\mathbf{D} = \epsilon^{(1)}\mathbf{E}, \tag{A.7a}$$

$$\mathbf{B} = \mu_m^{(1)}\mathbf{H}, \tag{A.7b}$$

we find by consistency of Eqs. (A.5a)–(A.7a) and (A.5b)–(A.7b) that

$$\epsilon^{(1)} = 1 + 4\pi\chi^{(1)}, \tag{A.8a}$$

$$\mu^{(1)} = 1 + 4\pi\chi_m^{(1)}. \tag{A.8b}$$

The fields $\mathbf{E}$ and $\mathbf{B}$ (rather than $\mathbf{D}$ and $\mathbf{H}$) are usually taken to constitute the fundamental electromagnetic fields. For example, the force on a particle of charge q moving at velocity $\mathbf{v}$ through an electromagnetic field is given by

$$\mathbf{F} = q\left(\mathbf{E} + \frac{\mathbf{v}}{c} \times \mathbf{B}\right). \tag{A.9}$$

Poynting's theorem can be derived from Maxwell's equations in the follow-ing manner. We begin with the vector identity

$$\nabla \cdot (\mathbf{E} \times \mathbf{H}) = \mathbf{H} \cdot (\nabla \times \mathbf{E}) - \mathbf{E} \cdot (\nabla \times \mathbf{H}) \tag{A.10}$$

and introduce expressions for $\nabla \times \mathbf{E}$ and $\nabla \times \mathbf{H}$ from the Maxwell equations (A.2a) and (A.2b), to obtain

$$\frac{c}{4\pi}\nabla \cdot (\mathbf{E} \times \mathbf{H}) + \frac{1}{4\pi}\left[\mathbf{H} \cdot \frac{\partial \mathbf{B}}{\partial t} + \mathbf{E} \cdot \frac{\partial \mathbf{D}}{\partial t}\right] = -\mathbf{J} \cdot \mathbf{E}. \tag{A.11}$$

Assuming for simplicity the case of a purely linear response, the second term on the left-hand side of this equation can be expressed as $\partial u/\partial t$, where

$$u = \frac{1}{8\pi}(\mathbf{E} \cdot \mathbf{D} + \mathbf{B} \cdot \mathbf{H}) \tag{A.12}$$

represents the energy density of the electromagnetic field. We also introduce the Poynting vector

$$\mathbf{S} = \frac{c}{4\pi}\mathbf{E} \times \mathbf{H}, \tag{A.13}$$

which gives the rate at which electromagnetic energy passes through a unit area whose normal is in the direction of $\mathbf{S}$. Equation (A.11) can then be written as

$$\nabla \cdot \mathbf{S} + \frac{\partial u}{\partial t} = -\mathbf{J} \cdot \mathbf{E}, \tag{A.14}$$

where $\mathbf{J} \cdot \mathbf{E}$ gives the rate per unit volume at which energy is lost to the field through Joule heating.

A wave equation for the electric field can be derived from Maxwell's equations, as described in Section 2.1, and for a linear, isotropic nonmagnetic (i.e., $\mu = 1$) medium that is free of sources has the form

$$-\nabla^2 \mathbf{E} + \frac{\epsilon^{(1)}}{c^2} \frac{\partial^2 \mathbf{E}}{\partial t^2} = 0. \tag{A.15}$$

This equation possesses solutions in the form of infinite plane waves, that is,

$$\mathbf{E} = \mathbf{E}_0 e^{i(\mathbf{k} \cdot \mathbf{r} - \omega t)} + \text{c.c.}, \tag{A.16}$$

where $\mathbf{k}$ and ω must be related by

$$k = n\omega/c \quad \text{where} \quad n = \sqrt{\epsilon^{(1)}} \text{ and } k = |\mathbf{k}|.$$

The magnetic field associated with this wave has the form

$$\mathbf{B} = \mathbf{B}_0 e^{i(\mathbf{k} \cdot \mathbf{r} - \omega t)} + \text{c.c.} \tag{A.17}$$

Note that, in accordance with the convention followed in the book, factors of $\frac{1}{2}$ are not included in these expressions. From Maxwell's equations, one can deduce that $\mathbf{E}_0$, $\mathbf{B}_0$, and $\mathbf{k}$ are mutually orthogonal and that the magnitudes of $\mathbf{E}_0$ and $\mathbf{B}_0$ are related by

$$n|\mathbf{E}_0| = |\mathbf{B}_0|. \tag{A.18}$$

In considering the energy relations associated with a time-varying field, it is useful to introduce a time-averaged Poynting vector $\langle \mathbf{S} \rangle$ and a time-averaged energy density $\langle u \rangle$. Through use of Eqs. (A.16)–(A.18), we find that these quantities are given by

$$\langle \mathbf{S} \rangle = \frac{nc}{2\pi} |E_0|^2 \hat{\mathbf{k}}, \tag{A.19a}$$

$$\langle u \rangle = \frac{n^2}{2\pi} |E_0|^2, \tag{A.19b}$$

where $\hat{\mathbf{k}}$ is a unit vector in the $\mathbf{k}$ direction. In this book the magnitude of the time-averaged Poynting vector is called the intensity $I = |\langle \mathbf{S} \rangle|$ and is given by $I = (nc/2\pi)|E_0|^2$.

References

J. D. Jackson, *Classical Electrodynamics*, 2nd ed., Wiley, New York, 1975.

J. B. Marion and M. A. Heald, *Classical Electromagnetic Radiation*, Academic Press, New York, 1980.

E. M. Purcell, *Electricity and Magnetism*, McGraw-Hill, New York, 1965.

Appendix B.   Systems of Units in Nonlinear Optics

There are several different systems of units that are commonly used in nonlinear optics. In this appendix we describe these different systems and show how to convert among them. For simplicity we restrict the discussion to a medium with instantaneous response so that the nonlinear susceptibilities can be taken to be dispersionless. Clearly the rules derived here for conversion among the systems of units are the same for a dispersive medium.

In the gaussian system of units, the polarization $\tilde{P}(t)$ is related to the field strength $\tilde{E}(t)$ by the equation

$$\tilde{P}(t) = \chi^{(1)} \tilde{E}(t) + \chi^{(2)} \tilde{E}^2(t) + \chi^{(3)} \tilde{E}^3(t) + \cdots. \tag{B.1}$$

In the gaussian system, all of the fields $\tilde{E}$, $\tilde{P}$, $\tilde{D}$, $\tilde{B}$, $\tilde{H}$, and $\tilde{M}$ have the same units; in particular, the units of $\tilde{P}$ and $\tilde{E}$ are given by

$$[\tilde{P}] = [\tilde{E}] = \frac{\text{statvolt}}{\text{cm}} = \frac{\text{statcoulomb}}{\text{cm}^2} = \left(\frac{\text{erg}}{\text{cm}^3}\right)^{1/2}. \tag{B.2}$$

Consequently, we see from Eq. (B.1) that the dimensions of the susceptibilities are as follows:

$$\chi^{(1)} \text{ is dimensionless}, \tag{B.3a}$$

$$\left[\chi^{(2)}\right] = \left[\frac{1}{\tilde{E}}\right] = \frac{\text{cm}}{\text{statvolt}} = \left(\frac{\text{erg}}{\text{cm}^3}\right)^{-1/2}, \tag{B.3b}$$

$$\left[\chi^{(3)}\right] = \left[\frac{1}{\tilde{E}^2}\right] = \frac{\text{cm}^2}{\text{statvolt}^2} = \left(\frac{\text{erg}}{\text{cm}^3}\right)^{-1}. \tag{B.3c}$$

The units of the nonlinear susceptibilities are often not stated explicitly in the gaussian system of units; one rather simply states that the value is given in electrostatic units (esu).

There are two different conventions in use regarding the units of the susceptibilities in the MKS system. The most common convention is to replace Eq. (B.1) by

$$\tilde{P}(t) = \epsilon_0 \left[\chi^{(1)} \tilde{E}(t) + \chi^{(2)} \tilde{E}^2(t) + \chi^{(3)} \tilde{E}^3(t) + \cdots\right] \tag{B.4}$$

where

$$\epsilon_0 = 8.85 \times 10^{-22} \text{ F/m} \tag{B.5}$$

denotes the permittivity of free space. Since the units of $\tilde{P}$ and $\tilde{E}$ in the MKS system are

$$[\tilde{P}] = \frac{C}{m^2}, \tag{B.6a}$$

$$[\tilde{E}] = \frac{V}{m}, \tag{B.6b}$$

and since 1 farad is equal to 1 coulomb per volt, it follows that the units of the susceptibilities are as follows:

$$\chi^{(1)} \text{ is dimensionless}, \tag{B.7a}$$

$$\left[\chi^{(2)}\right] = \left[\frac{1}{\tilde{E}}\right] = \frac{m}{V}, \tag{B.7b}$$

$$\left[\chi^{(3)}\right] = \left[\frac{1}{\tilde{E}^2}\right] = \frac{m^2}{V^2}. \tag{B.7c}$$

The other convention within the MKS system of units is to replace Eq. (B.1) with

$$\tilde{P}(t) = \epsilon_0 \chi^{(1)} \tilde{E}(t) + \chi^{(2)} \tilde{E}^2(t) + \chi^{(3)} \tilde{E}^3(t) + \cdots. \tag{B.8}$$

We shall refer to this as the alternative MKS definition in the following discussion. Since the units of $\tilde{P}$, $\tilde{E}$, and ϵ_0 are still given by Eqs. (B.6a–b), it follows that the dimensions of the susceptibilities are as follows:

$$\chi^{(1)} \text{ is dimensionless}, \tag{B.9a}$$

$$\left[\chi^{(2)}\right] = \frac{C}{V^2}, \tag{B.9b}$$

$$\left[\chi^{(3)}\right] = \frac{Cm}{V^3}. \tag{B.9c}$$

Conversion among the Systems

In order to facilitate conversion among the three systems introduced above, we express the three defining relations (B.1), (B.4), and (B.8) in the following forms:

$$\tilde{P}(t) = \chi^{(1)} \tilde{E}(t) \left[1 + \frac{\chi^{(2)} \tilde{E}(t)}{\chi^{(1)}} + \frac{\chi^{(3)} \tilde{E}^2(t)}{\chi^{(1)}} + \cdots\right] \text{ (gaussian)}, \tag{B.1'}$$

$$\tilde{P}(t) = \epsilon_0 \chi^{(1)} \tilde{E}(t) \left[1 + \frac{\chi^{(2)} \tilde{E}(t)}{\chi^{(1)}} + \frac{\chi^{(3)} \tilde{E}^2(t)}{\chi^{(1)}} + \cdots\right] \text{ (MKS)}, \tag{B.4'}$$

$$\tilde{P}(t) = \epsilon_0 \chi^{(1)} \tilde{E}(t) \left[1 + \frac{\chi^{(2)} \tilde{E}(t)}{\epsilon_0 \chi^{(1)}} + \frac{\chi^{(3)} \tilde{E}^2(t)}{\epsilon_0 \chi^{(1)}} + \cdots \right] \quad \text{(MKS; alt).} \quad \text{(B.8$'$)}$$

The power series shown in square brackets must be identical in each of these equations. However, the values of $\tilde{E}$, $\chi^{(1)}$, $\chi^{(2)}$, and $\chi^{(3)}$ are different in different systems. In particular, from Eqs. (B.2) and (B.5) and the fact that 1 statvolt $= 300$ V, we find that

$$\tilde{E}(\text{MKS}) = 3 \times 10^4 \tilde{E}(\text{gaussian}). \quad \text{(B.10)}$$

To determine how the linear susceptibilities in the gaussian and MKS systems are related, we make use of the fact that for a linear medium the displacement is given in the gaussian system by

$$\tilde{D} = \tilde{E} + 4\pi \tilde{P} = \tilde{E}\left(1 + 4\pi \chi^{(1)}\right), \quad \text{(B.11a)}$$

and in the MKS system by

$$\tilde{D} = \epsilon_0 \tilde{E} + \tilde{P} = \epsilon_0 \tilde{E}\left(1 + \chi^{(1)}\right). \quad \text{(B.11b)}$$

We thus find that

$$\chi^{(1)}(\text{MKS}) = 4\pi \chi^{(1)}(\text{gaussian}). \quad \text{(B.12)}$$

Using Eqs. (B.10) and (B.11a–b), and requiring that the power series of Eqs. (B.1$'$), (B.4$'$), and (B.8$'$) be identical, we find that the nonlinear susceptibilities in our three systems of unit are related by

$$\begin{aligned}
\chi^{(2)}(\text{MKS, Eq. (B.4)}) &= \frac{4\pi}{3 \times 10^4} \chi^{(2)}(\text{gaussian}) \\
&= 4.189 \times 10^{-4} \chi^{(2)}(\text{gaussian}),
\end{aligned} \quad \text{(B.13)}$$

$$\begin{aligned}
\chi^{(2)}(\text{MKS; alt. Eq. (B.8)}) &= \frac{4\pi \epsilon_0}{3 \times 10^4} \chi^{(2)}(\text{gaussian}) \\
&= 3.71 \times 10^{-15} \chi^{(2)}(\text{gaussian}),
\end{aligned} \quad \text{(B.14)}$$

$$\begin{aligned}
\chi^{(3)}(\text{MKS, Eq. (B.4)}) &= \frac{4\pi}{(3 \times 10^4)^2} \chi^{(3)}(\text{gaussian}) \\
&= 1.40 \times 10^{-8} \chi^{(3)}(\text{gaussian}),
\end{aligned} \quad \text{(B.15)}$$

$$\begin{aligned}
\chi^{(3)}(\text{MKS; alt. Eq. (B.8)}) &= \frac{4\pi \epsilon_0}{(3 \times 10^4)^2} \chi^{(3)}(\text{gaussian}) \\
&= 1.24 \times 10^{-19} \chi^{(3)}(\text{gaussian}).
\end{aligned} \quad \text{(B.16)}$$

Appendix C.   Relationship between Intensity and Field Strength

In the gaussian system of units, the intensity associated with the field

$$\tilde{E}(t) = E e^{-i\omega t} + \text{c.c.} \qquad (C.1)$$

is

$$I = \frac{nc}{2\pi} |E|^2, \qquad (C.2)$$

where n is the refractive index, $c = 3 \times 10^{10}$ cm/sec is the speed of light in vacuum, I is measured in erg/cm^2 sec, and E is measured in statvolts/cm.

In the MKS system, the intensity of the field described by Eq. (B.1) is given by

$$I = 2n \left(\frac{\epsilon_0}{\mu_0} \right)^{1/2} |E|^2 = \frac{2n}{Z_0} |E|^2, \qquad (C.3)$$

where $\epsilon_0 = 8.85 \times 10^{-12}$ F/m, $\mu_0 = 4\pi \times 10^{-7}$ H/m, and $Z_0 = 377\ \Omega$. I is measured in W/m^2, and E is measured in V/m. Using these relations we can obtain the results shown in Table C.1. As a numerical example, a pulsed laser of modest energy might produce a pulse energy or $Q = 1$ mJ with a pulse duration of $T = 10$ nsec. The peak laser power would then be of the order of $P = Q/T = 100$ kW. If this beam is focused to a spot size of $w_0 = 100\ \mu$m, the pulse intensity will be $I = P/\pi w_0^2 \simeq 0.3$ GW/cm^2.

TABLE C.1

Conventional	CGS		MKS	
I	I (erg/cm^2 sec)	E (statvolt/cm)	I (W/m^2)	E (V/m)
1 W/cm^2	10^7	0.0458	10^4	1.37×10^3
1 kW/cm^2	10^{10}	1.45	10^7	4.34×10^4
1 MW/cm^2	10^{13}	45.8	10^{10}	1.37×10^6
1 GM/cm^2	10^{16}	1.45×10^3	10^{13}	4.34×10^7
1 TW/cm^2	10^{19}	4.85×10^4	10^{16}	1.37×10^9

Appendix D. Physical Constants

Constant	Symbol	Value	CGS[a]	MKS[a]
Speed of light in vacuum	c	2.998	10^{10} cm/sec	10^8 m/sec
Elementary charge	e	4.803	10^{-10} esu	
		1.602		10^{-19} C
Avogadro number	N_A	6.023	10^{23} mol	10^{23} mol
Electron rest mass	$m = m_e$	9.109	10^{-28} g	10^{-31} kg
Proton rest mass	m_p	1.673	10^{-24} g	10^{-27} kg
Planck constant	h	6.626	10^{-27} erg sec	10^{-34} J sec
	$\hbar = h/2\pi$	1.054	10^{-27} erg sec	10^{-34} J sec
Fine structure constant[b]	$\alpha = e^2/\hbar c$	1/137	—	—
Compton wavelength of electron	$\lambda_C = h/mc$	2.426	10^{-10} cm	10^{-12} m
Rydberg constant	$R_\infty = me^4/2\hbar^2$	1.09737	10^5 cm^{-1}	10^7 m^{-1}
Bohr radius	$a_0 = \hbar^2/me^2$	5.292	10^{-9} cm	10^{-11} m
Electron radius[b]	$r_e = e^2/mc^2$	2.818	10^{-13} cm	10^{-15} m
Bohr magneton[b]	$\mu_S = eh/2m_e c$	9.273	10^{-21} erg/G	10^{-24} J/T
		$\Rightarrow$	1.4 MHz/G	
Nuclear magneton[b]	$\mu_N = e\hbar/2m_p c$	5.051	10^{-24} erg/G	10^{-27} J/T
Gas constant	R	8.314	10^7 erg/K m	10^0 J/K mole
Volume, mole of ideal gas	V_0	2.241	10^4 cm^3	10^{-2} m^3
Boltzmann constant	k_B	1.381	10^{-16} erg/K	10^{-23} J/K
Stefan–Boltzmann constant	σ	5.670	10^{-5} erg/cm^2 sec K^4	10^{-8} W/m^2 K^4
Gravitational constant	G	6.670	10^{-8} dyne cm^2/g^2	10^{-11} N m^2/kg^2
Electron volt	eV	1.602	10^{-12} erg	10^{-19} J

[a] Abbreviations: C, coulombs; mol, molecules; g, grams; J, joules; N, newtons; G, gauss; T, teslas.
[b] Defining equation is shown in the gaussian CGS system of units.

Physical constants specific to the MKS system

Constant	Symbol[a]	Value[a]
Permittivity of free space	ϵ_0	8.85×10^{-12} F/m
Permeability of free space	μ_0	$4\pi \times 10^{-7}$ H/m
Velocity of light in free space	$(\epsilon_0\mu_0)^{-1/2} = c$	2.997×10^8 m/sec
Impedance of free space	$(\mu_0/\epsilon_0)^{1/2} = Z_0$	377 Ω

[a] Abbreviations: F, farad = coulomb/volt, H, henry = weber/ampere.

Conversion between the systems

1 m	$= 100$ cm
1 kg	$= 1000$ g
1 newton	$= 10^5$ dynes
1 joule	$= 10^7$ erg
1 coulomb	$= 2.998 \times 10^9$ statcoulomb
1 volt	$= 1/299.8$ statvolt
1 ohm	$= 1.139 \times 10^{-12}$ sec/cm
1 tesla	$= 10^4$ gauss[a]
1 farad	$= 0.899 \times 10^{12}$ cm
1 henry	$= 1.113 \times 10^{-12}$ sec^2/cm

[a] Here 1 tesla $= 1$ weber/m^2; 1 gauss $= 1$ oersted.

Index

C

Carbon disulfide, 193, 194, 256, 317, 387, 421, 458, 478, 556
Carbon tetrachloride, 50, 194, 387, 421
Cascaded optical nonlinearities, 125
Cauchy's theorem, 57
Causality, 54, 56, 59
Centrosymmetric media, 43, 49, 62
Chaos (in stimulated Brillouin scattering), 428
Chiral materials, 252, 257
Chiral nematic liquid crystal, 255
Circular polarization, 199
Closure condition of quantum mechanics, 242
Coherent buildup length, 76, 108
Collisonal dephasing, 144, 146, 150, 266, 268
Collision-induced resonances, 176, 185
Compressibility, 384, 386, 414–415
Commutator, quantum mechanical, 149, 152, 155, 162, 172
Compton wavelength, 555
Conduction band, 224
Constitutive relations, 562
Continuity equation, 436
Confocal parameter, 113–114
Contour integration, 57, 60
Constant-pump approximation, 85, 114
Contracted notation, 37, 488
Conversion efficiency, 93
Conversion between systems of units, 565–567, 570
Counterpropagating waves, 329, 411
Counterrotating term, 157
Coupled-amplitude equations,
  for difference-frequency generation, 84
  for forward four-wave mixing with two-level atoms, 303
  for photorefractive media, 505, 508
  for second-harmonic generation, 88
  for stimulated Brillouin and Rayleigh scattering, 442
  for Stokes and anti-Stokes waves in stimulated Raman scattering, 465
  for sum-frequency generation, 72, 79
  with quasi-phase-matching, 109
Cross correlation, 431
Cross coupling (contrasted with self coupling), 192
Crystal systems, 41
Cubic (crystal), 41, 45, 49, 51

D

Damping, quantum mechanical, phenomenological, 149
  of sound waves, 387
Debye-Huckel screening, 227, 504
Debye relaxation model, 352, 477
Degeneracy factor, 20, 191
Determinant (of matrix) 82, 321
Denominator function, 23
Density of final states, 526
Density matrix,
  diagonal elements of, 147, 263
  equation of motion for, 151–153, 276
  formulation of quantum mechanics, 144–151
  off-diagonal elements of, 147, 151
Dextrorotatory, 253
Dextrose, 252
Diamond, 194, 209, 221
Diamond structure, 49
Dielectric constant, 563
Dielectric permittivity, 563
  tensor, 486
Dielectric relaxation time, 498, 508
Difference-frequency generation, 6–9, 25, 84–86, 99, 123, 253
Diffusion, 392, 497, 520
  diffusion field strength, 504
Diffusion constant, 500, 520
Diffraction, 314, 539
  diffraction length, 539
Dipole dephasing rate (time T_2), 150, 160, 265–268
Dipoles, radiation from, 68
Dipole moment operator, matrix representation of, 151
Dipole moment, induced, 155, 160, 163, 266, 288, 297, 377
Dirac delta function, 526
Dirac notation, 146
Director (of liquid crystal), 257
Dispersion (of refractive index), 97, 534
  dispersion length, 539
Dispersion theory, linear, 158–161
Dispersionless medium, 2, 4, 70, 356, 538
Dispersive lineshape, 272
Dispersive medium, 71, 534
Dissipative medium, 71, 232
Donors, 499
Doppler broadening, 275

S

T